Mathematical Theories for Metamaterials

From Condensed Matter Theory to Subwavelength Physics

Conference Board of the Mathematical Sciences

CBMS

Regional Conference Series in Mathematics

Number 136

Mathematical Theories for Metamaterials

From Condensed Matter Theory to Subwavelength Physics

Habib Ammari
Bryn Davies
Erik Orvehed Hiltunen

Published for the
Conference Board of the Mathematical Sciences
by the

AMERICAN MATHEMATICAL SOCIETY
Providence, Rhode Island

NSF–CBMS Regional Conference on Mathematical Methods for Novel Metamaterials, held at Auburn University on May 21–25, 2024.

2020 *Mathematics Subject Classification.* Primary 35J05, 35C20, 35B34, 35P20, 35P25, 35B27, 74J20.

For additional information and updates on this book, visit
www.ams.org/bookpages/cbms-136

Library of Congress Cataloging-in-Publication Data

Names: Ammari, Habib author | Davies, Bryn (Mathematician) author | Orvehed Hiltunen, Erik, 1993– author

Title: Mathematical theories for metamaterials : from condensed matter theory to subwavelength physics / Habib Ammari, Bryn Davies, Erik Orvehed Hiltunen.

Description: Providence, Rhode Island : Published for the Conference Board of the Mathematical Sciences by the American Mathematical Society, [2026] | Series: Conference Board of the Mathematical Sciences. CBMS, Regional conference series in mathematics, 0160-7642 ; number 136 | Includes bibliographical references and index.

Identifiers: LCCN 2026017892 | ISBN 9781470485351 paperback | ISBN 9781470486525 ebook

Subjects: LCSH: Metamaterials–Mathematical models | Differential equations, Partial | Resonators | AMS: Partial differential equations – Elliptic equations and systems – Laplacian operator, reduced wave equation (Helmholtz equation), Poisson equation | Partial differential equations – Representations of solutions – Asymptotic expansions | Partial differential equations – Qualitative properties of solutions – Resonances | Partial differential equations – Spectral theory and eigenvalue problems – Asymptotic distribution of eigenvalues and eigenfunctions | Partial differential equations – Spectral theory and eigenvalue problems – Scattering theory | Partial differential equations – Qualitative properties of solutions – Homogenization; equations in media with periodic structure | Mechanics of deformable solids – Waves – Wave scattering

Classification: LCC TK7871.15.M48 A44 2026

LC record available at https://lccn.loc.gov/2026017892

♾ The paper used in this book is acid-free and falls within the guidelines
established to ensure permanence and durability.
Visit the AMS home page at `https://www.ams.org/`

10 9 8 7 6 5 4 3 2 1 31 30 29 28 27 26

Contents

Preface

The manipulation of waves at subwavelength scales is revolutionising nanotechnology. It has applications in wireless communications, super-resolution biomedical imaging, quantum computing, and more. The aim of these lectures is to present a coherent mathematical theory for subwavelength physics. This theory is derived from the first principles and provides a rigorous framework to explain the spectacular properties of metamaterials. We will consider a variety of Hermitian, non-Hermitian and time-modulated systems of subwavelength resonators and identify the phase transitions and degeneracies in the mathematical structures of those models which are responsible for the exotic phenomena that have been experimentally observed. We will focus on the design of structures whose localisation, confinement, and guiding properties are robust with respect to imperfections.

The main mathematical results will be illustrated numerically. Our analysis in these lectures will be based on *generalised capacitance matrix*, which is a powerful tool for characterising the subwavelength resonant modes of a system of high-contrast resonators. The inclusion of *long-range interactions* is one of its crucial features. The capacitance formulation provides a discrete approximation to the low-frequency part of the spectrum of the continuous PDE model, valid in the high-contrast asymptotic limit. By comparison with discrete approximations used in condensed matter physics, our approach allows us to transpose well-known quantum phenomena to classical wave physics. Comparisons with tight-binding models will elucidate the similarities and fundamental differences between the quantum and classical worlds (such as the presence of long-range interactions). In our subwavelength setting, this approximation is based on first-principles analysis and provides a natural starting point for both theoretical analysis and numerical simulation of wave localisation and manipulation at subwavelength scales.

Introduction

Subwavelength wave physics is concerned with wave interactions in materials with structures at very small scales. Its ultimate goal is to manipulate waves at subwavelength scales. Recent breakthroughs, such as the emergence of the field of metamaterials (*i.e.*, micro-structured materials with unusual properties), have allowed us to do this in a robust way that beats traditional diffraction limits. Our aim in these lectures is to survey recent studies of coupled subwavelength resonator systems using first-principles mathematical analysis and to reveal new fundamental insights into the mechanisms responsible for the peculiar features of wave localisation, scattering, and guiding at deep subwavelength scales.

The starting point for this analysis is that, in the subwavelength limit, resonant eigenmodes are described by the eigenvectors of the generalised capacitance matrix, which appears by perturbing the elements of the kernel of the limiting operator. This discrete formulation can be applied to a wide variety of settings, including both finite- and infinite-sized models and both Hermitian and non-Hermitian systems. As a result, we are able to describe many different exotic phenomena in subwavelength regimes. For non-Hermitian systems, we demonstrate large-scale effective parameters leading to exceptional point degeneracies. As a result, these systems exhibit high sensitivity to perturbations. For periodic systems, we show that these coupled resonator systems can exhibit localised and guided waves on very small length scales. Using the concept of topologically protected edge or interface modes, such a localisation can be made robust against structural imperfections. We also show that these periodic resonator arrays can support subwavelength bandgap opening, Dirac points, momentum bandgaps, Fano-type transmission anomalies, unidirectional reflectionless transmission, extraordinary transmission, and non-reciprocal wave propagation. By introducing random perturbations to these periodic systems, we shed new light on the phenomenon of Anderson localisation in systems with long-range interactions. The localisation can be characterised in terms of level repulsion and we identify a phase transition as an eigenmode swapping event and show that, at this phase transition, strong localisation is achievable with an abnormally small random perturbation.

The results presented in these lectures are based on a new generalisation of the capacitance matrix. This discrete approximation takes the form of a dense matrix operator with long-range off-diagonal terms and a multiplicative material parameter matrix factor. This can be contrasted with existing matrix models used in metamaterial modelling and analysis, which are typically based on tight-binding approximations from condensed matter theory and include only nearest-neighbour interaction terms (long-range interactions are neglected). Condensed matter physics is concerned with situations where quantum physics and many-body interactions play a key role to create new physical phenomena. The capacitance matrix formulation here agrees with the tight-binding approximation only in the case of dilute resonators or one-dimensional (ordinary differential equation) models. Even in the dilute case, long-range interactions between subwavelength resonators are relatively strong, and nearest-neighbour approximations are not applicable [**31**]. The capacitance matrix formalism, which encodes the long-range interactions in the system, is therefore analogous to the tight-binding approximation in the sense that it provides a discrete approximation for classical wave systems as the tight-binding approximation does for quantum ones. Other than the fact that long-range interactions between subwavelength resonators is an essential feature for localisation properties (particularly under random perturbations of material parameters), the other difference from traditional tight-binding models is that hybridised modes of the subwavelength resonance problem are not simply linear combinations of uncoupled modes (since they need to be constant at leading order inside each of the resonators). The highly nontrivial hybridisation in subwavelength resonance problems is captured concisely by the generalised capacitance matrix.

In these lectures, we show that many exotic localisation, guiding, and scattering phenomena are not particular to quantum systems, as conjectured in the seminal papers [**100, 142**]. Subwavelength resonances are the key to realise them in classical wave systems. We describe in detail the mathematical foundations of some of these phenomena, while for others, we simply set out the main results and arguments in support of them.

These lectures are divided into three parts. In the first part, we consider systems of finitely many resonators. In the second part, we investigate the case where the resonators are repeated periodically. The central question of the third part is to what extent the resonant spectra of infinite and truncated structures can be related. Some open questions are listed in the concluding chapter. The appendices contain all the material needed to understand the mathematical and numerical tools used to study subwavelength resonator systems.

This monograph was written for the 2024 NSF–CBMS conference on "Mathematical Methods for Novel Metamaterials", hosted by Auburn University on May 21–25, 2024. It is an expanded version of the first author's CBMS ten lectures devoted to the mathematical foundations of subwavelength physics. We hope that it will demonstrate the challenges, excitement, and opportunities for research at the interface of mathematics and wave physics to the reader.

The first author's warmest thanks go to Yanzhao Cao and Junshan Lin for organizing the CBMS lecture series at Auburn University. The conference provided participants (especially the early-career scientists present) with unique opportunities for discussions, debates, and networking. He also warmly thanks his doctoral students Silvio Barandun, Jinghao Cao, Thea Kosche, Liora Rueff, and Alexander Uhlmann for carefully reading a first version of the text and for suggesting many corrections and improvements.

Part 1

Finite systems of subwavelength resonators

CHAPTER 1

Subwavelength resonances

In this chapter, we introduce a functional analytic approach to characterise a finite system of subwavelength resonators. This yields a discrete approximation to subwavelength scattering and resonance problems in terms of the generalised capacitance matrix. We describe the properties of this generalised capacitance matrix and give leading-order asymptotic expressions for both resonant modes and scattered solutions in terms of its eigenvalues and eigenvectors, which are accompanied by precise error bounds. Our analysis uses a boundary integral approach and is based on the asymptotic perturbation theory of Gohberg and Sigal; see Appendices A.1 and B.2. We also compare the capacitance matrix formalism to the tight-binding approximation in condensed matter physics, from which it has some crucial differences.

1.1. Scattering problem

Here, we assume that $d \in \{2,3\}$, $N \in \mathbb{N}$ and $D_1, D_2, \dots, D_N \subset \mathbb{R}^d$ are disjoint, connected sets with boundaries in $C^{1,s}$ for some $0 < s < 1$. An example of the setting is sketched in Figure 1.1. We use v_i to denote the wave speed in resonator D_i, then $k_i = \omega/v_i$ is the wave number in D_i, where $\omega \in \mathbb{R}, \omega \neq 0$, is the frequency of the incoming wave. Similarly, the wave speed and wave number in the background medium are denoted by v and k. Then, assume that $D := \cup_{i=1}^N D_i$ is illuminated by some incident wave u_{in} and consider the scattering problem

$$
(1.1)\qquad \begin{cases} \Delta u + k^2 u = 0 & \text{in } \mathbb{R}^d \setminus \overline{D}, \\ \Delta u + k_i^2 u = 0 & \text{in } D_i, \text{ for } i = 1, \dots, N, \\ u|_+ - u|_- = 0 & \text{on } \partial D, \\ \delta_i \left.\frac{\partial u}{\partial \nu}\right|_+ - \left.\frac{\partial u}{\partial \nu}\right|_- = 0 & \text{on } \partial D_i \text{ for } i = 1, \dots, N, \\ u - u_{\text{in}} \text{ satisfies the Sommerfeld radiation condition,} \end{cases}
$$

where the Sommerfeld radiation condition says that

$$
(1.2)\qquad \lim_{|x|\to+\infty} |x|^{\frac{d-1}{2}} \left(\frac{\partial}{\partial |x|} - ik \right)(u - u_{\text{in}}) = 0, \quad \text{uniformly in all directions } x/|x|,
$$

and guarantees that the energy is radiated outwards by the scattered solution. Here, $\nu(x)$ is the outward unit normal at $x \in \partial D_i$ and u is the total field, while $u - u_{\text{in}}$ is the scattered field. We assume that the incident field satisfies $\Delta u_{\text{in}} + k^2 u_{\text{in}} = 0$ in $\mathbb{R}^d$ and $\nabla u_{\text{in}}\big|_D = \mathcal{O}(\omega)$. We also assume that all contrast parameters δ_i are small, while the wave speeds are of order 1 as $\delta \to 0$. In other words,

$$
(1.3)\qquad v, v_i = \mathcal{O}(1), \delta_i = \mathcal{O}(\delta) \qquad \text{for } i = 1, \dots, N,
$$

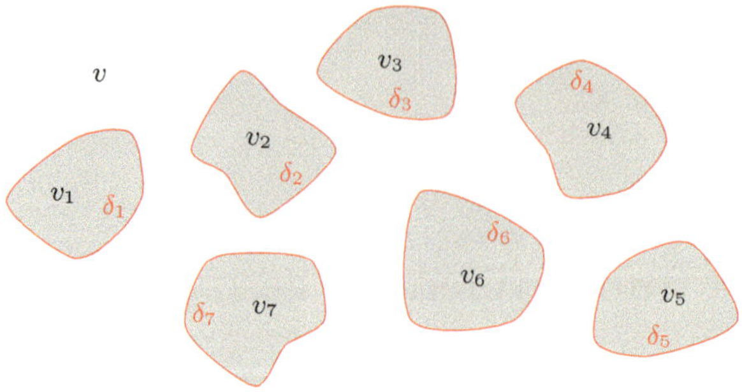

FIGURE 1.1. A finite collection of N resonators, with wave speeds v_i for $i = 1, \dots, N$, in a surrounding medium with wave speed v. The contrast between the ith resonator and the background is given by δ_i, where a small value of δ_i describes a large contrast.

where $0 < \delta \ll 1$. In order to concisely represent the different δ_i, we introduce the function $\widetilde{\delta}(x) \in L^2(\partial D)$ as

$$\widetilde{\delta}(x) = \delta_i \quad \text{for } x \in \partial D_i. \tag{1.4}$$

For $\omega \in \mathbb{R}, \omega \neq 0$, due to the Sommerfeld radiation condition (1.2), the problem (1.1) has a unique solution u. This follows from combining Rellich's lemma together with the unique continuation principle; see Appendix B.2.6.

We wish to characterise the solutions to (1.1) in terms of the system's subwavelength resonant modes. Before making a definition of a resonant mode being *subwavelength* as an asymptotic property in terms of δ, we first re-express the Sommerfeld radiation condition (1.2). To do so, we consider the ball B_R centred at the origin with a radius R large enough to contain D. We introduce the operator $\mathcal{T}_R^k : H^{1/2}(\partial B_R) \to H^{-1/2}(\partial B_R)$, which is given for $d = 2$ by

$$\mathcal{T}_R^k : \phi = \sum_{m \in \mathbb{Z}} \phi_m e^{\mathrm{i}m\theta} \mapsto \sum_{m \in \mathbb{Z}} z_m(k, R) \phi_m e^{\mathrm{i}m\theta},$$

and for $d = 3$ by

$$\mathcal{T}_R^k : \phi = \sum_{m=0}^{+\infty} \sum_{l=-m}^{m} \phi_m^l Y_m^l \mapsto \sum_{m=0}^{+\infty} z_m(k, R) \sum_{l=-m}^{m} \phi_m^l Y_m^l,$$

where

$$z_n(k, R) = \begin{cases} \frac{k (H_m^{(1)})'(kR)}{H_m^{(1)}(kR)} & \text{if } d = 2, \\ \frac{k (h_m^{(1)})'(kR)}{h_m^{(1)}(kR)} & \text{if } d = 3. \end{cases}$$

Here, θ is the angular variable, $\{Y_m^l\}$ are the spherical harmonics, and $\{H_m^{(1)}\}$ (respectively, $\{h_m^{(1)}\}$) are the Hankel functions of integer order (respectively, half-integer order); see Appendix B.2.1.

Then representing u in $\mathbb{R}^d \setminus B_R$ explicitly in terms of Hankel functions (see (B.45)), we can express the Sommerfeld radiation condition (1.2) as follows [**133**]:

$$\frac{\partial u}{\partial \nu} - \mathcal{T}_R^k[u] = \frac{\partial u_{\text{in}}}{\partial \nu} - \mathcal{T}_R^k[u_{\text{in}}] \quad \text{on } \partial B_R.$$

It then follows that the scattering problem (1.1) is equivalent to

$$(1.5)\qquad \begin{cases} \Delta u + k^2 u = 0 & \text{in } B_R \setminus \overline{D}, \\ \Delta u + k_i^2 u = 0 & \text{in } D_i, \text{ for } i = 1, \dots, N, \\ u|_+ - u|_- = 0 & \text{on } \partial D, \\ \delta_i \frac{\partial u}{\partial \nu}\Big|_+ - \frac{\partial u}{\partial \nu}\Big|_- = 0 & \text{on } \partial D_i \text{ for } i = 1, \dots, N, \\ \frac{\partial u}{\partial \nu} - \mathcal{T}_R^k[u] = \frac{\partial u_{\text{in}}}{\partial \nu} - \mathcal{T}_R^k[u_{\text{in}}] & \text{on } \partial B_R. \end{cases}$$

We are now ready to define subwavelength resonant frequencies and modes.

DEFINITION 1.1 (Subwavelength resonant frequency). *Given $\delta > 0$, a subwavelength resonant frequency $\omega = \omega(\delta) \in \mathbb{C}$ is defined to be such that*

(i) *there exists a nontrivial solution to (1.5) with $u_{\text{in}} \equiv 0$, known as an associated resonant mode;*

(ii) *ω depends continuously on δ and satisfies $\omega \to 0$ as $\delta \to 0$.*

Note that Definition 1.1 is independent of R. Note also that a subwavelength resonant frequency must have a negative imaginary part.

In order to understand the subwavelength resonance problem (*i.e.*, the problem of finding the subwavelength resonant frequencies), we first reframe problem (1.1) (or equivalently, (1.5)) as an operator equation using layer potential techniques.

1.2. Boundary integral formulation

The main tool that allows us to reframe problem (1.1) as an operator equation is the Helmholtz single layer potential; see Appendix B.2.4. This is an operator $\mathcal{S}_D^\omega : L^2(\partial D) \to H^1_{\text{loc}}(\mathbb{R}^d)$ that is defined as

$$(1.6)\qquad \mathcal{S}_D^\omega[\varphi](x) = \int_{\partial D} G^\omega(x-y)\varphi(y)\,\mathrm{d}\sigma(y), \quad x \in \mathbb{R}^d, \ \varphi \in L^2(\partial D),$$

where $H^1_{\text{loc}}(\mathbb{R}^d)$ is the set of functions in $H^1(A)$ (the usual Sobolev space of square-integrable functions whose weak derivative is square-integrable) for all compact subsets $A \subset \mathbb{R}^d$, while G^ω is the Helmholtz Green function, given by

$$(1.7)\qquad G^\omega(x) = \begin{cases} -\frac{\mathrm{i}}{4} H_0^{(1)}(\omega|x|), & d = 2, \\ -\frac{1}{4\pi|x|} e^{\omega|x|}, & d = 3, \end{cases} \qquad x \neq 0, \ \mathrm{Re}(\omega) > 0.$$

Here, $H_0^{(1)}$ is the Hankel function of the first kind and order zero. In the case $\omega = 0$, G^0 is the Laplace Green function given by (see Appendix B.2.2)

$$(1.8)\qquad G^0(x) = \begin{cases} \frac{1}{2\pi} \ln|x|, & d = 2, \\ -\dfrac{1}{4\pi|x|}, & d = 3, \end{cases} \qquad x \neq 0.$$

The single layer potential is useful because it allows us to seek solutions to (1.1) of the form

$$(1.9)\qquad u(x) = \begin{cases} \mathcal{S}_D^k[\phi](x) & x \in \mathbb{R}^d \setminus \overline{D}, \\ \mathcal{S}_D^{k_i}[\psi](x) & x \in D_i, \end{cases}$$

where $\phi, \psi \in L^2(\partial D)$ are density functions that need to be found. Here, we make the identification $L^2(\partial D) = L^2(\partial D_1) \times \ldots \times L^2(\partial D_N)$ and decompose the single layer potential as

$$\mathcal{S}_D^k = \begin{pmatrix} \mathcal{S}_{D_1}^k & \mathcal{S}_{D_2}^k\big|_{\partial D_1} & \cdots & \mathcal{S}_{D_N}^k\big|_{\partial D_1} \\ \mathcal{S}_{D_1}^k\big|_{\partial D_2} & \mathcal{S}_{D_2}^k & \cdots & \mathcal{S}_{D_N}^k\big|_{\partial D_2} \\ \vdots & \vdots & \ddots & \vdots \\ \mathcal{S}_{D_1}^k\big|_{\partial D_N} & \mathcal{S}_{D_2}^k\big|_{\partial D_N} & \cdots & \mathcal{S}_{D_N}^k \end{pmatrix},$$

where $\mathcal{S}_{D_i}^k\big|_{\partial D_j}$ is the evaluation of $\mathcal{S}_{D_i}^k$ on ∂D_j. The value of this representation is that a solution of the form (1.9) necessarily satisfies the Helmholtz equations and the Sommerfeld radiation condition in problem (1.1). In addition, the different wave numbers k_i inside D_i have been taken into account using different single layer potentials $\mathcal{S}_D^{k_i}$. We can collectively represent these single layer potentials through a single operator $\widetilde{\mathcal{S}}_D^\omega$, defined piecewise for $i = 1, \ldots, N$ as

$$\widetilde{\mathcal{S}}_D^\omega[\varphi](x) = \mathcal{S}_D^{k_i}[\varphi](x), \quad x \in D_i, \ \varphi \in L^2(\partial D). \tag{1.10}$$

We emphasise that $\widetilde{\mathcal{S}}_D^0 = \mathcal{S}_D^0$. Now, it only remains to find ψ and ϕ such that the transmission conditions on ∂D are fulfilled. This can be achieved through the introduction of an additional integral operator, the Neumann-Poincaré operator associated to D. This is an operator $\mathcal{K}_D^{\omega,*}$ on $L^2(\partial D)$ that is defined as

$$\mathcal{K}_D^{\omega,*}[\varphi](x) = \int_{\partial D} \frac{\partial}{\partial \nu(x)} G^\omega(x-y)\varphi(y)\, \mathrm{d}\sigma(y), \quad x \in \partial D, \ \varphi \in L^2(\partial D). \tag{1.11}$$

We define $\widetilde{\mathcal{K}}_D^{\omega,*}$ in the same spirit as $\widetilde{\mathcal{S}}_D^\omega$, namely,

$$\widetilde{\mathcal{K}}_D^{\omega,*}[\varphi](x) = \mathcal{K}_D^{k_i,*}[\varphi](x), \quad x \in \partial D_i, \ \varphi \in L^2(\partial D). \tag{1.12}$$

We are now able to describe how $\mathcal{S}_D^\omega$ and its normal derivative behave on ∂D. In particular, from Appendix B.2.4, it holds that for any $\varphi \in L^2(\partial D)$

$$\mathcal{S}_D^\omega[\varphi]\big|_+ = \mathcal{S}_D^\omega[\varphi]\big|_-, \tag{1.13}$$

and

$$\frac{\partial}{\partial \nu}\mathcal{S}_D^\omega[\varphi]\Big|_\pm = \left(\pm\frac{1}{2}I + \mathcal{K}_D^{\omega,*}\right)[\varphi], \tag{1.14}$$

where the subscripts + and − denote taking the limit from outside and inside the boundary ∂D, respectively. With abuse of notation, we use the same notation for $\mathcal{S}_D^\omega : L^2(\partial D) \to H^1_{\mathrm{loc}}(\mathbb{R}^d)$ as for its trace on ∂D.

With the so-called *jump conditions* (1.13) and (1.14) in hand, we can derive the following lemma, which characterises the subwavelength resonance problem as a boundary integral equation (see Proposition B.20 and Theorem B.21).

LEMMA 1.2. *The subwavelength resonance problem is equivalent to finding $\omega(\delta)$ such that $\omega(\delta) \to 0$ as $\delta \to 0$ and such that there exists a nontrivial pair of densities $(\phi, \psi) \in L^2(\partial D) \times L^2(\partial D)$ satisfying*

$$\mathcal{A}(\omega, \delta)\begin{pmatrix}\psi \\ \phi\end{pmatrix} = \begin{pmatrix}0 \\ 0\end{pmatrix}, \tag{1.15}$$

where the operator $\mathcal{A}(\omega,\delta) : L^2(\partial D) \times L^2(\partial D) \to H^1(\partial D) \times L^2(\partial D)$ *is defined as*

$$\mathcal{A}(\omega,\delta) = \begin{pmatrix} \widetilde{\mathcal{S}}_D^{\omega} & -\mathcal{S}_D^k \\ -\frac{1}{2}I + \widetilde{\mathcal{K}}_D^{\omega,*} & -\widetilde{\delta}\left(\frac{1}{2}I + \mathcal{K}_D^{k,*}\right) \end{pmatrix}, \tag{1.16}$$

where, as in (1.4), $\widetilde{\delta}(x) = \delta_i$ *for* $x \in \partial D_i$.

Turning to the scattering problem (1.1), the following result holds; see Theorem B.21.

LEMMA 1.3. *For real* $\omega \neq 0$ *such that* k_i^2 *is not a Dirichlet eigenvalue for* $-\Delta$ *on* D_i, $i = 1, \ldots, N$, *the solution* u *of the scattering problem* (1.1) *can be represented using the single layer potentials* $\mathcal{S}_D^k$ *and* $\widetilde{\mathcal{S}}_D^{\omega}$ *as follows:*

$$u(x) = \begin{cases} u_{\text{in}}(x) + \mathcal{S}_D^k[\phi](x), & x \in \mathbb{R}^3 \setminus \overline{D}, \\ \widetilde{\mathcal{S}}_D^{\omega}[\psi](x), & x \in D, \end{cases} \tag{1.17}$$

where the pair $(\phi,\psi) \in L^2(\partial D) \times L^2(\partial D)$ *is the unique solution to*

$$\mathcal{A}(\omega,\delta)\begin{pmatrix}\psi \\ \phi\end{pmatrix} = \begin{pmatrix} u_{\text{in}} \\ \widetilde{\delta}\frac{\partial u_{\text{in}}}{\partial \nu}\end{pmatrix} \quad \text{on } \partial D. \tag{1.18}$$

Lemma 1.2 shows that finding a subwavelength resonant frequency of the system is equivalent to finding, for a given δ, a characteristic value ω of $\mathcal{A}(\omega,\delta)$ that is such that $\omega(\delta) \to 0$ as $\delta \to 0$. Our approach to finding such solutions is to consider perturbations of the elements of the kernel of $\mathcal{A}(0,0)$. We will see that this space has dimension equal to the number of distinct resonators in the structure. Once we understand $\operatorname{Ker}\mathcal{A}(0,0)$, we can characterise the characteristic values of $\mathcal{A}(\omega,\delta)$ for small ω and δ as perturbations of this space. Our analysis is based on the asymptotic perturbation theory of Gohberg and Sigal [**37, 74, 92**] and allows us both to prove the existence of subwavelength resonant modes (satisfying Definition 1.1) and to derive asymptotic formulas for their values.

As will be shown in Section 1.5, using the subwavelength resonant modes, Lemma 1.3 yields a modal decomposition for solutions to the scattering problem with ω in a neighbourhood of 0.

1.3. Functional analytic approach

In this section, we will explore the use of a functional analytic approach to study scattering by a high-contrast material inclusion. This approach was first developed in this setting by [**34**] but, as we shall see, has since been developed to cover a variety of different applications. It has also been used to describe subwavelength resonance in a variety of different physical settings. For instance, it was used to characterise a system of subwavelength Helmholtz resonators in [**54**], plasmonic particles in [**51**] and high-contrast dielectric resonators in [**19, 50**] (see also Appendix C.3).

1.3.1. Abstract capacitance matrix. Let $\mathcal{H}$ and $\mathcal{H}'$ be two Hilbert spaces. We consider an operator-valued function $\omega \mapsto \mathcal{A}(\omega,\delta) \in \mathcal{L}(\mathcal{H},\mathcal{H}')$, where $\mathcal{L}(\mathcal{H},\mathcal{H}')$ is the space of bounded linear operators from $\mathcal{H}$ into $\mathcal{H}'$. We assume that $\mathcal{A}(\omega,\delta)$ is Fredholm of index zero and is holomorphic with respect to ω and δ.

At $\delta = 0$, we assume that the operator $\mathcal{A}(\omega, 0)$ has a characteristic value $\omega = 0$ of multiplicity $2N$, admitting the following pole-pencil decomposition:

$$\mathcal{A}(\omega,0)^{-1} = \frac{K}{\omega^2} + \mathcal{R}(\omega), \qquad \text{for} \quad K = \sum_{i=1}^{N} \langle \Phi_i, \cdot \rangle \Psi_i, \tag{1.19}$$

where $\operatorname{Ker} \mathcal{A}(0,0) = \operatorname{span}\{\Psi_j\}$, $\operatorname{Ker} \mathcal{A}^*(0,0) = \operatorname{span}\{\Phi_j\}$ and $\mathcal{R}$ is holomorphic for ω in a neighbourhood of 0. Here, $\mathcal{A}^*$ denotes the adjoint of $\mathcal{A}$. Moreover, we assume that $\mathcal{A}(\omega, \delta)$, for small but nonzero δ, satisfies

$$\mathcal{A}(\omega,\delta) = \mathcal{A}(\omega,0) + \mathcal{L}(\omega,\delta),$$

for some operator $\mathcal{L}$ satisfying (in corresponding operator norm) $\|\mathcal{L}\| = \mathcal{O}(\delta)$ uniformly for ω in a neighbourhood of 0.

By the generalised Rouché theorem (Theorem A.16), it follows that $\mathcal{A}(\cdot, \delta)$ has $2N$ characteristic values in a small neighbourhood of 0. In order to derive asymptotic formulas (in δ) of these characteristic values, we solve the equation $\mathcal{A}(\omega,\delta)[\Phi] = 0$. Multiplying with $\mathcal{A}(\omega,0)^{-1}$, we have

$$0 = \mathcal{A}(\omega,0)^{-1}\mathcal{A}(\omega,\delta)[\Phi] = \mathcal{A}(\omega,0)^{-1}\left(\mathcal{A}(\omega,0) + \mathcal{L}\right)[\Phi] = \left(I + \frac{K\mathcal{L}}{\omega^2} + \mathcal{R}\mathcal{L}\right)[\Phi].$$

Defining $\mathcal{B}(\omega,\delta) = \omega^2\mathcal{R}(\omega)\mathcal{L}(\omega,\delta)$ yields

$$\left(\omega^2 I + K\mathcal{L} + \mathcal{B}\right)[\Phi] = 0. \tag{1.20}$$

The characteristic values of $\mathcal{A}(\omega,\delta)$ are therefore determined by (1.20), which is generally a nonlinear eigenvalue problem since $\mathcal{L}$ and $\mathcal{B}$ depend on ω. For small ω, we have $\|\mathcal{B}\| = \mathcal{O}(\omega^2\delta)$ uniformly for ω and δ around 0. Similarly, we have $\mathcal{L} = \mathcal{L}_0 + \widehat{\mathcal{L}}$, where $\mathcal{L}_0$ is independent of ω and $\widehat{\mathcal{L}} = \mathcal{O}(\omega\delta)$. Therefore, the characteristic values of $\mathcal{A}(\omega,\delta)$ in a small neighbourhood of 0 are approximated by $\pm$ the square roots of the eigenvalues of the finite-rank operator $-K\mathcal{L}_0$ whose restriction to $\operatorname{Ker}\mathcal{A}(0,0)$ is given by the N by N matrix with entries

$$\mathcal{C}_{ij} = -\langle \Phi_i, \mathcal{L}_0[\Psi_j] \rangle.$$

Then the characteristic values of $\mathcal{A}(\omega,\delta)$ satisfy

$$\omega_n = \pm\sqrt{\lambda_n} + \mathcal{O}(\delta),$$

where λ_n are the eigenvalues of $\mathcal{C}$. Note that ω_n are of order $\sqrt{\delta}$ since $\|\mathcal{L}_0\| = \mathcal{O}(\delta)$.

Assume that instead of (1.19) we have the following pole-pencil decomposition:

$$\mathcal{A}(\omega,0)^{-1} = \frac{K + \omega\mathcal{R}^{(1)}}{\omega^2} + \mathcal{R}^{(2)}(\omega), \tag{1.21}$$

where $\mathcal{R}^{(1)}$ is independent of ω and $\mathcal{R}^{(2)}$ is holomorphic in ω in a neighbourhood of 0. We also assume that

$$\mathcal{A}(\omega,\delta) = \mathcal{A}(\omega,0) + \mathcal{L}(\omega,\delta), \quad \mathcal{L} = \mathcal{L}_0 + \widehat{\mathcal{L}}, \tag{1.22}$$

where $\|\mathcal{L}\| = \mathcal{O}(\delta)$, $\|\widehat{\mathcal{L}}\| = \mathcal{O}(\omega^2\delta)$ uniformly for ω in a neighbourhood of 0.

Then, analogously to (1.20), we have

$$\left(\omega^2 I + K\mathcal{L}_0 + \omega\mathcal{R}^{(1)}\mathcal{L}_0 + \mathcal{B}\right)[\Phi] = 0,$$

where $\|\mathcal{B}\| = \mathcal{O}(\omega^2\delta)$ uniformly for ω and δ around 0. Therefore, under the assumption that all the eigenvalues λ_n of $\mathcal{C}$ are simple, the characteristic values of $\mathcal{A}(\omega,\delta)$ are approximated as follows:

$$\omega_n = \pm\sqrt{\lambda_n} + \langle \mathcal{R}^{(1)}\mathcal{L}_0[\mathbf{v}_n], \mathbf{v}_n\rangle + \mathcal{O}(\delta^{3/2}), \tag{1.23}$$

where $\mathbf{v}_n$ are the normalised eigenvectors associated to the eigenvalues λ_n. In (1.23), the correction term is of order δ since $\|\mathcal{L}_0\| = \mathcal{O}(\delta)$.

1.3.2. Capacitance matrix formulation of the subwavelength resonance problem. The approach outlined in Section 1.3.1 can now be applied to Lemma 1.2 to prove the existence of subwavelength resonances, as defined in Definition 1.1 for a finite collection of N resonators in $d = 3$, and derive their asymptotic behaviour as $\delta \to 0$. Using the abstract setting in Section 1.3.1, we can derive a capacitance formulation of subwavelength resonances in three dimensions.

Let $d = 3$ and $\mathcal{H} = L^2(\partial D) \times L^2(\partial D), \mathcal{H}' = H^1(\partial D) \times L^2(\partial D)$. The idea here is to study the kernel of $\mathcal{A}(0,0) \in \mathcal{L}(\mathcal{H},\mathcal{H}')$, where

$$\mathcal{A}(0,0) = \begin{pmatrix} \mathcal{S}_D^0 & -\mathcal{S}_D^0 \\ -\frac{1}{2}I + \mathcal{K}_D^{0,*} & 0 \end{pmatrix}, \tag{1.24}$$

and then understand how $\operatorname{Ker}\mathcal{A}(0,0)$ is perturbed when δ and ω are nonzero. The following lemma describes the two operators that appear in $\mathcal{A}(0,0)$, as given in (1.24); see Lemmas B.5 and B.6.

LEMMA 1.4. *Consider a system of N subwavelength resonators $D = D_1 \cup \cdots \cup D_N$ in $\mathbb{R}^3$. Then,*

(i) *the Laplace single layer potential $\mathcal{S}_D^0 : L^2(\partial D) \to H^1(\partial D)$ is invertible;*
(ii) $\operatorname{Ker}(-\frac{1}{2}I + \mathcal{K}_D^{0,*}) = \operatorname{span}\{\psi_1, \psi_2, \ldots, \psi_N\}$, *where*

$$\psi_i := (\mathcal{S}_D^0)^{-1}[\chi_{\partial D_i}], \tag{1.25}$$

and $\chi_{\partial D_i}$ denotes the characteristic function of ∂D_i, for $i = 1, \ldots, N$.

From Lemma 1.4 we can see that $\mathcal{A}(0,0)$ has a N-dimensional kernel. Therefore, $\omega = 0$ is a characteristic value of $\mathcal{A}(\omega, 0)$. Due to symmetry, the multiplicity of $\omega = 0$ is, in fact, $2N$. When the material parameters are real, it is easy to see that $\overline{\mathcal{A}(\omega,\delta)} = \mathcal{A}(-\overline{\omega},\delta)$, from which we can see that the subwavelength resonant frequencies will be symmetric with respect to the imaginary axis, in the sense described in Lemma 1.5.

LEMMA 1.5. *The set of subwavelength resonant frequencies is symmetric in the imaginary axis. In particular, if $\delta_i, v_i \in \mathbb{R}$ for all $i = 1, \ldots, N$, and if ω is such that* (1.15) *is satisfied for some nonzero $\phi, \psi \in L^2(\partial D)$, then it will also hold that*

$$\mathcal{A}(-\overline{\omega},\delta)\begin{pmatrix} \overline{\psi} \\ \overline{\phi} \end{pmatrix} = \begin{pmatrix} 0 \\ 0 \end{pmatrix}.$$

With Lemma 1.5 in mind, we will subsequently state results only for the subwavelength resonant frequencies with non-negative real parts.

Let the operator $\mathcal{A} \in \mathcal{L}(\mathcal{H},\mathcal{H}')$ be given by (1.16). We then have (1.22) with

$$\mathcal{A}(\omega,0) = \begin{pmatrix} \widetilde{\mathcal{S}}_D^{\omega} & -\mathcal{S}_D^k \\ -\frac{1}{2}I + \widetilde{\mathcal{K}}_D^{\omega,*} & 0 \end{pmatrix}, \quad \mathcal{L}(\omega,\delta) = \begin{pmatrix} 0 & 0 \\ 0 & -\widetilde{\delta}\left(\frac{1}{2}I + \mathcal{K}_D^{k,*}\right) \end{pmatrix},$$

and

$$\mathcal{L}_0(\delta) = \begin{pmatrix} 0 & 0 \\ 0 & -\widetilde{\delta}\left(\frac{1}{2}I + \mathcal{K}_D^{0,*}\right) \end{pmatrix}.$$

Note that $\|\mathcal{L}_0\| = \mathcal{O}(\delta)$ and from (B.59) $\|\mathcal{L} - \mathcal{L}_0\| = \mathcal{O}(\omega^2\delta)$. Moreover, from Proposition B.25, it can be shown that $\mathcal{A}(\omega, 0)^{-1}$ satisfies (1.21) where

$$\Phi_i = -\frac{v_i^2}{|D_i|}\begin{pmatrix} 0 \\ \chi_{\partial D_i} \end{pmatrix}, \qquad \Psi_j = \begin{pmatrix} \psi_j \\ \psi_j \end{pmatrix}, \quad \psi_j = (\mathcal{S}_D^0)^{-1}[\chi_{\partial D_j}],$$

and $\mathcal{R}^{(1)}$ is given by

$$\mathcal{R}^{(1)} = \begin{pmatrix} \mathcal{R}_{11}^{(1)} & \mathcal{R}_{12}^{(1)} \\ 0 & 0 \end{pmatrix}$$

with

$$\mathcal{R}_{11}^{(1)} := \frac{v_i}{|D_i|}(\mathcal{S}_D^0)^{-1}\mathcal{S}_{D,1}(\mathcal{S}_D^0)^{-1}\big[\chi_{\partial D_i}\big]\langle \chi_{\partial D_i}, \cdot\rangle$$

and

$$\mathcal{R}_{12}^{(1)} := \frac{v}{|D_i|}(\mathcal{S}_D^0)^{-1}\mathcal{S}_{D,1}(\mathcal{S}_D^0)^{-1}\big[\chi_{\partial D_i}\big]\langle \chi_{\partial D_i}, \cdot\rangle.$$

Combining Lemma 1.4 together with Proposition B.25, the following result holds by applying the generalised Rouché theorem (Theorem A.16).

THEOREM 1.6. *Consider a system of N subwavelength resonators in $\mathbb{R}^d$ for $d \in \{2, 3\}$. For sufficiently small $\delta > 0$, there exist N subwavelength resonant frequencies $\omega_1(\delta), \dots, \omega_N(\delta)$ with non-negative real parts.*

On the other hand, from this it is straightforward to compute

$$\mathcal{C}_{ij} = -\frac{\delta_i v_i^2}{|D_i|}\langle \chi_{\partial D_i}, \psi_j\rangle.$$

We then make the following definitions.

DEFINITION 1.7 (Capacitance matrix). *For a system of $N \in \mathbb{N}$ resonators $D_1, \dots, D_N$ in $\mathbb{R}^3$ we can define the capacitance matrix $C = (C_{ij}) \in \mathbb{R}^{N\times N}$ to be the square matrix given by*

$$C_{ij} = -\int_{\partial D_i} (\mathcal{S}_D^0)^{-1}[\chi_{\partial D_j}]\,\mathrm{d}\sigma, \quad i, j = 1, \dots, N. \tag{1.26}$$

Due to the different material parameters inside each resonator we introduce the generalised capacitance matrix, which is the main quantity we use to describe the subwavelength resonances.

DEFINITION 1.8 (Generalised capacitance matrix). *For a system of $N \in \mathbb{N}$ resonators $D_1, \dots, D_N$ in $\mathbb{R}^3$, we can define the generalised capacitance matrix, denoted by $\mathcal{C} = (\mathcal{C}_{ij}) \in \mathbb{C}^{N\times N}$, to be the square matrix given by*

$$\mathcal{C}_{ij} = \frac{\delta_i v_i^2}{|D_i|} C_{ij}, \quad i, j = 1, \dots, N, \tag{1.27}$$

or equivalently,

$$\mathcal{C} = VC, \qquad V = \begin{pmatrix} \frac{v_1^2\delta_1}{|D_1|} & & \\ & \ddots & \\ & & \frac{v_N^2\delta_N}{|D_N|} \end{pmatrix}.$$

We can now show the following theorem by using the abstract approach described in Section 1.3.1.

THEOREM 1.9. *Let $d = 3$. Consider a system of N subwavelength resonators in $\mathbb{R}^3$. As $\delta \to 0$, the N subwavelength resonant frequencies (with non-negative real parts) satisfy the asymptotic formula*

$$\omega_n = \sqrt{\lambda_n} + \mathcal{O}(\delta), \quad n = 1, \ldots, N, \tag{1.28}$$

where $\{\lambda_n : n = 1, \ldots, N\}$ are the eigenvalues of the generalised capacitance matrix $\mathcal{C} \in \mathbb{C}^{N \times N}$, which satisfy $\lambda_n = \mathcal{O}(\delta)$ as $\delta \to 0$.

We refer to Theorem 1.31 for an explicit form of the term of order δ in the case where the material parameters on each resonator are the same. Note that the use of (1.23) leads to a higher-order correction in formula (1.28).

REMARK 1.10. *The assumption that the dimension is $d = 3$ in Definitions 1.7 and 1.8 is important as the Laplace single layer potential $\mathcal{S}_D^0$ is invertible in this case. As will be seen in Section 1.8, this is generally not the case when $d = 2$, which means that the corresponding version of Theorem 1.9 is slightly less elegant.*

COROLLARY 1.11. *Let $d = 3$. Let $\mathbf{v}_n$ be a normalised eigenvector of $\mathcal{C}$ associated with the eigenvalue λ_n. Then a normalised resonant mode u_n associated to the resonant frequency ω_n is given, as $\delta \to 0$, by*

$$u_n(x) = \begin{cases} \mathbf{v}_n \cdot \mathbf{S}_D^{\omega_n/v}(x) + \mathcal{O}(\delta^{1/2}), & x \in \mathbb{R}^3 \setminus \overline{D}, \\ \mathbf{v}_n \cdot \mathbf{S}_D^{\omega_n/v_i}(x) + \mathcal{O}(\delta^{1/2}), & x \in D_i, \end{cases} \tag{1.29}$$

where $S_D^{\omega_n/v} : \mathbb{R}^3 \to \mathbb{C}^N$ is the vector-valued function given by

$$\mathbf{S}_D^{\omega_n/v}(x) = \begin{pmatrix} \mathcal{S}_D^{\omega_n/v}[\psi_1](x) \\ \vdots \\ \mathcal{S}_D^{\omega_n/v}[\psi_N](x) \end{pmatrix},$$

with $\psi_i := (\mathcal{S}_D^0)^{-1}[\chi_{\partial D_i}]$ for $i = 1, \ldots, N$, and $S_D^{\omega_n/v_i}$ is defined analogously.

REMARK 1.12. *The capacitance matrix is defined solely in terms of the kernel of the integral operators $\mathcal{A}(0,0)$ and $\mathcal{A}^*(0,0)$. In fact, the adjoint $\mathcal{K}_D^0$ of the Neumann-Poincaré operator $\mathcal{K}_D^{0,*}$ satisfies*

$$\mathrm{Ker}\,(-\frac{1}{2}I + \mathcal{K}_D^0) = \mathrm{span}\{\chi_{\partial D_1}, \chi_{\partial D_2}, \ldots, \chi_{\partial D_N}\}.$$

Then,

$$C_{ij} = -\left\langle \chi_{\partial D_i}, \psi_j \right\rangle,$$

where, as in Lemma 1.4, $\{\psi_1, \ldots, \psi_N\}$ is a basis for $\mathrm{Ker}\,(-\frac{1}{2}I + \mathcal{K}_D^{0,*})$. In Section 1.3.1, we have used these ideas to define the generalised capacitance matrix purely in terms of the integral operator $\mathcal{A}$, thus providing a general method for studying subwavelength resonance systems.

1.4. Properties of the capacitance matrix

Through Theorem 1.9 and Corollary 1.11, we have reduced the subwavelength resonance problem to a matrix eigenproblem for the generalised capacitance matrix $\mathcal{C}$. We now wish to understand the properties of $\mathcal{C}$.

LEMMA 1.13. *For $i = 1, \ldots, N$, let V_i be defined as the solution to the exterior boundary value problem*

$$
\begin{cases} \Delta V_i = 0 & \text{in } \mathbb{R}^3 \setminus \overline{D}, \\ V_i = \delta_{ij} & \text{on } \partial D_j, \text{ for } j = 1, \ldots, N, \\ V_i(x) = \mathcal{O}\left(|x|^{-1}\right) & \text{as } |x| \to +\infty, \end{cases} \tag{1.30}
$$

where δ_{ij} is the Kronecker delta. Then, the capacitance coefficients, defined in Definition 1.7, are given by

$$
C_{ij} = \int_{\mathbb{R}^3 \setminus D} \nabla V_i \cdot \nabla V_j \, \mathrm{d}x \quad \text{for } i, j = 1, \ldots, N.
$$

PROOF. From item (i) in Lemma B.6, which is true for D with N connected components, it follows that the unique solution V_i to (1.30) is given by

$$
V_i = \mathcal{S}_D^0[(\mathcal{S}_D^0)^{-1}[\chi_{\partial D_i}]].
$$

Then by definition,

$$
\begin{aligned} C_{ij} &= -\int_{\partial D_i} (\mathcal{S}_D^0)^{-1}[\chi_{\partial D_j}] \, \mathrm{d}\sigma \\ &= -\int_{\partial D_i} \chi_{\partial D_i}(\frac{1}{2}I + \mathcal{K}_D^{0,*})(\mathcal{S}_D^0)^{-1}[\chi_{\partial D_j}] \, \mathrm{d}\sigma \\ &= -\int_{\partial D_i} V_i \frac{\partial V_j}{\partial \nu} \, \mathrm{d}\sigma \\ &= \int_{\mathbb{R}^3 \setminus D} \nabla V_i \cdot \nabla V_j \, \mathrm{d}x, \end{aligned} \tag{1.31}
$$

after integrating by parts. Here, we have used the jump relations (1.14) and the fact that $V_i = 1$ on ∂D_i. □

From Lemma 1.13 emerges a slightly different explanation for why the capacitance matrix approximation works. In the limiting case, when $\delta_i = 0$ for all i, the scattering problem (1.1) with $u_{\text{in}} \equiv 0$ is reduced to a Neumann boundary value problem in the interior of D and a Dirichlet boundary value problem in the exterior of D. When $\omega = 0$, the interior Neumann problem is solved by constant functions, which means that the solution of the exterior Dirichlet boundary value problem is a linear combination of the functions $V_1, \ldots, V_N$. The projection onto this finite-dimensional space yields a leading-order approximation of the solution as $\omega, \delta \to 0$, in the form of an eigenvalue problem for the generalised capacitance matrix. Lemma 1.13 is also useful as it allows us to immediately see the symmetry of the capacitance matrix.

LEMMA 1.14. *The capacitance matrix C is symmetric and positive definite.*

PROOF. Let $\xi = (\xi_1, \ldots, \xi_N)^\top \in \mathbb{R}^N$ and let $V = \sum_{i=1}^N \xi_i V_i$. Lemma 1.13 yields

$$
\sum_{i,j=1}^N \xi_i C_{ij} \xi_j = \int_{\mathbb{R}^3 \setminus D} |\nabla V|^2 \, \mathrm{d}x.
$$

Moreover, $\nabla V = 0$ implies that $V = 0$ in $\mathbb{R}^3 \setminus D$ and thus $\xi = 0$. □

Applying the maximum principle to the solutions of (1.30) yields the following results [**84**].

LEMMA 1.15. *The extra-diagonal entries of C are negative: $C_{ij} < 0$ for any $1 \leq i \neq j \leq N$. Moreover, C is diagonally dominant:*

$$C_{ii} > \sum_{j \neq i} |C_{ij}|, \quad \forall\, 1 \leq i \leq N.$$

PROOF. The maximum principle implies the inequality $0 \leq V_i(x) \leq 1$. Hence, it is necessary that $\nabla V_i \cdot \nu > 0$ on ∂D_j with $j \neq i$ and $\nabla V_i \cdot \nu < 0$ on ∂D_i. Then from Lemma 1.13, it follows that $C_{ij} < 0$ for any $1 \leq i \neq j \leq N$. Next, let $V = \sum_{i=1}^N V_i$. By the maximum principle, it holds $0 \leq V(x) \leq 1$ in $\mathbb{R}^3 \setminus D$, which implies $\nabla V \cdot \nu < 0$ on ∂D. Therefore, the following inequality holds for any $1 \leq i \leq N$:

$$0 < -\int_{\partial D_i} \frac{\partial V}{\partial \nu}\, \mathrm{d}\sigma = -\sum_{j=1}^N \int_{\partial D_i} \frac{\partial V_j}{\partial \nu}\, \mathrm{d}\sigma = \sum_{j=1}^N C_{ij} = C_{ii} - \sum_{j \neq i} |C_{ij}|.$$

□

The symmetry and positive definiteness of the capacitance matrix C is useful for understanding the properties of the generalised capacitance matrix $\mathcal{C}$, which is the product of C with a diagonal matrix containing the weights $\delta_i v_i^2/|D_i|$. In the case where $\delta_i v_i^2 > 0$, this diagonal matrix is positive definite, so, for example, we have the following lemma, which can be proved using the fact that C is always real symmetric.

LEMMA 1.16. *If $\delta_i v_i^2$, for $i = 1, \ldots, N$, are real-valued positive numbers, then the generalised capacitance matrix $\mathcal{C} \in \mathbb{C}^{N \times N}$ has N linearly independent eigenvectors.*

REMARK 1.17. *We will see later that nonzero imaginary parts of $\delta_i v_i^2$ can be used to model damping and amplification in the system. In this case, we can create* exceptional points *where eigenvalues and eigenvectors coincide and $\mathcal{C}$ is not diagonalisable.*

Explicit formulas for capacitance coefficients are generally beyond reach. However, by making some additional assumptions, we can make the capacitance matrix easier to understand. For example, spherical resonators can be easily described using appropriate radial coordinate systems. In this way, we can see that if D is a single sphere of radius r then its capacitance is given by

$$\mathrm{Cap}_D := -\int_{\partial D} (\mathcal{S}_D^0)^{-1}[\chi_{\partial D}]\, \mathrm{d}\sigma = 4\pi r.$$

Similarly, if we consider the case of two spherical resonators, as depicted in Figure 1.2, then we can use a bispherical coordinate system to derive explicit formulas for the capacitance coefficients [**30**].

LEMMA 1.18. *Suppose that $D = D_1 \cup D_2$ consists of two spheres of radius r_1 and r_2 separated by a distance d. Define the quantities α, ξ_1 and ξ_2 as*

$$\alpha = \frac{\sqrt{d(2r_1+d)(2r_2+d)(2r_1+2r_2+d)}}{2(r_1+r_2+d)} \quad \text{and} \quad \xi_i = \sinh^{-1}\left(\frac{\alpha}{r_i}\right), \text{ for } i = 1, 2.$$

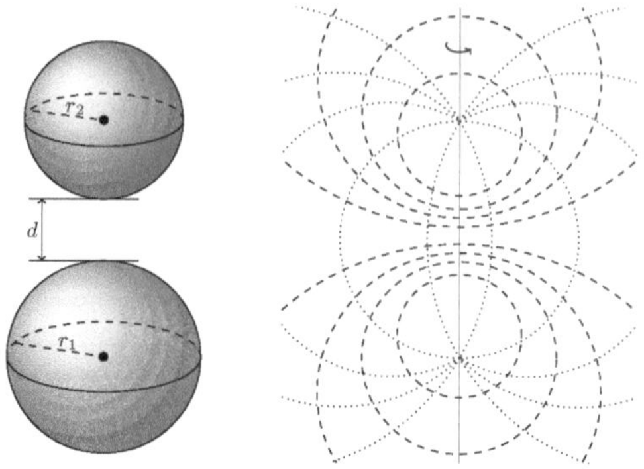

FIGURE 1.2. A system of two spherical resonators can be described using bispherical coordinates. Such a coordinate system is convenient since the boundaries of the spheres lie on level sets and the capacitance coefficients can be calculated explicitly.

Then,

$$C_{11} = 8\pi\alpha \sum_{n=0}^{+\infty} \frac{e^{(2n+1)\xi_2}}{e^{(2n+1)(\xi_1+\xi_2)} - 1}, \qquad C_{22} = 8\pi\alpha \sum_{n=0}^{+\infty} \frac{e^{(2n+1)\xi_1}}{e^{(2n+1)(\xi_1+\xi_2)} - 1},$$

$$C_{12} = C_{21} = -8\pi\alpha \sum_{n=0}^{+\infty} \frac{1}{e^{(2n+1)(\xi_1+\xi_2)} - 1}.$$

In the case of larger systems of resonators, we cannot hope to find such concise representations for the capacitance coefficients. However, a very useful property is that if we rescale a connected domain $D \subset \mathbb{R}^3$ by some factor $\varepsilon > 0$, then a scaling argument can be used to see that $\mathrm{Cap}_{\varepsilon D} = \varepsilon \mathrm{Cap}_D$. With this in mind, we can obtain explicit expressions in the case where the resonators are small compared to the distance between them.

LEMMA 1.19. *We assume that the resonators can be obtained by rescaling fixed domains $B_1, \dots, B_N$ as follows:*

$$D_i = \varepsilon B_i + z_i, \tag{1.32}$$

for some small parameter $\varepsilon > 0$. Then we have the following asymptotics of the capacitance matrix C_{ij} as $\varepsilon \to 0$:

$$C_{ii} = \varepsilon \mathrm{Cap}_{B_i} + \mathcal{O}(\varepsilon^3), \tag{1.33}$$

$$C_{ij} = -\frac{\varepsilon^2 \mathrm{Cap}_{B_i} \mathrm{Cap}_{B_j}}{4\pi |z_i - z_j|} + \mathcal{O}(\varepsilon^3) \quad \textit{for } i \neq j. \tag{1.34}$$

PROOF. Recall that the capacitance matrix coefficients C_{ij} can be written as

$$C_{ij} = -\int_{\partial D_i} (\mathcal{S}_D^0)^{-1}[\chi_{\partial D_j}] \, \mathrm{d}\sigma.$$

We shall compute the asymptotics of $(\mathcal{S}_D^0)^{-1}$ for small ε. Let us write the single layer potential $\mathcal{S}_D^0$ in a matrix form as

$$\mathcal{S}_D^0 = \begin{pmatrix} \mathcal{S}_{D_1}^0 & \cdots & \mathcal{S}_{D_N}^0|_{\partial D_1} \\ \vdots & \ddots & \vdots \\ \mathcal{S}_{D_1}^0|_{\partial D_N} & \cdots & \mathcal{S}_{D_N}^0 \end{pmatrix},$$

and then decompose it as

$$\mathcal{S}_D^0 = \begin{pmatrix} \mathcal{S}_{D_1}^0 & & \\ & \ddots & \\ & & \mathcal{S}_{D_N}^0 \end{pmatrix} + \begin{pmatrix} 0 & \mathcal{S}_{D_2}^0|_{\partial D_1} & \cdots & \mathcal{S}_{D_N}^0|_{\partial D_1} \\ \mathcal{S}_{D_1}^0|_{\partial D_2} & \ddots & & \mathcal{S}_{D_N}^0|_{\partial D_2} \\ \vdots & & \ddots & \vdots \\ \vdots & & \ddots & \mathcal{S}_{D_N}^0|_{\partial D_{N-1}} \\ \mathcal{S}_{D_1}^0|_{\partial D_N} & \cdots & \cdots & 0 \end{pmatrix}$$

$$:= S_I + S_{II}.$$

In order to keep the order of the norms in $L^2(\partial D)$ and $H^1(\partial D)$ constant as $\varepsilon \to 0$, we let $\mathbb{L}$ and $\mathbb{H}$, respectively, denote the spaces $L^2(\partial D)$ and $H^1(\partial D)$ along with the inner products

$$\langle \cdot, \cdot \rangle_{\mathbb{L}} = \frac{1}{|\partial D|} \langle \cdot, \cdot \rangle_{L^2(\partial D)}, \qquad \langle \cdot, \cdot \rangle_{\mathbb{H}} = \frac{1}{|\partial D|} \langle \cdot, \cdot \rangle_{H^1(\partial D)}.$$

Then, for a fixed $\widetilde{\varphi} \in L^2(\partial B)$, if we define $\varphi \in L^2(\varepsilon \partial B)$ as $\varphi(x) = \widetilde{\varphi}(\varepsilon^{-1}x)$ we have $\|\varphi\|_{\mathbb{L}} = \mathcal{O}(1)$ as $\varepsilon \to 0$.

Next, we estimate the operator norms of S_I and S_{II} in the space $\mathcal{L}(\mathbb{L}^2, \mathbb{H}^2)$ of bounded linear operators from $\mathbb{L}^2$ into $\mathbb{H}^2$. We first handle the operator S_I. By the scaling property $\mathcal{S}_{\varepsilon B}^0[\varphi] = \varepsilon \mathcal{S}_B^0[\widetilde{\varphi}]$, it can be shown that

$$\|\mathcal{S}_{D_j}^0\|_{\mathcal{L}(\mathbb{L},\mathbb{H})} \lesssim \varepsilon, \qquad \|(\mathcal{S}_{D_j}^0)^{-1}\|_{\mathcal{L}(\mathbb{H},\mathbb{L})} \lesssim \varepsilon^{-1}, \quad j = 1, \ldots, N,$$

which implies that

$$\|S_I\|_{\mathcal{L}(\mathbb{L}^2,\mathbb{H}^2)} \lesssim \varepsilon, \quad \|S_I^{-1}\|_{\mathcal{L}(\mathbb{H}^2,\mathbb{L}^2)} \lesssim \varepsilon^{-1}. \tag{1.35}$$

Here, the notation $A \lesssim B$ means that there exists a constant K independent of ε such that $A \leq KB$ for all small enough ε.

Let $\nabla_{\partial D}$ denote the surface gradient on ∂D. We can show that

$$\begin{aligned} \mathcal{S}_{D_j}^0[\varphi]|_{\partial D_i}(x) &= -\frac{\chi_{\partial D_i}(x)}{4\pi|z_i - z_j|} \int_{\partial D_j} \varphi(y)\, \mathrm{d}\sigma(y) + \mathcal{O}(\varepsilon^3 \|\varphi\|_{\mathbb{L}}), \\ \nabla_{\partial D} \mathcal{S}_{D_j}^0[\varphi]|_{\partial D_i}(x) &= \mathcal{O}(\varepsilon^3 \|\varphi\|_{\mathbb{L}}), \end{aligned} \tag{1.36}$$

and $\|\mathcal{S}_{D_j}^0|_{\partial D_i}\|_{\mathcal{L}(\mathbb{L},\mathbb{H})} \lesssim \varepsilon^2$. These estimates imply that

$$\|S_{II}\|_{\mathcal{L}(\mathbb{L}^2,\mathbb{H}^2)} \lesssim \varepsilon^2. \tag{1.37}$$

We now compute the asymptotic behaviour of $(\mathcal{S}_D^0)^{-1}$. We use the definition $\psi_j := S_I^{-1}[\chi_{\partial D_j}]$ and introduce the capacitance of each individual resonator D_j as $\mathrm{Cap}_{D_j} := -\int_{\partial D_j} \psi_j \, \mathrm{d}\sigma$. Note that $\mathrm{Cap}_{D_j} = \varepsilon \mathrm{Cap}_{B_j}$. Since we know from (1.35) and (1.37) that $S_I^{-1}S_{II} = \mathcal{O}(\varepsilon)$ in the operator norm, applying the Neumann series gives

$$
\begin{aligned}
(\mathcal{S}_D^0)^{-1}[\chi_{\partial D_j}] &= (S_I + S_{II})^{-1}[\chi_{\partial D_j}] \\
&= (I + S_I^{-1}S_{II})^{-1}S_I^{-1}[\chi_{\partial D_j}] \\
&= (I - S_I^{-1}S_{II})[\psi_j] + \mathcal{O}(\varepsilon), \qquad (1.38)
\end{aligned}
$$

where we have used the fact that

$$\|\psi_j\|_{\mathbb{L}} = \|\mathcal{S}_I^{-1}[\chi_{\partial D_j}]\|_{\mathbb{L}} \lesssim \varepsilon^{-1}.$$

Then, from (1.36), we obtain the expansions

$$
\begin{aligned}
S_{II}[\psi_i]|_{\partial D_i} &= \mathcal{O}(\varepsilon^2), \\
S_{II}[\psi_j]|_{\partial D_i} &= \frac{\varepsilon \mathrm{Cap}_{B_j}}{4\pi|z_i - z_j|}\chi_{\partial D_i} + \mathcal{O}(\varepsilon^2) \quad \text{for } j \neq i.
\end{aligned}
$$

We are ready to compute the capacitance matrix. From (1.38), we have

$$
\begin{aligned}
C_{ii} &= -\int_{\partial D_i} (\mathcal{S}_D^0)^{-1}[\chi_{\partial D_i}] \,\mathrm{d}\sigma = -\int_{\partial D_i} (I - S_I^{-1}S_{II})[\psi_i] \,\mathrm{d}\sigma - \int_{\partial D_i} \mathcal{O}(\varepsilon) \,\mathrm{d}\sigma \\
&= -\int_{\partial D_i} \psi_i \,\mathrm{d}\sigma + \mathcal{O}(\varepsilon^3) \\
&= \mathrm{Cap}_{D_i} + \mathcal{O}(\varepsilon^3) \\
&= \varepsilon \mathrm{Cap}_{B_i} + \mathcal{O}(\varepsilon^3).
\end{aligned}
$$

The expression for C_{ij} for $j \neq i$ can be derived in the same way. □

The following lemma follows from Lemma 1.19 with appropriate scaling arguments, where we fix the resonators and scale the distances between them [**27**].

LEMMA 1.20. *For $j = 1, \ldots, N$, let B_j be fixed bounded subsets of $\mathbb{R}^3$ with boundary in $C^{1,s}$ for some $0 < s < 1$. Then consider a dilute system of N subwavelength resonators with size of order ε, given by*

$$D = \bigcup_{j=1}^{N} \left(B_j + \varepsilon^{-1} z_j\right),$$

where $0 < \varepsilon \ll 1$ and $z_j \in \mathbb{R}^3$ are fixed vectors that describe the relative position of each resonator. In the limit as $\varepsilon \to 0$, the asymptotic behaviour of the capacitance matrix is given by

$$
C_{ij} = \begin{cases} \mathrm{Cap}_{B_i} + \mathcal{O}(\varepsilon^2), & i = j, \\ -\frac{\varepsilon \mathrm{Cap}_{B_i}\mathrm{Cap}_{B_j}}{4\pi|z_i - z_j|} + \mathcal{O}(\varepsilon^2), & i \neq j. \end{cases}
$$

Finally, we prove that the capacitance coefficients associated with two given resonators can only increase if we add additional resonators.

LEMMA 1.21. *Let $\mathcal{D}_N$ and $\mathcal{D}_{N+1}$ be collections of N and $N+1$ resonators such that $\mathcal{D}_{N+1} = \mathcal{D}_N \cup D_{N+1}$. For all $1 \leq i, j \leq N$, we then have*

$$(C_N)_{ij} \leq (C_{N+1})_{ij}.$$

PROOF. Let $V_{N,j}$ and $V_{N+1,j}$, respectively, be the unique solutions to the problems

$$\begin{cases} \Delta V_{N,j} = 0, & x \in \mathbb{R}^3 \setminus \mathcal{D}_N, \\ V_{N,j}(x) = \chi_{\partial D_j}(x), & x \in \partial \mathcal{D}_N, \\ V_{N,j}(x) = \mathcal{O}(|x|^{-1}), & |x| \to +\infty, \end{cases} \qquad \begin{cases} \Delta V_{N+1,j} = 0, & x \in \mathbb{R}^3 \setminus \mathcal{D}_{N+1}, \\ V_{N+1,j}(x) = \chi_{\partial D_j}(x), & x \in \partial \mathcal{D}_{N+1}, \\ V_{N+1,j}(x) = \mathcal{O}(|x|^{-1}), & |x| \to +\infty. \end{cases}$$

From the maximum principle, we observe that $0 < V_{N,j}(x), V_{N+1,j}(x) < 1$. Then we define

$$u(x) = V_{N,j}(x) - V_{N+1,j}(x).$$

Note that u satisfies

$$\begin{cases} \Delta u = 0, & x \in \mathbb{R}^3 \setminus \mathcal{D}_{N+1}, \\ u(x) = 0, & x \in \partial \mathcal{D}_N, \\ 0 < u(x) < 1, & x \in \partial D_{N+1}, \\ u(x) = \mathcal{O}(|x|^{-1}), & |x| \to +\infty. \end{cases}$$

Again, by the maximum principle we have $0 < u(x) < 1$ in $\mathbb{R}^3 \setminus \mathcal{D}_{N+1}$. Together with the boundary condition $u(x) = 0$ on $\partial \mathcal{D}_N$, this implies that

$$\frac{\partial u}{\partial \nu} \geq 0, \quad x \in \partial \mathcal{D}_N.$$

From

$$(C_N)_{ij} = -\int_{\partial D_i} \frac{\partial V_{N,j}}{\partial \nu} \, \mathrm{d}\sigma,$$

we then have

$$(C_{N+1})_{ij} - (C_N)_{ij} = \int_{\partial D_i} \frac{\partial u}{\partial \nu} \, \mathrm{d}\sigma \geq 0,$$

which proves the claim. □

REMARK 1.22. *Lemma* 1.21 *implies that the diagonal capacitance coefficients will always increase when adding additional resonators. In the physical situation of electrostatics, this result is intuitive: the self-capacitance of a conductor can only increase if additional conductors are introduced. Furthermore, since* $(C_N)_{ij} < 0$ *for* $i \neq j$, *Lemma* 1.21 *also shows that*

$$|(C_N)_{ij}| \lesssim \frac{1}{d_{ij}}, \tag{1.39}$$

where d_{ij} *is the distance between* D_i *and* D_j.

REMARK 1.23. *It is interesting to compare and contrast the capacitance formulation with the tight-binding approximation that is commonly used in quantum mechanical settings* [**1, 80, 163**]. *Both formulations construct matrix eigenvalue problems as discrete approximations to continuous differential equations. However, as will be shown in Section* 1.10, *a crucial difference is that the generalised capacitance matrix accounts for strong interactions between the resonators. The analogy with the tight-binding model is closer when the resonators are dilute. In this case, we see from Lemma* 1.20 *that* C_{ij} *depends only on the* i^{th} *and* j^{th} *resonator and is unaffected by the remaining resonators to leading-order. Moreover, in the dilute regime, the eigenmodes of the system can be approximated by a linear combination* *of the eigenmodes of the individual resonators. This property, which is a key assumption in the tight-binding approximation, does not hold in the case of nondilute subwavelength resonators.*

REMARK 1.24. *A matrix that can be expressed as $sI - A$, where $A = (a_{ij})$ with $a_{ij} \geq 0$ for all i, j and $s \geq \sigma(A)$, the spectral radius of A, is called M-matrix. In view of Lemma 1.15, the capacitance matrix C is a real symmetric nonsingular M-matrix. Hence, C has all the properties of M-matrices* [**87**, **139**, **140**, **168**]. *In particular, all the principal minors of C are positive and C^{-1} is also an M-matrix.*

1.5. Modal decompositions

The solution to the resonance problem, given in Theorem 1.9 and Corollary 1.11, can be used to understand the scattering behaviour of D. That is, we can use an expansion in terms of the resonant modes $u_1, \ldots, u_N$ (*i.e.*, a modal decomposition) to express the scattered field when D is illuminated by some incident wave u_{in}. The next result shows the modal decomposition approximation of the scattered field [**21**, **84**].

THEOREM 1.25. *Let $\mathcal{V}$ be the matrix of eigenvectors of $\mathcal{C}$. If $\omega = \mathcal{O}(\sqrt{\delta})$ as $\delta \to 0$ and $|\omega - \omega_i| > K\sqrt{\delta}$ for $i = 1, \ldots, N$, for some constant $K > 0$, then the solution to the scattering problem* (1.1) *can be written, uniformly for x in compact subsets of $\mathbb{R}^3$, as*

$$u(x) - u_{\text{in}}(x) = \sum_{n=1}^{N} a_n u_n(x) - \mathcal{S}_D^k \left[\left(\mathcal{S}_D^k\right)^{-1} [u_{\text{in}}] \right](x) + \mathcal{O}(\sqrt{\delta}), \tag{1.40}$$

for coefficients $a_n = a_n(\omega)$ that satisfy up to $\mathcal{O}(\delta^{3/2})$ the following system of linear equations:

$$\mathcal{V} \begin{pmatrix} \omega^2 - \omega_1^2 & & \\ & \ddots & \\ & & \omega^2 - \omega_N^2 \end{pmatrix} \begin{pmatrix} a_1 \\ \vdots \\ a_N \end{pmatrix} = \begin{pmatrix} \frac{\delta_1 v_1^2}{|D_1|} \int_{\partial D_1} (\mathcal{S}_D^0)^{-1}[u_{\text{in}}] \,\mathrm{d}\sigma \\ \vdots \\ \frac{\delta_N v_N^2}{|D_N|} \int_{\partial D_N} (\mathcal{S}_D^0)^{-1}[u_{\text{in}}] \,\mathrm{d}\sigma \end{pmatrix}.$$

PROOF. From Lemma 1.3, the solution u to (1.1) can be represented as

$$u(x) = \begin{cases} u_{\text{in}}(x) + \mathcal{S}_D^k[\phi](x), & x \in \mathbb{R}^3 \setminus \overline{D}, \\ \mathcal{S}_D^{k_i}[\psi](x), & x \in D_i, \ i = 1, \ldots, N, \end{cases} \tag{1.41}$$

for some surface potentials $(\phi, \psi) \in L^2(\partial D) \times L^2(\partial D)$, which must satisfy

$$\mathcal{S}_D^{k_i}[\psi] - \mathcal{S}_D^k[\phi] = u_{\text{in}} \quad \text{on } \partial D_i,$$
$$\left(-\frac{1}{2}I + \mathcal{K}_D^{k_i,*}\right)[\psi] - \delta_i \left(\frac{1}{2}I + \mathcal{K}_D^{k,*}\right)[\phi] = \delta_i \frac{\partial u_{\text{in}}}{\partial \nu} \quad \text{on } \partial D_i,$$

for $i = 1, \ldots, N$. From the asymptotic expansions (B.55) and (B.58) and the assumption that $\nabla u_{\text{in}}\big|_D = \mathcal{O}(\omega)$, we have that

$$\mathcal{S}_D^0[\psi - \phi] = u_{\text{in}} + \mathcal{O}(\omega) \quad \text{on } \partial D, \tag{1.42}$$
$$\left(-\frac{1}{2}I + \mathcal{K}_D^{0,*} + \frac{\omega^2}{v_i^2}\mathcal{K}_{D,2}\right)[\psi] - \delta_i \left(\frac{1}{2}I + \mathcal{K}_D^{0,*}\right)[\phi] = \mathcal{O}(\delta\omega + \omega^3) \quad \text{on } \partial D_i.$$

From (1.42) and the fact that $\mathcal{S}_D^0 : L^2(\partial D) \to H^1(\partial D)$ is invertible (see Lemmas 1.4 and B.6) we can see that

$$\phi = \psi - (\mathcal{S}_D^0)^{-1}[u_{\text{in}}] + \mathcal{O}(\omega). \tag{1.43}$$

Thus, we are left with the equations

$$\begin{aligned}\left(-\frac{1}{2}I + \mathcal{K}_D^{0,*} + \frac{\omega^2}{v_i^2}\mathcal{K}_{D,2} - \delta_i\left(\frac{1}{2}I + \mathcal{K}_D^{0,*}\right)\right)[\psi] &= -\delta_i\left(\frac{1}{2}I + \mathcal{K}_D^{0,*}\right)(\mathcal{S}_D^0)^{-1}[u_{\text{in}}] \\ &\quad + \mathcal{O}(\delta\omega + \omega^3),\end{aligned} \tag{1.44}$$

on ∂D_i, $i = 1, \ldots, N$. Integrating (1.44) over ∂D_i, and using Lemma B.13 gives us that

$$-\omega^2\int_{D_i} \mathcal{S}_D^0[\psi]\, \mathrm{d}x - v_i^2\delta_i\int_{\partial D_i} \psi\, \mathrm{d}\sigma = -v_i^2\delta_i\int_{\partial D_i} (\mathcal{S}_D^0)^{-1}[u_{\text{in}}]\, \mathrm{d}\sigma + \mathcal{O}(\delta\omega + \omega^3).$$

At leading-order, (1.44) says that $\left(-\frac{1}{2}I + \mathcal{K}_D^{0,*}\right)[\psi] = 0$ so, since ψ_j, $j = 1, \ldots, N$, defined in Lemma 1.4, form a basis for Ker $\left(-\frac{1}{2}I + \mathcal{K}_D^{0,*}\right)$, the solution can be written as

$$\psi = q_1\psi_1 + \ldots + q_N\psi_N + \mathcal{O}(\omega^2 + \delta), \tag{1.45}$$

for constants $q_1, \ldots, q_N = \mathcal{O}(1)$. Making this substitution, we reach the problem

$$(\mathcal{C} - \omega^2 I)\begin{pmatrix} q_1 \\ \vdots \\ q_N \end{pmatrix} = -\begin{pmatrix} \frac{v_1^2\delta_1}{|D_1|}\int_{\partial D_1}(\mathcal{S}_D^0)^{-1}[u_{\text{in}}]\, \mathrm{d}\sigma \\ \vdots \\ \frac{v_N^2\delta_N}{|D_N|}\int_{\partial D_N}(\mathcal{S}_D^0)^{-1}[u_{\text{in}}]\, \mathrm{d}\sigma \end{pmatrix} + \mathcal{O}(\delta\omega + \omega^3). \tag{1.46}$$

In view of Corollary 1.11, the result now follows from (1.41) combined with the expressions for ϕ, ψ in (1.43), (1.45) and (1.46). □

REMARK 1.26. *The term* $\mathcal{S}_D^k\left[\left(\mathcal{S}_D^k\right)^{-1}[u_{\text{in}}]\right](x)$ *in Theorem* 1.25 *is equal to* $u_{\text{in}}(x)$ *if* $x \in D$ *but not for* x *outside of the resonators.*

REMARK 1.27. *Theorems* 1.9 *and* 1.25 *have recently been reproduced using a different approach that is based on a variational formulation of* (1.1) *and the Dirichlet-to-Neumann operator associated to the Helmholtz equation in an exterior domain. This approach was first introduced in* [**86**] *for single resonators and then extended in* [**48**] *to systems of resonators.*

1.6. Higher-order approximations

The arguments used to derive the asymptotic formula in Theorem 1.9 can be continued to higher orders [**21, 84**]. For simplicity, we assume that the material parameters of each resonator are the same: $v_i = v_r$, and $\delta_i = \delta$, for $i = 1, \ldots, N$, and let $k_r = \omega/v_r$. Then the invertibility of $\mathcal{A}(\omega, \delta)$ in (1.16) is equivalent to that of the operator $\widetilde{\mathcal{A}} : L^2(\partial D) \to L^2(\partial D)$ defined by

$$\widetilde{\mathcal{A}}(\omega, \delta) := -\frac{1}{2}I + \mathcal{K}_D^{k_r,*} - \delta\left(\frac{1}{2}I + \mathcal{K}_D^{k,*}\right)(\mathcal{S}_D^k)^{-1}\mathcal{S}_D^{k_r}. \tag{1.47}$$

The operator $\widetilde{\mathcal{A}}(\omega,\delta)$ is holomorphic in ω. Indeed, by using the low-frequency asymptotic expansions for layer potentials in Appendix B.2.5, it is easy to see that (1.47) can be rewritten as

$$\widetilde{\mathcal{A}}(\omega,\delta) = -\frac{1}{2}I + \mathcal{K}_D^{0,*} + \omega^2\mathcal{B}_1(\omega) + \delta\mathcal{B}_2(\omega),$$

where $\mathcal{B}_1(\omega)$ and $\mathcal{B}_2(\omega)$ are the holomorphic operators in ω defined by

$$\mathcal{B}_1(\omega) := \sum_{p=2}^{+\infty} \frac{\omega^p}{v_r^p}\mathcal{K}_{D,p}^*, \qquad \mathcal{B}_2(\omega) := -\left(\frac{1}{2}I + \mathcal{K}_D^{k,*}\right)(\mathcal{S}_D^k)^{-1}\mathcal{S}_D^{k_r}. \tag{1.48}$$

Next, we introduce a new basis of functions $(\phi_i)_{1\leq i\leq N}$ of $\operatorname{Ker}(-\frac{1}{2}I + \mathcal{K}_D^{0,*})$ defined by

$$\phi_i := -\sum_{j=1}^{N}(C^{-1})_{ij}\psi_j, \qquad 1 \leq i \leq N, \tag{1.49}$$

where C is the capacitance matrix introduced in (1.26). Definition (1.49) ensures the property

$$\int_{\partial D_i} \phi_j \, \mathrm{d}\sigma = \delta_{ij} \text{ for any } 1 \leq i,j \leq N. \tag{1.50}$$

Let $\mathcal{P} : L^2(\partial D) \to L^2(\partial D)$ be the unique projection operator satisfying

$$\operatorname{Ker}\mathcal{P} = L_0^2(\partial D) \text{ and } \operatorname{Range}\mathcal{P} = \operatorname{Ker}(-\frac{1}{2}I + \mathcal{K}_D^{0,*}).$$

Then, for any $\phi \in L^2(\partial D)$, $\mathcal{P}[\phi]$ reads explicitly

$$\mathcal{P}[\phi] = \sum_{i=1}^{N}\left(\int_{\partial D_i} \phi \, \mathrm{d}\sigma\right)\phi_i. \tag{1.51}$$

PROPOSITION 1.28. *The operator $\widetilde{\mathcal{A}}(\omega,\delta)$ defined in (1.47) can be decomposed as*

$$\widetilde{\mathcal{A}}(\omega,\delta) = \widetilde{\mathcal{L}}_0 - \mathcal{P} + \omega^2\mathcal{B}_1(\omega) + \delta\mathcal{B}_2(\omega), \tag{1.52}$$

where $\widetilde{\mathcal{L}}_0 := -\frac{1}{2}I + \mathcal{K}_D^{0,*} + \mathcal{P}$ is an invertible Fredholm operator on $L^2(\partial D)$. The inverse of $\widetilde{\mathcal{L}}_0$ reads explicitly for $\phi \in L^2(\partial D)$ as follows:

$$\widetilde{\mathcal{L}}_0^{-1}[\phi] = \left(-\frac{1}{2}I + \mathcal{K}_D^{0,*}\right)^{-1}\left(\phi - \sum_{i=1}^{N}\left(\int_{\partial D_i} \phi \, \mathrm{d}\sigma\right)\phi_i\right) + \sum_{i=1}^{N}\left(\int_{\partial D_i} \phi \, \mathrm{d}\sigma\right)\phi_i, \tag{1.53}$$

where $(-\frac{1}{2}I + \mathcal{K}_D^{0,*})^{-1}$ is the inverse of the operator $-\frac{1}{2}I + \mathcal{K}_D^{0,*} : L_0^2(\partial D) \to L_0^2(\partial D)$. Furthermore, the following properties hold true:

(i) *$\mathcal{P}[\phi_i] = \widetilde{\mathcal{L}}_0[\phi_i] = \phi_i$ for any $1 \leq i \leq N$;*
(ii) *$\int_{\partial D_i} \widetilde{\mathcal{L}}_0^{-1}[\phi] \, \mathrm{d}\sigma = \int_{\partial D_i} \phi \, \mathrm{d}\sigma$ for any $1 \leq i \leq N$ and $\phi \in L^2(\partial D_i)$;*
(iii) *$\phi = (\phi - \mathcal{P}[\phi]) + \mathcal{P}[\phi]$ is the direct sum decomposition of $\phi \in L^2(\partial D)$ on $L_0^2(\partial D) \oplus \operatorname{Ker}(-\frac{1}{2}I + \mathcal{K}_D^{0,*})$.*

The decomposition (1.52) reads

$$\widetilde{\mathcal{A}}(\omega,\delta) = \mathcal{G}(\omega,\delta) - \mathcal{P}, \tag{1.54}$$

where $\mathcal{G}(\omega,\delta)$ is the operator

$$\mathcal{G}(\omega,\delta) := \widetilde{\mathcal{L}}_0 + \omega^2\mathcal{B}_1(\omega) + \delta\mathcal{B}_2(\omega).$$

Since $\widetilde{\mathcal{L}}_0$ is invertible, $\mathcal{G}(\omega,\delta) : L^2(\partial D) \to L^2(\partial D)$ is a holomorphic invertible Fredholm operator, whose inverse can be easily computed thanks to a Neumann series.

LEMMA 1.29. *The operator $\mathcal{G}(\omega,\delta)$ is invertible for sufficiently small ω and δ. More explicitly, the inverse reads*

$$(\mathcal{G}(\omega,\delta))^{-1} = \widetilde{\mathcal{L}}_0^{-1} - \widetilde{\mathcal{C}}(\omega,\delta), \tag{1.55}$$

where $\widetilde{\mathcal{C}}(\omega,\delta)$ is the operator of order $\mathcal{O}(\omega^2+\delta)$ defined by the following Neumann series:

$$\widetilde{\mathcal{C}}(\omega,\delta) := \sum_{p=1}^{+\infty}(-1)^{p+1}\widetilde{\mathcal{L}}_0^{-1}((\omega^2\mathcal{B}_1(\omega)+\delta\mathcal{B}_2(\omega))\widetilde{\mathcal{L}}_0^{-1})^p. \tag{1.56}$$

In (1.54), $\widetilde{\mathcal{A}}(\omega,\delta)$ is the sum of an easily invertible holomorphic operator and a constant finite-rank operator. This feature, together with Lemma 1.29, allows us to solve (1.18).

PROPOSITION 1.30. *Assume that the material parameters on each resonator are the same. Then, the operator $\mathcal{A}(\omega,\delta)$ defined in (1.16) is invertible if and only if the $N\times N$ matrix $A(\omega,\delta) \equiv (A(\omega,\delta)_{ij})_{1\le i,j\le N}$ defined by*

$$A(\omega,\delta)_{ij} := \int_{\partial D_i} \widetilde{\mathcal{C}}(\omega,\delta)[\phi_j]\, \mathrm{d}\sigma, \qquad 1\le i,j\le N, \tag{1.57}$$

is invertible. When it is the case, the solution (ϕ,ψ) to (1.18) reads

$$\begin{cases} \psi &= \sum_{i=1}^N x_i\mathcal{G}^{-1}(\omega,\delta)[\phi_i] + \mathcal{G}^{-1}(\omega,\delta)[f],\\ \phi &= \sum_{i=1}^N x_i(\mathcal{S}_D^k)^{-1}\mathcal{S}_D^{k_r}\mathcal{G}^{-1}(\omega,\delta)[\phi_i]\\ &\quad +(\mathcal{S}_D^k)^{-1}\mathcal{S}_D^{k_r}\mathcal{G}^{-1}(\omega,\delta)[f] - (\mathcal{S}_D^k)^{-1}[u_{\mathrm{in}}], \end{cases} \tag{1.58}$$

where $f \in L^2(\partial D)$ is the function

$$f := \delta\frac{\partial u_{\mathrm{in}}}{\partial\nu} - \delta\left(\frac{1}{2}I + \mathcal{K}_D^{k,*}\right)(\mathcal{S}_D^k)^{-1}[u_{\mathrm{in}}], \tag{1.59}$$

and where the coefficients $x := (x_i)_{1\le i\le N}$ are the solutions to the finite dimensional problem

$$A(\omega,\delta)x = F \text{ with } F := \left(\int_{\partial D_i}\mathcal{G}^{-1}(\omega,\delta)[f]\,\mathrm{d}\sigma\right)_{1\le i\le N}. \tag{1.60}$$

Inserting (1.58) into the integral representation formula (1.17) yields a modal expansion of the scattered field. Moreover, the N subwavelength resonant frequencies of the system are characterised as the characteristic values of the $N\times N$ matrix $A(\omega,\delta)$ of (1.57). Using the first explicit terms of the asymptotic expansion of $A(\omega,\delta)$ in ω,δ yields the following result [**21, 84**]. Higher-order terms in the asymptotic expansion of the subwavelength resonant frequencies in terms of δ can be obtained under a simplicity assumption on the eigenvalues of the capacitance matrix C [**84**]; see Appendix C.1. The term of order δ is given in the following theorem. Note that it is purely imaginary.

THEOREM 1.31. *Let $d = 3$. Consider a system of N subwavelength resonators in $\mathbb{R}^3$. Suppose that the material parameters are the same on each resonator, i.e., $v_1 = v_2 = \cdots = v_N$ and $\delta_1 = \delta_2 = \cdots = \delta_N$. As $\delta \to 0$, the N subwavelength resonant frequencies satisfy the asymptotic formula*

$$\omega_n = \sqrt{\lambda_n} - \mathrm{i}\tau_n + \mathcal{O}(\delta^{3/2}), \quad n = 1, \dots, N, \tag{1.61}$$

where λ_n for $n = 1, \dots, N$ are the eigenvalues of the generalised capacitance matrix $\mathcal{C}$ and τ_n are given by

$$\tau_n = \delta_1 \frac{v_1^2}{8\pi v} \frac{\mathbf{v}_n^\top C J C \mathbf{v}_n}{\|\mathbf{v}_n\|_D^2}, \tag{1.62}$$

with C being the capacitance matrix, J the $N \times N$ matrix of ones, $\mathbf{v}_n$ the eigenvector associated to λ_n, and the norm $\|x\|_D := \left(\sum_{i=1}^N |D_i| x_i^2\right)^{1/2}$ for $x \in \mathbb{R}^N$. Furthermore, for each $n = 1, \dots, N$, it holds that $\sqrt{\lambda_n} = \mathcal{O}(\delta^{1/2})$ and $\tau_n = \mathcal{O}(\delta)$ as $\delta \to 0$.

REMARK 1.32. *If the material parameters $v_1, \dots, v_N$ and $\delta_1, \dots, \delta_N$ are real, then λ_n and τ_n from Theorem 1.31 are all non-negative real numbers. This follows from the fact that the capacitance matrix C is symmetric and positive definite. Thus, in this case, the $\mathcal{O}(\delta^{1/2})$-term is the leading-order approximation of the real part, whereas the imaginary part appears at $\mathcal{O}(\delta)$ and accounts for the radiation effect.*

REMARK 1.33. *Due to the loss of energy (e.g. in the far field), the resonant frequencies will have negative imaginary parts when the material parameters are real. In many cases, $\tau_n = 0$ will hold for some n, which means that imaginary parts exhibit higher-order behaviour in δ. For example, the imaginary part of the second (dipole) frequency for a pair of identical resonators with real parameters is known to be $\mathcal{O}(\delta^2)$* [**39**].

REMARK 1.34. *The approach developed in this section can be used to obtain higher-order terms in the modal decomposition* (1.40) *of the scattered field; see* [**84**].

1.7. The cases of a single resonator and a dimer of resonators

When $N = 1$, the subwavelength resonant frequency ω_1 satisfies as δ goes to zero the asymptotic formula [**34**]

$$\omega_1 = \underbrace{\sqrt{\frac{\mathrm{Cap}_D}{|D|}} v_r \sqrt{\delta}}_{:=\omega_M} - \mathrm{i} \underbrace{\left(\frac{\mathrm{Cap}_D^2 v_r^2}{8\pi v |D|} \delta \right)}_{:=\tau_M} + \mathcal{O}(\delta^{\frac{3}{2}}), \tag{1.63}$$

where

$$\mathrm{Cap}_D := -\int_{\partial D} (\mathcal{S}_D^0)^{-1}[\chi_{\partial D}] \,\mathrm{d}\sigma \tag{1.64}$$

is the capacity of D. Moreover, the following monopole approximation of the scattered field $u^s := u - u_{\text{in}}$ for ω near ω_M holds [**34**]:

$$u^s(x) = g(\omega, \delta, D)(1 + o(1))u_{\text{in}}(0)G^k(x), \tag{1.65}$$

with the origin $0 \in D$ and the scattering coefficient g given by

$$g(\omega, \delta, D) = \frac{\text{Cap}_D}{1 - (\frac{\omega_M}{\omega})^2 + \mathrm{i}\gamma_M}, \tag{1.66}$$

where the damping constant γ_M is given by

$$\gamma_M := \frac{\omega(v + v_r)\text{Cap}_D}{8\pi v v_r} - \frac{(v - v_r)}{v}\frac{\delta v_r \text{Cap}_D^2}{8\pi|D|\omega}. \tag{1.67}$$

Formulas (1.65)–(1.66) show the scattering enhancement near ω_M.

In order to characterise the resonances of a dimer, we need the following straightforward generalisation of Lemma B.13.

LEMMA 1.35. *Assume that D has two connected components D_1 and D_2. Then for any $\varphi \in L^2(\partial D)$ we have, for $i = 1, 2$,*

$$\int_{\partial D_i} \mathcal{K}_{D,2}[\varphi]\, \mathrm{d}\sigma = -\int_{D_i} \mathcal{S}_D^0[\varphi]\, \mathrm{d}x, \qquad \int_{\partial D_i} \mathcal{K}_{D,3}[\varphi]\, \mathrm{d}\sigma = \frac{\mathrm{i}|D_i|}{4\pi}\int_{\partial D} \varphi\, \mathrm{d}\sigma. \tag{1.68}$$

Assume that D_1 and D_2 are symmetric with respect to the origin 0 and let C_{ij}, for $i, j = 1, 2$, be defined by (1.26) for $N = 2$. Then, from Theorem 1.9, there are two subwavelength resonances with positive real part for the resonator dimer D and as $\delta \to 0$, by using Lemma 1.35 it follows that [**39**]

$$\omega_1 = \underbrace{\sqrt{(C_{11} + C_{12})v_r\sqrt{\delta}}}_{:=\omega_{M,1}} - \mathrm{i}\tau_1\delta + \mathcal{O}(\delta^{3/2}), \tag{1.69}$$

$$\omega_2 = \underbrace{\sqrt{(C_{11} - C_{12})v_r\sqrt{\delta}}}_{:=\omega_{M,2}} + \delta^{3/2}\widehat{\eta}_1 + \mathrm{i}\delta^2\widehat{\eta}_2 + \mathcal{O}(\delta^{5/2}), \tag{1.70}$$

where $\widehat{\eta}_1$ and $\widehat{\eta}_2$ are real numbers determined by D, v, and v_r and

$$\tau_1 = \frac{v_r^2}{4\pi v}(C_{11} + C_{12})^2.$$

The resonances ω_1 and ω_2 are, respectively, called the monopole and dipole hybridised resonances of the resonator dimer D. Note that from Lemma 1.15, $\omega_{M,1}$ is slightly smaller than $\omega_{M,2}$. Note also that Im $\omega_1 = \mathcal{O}(\delta)$ while Im $\omega_2 = \mathcal{O}(\delta^2)$.

On the other hand, the resonator dimer can be approximated as a point scatterer with resonant monopole and resonant dipole modes. For $\omega = \mathcal{O}(\delta^{1/2})$ and $\delta \to 0$, and $|x|$ being sufficiently large, we have [**39**]

$$\begin{aligned} u^s(x) &= \underbrace{g^0(\omega)u_{\text{in}}(0)G^k(x)}_{\text{monopole}} \\ &\quad + \underbrace{\nabla u_{\text{in}}(0) \cdot g^1(\omega)\nabla G^k(x)}_{\text{dipole}} + \mathcal{O}(\delta|x|^{-1}), \end{aligned} \tag{1.71}$$

where the scattering coefficients $g^0(\omega)$ and $g^1(\omega) = (g^1_{ij}(\omega))$ are given by

$$g^0(\omega) = \frac{C(1,1)}{1-\omega_1^2/\omega^2}(1+\mathcal{O}(\delta^{1/2})), \quad C(1,1) := C_{11}+C_{12}+C_{21}+C_{22}, \tag{1.72}$$

$$g^1_{ij}(\omega) = \int_{\partial D} (\mathcal{S}_D^0)^{-1}[x_i](y) y_j - \frac{\delta v_r^2}{\omega^2|D|(1-\omega_2^2/\omega^2)} P^2 \delta_{i1}\delta_{j1}. \tag{1.73}$$

Here,

$$P := \int_{\partial D} y_1(\psi_1 - \psi_2)(y)\, \mathrm{d}\sigma(y), \tag{1.74}$$

with ψ_i, for $i = 1, 2$, being defined by (1.25), and δ_{i1} and δ_{j1} being the Kronecker delta.

As shown in (1.65)–(1.66), around ω_M, a single resonator in the free space scatters waves with a greatly enhanced amplitude. If a second resonator is introduced, coupling interactions will occur, giving according to (1.71) a system that has both monopole and dipole resonant modes. Note that the eigenvalues of the capacitance matrix of a dimer of resonators have the vectors $\frac{1}{\sqrt{2}}\begin{pmatrix}1\\1\end{pmatrix}$ and $\frac{1}{\sqrt{2}}\begin{pmatrix}1\\-1\end{pmatrix}$ as (normalised) eigenvectors. The first one corresponds to the monopole eigenmode while the second one corresponds to the dipole eigenmode of the dimer system of resonators.

REMARK 1.36. *When D_1 and D_2 are identical spheres, the coefficients C_{11}, C_{12} and P can be computed explicitly using bispherical coordinates. Explicit formulas for C_{11} and C_{12} are given in Lemma 1.18. Using the same approach as in the derivation of C_{ij}, we can also obtain an explicit formula for P. It holds that*

$$P = -4\pi r\left(r+\frac{d}{2}\right) - 8\pi\alpha^2 \sum_{n=0}^{+\infty}(2n+1)e^{-(2n+1)\xi}\coth\left(\left(n+\frac{1}{2}\right)\xi\right),$$

where r is the radius of the spheres D_1, D_2 and d is the separating distance between them.

1.8. Two-dimensional models

Throughout Section 1.3, we have mainly considered the problem of a resonator system in $\mathbb{R}^3$. This was convenient for two reasons. First, for small frequencies, the Laplace single layer potential $\mathcal{S}_D^0$ approximates the Helmholtz single layer potential $\mathcal{S}_D^\omega$ in the leading order, in the sense that $\mathcal{S}_D^\omega = \mathcal{S}_D^0 + \mathcal{O}(\omega)$ in the operator norm as $\omega \to 0$. Second, the fact that $\mathcal{S}_D^0$ is invertible in three dimensions was central to our definition of the capacitance matrix. However, if we consider a Helmholtz problem in two dimensions, then we do not have either of these helpful properties. In two dimensions, $\mathcal{S}_D^0$ is generally not injective and the low-frequency expansion of $\mathcal{S}_D^\omega$ is given by

$$\mathcal{S}_D^\omega = \frac{1}{2\pi}\ln\left(\frac{1}{2}\omega e^{\gamma - \mathrm{i}\frac{\pi}{2}}\right) I_{\partial D} + \mathcal{S}_D^0 + \mathcal{O}(\omega^2 \ln\omega) \quad \text{as } \omega \to 0, \tag{1.75}$$

where $I_{\partial D}$ is the map defined as $I_{\partial D}[\varphi] = \int_{\partial D}\varphi\, \mathrm{d}\sigma$ for $\varphi \in L^2(\partial D)$ and $\gamma = \lim_{n\to+\infty}(\sum_{k=1}^n \frac{1}{k} - \ln n) \approx 0.577\ldots$ is the Euler–Mascheroni constant. That is, the leading-order term in the expansion of $\mathcal{S}_D^\omega$ has a $\ln\omega$ singularity as $\omega \to 0$.

The invertibility of the Laplace single layer potential in two dimensions can be readily fixed. Let $L^2_0(\partial D)$ be the mean zero subspace of $L^2(\partial D)$ defined as

$$L^2_0(\partial D) = \left\{ f \in L^2(\partial D) \ : \ \int_{\partial D} f \ \mathrm{d}\sigma = 0 \right\}.$$

Then, we see that the Laplace single layer potential is well behaved on $L^2_0(\partial D)$ [**20**].

LEMMA 1.37. *Let $d = 2$. The Laplace single layer potential $\mathcal{S}_D^0$ is invertible from $L^2_0(\partial D)$ onto its image.*

With this in hand, we can show that an invertible version of $\mathcal{S}_D^0$ can be defined by adding a term proportional to the integral operator $I_{\partial D}[\varphi] = \int_{\partial D} \varphi \ \mathrm{d}\sigma$. For example, we have the following lemma [**20**].

LEMMA 1.38. *Let $d = 2$ and $I_{\partial D}$ be the integral map given by $I_{\partial D}[\varphi] = \int_{\partial D} \varphi \mathrm{d}\sigma$. For any $\omega \in \mathbb{C} \setminus \{z \in \mathbb{C} : z = \mathrm{i}y$ for some $y \leq 0\}$, the operator $\widehat{\mathcal{S}}_D^\omega : L^2(\partial D) \to L^2(\partial D)$, defined as*

$$\widehat{\mathcal{S}}_D^\omega = \frac{1}{2\pi} \ln \left(\frac{1}{2} \omega e^{\gamma - \mathrm{i}\frac{\pi}{2}} \right) I_{\partial D} + \mathcal{S}_D^0,$$

is invertible.

Notice that $\widehat{\mathcal{S}}_D^\omega$ is nothing other than the leading-order approximation of $\mathcal{S}_D^\omega$ from (1.75). This means that, up to some technical modifications, we can repeat the argument used to derive Theorem 1.9 to obtain the following analogous result.

THEOREM 1.39. *Let $d = 2$. A system of N subwavelength resonators in $\mathbb{R}^2$ has N subwavelength resonant frequencies. Further to this, if for any ω and δ we define the $N \times N$-matrix $\mathcal{A}^{(2)}_{\omega,\delta}$ as*

$$(1.76)\quad \begin{aligned} (\mathcal{A}^{(2)}_{\omega,\delta})_{ij} &= \omega^2 \ln \omega + \left(\left(1 + \frac{c_1}{b_1} - \ln v_i \right) - \frac{\mathcal{S}_D^0[\psi_j]|_{\partial D_i}}{4 b_1 (\int_{\partial D} \psi_j)} \right) \omega^2 \\ &\quad - \frac{v_i^2 \delta_i}{4 b_1 |D_i|} \left(\frac{\int_{\partial D_i} \psi_j}{(\int_{\partial D} \psi_j)} + \frac{\ln(v/v_i)}{2\pi} \int_{\partial D_i} (\widehat{\mathcal{S}}_D^k)^{-1}[\chi_{\partial D}] \right), \end{aligned}$$

where $b_1 = -\frac{1}{8\pi}$ and $c_1 = b_1(\gamma - \ln 2 - 1 - \mathrm{i}\frac{\pi}{2})$, then the subwavelength resonant frequencies are such that the determinant of $\mathcal{A}^{(2)}_{\omega,\delta}$ vanishes at leading-order:

$$\det(\mathcal{A}^{(2)}_{\omega,\delta}) = \mathcal{O}(\omega^4 \ln \omega + \delta \omega^2 \ln \omega) \quad \text{as } \omega, \delta \to 0.$$

Notice, finally, that $\mathcal{S}_D^0[\psi_j]$ is constant in D_j since $\psi_j \in \mathrm{Ker}\,(-\frac{1}{2}I + \mathcal{K}_D^{0,*})$. Then,

$$-\frac{\mathcal{S}_D^0[\psi_j]|_{\partial D_j}}{\int_{\partial D} \psi_j}$$

is nothing else than $1/(2\pi)\times$ the logarithm of the capacity of D_j (see Appendix B.1.4). Furthermore, the matrix with entries

$$-\frac{\mathcal{S}_D^0[\psi_j]|_{\partial D_i}}{\int_{\partial D} \psi_j}$$

in (1.76) can be considered as the two-dimensional analogue of the capacitance matrix introduced in Definition 1.7.

1.9. Numerical approaches

The subwavelength resonant frequencies can be computed numerically in a variety of ways. For example, we can use the boundary integral formulation (1.15) to derive a discrete version of the problem. This can be achieved, for example, by discretising the boundary using boundary elements or a multipole expansion method in the case where the resonators are spherical; see Appendix B.6. To compute the resonant frequencies for a given value of δ, one needs to use a numerical root finding algorithm to find ω such that the boundary integral equation (1.15) is satisfied. The discrete version of the boundary integral operator (which depends nonlinearly on ω) will need to be recomputed at each step in this iterative algorithm.

Muller's method can be used as a numerical root-finding algorithm, following the methodology of [**37**, Section 1.6]. The open source codes for its implementation are provided in [**37, 149**]. We recall that Muller's method allows us to find (complex) roots of holomorphic functions by using quadratic interpolants. In Figure 1.3, we apply Muller's method to obtain the zeros of the eigenvalues of the discrete matrix approximation of the operator $\mathcal{A}(\omega, \delta)$ in (1.16), obtained by multipole expansion. Since $\mathcal{A}(\omega, \delta)$ is not invertible at ω for which one of these eigenvalues is zero, computing such roots ω then yields the desired subwavelength resonant frequencies.

The generalised capacitance matrix can be used to obtain accurate numerical approximations with a significant reduction in computational power. Provided that δ is sufficiently small, Theorems 1.9 and 1.31 give an approximation of the resonant frequencies that is sufficiently accurate for many purposes. In Figure 1.3, we show the resonant frequencies of a system of ten spherical resonators computed using both the full multipole expansion method and the asymptotic approximation in terms of the capacitance matrix. The values derived from the eigenvalues of the capacitance matrix give a good approximation and require just 0.02 seconds of computation time, compared to the 41 seconds required for the full multipole computations, on the same desktop computer. If greater precision than that of the capacitance matrix approximation is required, then the values derived from the capacitance matrix can be used as initial values for root-finding algorithms to reduce computational time.

1.10. Tight-binding approximation

1.10.1. Tight-binding model in condensed matter physics. In the field of condensed matter theory in some settings a useful assumption is the tight-binding approximation. It consists of assuming that the major interaction of the system comes from self-interaction. The approximating modes are constructed by taking linear combinations of the modes obtained by only considering the self-interaction of a single atom at the time. This can be proven to be a good approximation for the Schrödinger equation when the electrons are tightly bonded to the atom's nuclei in terms of eigenvalues (see, for instance, [**80, 129, 130**]). To illustrate this, following [**155**], we consider the Schrödinger equation

$$\Big(H_{\mathrm{at}} + V\Big)\Psi = E\Psi,$$

where $H_{\mathrm{at}} = \sum_i H_i$ with H_i being the Hamiltonian of the single atom i, V is a potential that describes interactions between the atoms, and E is the energy.

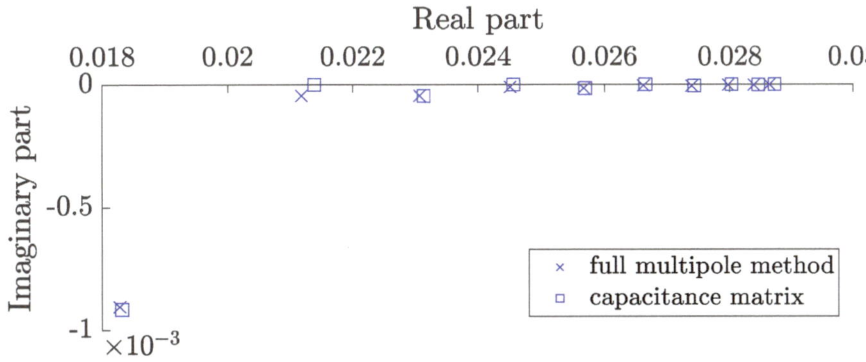

FIGURE 1.3. The subwavelength resonant frequencies of a system of ten spherical resonators. We compare the values computed using the multipole expansion method to discretise the full boundary integral equation and the values computed using the capacitance matrix. The computations using the full multipole expansion method took 41 seconds while the approximations from the capacitance matrix took just 0.02 seconds, on the same computer. Each resonator has unit radius and we use $\delta = 1/5000$.

In the tight-binding model, the wave function Ψ is expanded as a sum over atomic wave functions of the form

$$\Psi(x) = \sum_i \sum_n a_i^{(n)} \phi_i^{(n)}(x),$$

where $\phi_i^{(n)}$ is the atomic wave function on the atomic site i corresponding to the energy $e_i^{(n)}$ at the nth atomic level. Indeed, we assume that $\phi_i^{(n)} = \phi^{(n)}(x - z_i)$, where z_i is the position of the atom i,

$$H_j \phi_i^{(n)} = e_i^{(n)} \phi_i^{(n)} \delta_{ij},$$

and

$$\int \phi^{(n)}(x - z_i)\overline{\phi^{(m)}}(x - z_j)\, \mathrm{d}x = \delta_{ij}\delta_{nm}. \tag{1.77}$$

The Schrödinger equation becomes a matrix equation for the amplitudes $a_i^{(n)}$. This equation is further simplified by considering only one atomic level for each atom. Then, by using orthogonality of the atomic wave functions (1.77), we get an equation of the form

$$e_i a_i + \sum_j a_j \underbrace{\int V(x)\phi(x - z_i)\overline{\phi}(x - z_j)\, \mathrm{d}x}_{:=V_{ij}} = E a_i, \tag{1.78}$$

where e_i is the atomic energy level on the site i and V_{ij} is the matrix element of the Hamiltonian between the atomic sites i and j.

The tight-binding model (1.78) is often coupled with a nearest-neighbour approximation. Assuming that $V_{ii} = 0$ and $V_{ij} = 0$ for $|i - j| > 1$, we arrive at the

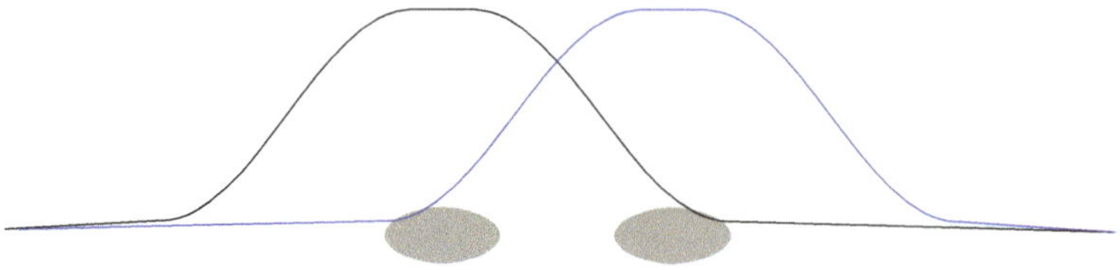

FIGURE 1.4. Modes of two different single resonators. The grey shaded areas represent the two resonators, and different colours have been used for the two different modes relative to each resonator separately. Observe that each mode is approximately constant inside the corresponding resonator.

widely used tridiagonal matrix model of the form

$$\underbrace{\begin{pmatrix} e_1 & V_{12} & & \\ V_{21} & e_2 & V_{23} & \\ & \ddots & \ddots & \ddots \\ & & V_{N(N-1)} & e_N \end{pmatrix}}_{:=H_{\text{tb}}} \begin{pmatrix} a_1 \\ a_2 \\ \vdots \\ a_N \end{pmatrix} = E \begin{pmatrix} a_1 \\ a_2 \\ \vdots \\ a_N \end{pmatrix}. \tag{1.79}$$

Note that under the additional assumptions $V_{ij} < 0$ and $e_i + \sum_j V_{ij} > 0$, the Hamiltonian matrix H_{tb} is a real nonsingular M-matrix.

1.10.2. Comparison between the tight-binding and capacitance matrix formulations. In this section, our goal is to compare the tight-binding formulation to the capacitance matrix formulation. Assuming that all the resonators are identical, we point out similarities and differences between the two approximations. To enhance this comparison, we recast the eigenvalue problem for the capacitance matrix C into the Hamiltonian form

$$\mathrm{i}\Phi'(t) = H\Phi(t), \tag{1.80}$$

where the Hamiltonian

$$H := \sqrt{\frac{\delta}{|D|}} v_r \begin{pmatrix} 0 & \sqrt{C} \\ \sqrt{C} & 0 \end{pmatrix}. \tag{1.81}$$

Here, the resonators are $D_i = D + z_i$, for $i = 1, \dots, N$, δ is the contrast parameter, and $v_r > 0$ is the wave speed inside the resonators. Note that the eigenvalues of H are the square roots with sign of the eigenvalues of the generalised capacitance matrix $\mathcal{C} := \sqrt{\delta/|D|} v_r C$.

The main fundamental difference between the tight-binding and capacitance formulations, aside from the fact that they were conceived to model different physical problems, is the following. As proven in Corollary 1.11, subwavelength modes need to be almost constant inside the resonators. As these modes will not, in general, be constant elsewhere, taking linear combinations of modes would contradict the almost constant nature of the true modes in the resonators. For a visual intuition of this fact, we refer to Figures 1.4 and 1.5.

Therefore, for a tight-binding approximation to work in the classical wave setting, we will need to assume the size of the resonators to be asymptotically small compared to the distance between them. We let $D_i = \varepsilon B + z_i$ for $i = 1, \dots, N$.

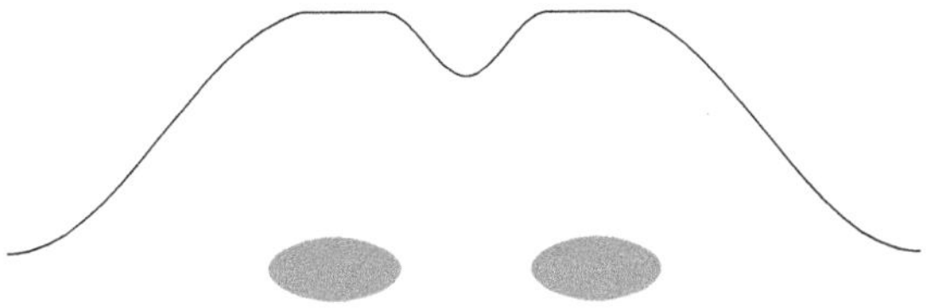

(A) Example of a mode for two coupled resonators. Observe that the mode is approximately constant inside both resonators.

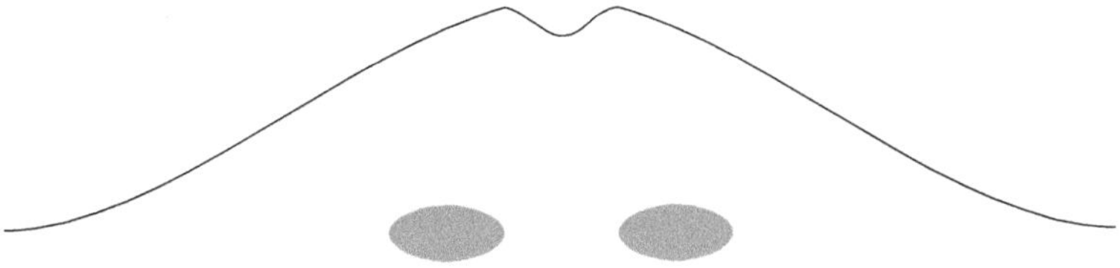

(B) Linear combination of the modes of the single resonators.

FIGURE 1.5. Comparison between a true mode (a) and a tight-binding-type approximant (b). Unless the resonators are asymptotically small, the linear combination in (b) is not constant inside the resonators and is not a suitable approximant of the mode sketched in (a).

From Lemma 1.19, it can be seen that in the dilute regime, up to first order in ε, the diagonal entries of the capacitance matrix C depend only on one resonator and represent the self-interaction of a particular resonator. The off-diagonal entries represent the mutual interactions between two resonators and, in the dilute regime, the entry C_{ij} depends solely on the ith and the jth resonator. Moreover, mutual interactions are represented by lower-order terms.

The requirements for a tight-binding approximation of (1.80) to be a good approximation are the following:

- Its eigenvalues are equal to the ones of the full Hamiltonian matrix H defined in (1.81), up to first-order in ε;
- The diagonal entries of the tight-binding approximant Hamiltonian should represent the magnitude of the modes related to self-interaction of single resonators. The off-diagonal entries of the approximant should represent the mutual interactions between the resonators. In accordance with the capacitance matrix in the dilute regime, these entries should be of higher order in ε compared to the diagonal entries. Furthermore, we will see that the off-diagonal terms decay too slowly to be omitted. Therefore, there will also be long-range interactions;
- In order to retain the symmetry with respect to zero of the eigenvalues of H, our approximant should not include any interaction between the positive modes and the negative ones.

The approximant will be represented as the sum of two matrices. We let $\widetilde{H}$ be the zero-order approximation which will denote the self-interaction of the resonators. The matrix Δ will account for the interactions and therefore will be of higher order in ε. Our approximant H_{tb} will be of the form $\widetilde{H} + \Delta$.

Our next step to determine these matrices will be to compute the eigenvalues of the full Hamiltonian H.

1.10.2.1. *Two-resonator case.* First, we consider the two-resonator case and compute the characteristic polynomial of H. By noticing that $H - \lambda \mathrm{I}_4$ can be viewed as a 2×2 block matrix whose blocks commute with each other, we may compute the block version of the determinant and turn the eigenvalue problem for H into the one for $\mathcal{C}$:

$$0 = \det(-\lambda \mathrm{I}_4 + H) = \begin{vmatrix} -\lambda \mathrm{I}_2 & \sqrt{\mathcal{C}} \\ \sqrt{\mathcal{C}} & -\lambda \mathrm{I}_2 \end{vmatrix} = \det\left(\lambda^2 \mathrm{I}_2 - \mathcal{C}\right). \tag{1.82}$$

Here, the notation $|A|$ for a matrix A is also used to denote the determinant of A. To compute the eigenvalues of $\mathcal{C}$, we will use the diluteness assumption. Matrices of the form

$$\begin{pmatrix} \alpha & \beta \\ \beta & \alpha \end{pmatrix},$$

where α, $\beta \in \mathbb{C}$, have $\alpha \pm \beta$ as two eigenvalues. From Lemma 1.19, $\mathcal{C}$ has this form and therefore, we can use this to compute its eigenvalues and find that

$$\lambda^2 = \delta \frac{\mathrm{Cap}_B}{|B|} v_r \pm \delta\varepsilon \frac{(\mathrm{Cap}_B)^2}{4\pi |z_1 - z_2||B|} v_r + \mathcal{O}(\varepsilon^2), \tag{1.83}$$

where z_1 and z_2 are the positions of the centres of the two resonators. To keep track of the four eigenvalues, we define

$$\begin{cases} \lambda_{11} := \sqrt{\delta \frac{\mathrm{Cap}_B}{|B|} v_r} \left(1 + \varepsilon \frac{\mathrm{Cap}_B}{8\pi|z_1 - z_2|} + \mathcal{O}(\varepsilon^2)\right), \\ \lambda_{12} := \sqrt{\delta \frac{\mathrm{Cap}_B}{|B|} v_r} \left(1 - \varepsilon \frac{\mathrm{Cap}_B}{8\pi|z_1 - z_2|} + \mathcal{O}(\varepsilon^2)\right), \\ \lambda_{21} := -\sqrt{\delta \frac{\mathrm{Cap}_B}{|B|} v_r} \left(1 + \varepsilon \frac{\mathrm{Cap}_B}{8\pi|z_1 - z_2|} + \mathcal{O}(\varepsilon^2)\right), \\ \lambda_{22} := -\sqrt{\delta \frac{\mathrm{Cap}_B}{|B|} v_r} \left(1 - \varepsilon \frac{\mathrm{Cap}_B}{8\pi|z_1 - z_2|} + \mathcal{O}(\varepsilon^2)\right). \end{cases} \tag{1.84}$$

Following [**51, 173**], we intend to find a tight-binding approximation for H. The choice that proved to fit better with the criteria listed above is the following:

$$\widetilde{H} := \sqrt{\delta \frac{\mathrm{Cap}_B}{|B|} v_r} \operatorname{diag}(1, 1, -1, -1). \tag{1.85}$$

It is clear from formula (1.84) that the eigenvalues of $\widetilde{H}$ coincide with those of H at zero order in ε. Hence, the first of the criteria listed above is satisfied.

We would now like to find a 4×4 matrix Δ such that $\widetilde{H} + \Delta$ has, up to first order in ε, the same eigenvalues as H. In line with the three requirements listed above, our choice of Δ is

$$\Delta = \begin{pmatrix} \Delta_1 & 0 \\ 0 & \Delta_2 \end{pmatrix},$$

where Δ_1 and Δ_2 are two 2×2 matrices of the form

$$\Delta_1 = \begin{pmatrix} 0 & a \\ a & 0 \end{pmatrix}, \qquad \Delta_2 = \begin{pmatrix} 0 & b \\ b & 0 \end{pmatrix} \tag{1.86}$$

with a, $b \in \mathbb{R}$ being two parameters to be determined. This way, Δ represents the interactions between the resonators and does not couple the negative modes with the positive ones. In this particular case of two resonators, the matrix $\widetilde{H} + \Delta$ is tridiagonal. A nearest-neighbour approximation holds trivially in this case. The parameters a and b can be viewed as the "hopping parameters" of the system.

To determine a and b, let us look at the eigenvalues of $\widetilde{H}+\Delta$. We have

$$
\begin{aligned}
0 &= \det\left(\widetilde{H}+\Delta-\widetilde{\lambda}\mathrm{I}_4\right)\\
&= \begin{vmatrix} \sqrt{\delta\frac{\mathrm{Cap}_B}{|B|}}v_r-\widetilde{\lambda} & a \\ a & \sqrt{\delta\frac{\mathrm{Cap}_B}{|B|}}v_r-\widetilde{\lambda}\end{vmatrix}\begin{vmatrix} -\sqrt{\delta\frac{\mathrm{Cap}_B}{|B|}}v_r-\widetilde{\lambda} & b \\ b & -\sqrt{\delta\frac{\mathrm{Cap}_B}{|B|}}v_r-\widetilde{\lambda}\end{vmatrix}\\
&= \left[\left(\sqrt{\delta\frac{\mathrm{Cap}_B}{|B|}}v_r-\widetilde{\lambda}\right)^2-a^2\right]\left[\left(-\sqrt{\delta\frac{\mathrm{Cap}_B}{|B|}}v_r-\widetilde{\lambda}\right)^2-b^2\right],
\end{aligned}
$$

from which we obtain

$$
\begin{aligned}
\widetilde{\lambda}_{11} &:= \sqrt{\delta\frac{\mathrm{Cap}_B}{|B|}}v_r+a, & \widetilde{\lambda}_{12} &:= \sqrt{\delta\frac{\mathrm{Cap}_B}{|B|}}v_r-a,\\
\widetilde{\lambda}_{21} &:= -\sqrt{\delta\frac{\mathrm{Cap}_B}{|B|}}v_r-b, & \widetilde{\lambda}_{22} &:= -\sqrt{\delta\frac{\mathrm{Cap}_B}{|B|}}v_r+b.
\end{aligned}
\tag{1.87}
$$

By comparing the approximate eigenvalues (1.87) to the eigenvalues of the full matrix H given in (1.84), we obtain

$$
a=\varepsilon\frac{\mathrm{Cap}_B}{8\pi|z_1-z_2|}, \qquad b=\varepsilon\frac{\mathrm{Cap}_B}{8\pi|z_1-z_2|}.
$$

Plugging the values of a and b into formulas (1.87), we get

$$
\begin{aligned}
\widetilde{\lambda}_{11} &:= \sqrt{\delta\frac{\mathrm{Cap}_B}{|B|}}v_r+\varepsilon\frac{\mathrm{Cap}_B}{8\pi|z_1-z_2|},\\
\widetilde{\lambda}_{12} &:= \sqrt{\delta\frac{\mathrm{Cap}_B}{|B|}}v_r-\varepsilon\frac{\mathrm{Cap}_B}{8\pi|z_1-z_2|},\\
\widetilde{\lambda}_{21} &:= -\sqrt{\delta\frac{\mathrm{Cap}_B}{|B|}}v_r-\varepsilon\frac{\mathrm{Cap}_B}{8\pi|z_1-z_2|},\\
\widetilde{\lambda}_{22} &:= -\sqrt{\delta\frac{\mathrm{Cap}_B}{|B|}}v_r+\varepsilon\frac{\mathrm{Cap}_B}{8\pi|z_1-z_2|}.
\end{aligned}
\tag{1.88}
$$

Finally, the approximant matrix reads

$$
\widetilde{H}+\Delta=\begin{pmatrix}
\sqrt{\delta\frac{\mathrm{Cap}_B}{|B|}}v_r & \varepsilon\frac{\mathrm{Cap}_B}{8\pi|z_1-z_2|} & 0 & 0\\
\varepsilon\frac{\mathrm{Cap}_B}{8\pi|z_1-z_2|} & \sqrt{\delta\frac{\mathrm{Cap}_B}{|B|}}v_r & 0 & 0\\
0 & 0 & -\sqrt{\delta\frac{\mathrm{Cap}_B}{|B|}}v_r & \varepsilon\frac{\mathrm{Cap}_B}{8\pi|z_1-z_2|}\\
0 & 0 & \varepsilon\frac{\mathrm{Cap}_B}{8\pi|z_1-z_2|} & -\sqrt{\delta\frac{\mathrm{Cap}_B}{|B|}}v_r
\end{pmatrix}.
$$

We note that the approximant matrix, up to second order in ε, is similar to the starting Hamiltonian.

1.10.2.2. *Case of an arbitrary number of resonators.* In this final subsection, we outline the case of an arbitrary number of resonators. We proceed analogously to the two-resonator case. To do so, we compute the eigenvalues of H up to first order in ε. We first notice that

$$0 = \det(-\lambda \mathrm{I}_{2N} + H) = \begin{vmatrix} -\lambda \mathrm{I}_N & \sqrt{\mathcal{C}} \\ \sqrt{\mathcal{C}} & -\lambda \mathrm{I}_N \end{vmatrix} = \det\left(\lambda^2 \mathrm{I}_N - \mathcal{C}\right), \tag{1.89}$$

since the submatrices shown in the computation all commute with each other. Therefore, the problem of computing the eigenvalues of H is equivalent to the problem of computing the eigenvalues of $\mathcal{C}$. Let us therefore compute the eigenvalues of $\mathcal{C}$. From Lemma 1.19, it follows that $\mathcal{C}$ is of the form

$$\mathcal{C} = \delta \frac{\mathrm{Cap}_B}{|B|} v_r^2 \left(\mathrm{I}_N + \varepsilon A\right) + \mathcal{O}(\varepsilon^2),$$

where A is the $N \times N$ matrix whose $(j,k)^{\text{th}}$ entry is

$$A_{jk} = \begin{cases} \frac{\mathrm{Cap}_B}{4\pi |z_j - z_k|} & \text{if } j \neq k, \\ 0 & \text{if } j = k. \end{cases}$$

As $\mathcal{C}$ is diagonalisable, it follows that its eigenvalues satisfy

$$\lambda_j^2 = \delta \frac{\mathrm{Cap}_B}{|B|} v_r^2 \left(1 + \varepsilon a_j\right) + \mathcal{O}(\varepsilon^2),$$

where $1 \leq j \leq N$ and a_j is the jth eigenvalue of A. Consequently, the eigenvalues of H are

$$\lambda_j = \pm \sqrt{\delta \frac{\mathrm{Cap}_B}{|B|}} v_r \left(1 + \varepsilon \frac{a_j}{2}\right) + \mathcal{O}(\varepsilon^2). \tag{1.90}$$

To choose an approximant, as in the case for $N = 2$, we define

$$\widetilde{H} := \sqrt{\delta \frac{\mathrm{Cap}_B}{|B|}} v_r \operatorname{diag}(\underbrace{1,\ 1,\ \ldots,\ 1}_{N}, \underbrace{-1,\ -1,\ \ldots,\ -1}_{N}).$$

As in Section 1.10.2.1, the higher-order coupling matrix Δ will, in general, be a 2×2 block diagonal, so that no coupling occurs between the negative modes and the positive ones. We therefore seek Δ of the form

$$\Delta = \begin{pmatrix} \Delta_1 & 0 \\ 0 & \Delta_2 \end{pmatrix},$$

where Δ_1, Δ_2 are symmetric, off-diagonal matrices. Comparing the eigenvalues of $\widetilde{H} + \Delta$ to λ_j, given in (1.90), we obtain that (up to similarity transformations)

$$\Delta_1 = \Delta_2 = \frac{\varepsilon}{2} \sqrt{\delta \frac{\mathrm{Cap}_B}{|B|}} v_r A.$$

It is evident that the two diagonal blocks Δ_1 and Δ_2 are dense and the tight-binding approximant is not tridiagonal. In other words, long-range interactions cannot be ignored, and nearest-neighbour approximations are in general not appropriate. This is one of the crucial differences between the capacitance matrix formulation in subwavelength physics and the tight-binding approximation in condensed matter physics.

CHAPTER 2

Exceptional points in non-Hermitian systems of subwavelength resonators

The behaviour of subwavelength resonators, which interact strongly with waves at subwavelength scales, can be greatly influenced by even very small perturbations. As a result, subwavelength resonators are ideal building blocks for designing small and powerful sensors capable of detecting the presence of small particles such as viruses and nanoparticles. These devices rely on measuring the shifts in the structure's resonant frequencies caused by the perturbations [**143, 161, 162**]. The shift in the resonant frequencies typically scales in proportion to the size of the perturbation, meaning that the shift is very small and difficult to detect when the perturbation is small. However, this weakness can be overcome through the use of structures with *exceptional points*.

An exceptional point is a point in parameter space at which two or more eigenvalues, and also the corresponding eigenvectors, coincide [**102, 128**]. A deep degeneracy of this nature gives rise to structures with remarkable properties. In particular, if a perturbation of order η is made to a structure with an Nth-order exceptional point (that is, one at which N eigenvalues and eigenvectors coincide) then the eigenvalues can experience perturbations of order $\eta^{1/N}$; see Appendix C.2. Thus, when trying to detect small perturbations, the measurable response will be relatively large, offering a means to design enhanced sensing systems [**66, 103, 167**].

Exceptional points arise in non-Hermitian systems. They can be created by introducing gain and loss, which can be represented by the imaginary parts of complex-valued material parameters. As is common in the field, we will consider structures with *parity–time ($\mathcal{PT}$) symmetry* [**102, 128**]. The exceptional points in such systems originate from the fact that the spectrum of a $\mathcal{PT}$-symmetric operator is symmetric under complex conjugation. Exceptional points are the transition points when pairs of real eigenvalues coalesce and change to be complex conjugate pairs. They are a consequence of balanced symmetries in the system, which cause the eigenvectors to align. The motivation for imposing parity–time symmetry is that the system already has some underlying symmetry, making them a natural setting to search for exceptional points.

In this chapter, we study the occurrence of exceptional points in structures composed of subwavelength resonators. The sources of energy gain and loss are represented through complex-valued material parameters. We choose the distribution of the material parameters and the geometry so that the system is $\mathcal{PT}$-symmetric. We search for high-order asymptotic exceptional points in suitably large arrays. Our analysis reveals the range of different configurations that can give rise to such exceptional points and provides efficient techniques to compute them. The results presented in this chapter are from [**27, 28**].

2.1. Exceptional points

Let $N \in \mathbb{N}$ and let $D_1, D_2, \ldots, D_N \subset \mathbb{R}^3$ be disjoint connected sets with boundaries in $C^{1,s}$ for some $0 < s < 1$. As in Chapter 1, we use v_i and δ_i to respectively denote the wave speed and the contrast parameter in resonator D_i. We assume that $v_i \in \mathbb{C}$, while the wave speed of the background medium v is assumed to be real. The imaginary parts can be interpreted as the magnitude of the gain or loss.

DEFINITION 2.1. *Let $\mathcal{C}$ be the generalised capacitance matrix introduced in Definition 1.8. We say that a system of $N \in \mathbb{N}$ resonators $D_1, D_2, \ldots, D_N$ in $\mathbb{R}^3$ admits an Nth-order exceptional point if there exists γ such that*

$$\det(\mathcal{C} - xI) = (\gamma - x)^N, \tag{2.1}$$

$$\dim \operatorname{Ker}(\mathcal{C} - \gamma I) = 1. \tag{2.2}$$

We will assume that the problem is parity–time symmetric in the sense that each resonator D_i can be uniquely associated to another resonator D_j (possibly with $i = j$) such that

$$D_i = \mathcal{P} D_j \quad \text{and} \quad v_i^2 \delta_i = \mathcal{T}(v_j^2 \delta_j), \tag{2.3}$$

where the parity operator $\mathcal{P} : \mathbb{R}^3 \to \mathbb{R}^3$ and the time-reversal operator $\mathcal{T} : \mathbb{C} \to \mathbb{C}$ are given, respectively, by

$$\mathcal{P}(x) = -x \quad \text{and} \quad \mathcal{T}(z) = \overline{z}. \tag{2.4}$$

The creation of Nth-order singularities in the generalised capacitance matrix, where the eigenvalues and the eigenvectors coincide, leads to the design of subwavelength resonant structures with higher-order resonant singularities, as shown in the following theorem.

THEOREM 2.2. *Consider the Helmholtz resonance problem (1.1) for the domain $D = D_1 \cup \cdots \cup D_N$, where $N \in \mathbb{N}$. For a set of material parameter values where an Nth-order exceptional point for the generalised capacitance matrix $\mathcal{C}$ occurs, there exist N resonant frequencies $\omega_1, \ldots, \omega_N$ and associated eigenmodes $u_1, \ldots, u_N$ such that for any $i, j \in \{1, \ldots, N\}$*

$$\omega_i = \omega_j + \mathcal{O}(\delta) \quad \text{as } \delta \to 0,$$

and for any $i, j \in \{1, \ldots, N\}$ there exists some $K \in \mathbb{C}$ such that

$$u_i = K u_j + \mathcal{O}(\delta) \quad \text{as } \delta \to 0.$$

REMARK 2.3. *Theorem 2.2 states that exceptional points occur for subwavelength resonator systems only in the leading order in δ. We do not expect the original differential problem (1.1) to support exact exceptional points, in the sense of exactly degenerate resonant frequencies and coalescent eigenmodes. The exceptional points studied here are related to the $\mathcal{PT}$-symmetry of the problem. Even under the assumption of symmetric gain and loss, the problem (1.1) is not $\mathcal{PT}$-symmetric, since the radiation condition swaps sign under complex conjugation. Nevertheless, the discrete approximation given by the generalised capacitance matrix is indeed $\mathcal{PT}$-symmetric.*

2.2. Exceptional points of order two

Let $N = 2$. We define $\delta := |\delta_1|$, and assume that

$$\delta \ll 1, \quad \delta_2 = \mathcal{O}(\delta), \quad v_i = \mathcal{O}(1).$$

The assumption that δ is small means that we are studying high-contrast resonators. We introduce the notation

$$\frac{v_1^2\delta_1}{|D_1|} := a + ib, \qquad \frac{v_2^2\delta_2}{|D_2|} := a - ib, \tag{2.5}$$

for positive real-valued parameters a and b. Here, $|D_i|$ denotes the volume of D_i. Under the sign convention fixed by (1.2), the parameter b represents the magnitude of the gain and loss, which are imposed on D_1 and D_2, respectively. Moreover, under the parity symmetry assumption, we have

$$|D_1| = |D_2| \text{ and } C_{11} = C_{22}$$

with C_{ij} being the entries of the capacitance matrix defined by (1.26).

For a system of two subwavelength resonators, we can find the eigenvalues and eigenvectors of the generalised capacitance matrix explicitly. Using these formulas, we can show that there exists an exceptional point for certain parameter values [**28**].

Let $\lambda_i, \mathbf{v}_i = \begin{pmatrix} v_i^1 \\ v_i^2 \end{pmatrix}$ denote an eigenpair of $\mathcal{C}$. From Theorem 1.9 together with (1.41) and (1.45), it follows that the subwavelength resonant frequencies and corresponding eigenmodes satisfy the asymptotic formulas

$$\omega_i = \omega_i^{(0)} + \mathcal{O}(\delta), \quad u_i = u_i^{(0)} + \mathcal{O}(\delta^{1/2}) \quad \text{as } \delta \to 0,$$

where

$$\omega_i^{(0)} := \sqrt{\lambda_i}, \quad u_i^{(0)} := v_i^1 S_1^\omega + v_i^2 S_2^\omega.$$

The eigenvalues of the matrix $\mathcal{C}$ are given by

$$\lambda_i = aC_{11} + (-1)^i\sqrt{a^2C_{12}^2 - b^2(C_{11}^2 - C_{12}^2)}, \tag{2.6}$$

where the parameters a and b were defined in (2.5). We then have the following theorem for the exceptional points of the dimer, which occur when $\lambda_1 = \lambda_2$.

THEOREM 2.4. *A $\mathcal{PT}$-symmetric pair of subwavelength resonators D_1 and D_2, that is, satisfying* (2.3)*, has an asymptotic exceptional point of order two with respect to δ in the subwavelength regime. In other words, there is a set of material parameters such that the eigenvalues and eigenvectors of the associated generalised capacitance matrix defined by* (1.27) *coincide. In particular, if*

$$\operatorname{Im}(v_1^2\delta_1) = b^*(a) := \frac{\operatorname{Re}(v_1^2\delta_1)C_{12}}{\sqrt{C_{11}^2 - C_{12}^2}},$$

then $\lambda_1 = \lambda_2$ and $\mathbf{v}_1 = K\mathbf{v}_2$ for some $K \in \mathbb{C}$, where $(\lambda_i, \mathbf{v}_i)$, $i = 1, 2$, are the eigenpairs of the generalised capacitance matrix $\mathcal{C}$. Further to this,

- *if $\operatorname{Im}(v_1^2\delta_1) < b^*$ then $\sqrt{\lambda_1}, \sqrt{\lambda_2}$ are real valued and $\sqrt{\lambda_1} \neq \sqrt{\lambda_2}$;*
- *if $\operatorname{Im}(v_1^2\delta_1) > b^*$ then $\sqrt{\lambda_1}, \sqrt{\lambda_2}$ are purely imaginary and $\sqrt{\lambda_1} \neq \sqrt{\lambda_2}$.*

PROOF. Combining Theorem 1.9 and formula (2.6), we find that

$$\frac{\operatorname{Im}(v_1^2\delta_1)}{\operatorname{Re}(v_1^2\delta_1)} = \frac{C_{12}}{\sqrt{C_{11}^2 - C_{12}^2}}$$

corresponds to the point $b^*(a)$ where $\mathcal{C}$ has a double eigenvalue corresponding to a one-dimensional eigenspace. The remaining statements are straightforward to check. □

LEMMA 2.5. *If $b \neq b^*$, the eigenmodes u_n corresponding to the resonant frequencies ω_n, for $n = 1, 2$, are given by*

$$u_n = v_n^1 S_1^\omega + v_n^2 S_2^\omega + \mathcal{O}(\delta^{1/2}),$$

as $\delta \to 0$, where

$$S_j^\omega(x) = \begin{cases} \mathcal{S}_D^{\omega_n/v}[\psi_j](x), & x \in \mathbb{R}^3 \setminus \overline{D}, \\ \mathcal{S}_D^{\omega_n/v_i}[\psi_j](x), & x \in D_i, i = 1, 2, \end{cases}$$

for $j = 1, 2$, with ψ_j being defined by (1.25) and $\mathbf{v}_n = \begin{pmatrix} v_n^1 \\ v_n^2 \end{pmatrix}$ are the eigenvectors of $\mathcal{C}$ given by

$$\mathbf{v}_n = \begin{pmatrix} -C_{12} \\ C_{11} - \mu_n \end{pmatrix}, \qquad \mu_n = \frac{\lambda_n}{(a + \mathrm{i}b)}.$$

The approximate nature of the asymptotic exceptional point predicted in Theorem 2.4 is demonstrated by Figure 2.1, where the subwavelength resonant frequencies of the full differential system are simulated directly using the multipole expansion method derived in Appendix B.6. Note that since the eigenmodes are asymptotically constant on each domain D_i, accurate and efficient computations can be achieved by approximating functions by only the first term in the multipole expansion. We can see that there is a critical value of the gain and loss (the imaginary parts of the material parameters) such that the eigenvalues coincide at leading order. Below this critical value, the leading-order parts of the resonant frequencies are real, and they form a purely imaginary conjugate pair above this value.

2.3. Exceptional points in the dilute regime

In order to produce higher-order subwavelength exceptional points, we need to study larger systems of resonators. In order for the capacitance matrix to be easier to work with in this case, we will make the additional assumption that the resonators are relatively far apart, meaning that we can use the dilute approximation given in Lemma 1.20. This means that we can efficiently analyse large systems of resonators in terms of this leading-order approximation of the capacitance matrix [**27**].

We will assume that $D_1, \ldots, D_N$ are all given by translations of some domain B which is parity-symmetric (*i.e.*, it satisfies $\mathcal{P}B = B$). Without loss of generality, we choose the length scale such that $\operatorname{Cap}_B = 4\pi$ (this holds *e.g.* if B is the unit sphere). For $\varepsilon > 0$, we define the resonators as

$$D_i = B - \left(i - \frac{N+1}{2}\right)(\varepsilon^{-1}, 0, 0).$$

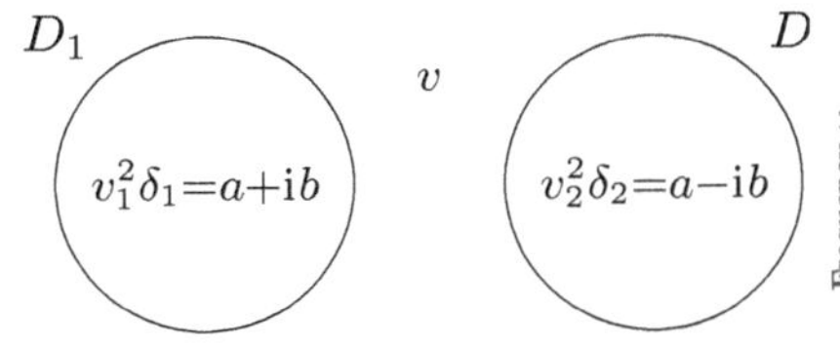

(A) A $\mathcal{PT}$-symmetric pair of spherical resonators.

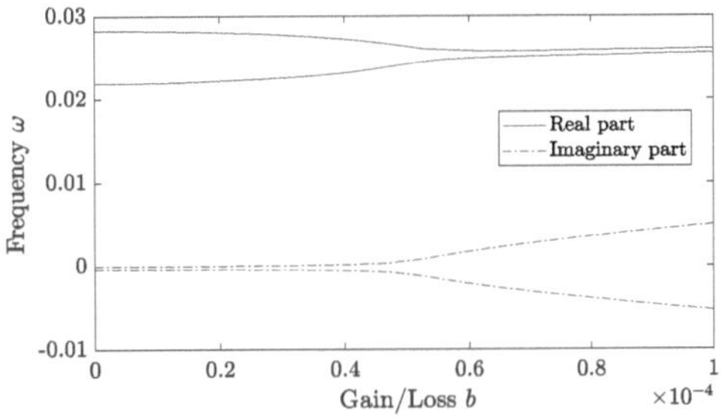

(B) The two subwavelength resonant frequencies.

FIGURE 2.1. The two subwavelength resonant frequencies of a pair of $\mathcal{PT}$-symmetric resonators can be simulated directly from the full differential system using the multipole expansion method. An asymptotic exceptional point occurs at $b^* \approx 0.5 \times 10^{-4}$, where the frequencies coincide at leading-order. For b smaller than b^*, the leading-order terms of the resonant frequencies are real, while for b larger than b^* they are purely imaginary and are the conjugate of one other, again to leading-order.

Notice that the resonators are equally spaced along the x-axis and that all the resonators are far away from each other when ε is small. Again, we define the contrast parameter $\delta := |\delta_1|$, which will be our asymptotic parameter. We assume that

$$\delta \ll 1,$$

and that in the asymptotic limit $\delta \to 0$ we have $\delta_i = \mathcal{O}(\delta)$ while $v_i = \mathcal{O}(1)$ for all $i = 1, \ldots, N$. We also define $a \in \mathbb{R}$ to be such that $\operatorname{Re}(v_1^2\delta_1) = \delta a$ and assume that $a \neq 0$. Then, using the diagonal matrix V with nonzero entries given by

$$V_{ii} = \frac{v_i^2\delta_i}{\delta a}, \quad i = 1, \ldots, N, \tag{2.7}$$

we define the matrix $\mathcal{C}_d$ as

$$\mathcal{C}_d = V \begin{pmatrix} 1 & -\varepsilon & -\varepsilon/2 & \cdots & -\varepsilon/(N-1) \\ -\varepsilon & 1 & -\varepsilon & \cdots & -\varepsilon/(N-2) \\ -\varepsilon/2 & -\varepsilon & 1 & \cdots & -\varepsilon/(N-3) \\ \vdots & \vdots & \vdots & \ddots & \vdots \\ -\varepsilon/(N-1) & -\varepsilon/(N-2) & -\varepsilon/(N-3) & \cdots & 1 \end{pmatrix}. \tag{2.8}$$

By Lemma 1.20, we see that $\mathcal{C}_d$ gives a dilute approximation of $\mathcal{C}$ in the sense that

$$\mathcal{C} = 4\pi a\delta \left(\mathcal{C}_d + \mathcal{O}(\varepsilon^2)\right).$$

Also, under this choice of D, we can follow the proof of [**28**, Theorem 2.3] to conclude that the error term in Theorem 1.9 holds uniformly in ε. Therefore, from Theorem 1.9, we get the following result.

THEOREM 2.6. *Let γ_n be the eigenvalues of $\mathcal{C}_d$ and $\mathbf{q}_n$ the corresponding eigenvectors. Then, for small ε and δ, we have the following asymptotic behaviour of ω_n*

and u_n*:*

$$\omega_n = \sqrt{\frac{4\pi a\delta\gamma_n}{|D_1|}} + \mathcal{O}(\delta + \delta^{1/2}\varepsilon^2), \tag{2.9}$$

$$u_n(x) = \mathbf{q}_n \cdot \mathbf{S}^{\omega_n}(x) + \mathcal{O}(\varepsilon^2 + \delta^{1/2}), \quad n = 1, \dots, N,$$

where $\mathbf{S}^{\omega_n} = (S_1^{\omega_n}, \dots, S_N^{\omega_n})^\top$ *with* $S_j^{\omega_n}$ *for* $j = 1, \dots, N$*, being defined by*

$$\mathbf{S}_j^{\omega_n}(x) = \begin{cases} \mathcal{S}_D^{\omega_n/v}[\psi_j](x), & x \in \mathbb{R}^3 \setminus \overline{D}, \\ \mathcal{S}_D^{\omega_n/v_i}[\psi_j](x), & x \in D_i, i = 1, \dots, N. \end{cases}$$

In (2.9)*, the error terms hold uniformly for* ε *and* δ *in neighbourhoods of* 0.

This theorem gives a discrete approximation of the resonant frequencies and eigenmodes, in terms of the eigenvalues and eigenvectors of $\mathcal{C}_d$. It shows that the exceptional points of $\mathcal{C}_d$ will be the asymptotic exceptional points of the full differential equation problem (1.1). We recall that an Nth-order exceptional point of $\mathcal{C}_d$ is a set of parameter values such that

$$\det(\mathcal{C}_d - xI) = (\gamma - x)^N \quad \text{and} \quad \dim \operatorname{Ker}(\mathcal{C}_d - \gamma I) = 1,$$

for some γ. In what follows, we will study these points by expanding the characteristic polynomial of $\mathcal{C}_d$ and matching the coefficients to those of $(\gamma - x)^N$.

2.3.1. A third-order exceptional point. In this section, we consider an array of three resonators and search for a third-order exceptional point. Observe that since N is odd, the centre resonator must have real material parameters in order to be $\mathcal{PT}$-symmetric. We introduce the notation

$$v_1^2\delta_1 := \delta a(1 + ib), \qquad v_2^2\delta_2 := \delta ac, \qquad v_3^2\delta_3 := \delta a(1 - ib),$$

for real-valued parameters a, b and c, such that $a, b, c = \mathcal{O}(1)$. In this case, the matrix $\mathcal{C}_d$ is given by

$$\mathcal{C}_d = \begin{pmatrix} 1 + ib & -(1 + ib)\varepsilon & -(1 + ib)\varepsilon/2 \\ -c\varepsilon & c & -c\varepsilon \\ -(1 - ib)\varepsilon/2 & -(1 - ib)\varepsilon & 1 - ib \end{pmatrix}.$$

Next, we shall show that the discrete model matrix $\mathcal{C}_d$ has an exceptional point of order 3.

The characteristic polynomial $P(x)$ of $\mathcal{C}_d$ is given by

$$P(x) = x^3 - (c + 2)x^2 + \left(1 + b^2 + 2c - \frac{\varepsilon^2}{4}(1 + b^2 + 8c)\right)x \\ - c(1 + b^2)\left(1 - \frac{9}{4}\varepsilon^2 - \varepsilon^3\right).$$

In order to have an exceptional point of order 3, we require that there is some γ such that

$$P(x) = (x - \gamma)^3 = x^3 - 3\gamma x^2 + 3\gamma^2 x - \gamma^3,$$

and $\dim \operatorname{Ker}(\mathcal{C}_d - \gamma I) = 1$. Comparing the above two expressions for $P(x)$, we see that b, c and ε should satisfy

$$3\gamma = c + 2, \tag{2.10}$$

$$3\gamma^2 = 1 + b^2 + 2c - \frac{\varepsilon^2}{4}(1 + b^2 + 8c), \tag{2.11}$$

$$\gamma^3 = c(1 + b^2)\left(1 - \frac{9}{4}\varepsilon^2 - \varepsilon^3\right). \tag{2.12}$$

LEMMA 2.7. *For any small $\varepsilon > 0$, there exist $b > 0$ and $c > 0$ satisfying* (2.11) *and* (2.12). *Moreover, we have*

$$b = b_1\varepsilon + \mathcal{O}(\varepsilon^2), \tag{2.13}$$

$$c = 1 + c_1\varepsilon + \mathcal{O}(\varepsilon^2), \tag{2.14}$$

where b_1 and c_1 are specified as roots of given polynomials. Therefore, for such b and c, the characteristic polynomial of $\mathcal{C}_d$ is given by

$$P(x) = (x - \gamma)^3 \quad \textit{with } \gamma = \frac{c+2}{3} = 1 + \mathcal{O}(\varepsilon).$$

PROOF. From (2.10) we have $\gamma = (c+2)/3$. Then, (2.11) can be written as

$$1 + b^2 = \left(1 - \frac{\varepsilon^2}{4}\right)^{-1}\left[\frac{(c+2)^2}{3} - 2(1 - \varepsilon^2)c\right]. \tag{2.15}$$

Then, substituting the above into (2.12), we get

$$\left(1 - \frac{\varepsilon^2}{4}\right)\frac{(c+2)^3}{27} = c\left(1 - \frac{9}{4}\varepsilon^2 - \varepsilon^3\right)\left[\frac{(c+2)^2}{3} - 2(1 - \varepsilon^2)c\right], \tag{2.16}$$

which is a cubic polynomial in c. If $\varepsilon = 0$, then $c = 1$ is a solution to (2.16). Since it is a regularly perturbed cubic equation by small $\varepsilon > 0$, (2.16) has a real root c satisfying

$$c = 1 + c_1\varepsilon + \mathcal{O}(\varepsilon^2).$$

Substituting the above expansion to (2.16), we can see that c_1 is the real root of the polynomial equation

$$c_1^3 + \frac{27}{4}c_1 - \frac{27}{8} = 0,$$

that is, $c_1 \approx 0.483\ldots$. Then, by (2.15), we can compute the expansion of b as

$$b = b_1\varepsilon + \mathcal{O}(\varepsilon^2),$$

where $b_1 = \sqrt{\frac{9}{4} + \frac{c_1^2}{3}} \approx 1.53\ldots$. □

We next show that, at the exceptional point, all the eigenvectors coalesce.

LEMMA 2.8. *For a given small $\varepsilon > 0$, let b and c be chosen as in Lemma* 2.7 *and let γ be the corresponding eigenvalue of $\mathcal{C}_d$. Then we have*

$$\dim \operatorname{Ker}(\mathcal{C}_d - \gamma I) = 1.$$

PROOF. Since $\gamma = (c+2)/3$, we have

$$\mathcal{C}_d - \gamma I = \begin{pmatrix} 1 + \mathrm{i}b - \frac{c+2}{3} & -(1+\mathrm{i}b)\varepsilon & -(1+\mathrm{i}b)\varepsilon/2 \\ -c\varepsilon & c - \frac{c+2}{3} & -c\varepsilon \\ -(1-\mathrm{i}b)\varepsilon/2 & -(1-\mathrm{i}b)\varepsilon & (1-\mathrm{i}b) - \frac{c+2}{3} \end{pmatrix}.$$

We will show that $\operatorname{rank}(\mathcal{C}_d - \gamma I) = 2$, which implies the conclusion. First of all, since γ is an eigenvalue, $\operatorname{rank}(\mathcal{C}_d - \gamma I) < 3$. We therefore only need to show that two of the column vectors of $\mathcal{C}_d - \gamma I$ are linearly independent. For small $\varepsilon > 0$, by (2.13) and (2.14), we have

$$\mathcal{C}_d - \gamma I = \varepsilon \begin{pmatrix} -\frac{c_1}{3} + \mathrm{i}b_1 & -1 & -1/2 \\ -1 & \frac{2c_1}{3} & -1 \\ -1/2 & -1 & -\frac{c_1}{3} - \mathrm{i}b_1 \end{pmatrix} + \mathcal{O}(\varepsilon^2). \tag{2.17}$$

Observe that both b_1 and c_1 are nonzero and real. From the asymptotic behaviour of (2.17), we see that the first and last column vectors are linearly independent for sufficiently small ε. □

By the above two lemmas, we see that the trimer has an asymptotic exceptional point of order 3.

THEOREM 2.9. *A $\mathcal{PT}$-symmetric system D of three dilute resonators has an asymptotic exceptional point of order 3 with respect to ε and δ at the resonant frequency ω^*, which is given as $\varepsilon, \delta \to 0$ by*

$$\omega^* = \sqrt{\frac{4\pi(3 + \varepsilon c_1)\operatorname{Re}(v_1^2\delta_1)}{3|D_1|}} + \mathcal{O}(\delta + \delta^{1/2}\varepsilon),$$

where c_1 is the real root of the polynomial equation $c_1^3 + \frac{27}{4}c_1 - \frac{27}{8} = 0$, (that is, $c_1 \approx 0.483\ldots$).

In Figure 2.2, the eigenvalues of $\mathcal{C}_d$ and the resonant frequencies of the full differential system are compared. Figure 2.2 shows, to leading-order and for $N = 3$, the resonant frequencies as the gain and loss increase from 0 and cross the exceptional points. The material parameters are $v_i^2\delta_i = a_i^* + \mathrm{i}\tau b_i^*$, for $0 \le \tau \le 2$, where $a_i^* + \mathrm{i}b_i^*$ corresponds to the material parameters at the asymptotic exceptional point.

2.3.2. Exceptional points of order four. We now seek fourth-order exceptional points in the case that D is an array of four $\mathcal{PT}$-symmetric resonators with material parameters given by

$$v_1^2\delta_1 := \delta a(1 + \mathrm{i}b), \quad v_2^2\delta_2 := \delta a(c + \mathrm{i}d), \quad v_3^2\delta_3 := \delta a(c - \mathrm{i}d), \quad v_4^2\delta_4 := \delta a(1 - \mathrm{i}b).$$

In this setting, the matrix $\mathcal{C}_d$, as defined in (2.8), is

$$\mathcal{C}_d = \begin{pmatrix} 1 + \mathrm{i}b & -(1 + \mathrm{i}b)\varepsilon & -(1 + \mathrm{i}b)\varepsilon/2 & -(1 + \mathrm{i}b)\varepsilon/3 \\ -(c + \mathrm{i}d)\varepsilon & c + \mathrm{i}d & -(c + \mathrm{i}d)\varepsilon & -(c + \mathrm{i}d)\varepsilon/2 \\ -(c - \mathrm{i}d)\varepsilon/2 & -(c - \mathrm{i}d)\varepsilon & c - \mathrm{i}d & -(c - \mathrm{i}d)\varepsilon \\ -(1 - \mathrm{i}b)\varepsilon/3 & -(1 - \mathrm{i}b)\varepsilon/2 & -(1 - \mathrm{i}b)\varepsilon & 1 - \mathrm{i}b \end{pmatrix},$$

and the characteristic polynomial $P(x) = \det(\mathcal{C}_d - xI)$ can be computed as

$$\begin{aligned} P(x) &= x^4 - 2(c + 1)x^3 \\ &\quad + \left(1 + b^2 + c(4 + c) + d^2 - \varepsilon^2\frac{1}{18}(2 + 2b^2 + 9c(5 + 2c) - 27bd + 18d^2)\right)x^2 \\ &\quad + \frac{1}{18}\left(9(c^2 + d^2)(-4 + 9\varepsilon^2 + 4\varepsilon^3) + c(1 + b^2)(-36 + \varepsilon^2(49 + 12\varepsilon))\right)x \\ &\quad + \frac{1}{144}(1 + b^2)(c^2 + d^2)\left(144 - \varepsilon^2(520 + 384\varepsilon + 23\varepsilon^2)\right). \end{aligned}$$

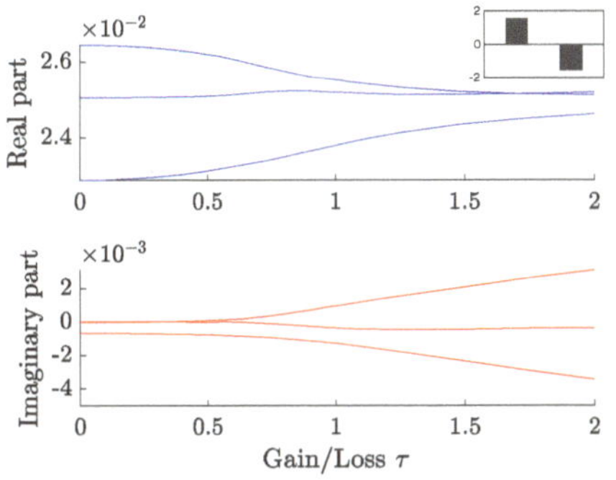

(A) The subwavelength resonant frequencies of the full differential problem.

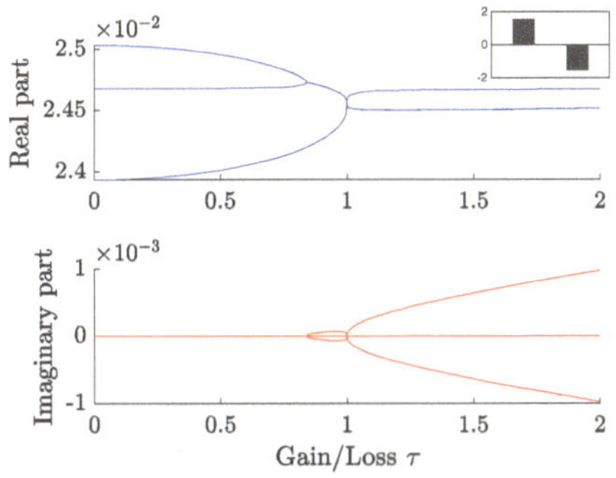

(B) The approximate subwavelength resonant frequencies, under the dilute approximation.

FIGURE 2.2. A $\mathcal{PT}$-symmetric system of three subwavelength resonators supports an asymptotic exceptional point, where the eigenvalues (and corresponding eigenvectors) of the dilute capacitance matrix coincide. We can compare the resonant frequencies of the full differential problem in (a), computed using the multipole expansion method, and the approximate frequencies in (b), computed using the dilute approximation of the generalised capacitance matrix. In both cases, the relative size of the imaginary parts are fixed and they are rescaled by some τ, which is such that the asymptotic exceptional point occurs at $\tau = 1$.

To obtain an exceptional point of order 4, we require that

$$P(x) = (x-\gamma)^4 = x^4 - 4\gamma x^3 + 6\gamma^2 x^2 - 4\gamma^3 x + \gamma^4,$$

for some γ. Comparing the two expressions for P, we see that $\gamma = (c+1)/2$ and that

(2.18)
$$\frac{3}{2}(c+1)^2 = 1 + b^2 + c(4+c) + d^2 - \frac{\varepsilon^2}{18}\left(2 + 2b^2 + 9c(5+2c) - 27bd + 18d^2\right),$$

(2.19)
$$\frac{1}{2}(c+1)^3 = -\frac{1}{18}\left(9(c^2+d^2)(-4+9\varepsilon^2+4\varepsilon^3) + c(1+b^2)(-36+49\varepsilon^2+12\varepsilon^3)\right),$$

(2.20)
$$\frac{1}{16}(c+1)^4 = \frac{1}{144}(1+b^2)(c^2+d^2)(144 - \varepsilon^2(520+384\varepsilon+23\varepsilon^2)).$$

We are interested in solutions to this system for small ε. At $\varepsilon = 0$, we have a unique solution given by

$$b = 0, \qquad c = 1, \qquad d = 0.$$

For small but nonzero ε, we make the ansatz

$$b = b_1\varepsilon + \mathcal{O}(\varepsilon^2), \qquad c = 1 + c_1\varepsilon + \mathcal{O}(\varepsilon^2), \qquad d = d_1\varepsilon + \mathcal{O}(\varepsilon^2).$$

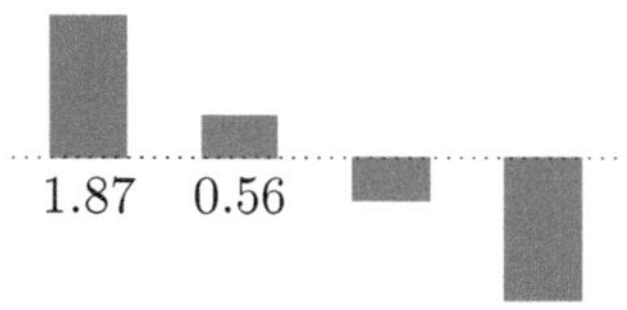

(A) Exceptional point satisfying $b_1 d_1 > 0$ and $|b_1| > |d_1|$, with $c_1 = 0.654$.

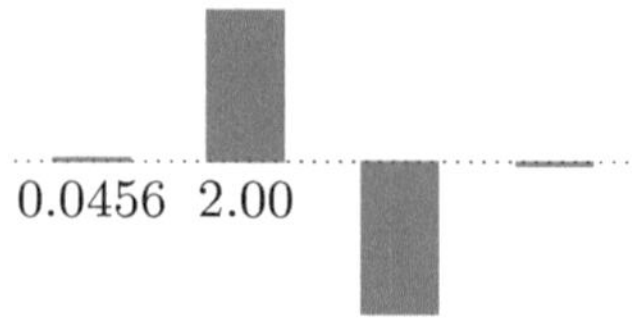

(B) Exceptional point satisfying $b_1 d_1 > 0$ and $|b_1| < |d_1|$, with $c_1 = -0.863$.

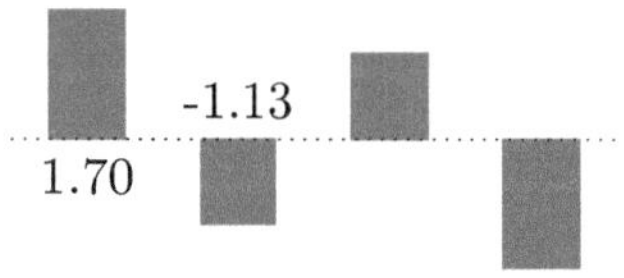

(C) Exceptional point satisfying $b_1 d_1 < 0$ and $|b_1| > |d_1|$, with $c_1 = 1.07$.

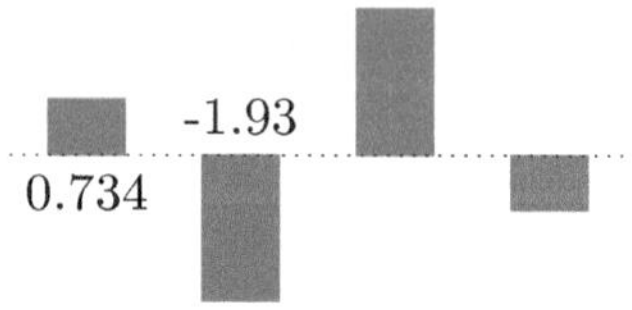

(D) Exceptional point satisfying $b_1 d_1 < 0$ and $|b_1| < |d_1|$, with $c_1 = -1.15$.

FIGURE 2.3. A system of four $\mathcal{PT}$-symmetric resonators supports four asymptotic exceptional points. Here, we plot the leading-order coefficients of the imaginary parts of the material parameters (the gain or loss) at each of the four exceptional points.

By substituting these expansions into (2.18)–(2.20), we find that

$$c_1^2 - 2(b_1^2 + d_1^2) + \frac{65}{9} = 0, \tag{2.21}$$

$$c_1^3 + c_1\left(\frac{49}{9} - 4b_1^2\right) + \frac{16}{3} = 0, \tag{2.22}$$

$$c_1^4 - 16(c_1^2 + d_1^2)\left(b_1^2 - \frac{1}{9}\right) + \frac{32}{3}c_1 + 16b_1^2 - 24b_1d_1 + \frac{23}{9} = 0. \tag{2.23}$$

Moreover, it is clear that if (b_1, c_1, d_1) is a solution, then $(-b_1, c_1, -d_1)$ is also a solution. Solving the above system numerically, we obtain 4 solutions up to this symmetry (or 8 solutions in total), presented in Figure 2.3.

The four different solutions can be described in terms of the relative magnitude and signs of b_1 and d_1: each solution corresponds to one of the four cases depending on if b_1 and d_1 have the same or opposite sign, and if b_1 or if d_1 is larger in magnitude. This is depicted in Figure 2.3. An exceptional point with the same qualitative features as in Figure 2.3a was previously observed in the setting of a Hamiltonian system in [**176**].

We have used formal asymptotics to approximate the continuously differentiable solutions to (2.18)–(2.20). Next, we show that any solution to this system is indeed a fourth-order asymptotic exceptional point, or in other words, all eigenvectors of $\mathcal{C}_d$ coalesce.

LEMMA 2.10. *For a given small $\varepsilon > 0$, let b, c and d be solutions to* (2.18)–(2.20) *and let γ be the corresponding eigenvalue of $\mathcal{C}_d$. Then, we have*

$$\dim \operatorname{Ker}(\mathcal{C}_d - \gamma I) = 1.$$

PROOF. We will show the equivalent statement that $\mathcal{C}_d - \gamma I$ has rank 3. Since γ is an eigenvalue of $\mathcal{C}_d$, the rank is at most 3. Moreover, since $\gamma = (c+1)/2$, we

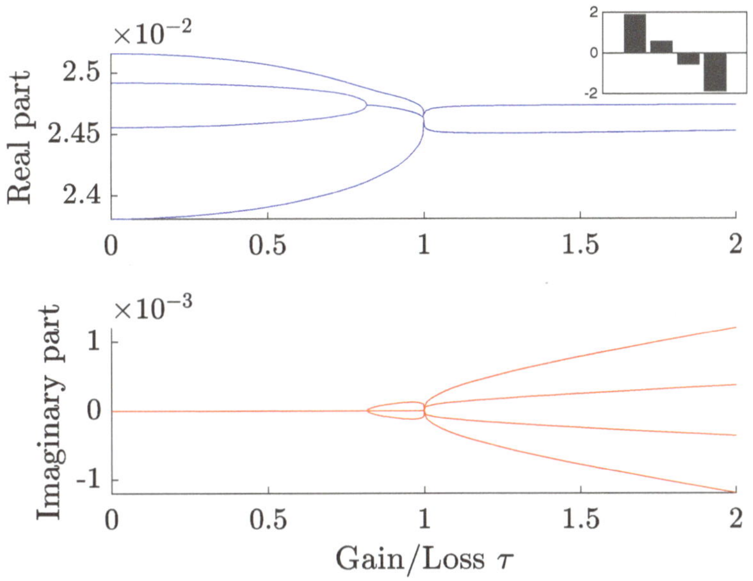

FIGURE 2.4. Evolution of the resonant frequencies, at leading-order, as gain and losses are introduced. Here, the imaginary parts b_i are rescaled by $\tau \in [0, 2]$, where $\tau = 1$ corresponds to an exceptional point of order four.

have for small ε,

$$\mathcal{C}_d - \gamma I = \varepsilon \begin{pmatrix} -\frac{c_1}{2} + \mathrm{i}b_1 & -1 & -1/2 & -1/3 \\ -1 & \frac{c_1}{2} + \mathrm{i}d_1 & -1 & -1/2 \\ -1/2 & -1 & \frac{c_1}{2} - \mathrm{i}d_1 & -1 \\ -1/3 & -1/2 & -1 & -\frac{c_1}{2} - \mathrm{i}b_1 \end{pmatrix} + \mathcal{O}(\varepsilon^2).$$

The determinant of the 3×3 upper right block is given by

$$\det \begin{pmatrix} -1 & -1/2 & -1/3 \\ \frac{c_1}{2} + \mathrm{i}d_1 & -1 & -1/2 \\ -1 & \frac{c_1}{2} - \mathrm{i}d_1 & -1 \end{pmatrix} = -\frac{1}{12}\left((c_1+3)^2 + 4d_1^2 + 2\right),$$

and is negative for any c_1 and d_1. Therefore, for small ε, the rank of $\mathcal{C}_d - \gamma I$ is at least 3. □

Figure 2.4 shows the evolution of the resonant frequencies as gain and loss are introduced. This example follows the same pattern as the one in Figure 2.2.

2.3.3. Exceptional points of arbitrary order. Here, we study exceptional points in larger resonator systems. We will consider a $\mathcal{PT}$-symmetric array with an even or odd number of resonators. In the case of an odd number, analogously to the third-order exceptional point, we assume that the centre resonator has no gain or loss.

To be specific, we consider an array of N resonators with material parameters given by

$$v_i^2 \delta_i = \delta a(a_i + \mathrm{i}b_i) \text{ for } i = 1, \dots, N,$$

for some $a_i, b_i \in \mathbb{R}$. As before, a is chosen so that $a_1 = 1$. The assumption of $\mathcal{PT}$ symmetry, specified in (2.3), means that in the respective case of even or odd number of resonators, we have

$$\begin{cases} a_i = a_{2n+1-i}, & b_i = -b_{2n+1-i}, & & N = 2n \text{ is even}, \\ a_i = a_{2n+2-i}, & b_i = -b_{2n+2-i}, & b_{n+1} = 0, & N = 2n+1 \text{ is odd}, \end{cases}$$

where $n \in \mathbb{N}$. With this notation, the dilute capacitance matrix $\mathcal{C}_d = ((\mathcal{C}_d)_{ij})$, as defined in (2.8), is the matrix with entries specified by

$$(\mathcal{C}_d)_{ij} = \begin{cases} a_i + \mathrm{i} b_i, & i = j, \\ -(a_i + \mathrm{i} b_i)\frac{\varepsilon}{|i-j|}, & i \neq j. \end{cases}$$

Again, to have an exceptional point of order N, we require that (2.1) and (2.2) hold for some γ. For general N, (2.1) is a system of N polynomial equations of order N in terms of the N unknown parameters $\gamma, b_1, a_2, b_2, \ldots, a_n, b_n$ and if N is odd, a_{n+1}. For N larger than 4, it is not possible to explicitly derive the solutions, so we will numerically study this system of equations in Section 2.3.3. However, under the assumption that (2.1) holds, we can derive asymptotic formulas for unknown parameters, analogously to (2.21)–(2.23).

We begin by observing the following simple equation for γ:

$$\gamma = \frac{1}{N}\sum_{i=1}^{N} a_i. \tag{2.24}$$

As $\varepsilon \to 0$, we also have the following result.

LEMMA 2.11. *Assume that there is a solution a_i, b_i, for $i = 1, \ldots, N$, to (2.1) that is continuous as $\varepsilon \to 0$. Then, we have*

$$a_i = 1 + \mathcal{O}(\varepsilon), \qquad b_i = \mathcal{O}(\varepsilon),$$

and consequently, $\gamma = 1 + \mathcal{O}(\varepsilon)$.

PROOF. As $\varepsilon \to 0$, denote the limiting values of a_i and b_i by $a_{i,0}$ and $b_{i,0}$, respectively. Since (2.1) has no terms of order $\mathcal{O}(\varepsilon)$, we find that

$$a_i = a_{i,0} + \mathcal{O}(\varepsilon), \qquad b_i = b_{i,0} + \mathcal{O}(\varepsilon).$$

We then have

$$\begin{aligned} \det(\mathcal{C}_d - xI) &= \prod_{i=1}^{N}(a_{i,0} + \mathrm{i} b_{i,0} - x) + \mathcal{O}(\varepsilon) \\ &= (\gamma_0 - x)^N + \mathcal{O}(\varepsilon), \end{aligned}$$

where, from (2.24),

$$\gamma_0 = \frac{1}{N}\sum_{i=1}^{N} a_{i,0}.$$

Since $\gamma_0 \in \mathbb{R}$, it follows that $b_{i,0} = 0$ for all $i = 1, \ldots, N$, and since $a_{1,0} = 1$, it follows that $a_{i,0} = 1$ for all $i = 1, \ldots, N$. □

Assume that (2.1) holds for some γ. Then, for small $\varepsilon > 0$, we have by Lemma 2.11 that

$$a_i = 1 + a_{i,1}\varepsilon + o(\varepsilon), \qquad b_i = b_{i,1}\varepsilon + o(\varepsilon), \qquad \gamma = 1 + \gamma_1\varepsilon + o(\varepsilon),$$

for some $a_{i,1}, b_{i,1}$ and γ_1 independent of ε. In this case, it holds that

$$\mathcal{C}_d = I + \varepsilon \mathcal{C}_{d,1} + o(\varepsilon)\,, \tag{2.25}$$

where

$$\mathcal{C}_{d,1} = \begin{pmatrix} a_{1,1} + \mathrm{i}b_{1,1} & -1 & -1/2 & \cdots & -1/(N-1) \\ -1 & a_{2,1} + \mathrm{i}b_{2,1} & -1 & \cdots & -1/(N-2) \\ -1/2 & -1 & a_{3,1} + \mathrm{i}b_{3,1} & \cdots & -1/(N-3) \\ \vdots & \vdots & \vdots & \ddots & \vdots \\ -1/(N-1) & -1/(N-2) & -1/(N-3) & \cdots & a_{1,1} - \mathrm{i}b_{1,1} \end{pmatrix}.$$

Then

$$\mathcal{C}_d - \gamma I = \varepsilon\left(\mathcal{C}_{d,1} - \gamma_1 I\right) + o(\varepsilon)\,.$$

Therefore, for γ to be an Nth-order asymptotic exceptional point of $\mathcal{C}_d$, we must have that γ_1 is an Nth-order exceptional point of $\mathcal{C}_{d,1}$. We can then obtain a system of polynomial equations describing the exceptional point (analogous to (2.21)–(2.23) but for general N) by expanding the characteristic polynomial of $\mathcal{C}_{d,1}$. Moreover, assuming that (2.1) holds for some γ, a simple way to prove that (2.2) holds is to check that $\mathcal{C}_{d,1}$ has a one-dimensional kernel.

At an exceptional point γ_1 of $\mathcal{C}_{d,1}$, it follows from Theorem 2.6 and (2.25) that the full system exhibits an asymptotic exceptional point with frequency given by

$$\omega^* = \sqrt{\frac{4\pi a\delta}{|D_1|}(1+\varepsilon\gamma_1) + \mathcal{O}(\delta) + \delta^{1/2}\, o(\varepsilon)}\,.$$

Figure 2.5 shows examples of symmetric families of gain/loss distributions (that is, the imaginary parts of the material parameters) corresponding to exceptional points of different orders. Figures 2.5a and 2.5b show roughly linearly growing distributions, while Figures 2.5c and 2.5d show alternating distributions.

Figure 2.6 shows the evolution of the resonant frequencies as gain and losses are introduced for exceptional points of orders 8 and 14. In these examples, we observe the same pattern as in Figures 2.2 and 2.4.

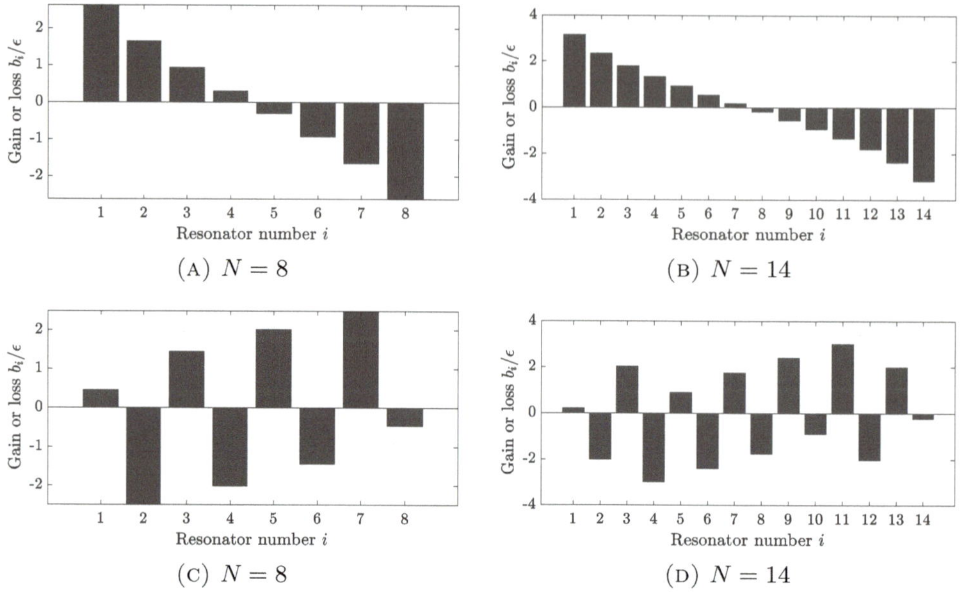

FIGURE 2.5. Higher-order asymptotic exceptional points can be found in larger arrays of resonators.

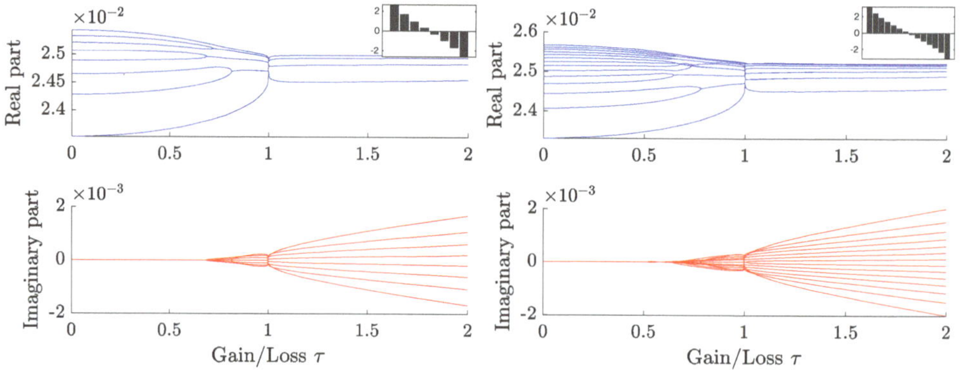

FIGURE 2.6. Patterns of exceptional points of order 8 and 14 as the gain/loss grows linearly.

CHAPTER 3

Effective medium theory for finite systems of weakly interacting subwavelength resonators

In this chapter, we derive an effective medium theory for wave propagation in finite but large systems of weakly interacting subwavelength resonators. We start with a multiple-scattering formulation of the scattering problem of an incident wave by a large number of identical small resonators. Under certain conditions on the configuration of the resonators, we establish an effective medium theory for the system of resonators, as the number of resonators tends to infinity. As a consequence, we show that just below the subwavelength resonant frequency ω_M of a single resonator given by (1.63), the effective medium obtained can have a high refractive index, while just above ω_M, it behaves as a diffusive medium. On the other hand, using dimers of resonators, under the assumption that their volume fraction is small and their orientation is uniformly distributed, we prove that both the effective material parameters can be negative at frequencies slightly higher than the dipole hybridised frequency $\omega_{M,2}$ for a single constituent dimer defined by (1.70). The results presented in this chapter are from [**33, 39, 56**]. See also [**83, 85**].

3.1. Problem formulation

Consider the scattering of waves by N identical resonators distributed in $\Omega \Subset \mathbb{R}^3$. The resonators are represented by

$$D^N := \cup_{1\leq j\leq N} D_j^N,$$

where $D_j^N = y_j^N + sB$ for $1 \leq j \leq N$ with y_j^N being the location, s being the characteristic size and B being the normalised resonator, which is a smooth and simply connected domain with size of order one. We denote by ρ_r and κ_r the material parameters within the resonator, which are different from the corresponding ρ and κ in the background medium $\mathbb{R}^3 \backslash D^N$. We let $\delta = \rho_r/\rho$, $k = \omega/v$, $k_r = \omega/v_r$, where $v = \sqrt{\rho/\kappa}$ and $v_r = \sqrt{\rho_r/\kappa_r}$.

We assume that $0 < \delta \ll 1, 0 < s \ll 1$, $N \gg 1$ and $\{y_j^N\} \subset \Omega$. Let u_{in} be the incident wave that we assume to be a plane wave for simplicity. The scattering can be modelled by the following system of equations:

$$\begin{cases} \Delta u^N + k^2 u^N = 0 \quad \text{in } \mathbb{R}^3 \backslash D^N, \\ \Delta u^N + k_r^2 u^N = 0 \quad \text{in } D^N, \\ u^N|_+ - u^N|_- = 0 \quad \text{on } \partial D^N, \\ \delta \left.\frac{\partial u^N}{\partial \nu}\right|_+ - \left.\frac{\partial u^N}{\partial \nu}\right|_- = 0 \quad \text{on } \partial D^N, \\ u^N - u_{\text{in}} \text{ satisfies the Sommerfeld radiation condition,} \end{cases} \tag{3.1}$$

where u^N is the total field and ω is the operating frequency.

As shown in Chapter 1, the solution u^N can be written as

$$u^N(x) = \begin{cases} u_{\text{in}} + \mathcal{S}_{D^N}^k[\phi^N], & x \in \mathbb{R}^3 \setminus \overline{D^N}, \\ \mathcal{S}_D^{k_r}[\psi^N], & x \in D^N, \end{cases} \tag{3.2}$$

for some surface potentials $\phi^N, \psi^N \in L^2(\partial D^N)$, where

$$\mathcal{S}_{D^N}^k[\phi^N] = \sum_{1 \le j \le N} \mathcal{S}_{D_j^N}^k[\phi_j^N],$$

$$\mathcal{S}_D^{k_r}[\psi^N] = \sum_{1 \le j \le N} \mathcal{S}_{D_j^N}^{k_r}[\psi_j^N],$$

with $\phi_j^N, \psi_j^N \in L^2(\partial D_j^N)$.

Using the jump relations for the single layer potentials, ϕ^N and ψ^N satisfy the following system of boundary integral equations:

$$\mathcal{A}^N(\omega, \delta)[\Psi^N] = F^N \quad \text{on } \partial D^N, \tag{3.3}$$

where

$$\mathcal{A}^N(\omega, \delta) = \begin{pmatrix} \mathcal{S}_{D^N}^{k_r} & -\mathcal{S}_{D^N}^k \\ -\frac{1}{2}I + \mathcal{K}_{D^N}^{k_r,*} & -\delta(\frac{1}{2}I + \mathcal{K}_{D^N}^{k,*}) \end{pmatrix},$$

$$\Psi^N = \begin{pmatrix} \psi^N \\ \phi^N \end{pmatrix}, \quad F^N = \begin{pmatrix} u_{\text{in}} \\ \delta \frac{\partial u_{\text{in}}}{\partial \nu} \end{pmatrix} \Bigg|_{\partial D^N}.$$

In terms of ϕ_j^N, ψ_j^N, this leads to a system of equations of the form

$$\mathcal{A}_{D_1,\ldots D_N} \begin{pmatrix} \psi_1^N \\ \phi_1^N \\ \vdots \\ \psi_N^N \\ \phi_N^N \end{pmatrix} = \begin{pmatrix} u_{\text{in}}\big|_{\partial D_1} \\ \frac{\partial u_{\text{in}}}{\partial \nu_1}\big|_{\partial D_1} \\ \vdots \\ u_{\text{in}}\big|_{\partial D_N} \\ \frac{\partial u_{\text{in}}}{\partial \nu_N}\big|_{\partial D_N} \end{pmatrix}, \tag{3.4}$$

where ν_i is the outward unit normal at $\partial D_i, i = 1, \ldots, N$. The operator $\mathcal{A}_{D_1,\ldots,D_N}$ has the block diagonal form

$$\mathcal{A}_{D_1,\ldots,D_N} = \begin{pmatrix} \mathcal{M}_1 & \mathcal{L}_{1,2} & \mathcal{L}_{1,3} & \cdots \\ \mathcal{L}_{2,1} & \mathcal{M}_2 & \mathcal{L}_{2,3} & \cdots \\ \vdots & \vdots & \vdots & \vdots \\ \mathcal{L}_{N,1} & \mathcal{L}_{N,2} & \cdots & \mathcal{M}_N \end{pmatrix},$$

where $\mathcal{M}_j$ is the self-interaction for the jth resonator, defined by

$$\mathcal{M}_j := \begin{pmatrix} \mathcal{S}_{D_j}^{k_r} & -\mathcal{S}_{D_j}^k \\ \frac{1}{\delta}\left(-\frac{1}{2}I + \mathcal{K}_{D_j}^{k_r,*}\right) & -\left(\frac{1}{2}I + \mathcal{K}_{D_j}^{k,*}\right) \end{pmatrix},$$

and, for $i \neq j$, $\mathcal{L}_{i,j}$ encodes the effect of the jth resonator D_j to the ith resonator:

$$\mathcal{L}_{i,j} = \begin{pmatrix} 0 & -\mathcal{S}_{D_i,D_j}^k \\ 0 & -\mathcal{L}_{D_i,D_j}^k \end{pmatrix}.$$

Here, we have introduced the operators $\mathcal{S}^k_{D_i,D_j} : L^2(\partial D_j) \to L^2(\partial D_i)$ and $\mathcal{L}^k_{D_i,D_j} : L^2(\partial D_j) \to L^2(\partial D_i)$, which are respectively defined by

$$\forall\, \varphi \in L^2(\partial D_j), \quad \mathcal{S}^k_{D_i,D_j}[\varphi] = \mathcal{S}^k_{D_j}[\varphi]\Big|_{\partial D_i} \quad \text{and} \quad \mathcal{L}^k_{D_i,D_j}[\varphi] = \frac{\partial}{\partial \nu_i}\mathcal{S}^k_{D_j}[\varphi]\Big|_{\partial D_i}.$$

3.2. Point interaction approximation

We are interested in the case where there is a large number N of small identical resonators distributed in a bounded domain Ω and the incident wave has a frequency near the subwavelength resonant frequency for an individual resonator. We recall from (1.63) that for a resonator given by $D_j^N = y_j^N + sB$, its corresponding subwavelength resonant frequency ω_M is approximately given by

$$\omega_M = \frac{1}{s}\sqrt{\frac{\mathrm{Cap}_B \delta}{|B|}}\, v_r. \tag{3.5}$$

We assume that the resonators are well separated, *i.e.*, their separation distance is large compared to their size and denote by $u_j^{\mathrm{in},N}$ the field incident on the jth resonator D_j, and by $u_j^{s,N}$ the field scattered from D_j. It is clear that

$$u_i^{\mathrm{in},N}(x) = u_{\mathrm{in}}(x) + \sum_{j\neq i} u_j^{s,N}(x). \tag{3.6}$$

Using (1.65), and the fact that the resonators are well separated, the scattered field $u_i^{s,N}(x)$ far from the ith resonator can be approximated by

$$u_i^{s,N}(x) = g(\omega,\delta,sB)G^k(x - y_i^N)u_i^{\mathrm{in},N}(y_i^N), \tag{3.7}$$

where the scattering coefficient g is defined by (1.66) with $D = sB$. By taking $x = y_j^N$ in the above equation and using (3.6), we obtain the following system of linear equations for $u_i^{\mathrm{in},N}(y_i^N)$:

$$u_i^{\mathrm{in},N}(y_i^N) + g(\omega,\delta,sB)\sum_{j\neq i} G^k(y_i^N - y_j^N)u_j^{\mathrm{in},N} = u_{\mathrm{in}}(y_i^N),$$

or equivalently,

$$M\begin{pmatrix} u_1^{\mathrm{in},N}(y_1^N) \\ \vdots \\ u_N^{\mathrm{in},N}(y_N^N)\end{pmatrix} = \begin{pmatrix} u_{\mathrm{in}}(y_1^N) \\ \vdots \\ u_{\mathrm{in}}(y_N^N)\end{pmatrix},$$

where M is the $N\times N$ matrix defined by

$$M_{ij} = \begin{cases} 1, & i = j, \\ g(\omega,\delta,sB)G^k(y_i^N - y_j^N), & i\neq j. \end{cases} \tag{3.8}$$

Finally, the field u^N is the sum of the incoming wave and all the waves scattered by the different resonators. Together with (3.7), we obtain that

$$u^N(x) = u_{\mathrm{in}}(x) + \sum_{1\leq i\leq N} g(\omega,\delta,sB)G^k(x - y_i^N)\sum_{1\leq j\leq N}(M^{-1})_{ij}u_{\mathrm{in}}(y_j^N), \tag{3.9}$$

where $(M^{-1})_{ij}$ denotes the $(ij)^{\mathrm{th}}$ component of the inverse matrix M^{-1}. Equation (3.9) is called the point interaction approximation of the solution u^N of (3.1) [**136, 138**]. It can be rigorously justified under the following assumptions on the frequency, the resonator volume fraction, and the configuration of the resonator distribution; see [**56**].

ASSUMPTION 3.1. *The frequency* $\omega = \mathcal{O}(1)$ *and is independent of* N. *Moreover,*

$$1-\Big(\frac{\omega_M}{\omega}\Big)^2 = \beta_0 \tag{3.10}$$

for some small constant β_0.

ASSUMPTION 3.2. *The following identity holds:*

$$sN = \Lambda, \tag{3.11}$$

where Λ *is a positive constant independent of* N.

ASSUMPTION 3.3. *The following conditions hold:*

$$\begin{cases} \min_{l\neq j} |y_l^N - y_j^N| \gtrsim N^{-\frac{1}{3}}, \\ s \ll N^{-\frac{1}{3}}. \end{cases}$$

ASSUMPTION 3.4. *There exists a function* $\widetilde{V} \in C^1(\overline{\Omega})$ *such that for any* $f \in C^{0,\alpha}(\Omega)$ *with* $0 < \alpha < 1$,

$$\max_{1\leq j\leq N}\left| \frac{1}{N}\sum_{l\neq j} \frac{1}{|y_l^N - y_j^N|} f(y_l^N) - \int_\Omega \frac{1}{|y - y_j^N|}\widetilde{V}(y) f(y)\,\mathrm{d}y \right| \lesssim \frac{1}{N^{\frac{\alpha}{3}}}\|f\|_{C^{0,\alpha}}. \tag{3.12}$$

Equation (3.10) in Assumption 3.1 controls the frequency deviation from ω_M. There are two cases depending on whether $\omega > \omega_M$ or $\omega < \omega_M$. In the former case, we have $\beta_0 > 0$, while in the latter case we have $\beta_0 < 0$. Moreover, from (3.5), $\omega = \mathcal{O}(1)$ implies that $s = \mathcal{O}(\delta^2)$. Assumption 3.2 is on the resonator volume fraction. It indicates that both the resonator size s and contrast parameter δ depend on N and in the limit when $N \to +\infty$, we have $s \to 0$ and $\delta = \mathcal{O}(s^2) \to 0$. In Assumption 3.3, we impose conditions on the resonator configuration by assuming that the size of each resonator is much smaller than the typical distance between neighbouring resonators. Assumption 3.4 is a condition on the regularity of the "sampling" points $\{y_j^N\}$. One can easily check that Assumptions 3.3 and 3.4 hold for periodically distributed y_j^N's.

3.3. Derivation of an effective potential

To pass to the limit $N \to +\infty$ in (3.9) under the assumptions we made in the previous section, we let $\Omega_N = \Omega \setminus \cup_{1\leq i\leq N} B(y_i^N, \sqrt{s})$ and prove that there exists some macroscopic field $u \in C^{1,\alpha}(\Omega)$ such that

$$u^N(x) \to u(x) \quad \text{for } x \in \Omega_N.$$

Here, the convergence is understood in the sense that for any $\varepsilon > 0$, there exists N_0 such that for all $N \geq N_0$, we have

$$\|u^N - u\|_{C^{1,\alpha}(\Omega_N)} \leq \varepsilon.$$

This implies in particular that

$$u_j^{\text{in},N}(y_j^N) \to u(y_j^N).$$

Therefore,

$$g(\omega,\delta,sB)G^k(x - y_j^N)u_j^{\text{in},N}(y_j^N) \to \frac{1}{N}\Lambda g(\omega,\delta,B)G^k(x-y_j^N)u(y_j^N).$$

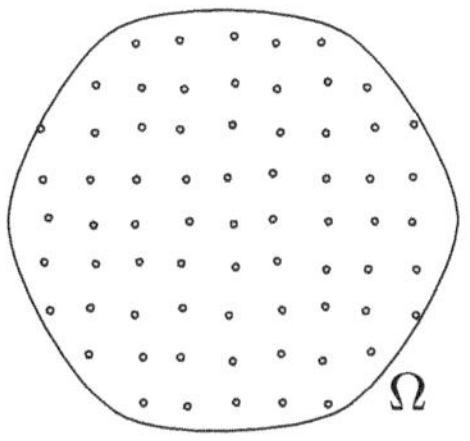

FIGURE 3.1. A system of weakly interacting single resonators.

On the other hand, for $x \in \Omega_N$, we have

$$u(x) = u_{\text{in}}(x) + \sum_{1 \leq j \leq N} g(\omega, \delta, sB) G^k(x - y_j^N) u_j^{\text{in},N}(y_j^N) + o(1)\,.$$

By letting $N \to +\infty$, we obtain

$$u(x) = u_{\text{in}}(x) + \frac{\Lambda \text{Cap}_B}{\beta_0} \int_{\Omega} \widetilde{V}(y) G^k(x-y) u(y)\, dy.$$

By applying the operator $\Delta + k^2$ to both sides of the above equation, we get

$$\left(\Delta + k^2 - \Lambda \text{Cap}_B / \beta_0 \widetilde{V}(x)\right) u(x) = 0 \quad \text{in } \Omega.$$

Therefore, when $-\Lambda \text{Cap}_B / \beta_0 \gg 1$ we obtain an effective medium with a high refractive index while when $-\frac{\Lambda \text{Cap}_B}{\beta_0} \ll -1$, the effective medium is diffusive.

To illustrate our results in this section, we consider a source at the origin surrounded by a system of uniformly distributed resonators, as shown in Figure 3.1. We assume that B is the unit sphere and denote by $G^{k_{\text{eff}}}(x)$ the Green function in the presence of the resonators at frequency ω. Then, from (1.65) and (1.67), it follows that

$$g(\omega, \delta, sB) = \frac{4\pi s}{1 - (\frac{\omega_M}{\omega})^2 + \mathrm{i}\gamma_M}, \qquad \omega_M = \frac{\sqrt{3\delta}}{s}, \qquad \gamma_M = s\omega.$$

Moreover, using the point interaction approximation (3.9), we obtain that

(3.13)
$$G^{k_{\text{eff}}}(x) := G^k(x) - 4\pi s \sum_{1 \leq i \leq N} \frac{1}{1 - (\frac{\omega_M}{\omega})^2 + \mathrm{i}s\omega} G^k(x - y_i^N) \sum_j (M^{-1})_{ij} G^k(y_j^N).$$

In Figure 3.2, we plot $|\operatorname{Im}(G^{k_{\text{eff}}}(x))|$ when no resonators are present, and in Figure 3.3, we plot $|\operatorname{Im}(G^{k_{\text{eff}}}(x))|$, for resonator clouds of increasing volume fraction $f := 4\pi s^3 N/3$. When the volume fraction reaches $f = 2 \times 10^{-4}$, we obtain a very sharp peak over the origin just below ω_M, indicating that the effective refractive index should be greatly enhanced in this frequency regime.

3.4. Systems of weakly interacting dimer resonators

Our aim in this section is to understand the mechanism behind the double-negative refractive index phenomenon in dilute dimer systems of subwavelength resonators.

We consider the scattering of an incident plane wave u_{in} by N identical resonator dimers with different orientations distributed in $\Omega \Subset \mathbb{R}^3$. The N identical resonator dimers are generated by scaling the normalised dimer D by a factor s,

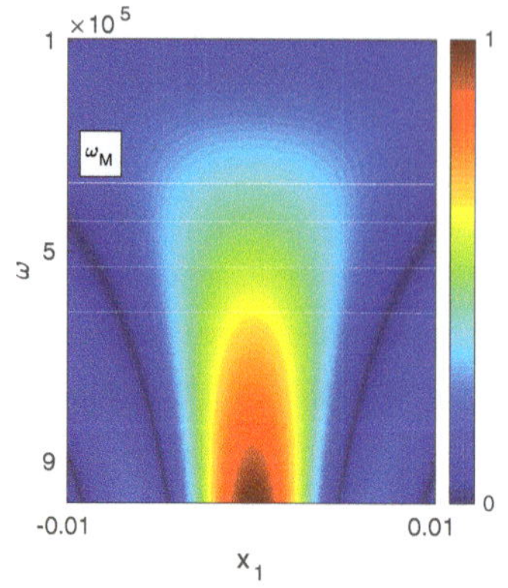

FIGURE 3.2. The imaginary part of the Green function when no resonators are present. The white line indicates ω_M.

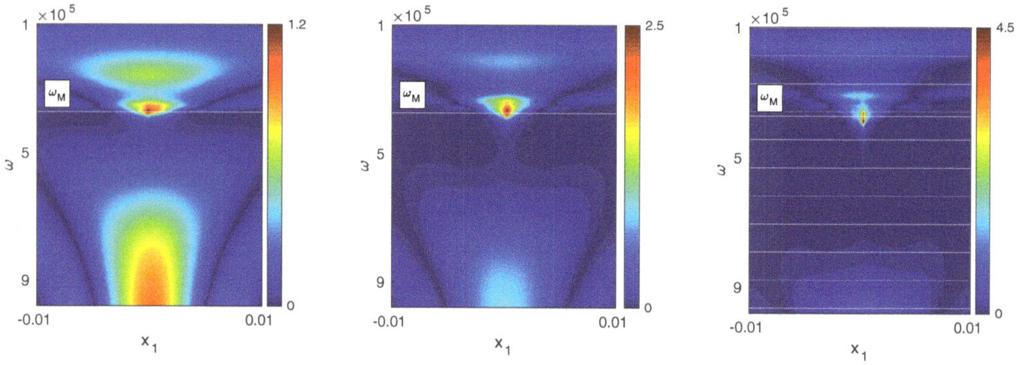

(A) $f = 0.5 \times 10^{-4}$ (B) $f = 1 \times 10^{-4}$ (C) $f = 2 \times 10^{-4}$

FIGURE 3.3. The imaginary part of the Green function averaged over 25 resonator cloud configurations for increasing resonator concentrations f. The results are normalised with respect to the background Green function in Figure 3.2. The function peaks over the source at the origin when the excitation frequency is very near ω_M, indicating enhanced effective refractive index.

and then rotating the orientation and translating the centre. More precisely, the resonator dimers occupy the domain

$$D^N := \cup_{1 \le j \le N} D_j^N,$$

where $D_j^N = y_j^N + sR_{d_j^N} D$ for $1 \le j \le N$, with y_j^N being the centre of the dimer D_j^N, s being the characteristic size, and $R_{d_j^N}$ being the rotation in $\mathbb{R}^3$ which aligns the dimer D_j^N in the direction d_j^N. Here, d_j^N is a unit length vector in $\mathbb{R}^3$.

We assume that $0 < s \ll 1$, $N \gg 1$ and $\{y_j^N : 1 \le j \le N\} \subset \Omega$. The scattering of waves by the resonator dimers can be modelled by (3.1), where D^N is the collection of resonator dimers. We make the same assumptions as in Assumptions 3.2, 3.3, and 3.4. We replace Assumption 3.1 with the following assumption.

ASSUMPTION 3.5. $\delta = \mu^2 s^2$ *for some positive number* $\mu > 0$ *and* $\omega = \omega_{M,2} + as^2$ *for some real number* $a \neq \mu^3 \widehat{\eta}_1$, *where* $\omega_{M,2}$ *and* $\widehat{\eta}_1$ *are defined by* (1.70).

Furthermore, we need the following assumption.

ASSUMPTION 3.6. *There exists a matrix-valued function* $\widetilde{B} \in C^1(\overline{\Omega})$ *such that for any* $f \in C^{0,\alpha}(\Omega)$ *with* $0 < \alpha < 1$,

$$\max_{1 \leq j \leq N} \left| \frac{1}{N} \sum_{l \neq j} (d_l^N \cdot \frac{1}{|y_l^N - y_j^N|}) d_l^N \cdot f(y_l^N) - \int_\Omega \widetilde{B}(y) \nabla_y (\frac{1}{|y - y_j^N|}) \cdot f(y) \, \mathrm{d}y \right|$$

$$\lesssim \frac{1}{N^{\frac{\alpha}{3}}} \|f\|_{C^{0,\alpha}}. \tag{3.14}$$

Assumption 3.5 is necessary to obtain an effective double negative medium. Recall that $\omega_{1,M} < \omega_{2,M}$ and note that for any frequency ω slightly higher than the monopole resonant frequency $\omega_{1,M}$ the effective potential is always negative. Then in order to make both the effective scalar and matrix material parameters negative, one needs ω to be very close to the dipole resonant frequency $\omega_{M,2}$.

On the other hand, note also that Assumption 3.6 is a higher condition on the regularity of the "sampling" points $\{y_j^N\}$ than the one in Assumption 3.4. It is needed because of the occurrence of dipolar resonant modes; see Section 1.7.

Now, we introduce the two constants

$$\widetilde{g}^0 = \frac{2(C_{11} + C_{12})}{1 - \omega_{M,1}^2 / \omega_{M,2}^2}, \quad \widetilde{g}^1 = \frac{\mu^2 v_r^2}{2|D|\omega_{M,2}(\mu^3 \widehat{\eta}_1 - a)} P^2,$$

where $\omega_{M,1}$ is defined by (1.69), and the two functions

$$M_1 = \begin{cases} I & \text{in } \mathbb{R}^3 \setminus \Omega, \\ I - \Lambda \widetilde{g}^1 \widetilde{B} & \text{in } \Omega, \end{cases}$$

and

$$M_2 = \begin{cases} k^2 & \text{in } \mathbb{R}^3 \setminus \Omega, \\ k^2 - \Lambda \widetilde{g}^0 \widetilde{V} & \text{in } \Omega. \end{cases}$$

The following result from [**39**] holds.

THEOREM 3.7. *Suppose that there exists a unique solution* u *to*

$$\nabla \cdot M_1(x) \nabla u(x) + M_2(x) u(x) = 0 \quad \text{in } \mathbb{R}^3, \tag{3.15}$$

such that $u - u_{\text{in}}$ *satisfies the Sommerfeld radiation condition at infinity. Then, under Assumptions* 3.2, 3.3, 3.4, *and* 3.5, *we have* $u^N(x) \to u(x)$ *uniformly for* $x \in \Omega_N$.

If the resonator dimers are distributed so that $\widetilde{B}$ is a positive matrix with $\widetilde{B}(x) \geq C > 0$ for some constant C for all $x \in \Omega$, then we see that for ω of the form $\omega = \omega_{M,2} + as^2$ with $a < \mu^3 \widehat{\eta}_1$, and sufficiently large Λ, both the matrix $I - \Lambda \widetilde{g}^1 \widetilde{B}$ and the scalar function $k^2 - \Lambda \widetilde{g}^0 \widetilde{V}$ are negative. Therefore, we obtain an effective double negative medium. On the other hand, for ω in between $\omega_{M,1}$ and $\omega_{M,2}$ but away from the dipolar resonance $\omega_{M,2}$, $\widetilde{g}^1$ may be small enough so that the matrix $I - \Lambda \widetilde{g}^1 \widetilde{B}$ is positive, while the matrix $k^2 - \Lambda \widetilde{g}^0 \widetilde{V}$ remains negative. Then the obtained effective medium has one effective single negative material parameter.

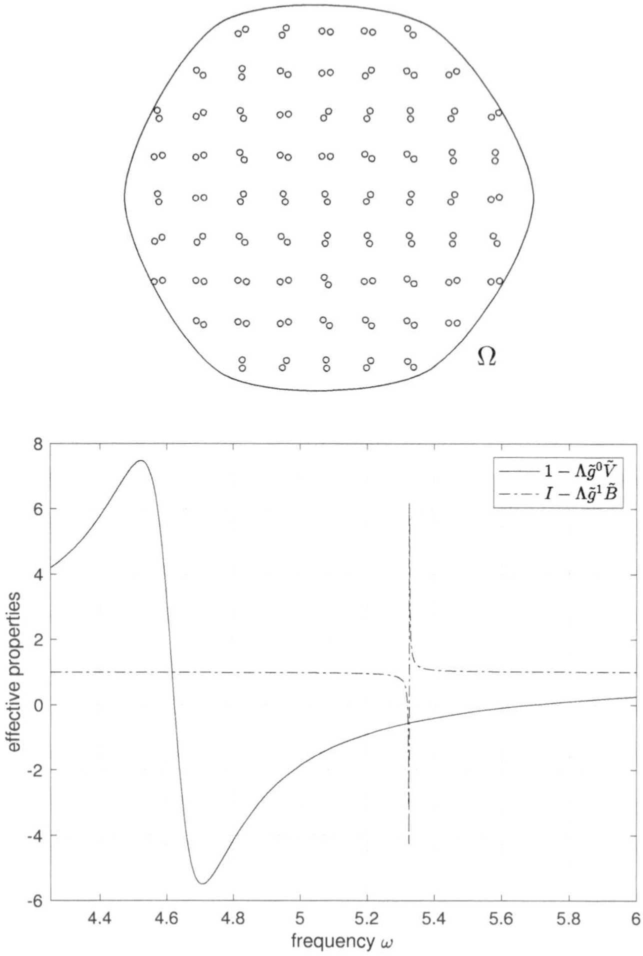

FIGURE 3.4. The effective properties of a system of weakly interacting dimer resonators.

To illustrate the phenomenon of double negative refractive index in dilute systems of dimer resonators, we consider a cubic array of identical spherical resonator dimers, as illustrated in Figure 3.4. We assume that each dimer is randomly oriented so that the unit direction vector D_j^N is uniformly distributed on the unit sphere. Figure 3.4 shows the two effective coefficients as functions of frequency. Clearly, there is a narrow region in which both coefficients are negative. So, double-negative refraction occurs in this frequency regime.

Part 2

Periodic systems of subwavelength resonators

CHAPTER 4

Hermitian systems of subwavelength resonators

In this chapter, we investigate the case where the resonators are repeated periodically, as illustrated in Figure 4.1. Periodicity can give rise to exotic properties, and these structures are often referred to as *metamaterials*. In order to describe resonance and scattering problems in periodic structures, we use methods analogous to those in Chapter 1. The Green function will need to be modified to account for periodicity, and we will introduce the concepts of generalised quasiperiodic and periodic capacitance matrices. These concepts can be applied to explain a variety of interesting physical phenomena. We particularly want to study Dirac degeneracies, near-zero effective material parameters, robust localisation at subwavelength scales, resonances in the first radiation continuum, bound states in the continuum, Fano-type transmission scattering, and Anderson-type localisation in the subwavelength regime.

4.1. Subwavelength resonance problem in periodic structures

Conceptually, there are three different cases depending on the relative dimensions of periodicity of the lattice d_l and of the ambient space d:

- $d - d_l = 0$, which we shall refer to as *crystal*. In this fully periodic case (either $d = d_l = 2$ or $d = d_l = 3$), the metamaterial has no boundary to the surrounding space;
- $d - d_l = 1$, which we will refer to as a *screen*. In this case, the structure consists of a thin sheet of resonators;
- $d - d_l = 2$, which we will refer to as a *chain*. There is one example of this case, namely $d = 3, d_l = 1$.

We assume that D, as defined in Chapter 1, is repeated in a periodic lattice Λ. We let $P_l : \mathbb{R}^d \to \mathbb{R}^{d_l}$ be the projection onto the first d_l coordinates and $P_\perp : \mathbb{R}^d \to \mathbb{R}^{d-d_l}$ be the projection onto the last $d - d_l$ coordinates.

We let $l_1, \ldots, l_{d_l} \in \mathbb{R}^d$ denote the lattice vectors generating the lattice Λ, in other words, such that

$$\Lambda := \{m_1 l_1 + \cdots + m_{d_l} l_{d_l} \mid m_i \in \mathbb{Z}\}.$$

For simplicity, we assume that $P_\perp l_i = 0$, which means that the lattice is aligned with the first d_l coordinate axes.

For a point $x \in \mathbb{R}^d$, we will use the notation $x = (x_l, x_0)$, where $x_l \in \mathbb{R}^{d_l}$ is the vector along the first d_l dimensions and $x_0 \in \mathbb{R}^{d-d_l}$. Denote by $Y \subset \mathbb{R}^d$ a fundamental domain of the given lattice. Explicitly, we take

$$Y := \{c_1 l_1 + \cdots + c_{d_l} l_{d_l} \mid 0 \leq c_1, \ldots, c_{d_l} \leq 1\}.$$

The dual lattice of Λ, denoted by Λ^*, is generated by $\alpha_1, \ldots, \alpha_{d_l}$ satisfying

$$\alpha_i \cdot l_j = 2\pi\delta_{ij} \text{ and } P_\perp \alpha_i = 0 \text{ for } i, j = 1, \ldots, d_l.$$

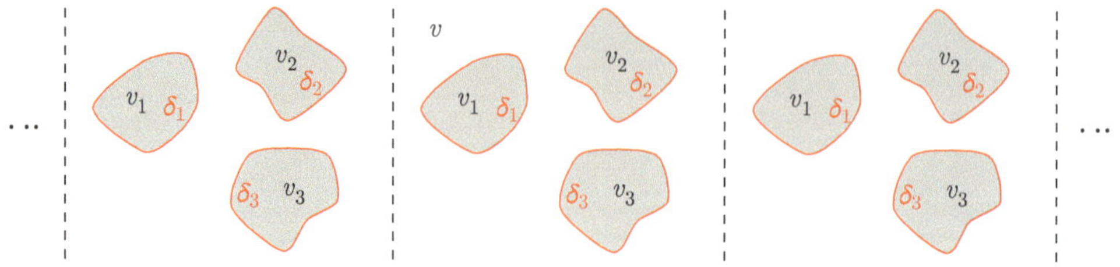

FIGURE 4.1. A periodic array of material inclusions. Here, three material inclusions (resonators) are sketched with periodicity in one dimension. Each interior has a different wave speed v_1, v_2, v_3 and the surrounding medium has a wave speed v. The contrast between the ith resonator and the background is given by δ_i, where a small value of δ_i describes a large contrast.

The *Brillouin zone* Y^* is defined as $Y^* := \left(\mathbb{R}^{d_l} \times \{\mathbf{0}\}\right)/\Lambda^*$, where $\mathbf{0}$ is the zero-vector in $\mathbb{R}^{d-d_l}$. We let $Y_l := Y \cap \{x_0 = 0\}$ and remark that Y^* can be written as $Y^* = Y_l^* \times \{\mathbf{0}\}$, where Y_l^* has the topology of a torus in d_l dimensions.

The periodically repeated i^{th} resonator $\mathcal{D}_i$ and the full periodic structure $\mathcal{D}$ are given, respectively, by

$$D_i^m = D_i + m, \qquad \mathcal{D}_i = \bigcup_{m\in\Lambda} D_i + m, \qquad \mathcal{D} = \bigcup_{i=1}^{N} \mathcal{D}_i. \tag{4.1}$$

With this notation in hand, we rewrite (1.1) (with $u_{\text{in}} = 0$) as follows:

$$\begin{cases} \Delta u + k^2 u = 0 & \text{in } \mathbb{R}^d \setminus \mathcal{D}, \\ \Delta u + k_i^2 u = 0 & \text{in } \mathcal{D}_i, \ i = 1, \ldots, N, \\ u|_+ - u|_- = 0 & \text{on } \partial\mathcal{D}, \\ \delta_i \dfrac{\partial u}{\partial \nu}\Big|_+ - \dfrac{\partial u}{\partial \nu}\Big|_- = 0 \text{ on } \partial\mathcal{D}_i, \ i = 1, \ldots, N, \\ u(x_l, x_0) \text{ satisfies the outgoing radiation condition as } |x_0| \to +\infty. \end{cases} \tag{4.2}$$

Here, as before, $k = \omega/v$ and $k_i = \omega/v_i$. Notice that while the set of material inclusions is compact in Chapter 1, it is not the case in the infinite periodic setting. Consequently, the radiation condition in (4.2) is no longer the one in (1.5). Moreover, it depends on the dimensionality. Note also that we only impose a radiation condition for x away from the periodic structure, and, in the fully periodic case, we do not assume any radiation condition. As we shall see, since there is no radiation condition in the first d_l coordinate dimensions, the spectrum σ of (4.2) is in general continuous. In order to effectively study this equation, we will use the *Floquet-Bloch* theory, which is outlined below; see also Appendix A.2.

4.1.1. Floquet-Bloch theory. A function $f(x) \in L^2(\mathbb{R}^d)$ is said to be α-quasiperiodic, with quasiperiodicity $\alpha \in Y^*$, if $e^{-\mathrm{i}\alpha\cdot x} f(x)$ is Λ-periodic. Given a function $f \in L^2(\mathbb{R}^d)$, the Floquet transform of f is defined as

$$\mathcal{U}[f](x, \alpha) := \sum_{m\in\Lambda} f(x - m) e^{\mathrm{i}\alpha\cdot m}, \quad x, \alpha \in \mathbb{R}^d. \tag{4.3}$$

$\mathcal{U}[f]$ is always α-quasiperiodic in x and periodic in α. The Floquet transform is an invertible map $\mathcal{U} : L^2(\mathbb{R}^d) \to L^2(Y \times Y^*)$, with inverse given by

$$\mathcal{U}^{-1}[g](x) = \frac{1}{|Y_l^*|} \int_{Y^*} g(x, \alpha)\, \mathrm{d}\alpha, \quad x \in \mathbb{R}^d,$$

where $g(x, \alpha)$ is extended quasiperiodically for x outside of the unit cell Y.

If we apply the Floquet transform to (4.2) we obtain, where $u^\alpha(x) := \mathcal{U}[u](x, \alpha)$, that

$$\begin{cases} \Delta u^\alpha + k^2 u^\alpha = 0 & \text{in } \mathbb{R}^d \setminus \mathcal{D}, \\ \Delta u^\alpha + k_i^2 u^\alpha = 0 & \text{in } \mathcal{D}_i,\ i = 1, \dots, N, \\ u^\alpha|_+ - u^\alpha|_- = 0 & \text{on } \partial\mathcal{D}, \\ \delta_i \frac{\partial u^\alpha}{\partial \nu}\Big|_+ - \frac{\partial u^\alpha}{\partial \nu}\Big|_- = 0 & \text{on } \partial\mathcal{D}_i,\ i = 1, \dots, N, \\ u^\alpha(x_l, x_0) \text{ is } \alpha\text{-quasiperiodic in } x_l, \\ u^\alpha(x_l, x_0) \text{ satisfies } \alpha\text{-quasiperiodic radiation condition as } |x_0| \to +\infty. \end{cases} \tag{4.4}$$

As mentioned above, the α-quasiperiodic radiation condition depends on the dimensionality. We refer, for instance, to [**37**, Chapter 4] for its explicit form. The spectrum σ of the original problem (4.2) is parameterised by the spectra $\sigma(\alpha)$, $\alpha \in Y^*$, of the problem (4.4), which in turn are known to consist of discrete values $\omega = \omega_i^\alpha$:

$$\sigma = \bigcup_{\alpha \in Y^*} \sigma(\alpha), \quad \sigma(\alpha) = \bigcup_{i=1}^{+\infty} \omega_i^\alpha.$$

DEFINITION 4.1 (Band function). *The resonant frequencies ω_i^α, seen as functions of α, are called band functions. The collection of band functions is called the band structure.*

DEFINITION 4.2 (Bandgap). *A bandgap of $\mathcal{D}$ is a connected component of $\mathbb{C}\setminus\sigma$. If the spectrum σ is real, we define a bandgap of $\mathcal{D}$ as a connected component of $\mathbb{R} \setminus \sigma$, which consists of an interval in $\mathbb{R}$.*

As in Chapter 1, we will focus on the subwavelength part of the spectrum, which are the resonant frequencies ω_i^α which tend to 0 as the contrast parameter $\delta \to 0$.

Using the tools presented in Appendix B.4, the capacitance matrix formulation developed in Theorem 1.9 for finite systems of resonators will be extended to infinitely periodic ones.

4.2. Subwavelength bandgap opening in a crystal

In this section, we consider a crystal of subwavelenth resonators and first show that a subwavelength bandgap opening may occur. In general, for bandgaps to occur, the period of the structure must be of the order of the wavelength and the contrast in the material parameters must be large [**107**]. This limits the use of crystalline structures in applications targeting low frequencies, because crystals would require impractically large geometries [**154**].

To demonstrate the opening of a subwavelength bandgap, we consider a periodic arrangement of resonators and exploit their subwavelength resonances. We provide a mathematical and numerical framework for analysing bandgap openings in subwavelength crystals at low frequencies. We derive a formula for the quasiperiodic subwavelength resonant frequencies of an arbitrarily shaped resonator in terms of the contrast between the material parameters, along with proving the existence of a subwavelength bandgap. Then we prove that the first Bloch eigenvalue achieves its maximum at the corner of the Brillouin zone. Moreover, by computing the asymptotic of the Bloch eigenmodes in the periodic structure near that critical frequency, we demonstrate that these eigenmodes can be decomposed into two parts: one part is slowly varying and satisfies a homogenised equation, while the other is periodic across each elementary crystal cell and is varying. These results extend the effective medium theory developed in Chapter 3 for finite resonator systems in the dilute regime to the nondilute case. However, they only describe the behaviour of the low-frequency part of the field. The results presented in this section are from [**38**, **49**].

4.2.1. Problem formulation. In this subsection, we let $d_l = d = 2$ or $d_l = d = 3$. Then we describe the crystal of subwavelength resonators under consideration. Let the lattice Λ be $\mathbb{Z}^d$. Assume that there is only one resonator in the unit cell, *i.e.*, the resonators occupy $\cup_{n\in\mathbb{Z}^d}(D+n)$ for a bounded and simply connected domain D with $\partial D \in C^{1,s}$ with $0 < s < 1$. We denote by ρ_r and κ_r the material parameters inside the resonators and by ρ and κ the corresponding parameters for the background medium and let $v = \sqrt{\kappa/\rho}$, $v_r = \sqrt{\kappa_r/\rho_r}$, $k = \omega/v$, $k_r = \omega/v_r$, and $\delta = \rho_r/\rho$, where ω is the frequency.

We let $\alpha \in Y^*$ and consider the following $\alpha-$quasiperiodic equation in the unit cell $Y = [-1/2, 1/2]^d$:

$$
\begin{cases}
\Delta u^\alpha + k^2 u^\alpha = 0 & \text{in} \quad Y\backslash\overline{D}, \\
\Delta u^\alpha + k_r^2 u^\alpha = 0 & \text{in} \quad D, \\
u^\alpha|_+ - u^\alpha|_- = 0 & \text{on } \partial D, \\
\delta \left.\dfrac{\partial u^\alpha}{\partial \nu}\right|_+ - \left.\dfrac{\partial u^\alpha}{\partial \nu}\right|_- = 0 & \quad \text{on } \partial D, \\
e^{-\mathrm{i}\alpha\cdot x} u^\alpha \text{ is periodic.}
\end{cases}
\tag{4.5}
$$

From Appendix A.2.2, since the problem is self-adjoint with compact resolvent, (4.5) has nontrivial solutions for discrete values of ω such as

$$0 \le \omega_1^\alpha \le \omega_2^\alpha \le \cdots$$

and we have the following band structure of propagating frequencies for the given periodic structure:

$$[0, \max_\alpha \omega_1^\alpha] \cup [\min_\alpha \omega_2^\alpha, \max_\alpha \omega_2^\alpha] \cup [\min_\alpha \omega_3^\alpha, \max_\alpha \omega_3^\alpha] \cup \cdots .$$

4.2.2. Subwavelength bandgaps. We use the quasiperiodic single layer potential introduced in Appendix B.4 to represent the solution to the scattering problem (4.5) in $Y \setminus \overline{D}$. We look for a solution u^α of (4.5) of the form:

$$
u^\alpha = \begin{cases}
\mathcal{S}_D^{\alpha,k}[\phi] & \text{in } Y \setminus \overline{D}, \\
\mathcal{S}_D^{\alpha,k_r}[\psi] & \text{in } D,
\end{cases}
\tag{4.6}
$$

for some surface potentials $\phi, \psi \in L^2(\partial D)$. Using the jump relations for the single layer potentials, one can show that (4.5) is equivalent to the boundary integral equation

$$\mathcal{A}^\alpha(\omega,\delta)[\Psi]=0, \tag{4.7}$$

where

$$\mathcal{A}^\alpha(\omega,\delta)=\begin{pmatrix}\mathcal{S}_D^{\alpha,k_r} & -\mathcal{S}_D^{\alpha,k}\\ -\frac{1}{2}I+(\mathcal{K}_D^{-\alpha,k_r})^* & -\delta(\frac{1}{2}+(\mathcal{K}_D^{-\alpha,k})^*)\end{pmatrix},\ \Psi=\begin{pmatrix}\psi\\ \phi\end{pmatrix}. \tag{4.8}$$

Again, we denote by $\mathcal{H}=L^2(\partial D)\times L^2(\partial D)$ and by $\mathcal{H}_1=H^1(\partial D)\times L^2(\partial D)$. It is clear that $\mathcal{A}^\alpha(\omega,\delta)$ is a bounded linear operator from $\mathcal{H}$ to $\mathcal{H}_1$, *i.e.*, $\mathcal{A}^\alpha(\omega,\delta)\in\mathcal{L}(\mathcal{H},\mathcal{H}_1)$. Moreover, we can check that the characteristic values of $\mathcal{A}^\alpha(\omega,\delta)$ can be written as

$$0\le\omega_1^\alpha\le\omega_2^\alpha\le\cdots.$$

We first look at the limiting case when $\delta=0$. The operator $\mathcal{A}^\alpha(\omega,\delta)$ is a perturbation of

$$\mathcal{A}^\alpha(\omega,0)=\begin{pmatrix}\mathcal{S}_D^{\alpha,k_r} & -\mathcal{S}_D^{\alpha,k}\\ -\frac{1}{2}I+(\mathcal{K}_D^{-\alpha,k_r})^* & 0\end{pmatrix}. \tag{4.9}$$

We see that ω_0 is a characteristic value of $\mathcal{A}^\alpha(\omega,0)$ if only if $(\omega_0/v_r)^2$ is a Neumann eigenvalue of D or $(\omega_0/v)^2$ is a Dirichlet eigenvalue of $Y\backslash D$ with α-quasiperiodic boundary conditions on ∂Y. Since zero is a Neumann eigenvalue of D, $\omega_0=0$ is a characteristic value for the holomorphic operator-valued function $\omega\mapsto\mathcal{A}^\alpha(\omega,0)$. By noting that there is a positive lower bound for the other Neumann eigenvalues of D and all the Dirichlet eigenvalues of $Y\backslash D$ with α-quasiperiodic boundary conditions on ∂Y, we can conclude the following result by the asymptotic Gohberg-Sigal theory in Appendix A.1.

LEMMA 4.3. *Fix $\alpha\in Y^*$. For any δ sufficiently small, there exists one and only one characteristic value $\omega_1^\alpha(\delta)$ in a neighbourhood of the origin in the complex plane to the holomorphic operator-valued function $\mathcal{A}^\alpha(\omega,\delta)$. Moreover, $\omega_1^\alpha(0)=0$, and ω_1^α depends on δ continuously.*

Lemma 4.3 can be proved analogously to Lemma 4.14.

4.2.2.1. *Asymptotic behaviour of ω_1^α.* In this subsection, we assume that $\alpha\neq0$. We define the α-quasiperiodic capacity $\mathrm{Cap}_{\alpha,D}$ by

$$\mathrm{Cap}_{\alpha,D}:=-\int_{\partial D}(\mathcal{S}_D^{\alpha,0})^{-1}[\chi_{\partial D}]. \tag{4.10}$$

Note that in the two-dimensional case (*i.e.*, $d_l=d=2$) the quasiperiodic single layer operator $\mathcal{S}_D^{\alpha,0}:L^2(\partial D)\to H^1(\partial D)$ is invertible. The following theorem from [**38**] holds. It can be proved analogously to Theorem 1.9.

THEOREM 4.4. *For $\alpha\neq0$ and sufficiently small δ, we have*

$$\omega_1^\alpha=\sqrt{\frac{\delta\mathrm{Cap}_{\alpha,D}}{|D|}}v_r+\mathcal{O}(\delta^{3/2}). \tag{4.11}$$

REMARK 4.5. *The leading-order term in* (4.11)

$$\omega_{M,\alpha}:=\sqrt{\frac{\delta\mathrm{Cap}_{\alpha,D}}{|D|}}v_r\to0\quad as\ \alpha\to0 \tag{4.12}$$

because $((1/2)I+(\mathcal{K}_D^{-\alpha,0})^*)(\mathcal{S}_D^{\alpha,0})^{-1}[\chi_{\partial D}]\to0$ and so $\mathrm{Cap}_{\alpha,D}\to0$ as $\alpha\to0$. Moreover, since $\delta\ll1$, $\omega_{M,\alpha}$ lies in a small neighbourhood of zero for all $\alpha\in Y^*$.

We define $\omega_1^* := \max_\alpha \omega_{M,\alpha}$. Then we deduce the following result regarding a subwavelength bandgap opening [**38**].

THEOREM 4.6. *For every $\varepsilon > 0$, there exist $\delta_0 > 0$ and $\widetilde{\omega} > \omega_1^* + \varepsilon$ such that*

$$[\omega_1^* + \varepsilon, \widetilde{\omega}] \subset [\max_\alpha \omega_1^\alpha, \min_\alpha \omega_2^\alpha] \tag{4.13}$$

for $\delta < \delta_0$.

PROOF. Using $\omega_1^0 = 0$ and the continuity of ω_1^α in α and δ, we get α_0 and δ_1 such that $\omega_1^\alpha < \omega_1^*$ for every $|\alpha| < \alpha_0$ and $\delta < \delta_1$. Following the derivation of (4.11), we can check that it is uniformly valid in α as far as $|\alpha| \geq \alpha_0$. Thus, there exists $\delta_0 < \delta_1$ such $\omega_1^\alpha \leq \omega_1^* + \varepsilon$ for $|\alpha| \geq \alpha_0$. We have shown that $\max_\alpha \omega_1^\alpha \leq \omega_1^* + \varepsilon$ for sufficiently small δ. To have $\min_\alpha \omega_2^\alpha > \omega_1^* + \varepsilon$ for small δ, it is enough to check that $\mathcal{A}^\alpha(\omega, \delta)$ has no small characteristic values other than ω_1^α. For α away from 0, we can see that it is true following the proof of Theorem 4.4. If $\alpha = 0$, we have

$$\mathcal{A}^\alpha(\omega, \delta) = \mathcal{A}^\alpha(\omega, 0) + \mathcal{O}(\delta), \tag{4.14}$$

near ω_2^0 with $\delta = 0$. Since $\omega_2^0 \neq 0$, we have $\omega_2^0(\delta) > \omega_1^* + \varepsilon$ for sufficiently small δ. Finally, using the continuity of ω_2^α in α, we obtain $\min_\alpha \omega_2^\alpha > \omega_1^* + \varepsilon$ for small δ. This completes the proof. □

We are now interested in the point in the Brillouin zone where the maximum of the first Bloch eigenvalue ω_1^α is achieved. It is clear that due to time reversal symmetry, we have

$$\omega_1^\alpha = \omega_1^{-\alpha}.$$

On the other hand,

$$\omega_1^\alpha = \omega_1^{\alpha + (2\pi, \ldots, 2\pi)}.$$

Therefore,

$$\omega_1^\alpha = \omega_1^{-\alpha + (2\pi, \ldots, 2\pi)}.$$

Then it follows that $(\pi, \ldots, \pi)$ is a critical point of ω_1^α.

As shown in [**49**], under a symmetry assumption on D, ω_1^α attains its maximum ω_1^* at the corner M of the Brillouin zone Y^* corresponding to the quasiperiodicity $(\pi, \ldots, \pi)$.

THEOREM 4.7. *Assume that D is symmetric with respect to $\{x_j = 0\}$ for $j = 1, \ldots, d$. Then both $\mathrm{Cap}_{\alpha,D}$ and ω_1^α attain their maxima at $\alpha^* = (\pi, \ldots, \pi)$.*

PROOF. Let $\mathcal{H}_\alpha$ be the closure of the set $C_\alpha^\infty(Y)$ in H^1, and let $\mathcal{V}_\alpha := \{v \in \mathcal{H}_\alpha, v = 1 \text{ on } \partial D\}$. From Lemma B.36, we have the following characterisation of $\mathrm{Cap}_{\alpha,D}$:

$$\mathrm{Cap}_{\alpha,D} = \min_{v \in \mathcal{V}_\alpha} \int_{Y \setminus D} |\nabla v|^2.$$

Let $\mathcal{V}_{\alpha,0} := \{v \in \mathcal{H}_\alpha : v = 1 \text{ on } \partial D,\ v = 0 \text{ on } \partial Y\}$. Then $\mathcal{V}_{\alpha,0} \subset \mathcal{V}_\alpha$ and

$$\mathrm{Cap}_{\alpha,D} = \min_{v \in \mathcal{V}_\alpha} \int_{Y \setminus D} |\nabla v|^2 \leq \min_{v \in \mathcal{V}_{\alpha,0}} \int_{Y \setminus D} |\nabla v|^2.$$

Now, let v_0 be a harmonic function satisfying

$$\begin{cases} v_0 = 1 & \text{on } \partial D, \\ v_0 = 0 & \text{on } \partial Y. \end{cases} \tag{4.15}$$

Since D is symmetric, v_0 is symmetric with respect to $\{x_j = 0\}$, $j = 1, \ldots, d$, and hence v_0 can be extended to an α^*-quasiperiodic function (antiperiodic in each direction). By the characterisation of $\mathrm{Cap}_{\alpha^*,D}$, we have

$$\int_{Y \setminus D} |\nabla v_0|^2 = \mathrm{Cap}_{\alpha^*,D}.$$

It then follows that

$$\mathrm{Cap}_{\alpha,D} \leq \min_{v \in \mathcal{Y}} \int_{Y \setminus D} |\nabla v|^2 \leq \int_{Y \setminus D} |\nabla v_0|^2 = \mathrm{Cap}_{\alpha^*,D}.$$

Thus, $\mathrm{Cap}_{\alpha,D}$ attains its maximum at $\alpha = \alpha^*$. The argument for ω_1^α is similar to the one for $\mathrm{Cap}_{\alpha,D}$. By symmetry, we see that $u_1^{\alpha^*}$ is symmetric with respect to $\{x_j = 0\}$, $j = 1, \ldots, d$. This, combined with antiperiodicity, yields that $u_1^{\alpha^*}$ is zero on ∂Y. Let

$$\mathcal{H}_{\alpha,0} := \big\{ u \in \mathcal{H}_\alpha : \ u = 0 \text{ on } \partial Y \big\}.$$

Then

$$(\omega_1^\alpha)^2 = \min_{u \in \mathcal{H}_\alpha} \frac{\int_Y \widetilde{\rho}^{-1} |\nabla v|^2}{\int_Y \widetilde{\kappa}^{-1} |v|^2} \leq \min_{u \in \mathcal{H}_{\alpha,0}} \frac{\int_Y \widetilde{\rho}^{-1} |\nabla v|^2}{\int_Y \widetilde{\kappa}^{-1} |v|^2} \leq \frac{\int_Y \widetilde{\rho}^{-1} |\nabla u_1^{\alpha^*}|^2}{\int_Y \widetilde{\kappa}^{-1} |u_1^{\alpha^*}|^2} = (\omega_1^{\alpha^*})^2,$$

where $1/\widetilde{\rho} = (1/\rho)\chi_{Y \setminus D} + (1/\rho_r)\chi_D$, and $1/\widetilde{\kappa}$ is defined analogously. Here, $\chi_{Y \setminus D}$ and χ_D are respectively the characteristic functions of $Y \setminus D$ and D. This completes the proof. □

We shall investigate the behaviour of ω_1^α near $\alpha = \alpha^*$. We let

$$G_1^{\widetilde{\alpha}}(x) = \sum_{m \in \mathbb{Z}^d} \frac{e^{\mathrm{i}(2\pi m + \alpha^*) \cdot x}}{k^2 - |2\pi m + \alpha^*|^2} \left(\frac{|\widetilde{\alpha}|^2}{k^2 - |2\pi m + \alpha^*|^2} + \frac{4((2\pi m + \alpha^*) \cdot \widetilde{\alpha})^2}{(k^2 - |2\pi m + \alpha^*|^2)^2} \right),$$

and define the boundary integral operator

$$\mathcal{S}_1^{\widetilde{\alpha}}[\varphi](x) := \int_{\partial D} G_1^{\widetilde{\alpha}}(x - y) \varphi(y) \, d\sigma(y).$$

The following lemma holds.

LEMMA 4.8. *We have*

$$\mathcal{S}_D^{\alpha^* + \varepsilon \widetilde{\alpha}, 0} = e^{\mathrm{i}\varepsilon \widetilde{\alpha} \cdot x} \left(\mathcal{S}_D^{\alpha^*,0} + \varepsilon^2 \mathcal{S}_1^{\widetilde{\alpha}} + \mathcal{O}(\varepsilon^4) \right),$$

where the $\mathcal{O}(\varepsilon^4)$ term is an operator from $L^2(\partial D)$ to $H^1(\partial D)$ whose operator norm is of order ε^4.

PROOF. Since

$$\begin{aligned} &\frac{1}{k^2 - |2\pi m + \alpha^* + \varepsilon \widetilde{\alpha}|^2} \\ &= \frac{1}{k^2 - |2\pi m + \alpha^*|^2} \left(1 + \frac{\varepsilon^2 |\widetilde{\alpha}|^2}{k^2 - |2\pi m + \alpha^*|^2} + \frac{4\varepsilon^2 ((2\pi m + \alpha^*) \cdot \widetilde{\alpha})^2}{(k^2 - |2\pi m + \alpha^*|^2)^2} \right) + \mathcal{O}(\varepsilon^3 |\widetilde{\alpha}|^3), \end{aligned}$$

using the symmetry in the summation, we can derive that

$$G^{\alpha^*+\varepsilon\widetilde{\alpha},k}(x) = e^{\mathrm{i}\varepsilon\widetilde{\alpha}\cdot x} \sum_{m\in\mathbb{Z}^d} \frac{e^{\mathrm{i}(2\pi m+\alpha^*)\cdot x}}{k^2-|2\pi m+\alpha^*+\varepsilon\widetilde{\alpha}|^2}$$

$$= e^{\mathrm{i}\varepsilon\widetilde{\alpha}\cdot x} \sum_{m\in\mathbb{Z}^d} \frac{e^{\mathrm{i}(2\pi m+\alpha^*)\cdot x}}{k^2-|2\pi m+\alpha^*|^2}\left(1+\frac{\varepsilon^2|\widetilde{\alpha}|^2}{k^2-|2\pi m+\alpha^*|^2}+\frac{4\varepsilon^2((2\pi m+\alpha^*)\cdot\widetilde{\alpha})^2}{(k^2-|2\pi m+\alpha^*|^2)^2}\right)$$
$$+\mathcal{O}(\varepsilon^4|\widetilde{\alpha}|^4)$$
$$= e^{\mathrm{i}\varepsilon\widetilde{\alpha}\cdot x}\left(G^{\alpha^*,k}(x)+\varepsilon^2 G_1^{\widetilde{\alpha}}(x)\right)+\mathcal{O}(\varepsilon^4|\widetilde{\alpha}|^4).$$

The lemma follows. □

Let $\Lambda_D^{\widetilde{\alpha}}$ be the quadratic function in $\widetilde{\alpha}$ defined by

$$\Lambda_D^{\widetilde{\alpha}} := \frac{1}{2}\int_{\partial D}\Bigg[(\widetilde{\alpha}\cdot y)^2\left(\mathcal{S}_D^{\alpha^*,0}\right)^{-1}[\chi_{\partial D}]+\left(\mathcal{S}_D^{\alpha^*,0}\right)^{-1}[(\widetilde{\alpha}\cdot x)^2]$$
$$-2\left(\mathcal{S}_D^{\alpha^*,0}\right)^{-1}\mathcal{S}_1^{\widetilde{\alpha}}\left(\mathcal{S}_D^{\alpha^*,0}\right)^{-1}[\chi_{\partial D}]\Bigg]\,\mathrm{d}\sigma(y).$$

LEMMA 4.9. *For every small $\varepsilon>0$, it holds that*

$$\mathrm{Cap}_{\alpha^*+\varepsilon\widetilde{\alpha},D}=\mathrm{Cap}_{\alpha^*,D}+\varepsilon^2\Lambda_D^{\widetilde{\alpha}}+\mathcal{O}(\varepsilon^4).$$

Moreover, $\Lambda_D^{\widetilde{\alpha}}$ is a negative semidefinite quadratic function of $\widetilde{\alpha}$.

PROOF. Recall that

$$\mathrm{Cap}_{\alpha^*+\varepsilon\widetilde{\alpha},D}=-\int_{\partial D}\left(\mathcal{S}_D^{\alpha^*+\varepsilon\widetilde{\alpha},0}\right)^{-1}[\chi_{\partial D}](y)\,\mathrm{d}\sigma(y).$$

We solve on ∂D

$$1=\mathcal{S}_D^{\alpha^*+\varepsilon\widetilde{\alpha},0}[\phi](x)$$

for ϕ. Since

$$\mathcal{S}_D^{\alpha^*+\varepsilon\widetilde{\alpha},0}=e^{\mathrm{i}\varepsilon\widetilde{\alpha}\cdot x}\left(\mathcal{S}_D^{\alpha^*,0}+\varepsilon^2\mathcal{S}_1^{\widetilde{\alpha}}+\mathcal{O}(\varepsilon^4)\right),$$

we get

$$\phi(y)=e^{\mathrm{i}\widetilde{\alpha}\cdot\varepsilon y}\left(\left(\mathcal{S}_D^{\alpha^*,0}\right)^{-1}-\varepsilon^2\left(\mathcal{S}_D^{\alpha^*,0}\right)^{-1}\mathcal{S}_1^{\widetilde{\alpha}}\left(\mathcal{S}_D^{\alpha^*,0}\right)^{-1}\right)[e^{-\mathrm{i}\widetilde{\alpha}\cdot\varepsilon x}](y)+\mathcal{O}(\varepsilon^4)$$
$$=e^{\mathrm{i}\widetilde{\alpha}\cdot\varepsilon y}\Bigg[\left(\mathcal{S}_D^{\alpha^*,0}\right)^{-1}[\chi_{\partial D}]-\varepsilon^2\Bigg(\left(\mathcal{S}_D^{\alpha^*,0}\right)^{-1}[(\widetilde{\alpha}\cdot x)^2/2]$$
$$-\left(\mathcal{S}_D^{\alpha^*,0}\right)^{-1}\mathcal{S}_1^{\widetilde{\alpha}}\left(\mathcal{S}_D^{\alpha^*,0}\right)^{-1}[\chi_{\partial D}]\Bigg)\Bigg](y)+\mathcal{O}(\varepsilon^4).$$

Using the expansion

$$e^{\mathrm{i}\widetilde{\alpha}\cdot\varepsilon y}=1+\mathrm{i}\varepsilon\widetilde{\alpha}\cdot y-\frac{1}{2}\varepsilon^2(\widetilde{\alpha}\cdot y)^2-\mathrm{i}\frac{\varepsilon^3}{6}(\widetilde{\alpha}\cdot y)^3+\mathcal{O}(\varepsilon^4),$$

and the symmetry in the integration, we have

$$\begin{aligned}\mathrm{Cap}_{\alpha^*+\varepsilon\widetilde{\alpha},D} &= -\int_{\partial D} \phi(y)\ \mathrm{d}\sigma(y)\\ &= \mathrm{Cap}_{\alpha^*,D} + \varepsilon^2 \Lambda_D^{\widetilde{\alpha}} + \mathcal{O}(\varepsilon^4).\end{aligned}$$

Finally, since $\mathrm{Cap}_{\alpha,D}$ attains its maximum at $\alpha = \alpha^*$, $\Lambda_D^{\widetilde{\alpha}}$ is a negative semidefinite quadratic function of $\widetilde{\alpha}$. This completes the proof. □

We now introduce a matrix $(\lambda_{ij})_{i,j=1,2,3}$ such that

$$\frac{v_r^2}{|D|}\Lambda_D^{\widetilde{\alpha}} := -\sum_{1\le i,j\le d} \lambda_{ij}\widetilde{\alpha}_i\widetilde{\alpha}_j. \tag{4.16}$$

It is clear that the matrix $(\lambda_{ij})_{i,j=1,2,3}$ is symmetric and positive semidefinite.

LEMMA 4.10. *We have*

$$(\omega_1^{\alpha^*+\varepsilon\widetilde{\alpha}}(\delta))^2 = \delta\omega_{M,\alpha^*}^2 - \varepsilon^2\delta \sum_{1\le i,j\le d} \lambda_{ij}\widetilde{\alpha}_j\widetilde{\alpha}_j + \lambda_0(\alpha^*)\delta^2 + \mathcal{O}(\varepsilon^4\delta + \varepsilon^2\delta^2 + \delta^3).$$

PROOF. Recall from Theorem 4.4 that the following asymptotic formula of ω_1^α holds:

$$(\omega_1^\alpha(\delta))^2 = \frac{\delta v_r^2 \mathrm{Cap}_{\alpha,D}}{|D|} + \mathcal{O}(\delta^2) := \delta(\omega_{M,\alpha})^2 + \mathcal{O}(\delta^2), \tag{4.17}$$

where $\omega_{M,\alpha}$ is defined in (4.12), and the $\mathcal{O}(\delta^2)$ term is uniformly of order δ^2 with respect to α away from 0. A further asymptotic expansion implies that the $\mathcal{O}(\delta^2)$-term can be further decomposed as

$$\mathcal{O}(\delta^2) = \lambda_0(\alpha)\delta^2 + \mathcal{O}(\delta^3),$$

where $\mathcal{O}(\delta^3)$ denotes some uniformly bounded remaining term. Since α^* is a critical point of ω_1^α regardless of δ, we have $\lambda_0(\alpha) = \lambda_0(\alpha^*) + \mathcal{O}(|\alpha-\alpha^*|^2)$.

Finally, we obtain that

$$(\omega_{M,\,\alpha^*+\varepsilon\widetilde{\alpha}})^2 = \frac{v_r^2\mathrm{Cap}_{\alpha^*+\varepsilon\widetilde{\alpha},D}}{|D|} = (\omega_{M,\alpha^*})^2 + \frac{\varepsilon^2 v_r^2}{|D|}\Lambda_D^{\widetilde{\alpha}} + \mathcal{O}(\varepsilon^4).$$

This completes the proof. □

4.2.2.2. *High-frequency homogenisation of the Bloch eigenmodes.* In [**49**], an asymptotic expansion of the Bloch eigenmodes near the critical frequency ω_1^* has also been computed. It is proved that these eigenmodes can be decomposed into two parts: one slowly varies and satisfies a homogenised equation, while the other is periodic across each elementary cell and varies at the microscopic scale.

To simplify the presentation, we assume that $v = v_r = 1$. We let u_1^α be the α-quasiperiodic propagating wave mode corresponding to ω_1^α, which is given by

$$u_1^\alpha = \begin{cases} \mathcal{S}_D^{\alpha,\omega_1^\alpha}\left(\mathcal{S}_D^{\alpha,0}\right)^{-1}[\chi_{\partial D}] + \mathcal{O}(\delta^{1/2}) & \text{in } Y\setminus D,\\ \mathcal{S}_D^{\alpha,\omega_1^\alpha}\left(\mathcal{S}_D^{0}\right)^{-1}[\chi_{\partial D}] + \mathcal{O}(\delta^{1/2}) & \text{in } D.\end{cases} \tag{4.18}$$

Since we have

$$\mathcal{S}_D^{\alpha,\omega_1^\alpha}\left(\mathcal{S}_D^{0}\right)^{-1}[\chi_{\partial D}] \approx \mathcal{S}_D^{\alpha,\omega_1^\alpha}\left(\mathcal{S}_D^{\alpha,0}\right)^{-1}[\chi_{\partial D}]$$

in D up to a remainder of order of $\mathcal{O}(\delta^{1/2})$, u_1^α can be approximated in Y by

$$u_1^\alpha = \mathcal{S}_D^{\alpha,\omega_1^\alpha}\left(\mathcal{S}_D^{\alpha,0}\right)^{-1}[\chi_{\partial D}] + \mathcal{O}(\delta^{1/2}). \tag{4.19}$$

We now consider the asymptotic behaviours of the Bloch eigenmodes in a crystal with period s. We first present the relation between the Bloch eigenvalues and eigenmodes at different scales, which follows from a scaling argument.

LEMMA 4.11. *Let ω_1^α and u_1^α be the Bloch eigenvalue and eigenmode of the crystal with period* 1, *then the crystal with period s has Bloch eigenvalue*

$$\omega_{1,s}^{\alpha/s} = \frac{1}{s}\omega_1^\alpha$$

and eigenmode

$$u_{1,s}^{\alpha/s}(x) = u_1^\alpha\left(\frac{x}{s}\right).$$

As a consequence, we see that

$$\omega_*^s = \max_\alpha \omega_{1,s}^\alpha = \frac{1}{s}\max_\alpha \omega_1^\alpha = \frac{1}{s}\omega_1^{\alpha^*}.$$

Moreover, the maximum is attained at the corner point α^*/s of the Brillouin zone Y_s^*. We are interested in the behaviour of the Bloch eigenmode $u_{1,s}^\alpha$ for α near α^*/s. The following result holds.

LEMMA 4.12. *We have*

$$u_{1,s}^{\alpha^*/s+\widetilde{\alpha}}(x) = e^{\mathrm{i}\widetilde{\alpha}\cdot x} S\left(\frac{x}{s}\right) + \mathcal{O}(s^2 + \delta^{1/2}), \tag{4.20}$$

where the function S is defined by

$$S(x) = \mathcal{S}_D^{\alpha^*,0}\left[\left(\mathcal{S}_D^{\alpha^*,0}\right)^{-1}[\chi_{\partial D}]\right](x), \quad x \in Y. \tag{4.21}$$

PROOF. Note that $u_{1,s}^{\alpha^*/s+\widetilde{\alpha}}(x) = u_1^{\alpha^*+s\widetilde{\alpha}}(x/s)$. By (4.19), we have

$$u_1^{\alpha^*+s\widetilde{\alpha}}(x/s) = \mathcal{S}_D^{\alpha^*+s\widetilde{\alpha},\omega_1^{\alpha^*+s\widetilde{\alpha}}}\left[\left(\mathcal{S}_D^{\alpha^*+s\widetilde{\alpha},0}\right)^{-1}[\chi_{\partial D}]\right]\left(\frac{x}{s}\right) + \mathcal{O}(\delta^{1/2}).$$

Since $\omega_1^{\alpha^*+s\widetilde{\alpha}} = \mathcal{O}(\delta^{1/2})$, using (B.98), we further obtain

$$u_1^{\alpha^*+s\widetilde{\alpha}}(x/s) = \mathcal{S}_D^{\alpha^*+s\widetilde{\alpha},0}\left[\left(\mathcal{S}_D^{\alpha^*+s\widetilde{\alpha},0}\right)^{-1}[\chi_{\partial D}]\right]\left(\frac{x}{s}\right) + \mathcal{O}(\delta^{1/2}). \tag{4.22}$$

On the other hand, in the proof of Lemma 4.9, we have showed that

$$\left(\mathcal{S}_D^{\alpha^*+s\widetilde{\alpha},0}\right)^{-1}[\chi_{\partial D}]\left(\frac{y}{s}\right) = e^{\mathrm{i}\widetilde{\alpha}\cdot y}\left(\mathcal{S}_D^{\alpha^*,0}\right)^{-1}[\chi_{\partial D}]\left(\frac{y}{s}\right) + \mathcal{O}(s^2). \tag{4.23}$$

It follows from (4.22) and (4.23) that

$$\begin{aligned}
u_1^{\alpha^*+s\widetilde{\alpha}}(x/s) &= \int_{s\partial D} e^{\mathrm{i}\widetilde{\alpha}\cdot(x-y)} G^{\alpha^*,0}\left(\frac{x-y}{s}\right) e^{\mathrm{i}\widetilde{\alpha}\cdot y}\left(\left(\mathcal{S}_D^{\alpha^*,0}\right)^{-1}[\chi_{\partial D}]\right)\left(\frac{y}{s}\right) s^{-2}\, \mathrm{d}\sigma(y) \\
&\quad + \mathcal{O}(s^2 + \delta^{1/2}) \\
&= e^{\mathrm{i}\widetilde{\alpha}\cdot x}\mathcal{S}_D^{\alpha^*,0}\left[\left(\mathcal{S}_D^{\alpha^*,0}\right)^{-1}[\chi_{\partial D}]\right]\left(\frac{x}{s}\right) + \mathcal{O}(s^2 + \delta^{1/2}).
\end{aligned}$$

Then the lemma follows. □

We note that the function S defined by (4.21) is a piecewise harmonic function with $S = 1$ on ∂D and $S = 0$ on ∂Y. Hence, $S(x/s)$ varies on the small scale s and describes the microscopic behaviour of $u_{1,s}^{\alpha^*/s+\widetilde{\alpha}}$. On the other hand, the function $e^{\mathrm{i}\widetilde{\alpha}\cdot x}$ represents the *macroscopic behaviour* (or the plane wave envelope) of $u_{1,s}^{\alpha^*/s+\widetilde{\alpha}}$. We now derive a homogenised equation for this macroscopic field near the critical frequency ω_*^s.

We first recall that

$$\omega_1^{\alpha^*} = C\sqrt{\delta} + \mathcal{O}(\delta^{3/2})$$

for some positive constant C of order one which depends on D and α^*. In order to keep the critical frequency ω_*^s at a fixed order when the cell size s tends to zero, we assume that $\delta = \mu s^2$ for some positive constant μ. As a result, we have

$$\max_{\alpha} \omega_{1,s}^{\alpha} = \omega_*^s = \frac{1}{s}\omega_1^{\alpha^*} = \mathcal{O}(1).$$

Now, suppose that ω is near ω_*^s. We need to find the corresponding Bloch eigenmodes, or $\widetilde{\alpha}$ so that

$$\omega = \omega_{1,s}^{\alpha^*/s+\widetilde{\alpha}}.$$

Using Lemmas 4.10 and 4.11, we have

$$\omega_*^2 - \omega^2 = \delta \sum_{1\leq i,j\leq d} \lambda_{ij}\widetilde{\alpha}_i\widetilde{\alpha}_j + \mathcal{O}(s^4), \tag{4.24}$$

where $\omega_* := \omega_*^s$.

Let

$$\omega_*^2 - \omega^2 = \beta\delta$$

for some constant β. Then we have

$$\sum_{1\leq i,j\leq d} \lambda_{ij}\widetilde{\alpha}_i\widetilde{\alpha}_j = \beta + \mathcal{O}(s^2),$$

which shows that all the solutions $\widetilde{\alpha}$ lie approximately on the ellipse or ellipsoid defined by

$$\Big\{\alpha = (\alpha_1, \ldots, \alpha_d) : \sum_{1\leq i,j\leq d} \lambda_{ij}\alpha_i\alpha_j = \beta\Big\}.$$

It also implies that the plane wave $\widetilde{u}(x) := e^{\mathrm{i}\widetilde{\alpha}\cdot x}$ satisfies to a leading-order term

$$\sum_{1\leq i,j\leq d} \lambda_{ij}\partial_{ij}^2\widetilde{u}(x) + \beta\widetilde{u}(x) = 0. \tag{4.25}$$

We now consider two cases. Case I, $\beta > 0$, and Case II, $\beta < 0$.

In the first case, $\widetilde{\alpha}$ is a well-defined vector in $\mathbb{R}^d$, which justifies the existence of true Bloch eigenmodes. However, in the second case $\widetilde{\alpha}$ is a purely imaginary vector, and hence is not a Bloch eigenmode. The associated function $e^{\mathrm{i}\widetilde{\alpha}\cdot x}$ either grows or decays exponentially along a certain direction. This justifies that a bandgap opens at the critical frequency ω_*.

To conclude, we have the following main result on the high-frequency homogenisation theory of subwavelength resonator crystals.

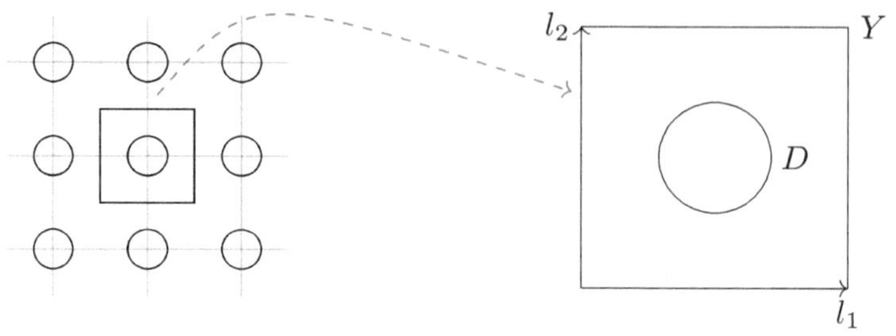

FIGURE 4.2. Illustration of the square lattice crystal and quantities in the fundamental domain Y.

THEOREM 4.13. *For frequencies in a small neighbourhood of maximum of the first Bloch eigenvalue, say $\omega_*^2 - \omega^2 = \mathcal{O}(s^2)$, the following asymptotic of the Bloch eigenmode $u_{1,s}^{\alpha^*/s+\widetilde{\alpha}}$ holds:*

$$u_{1,s}^{\alpha^*/s+\widetilde{\alpha}}(x) = e^{\mathrm{i}\widetilde{\alpha}\cdot x} S\left(\frac{x}{s}\right) + \mathcal{O}(s),$$

where the macroscopic field $e^{\mathrm{i}\widetilde{\alpha}\cdot x}$ satisfies the following equation:

$$\sum_{1\leq i,j\leq d} \lambda_{ij}\partial_i\partial_j\widetilde{u}(x) + \frac{\omega_*^2 - \omega^2}{\delta}\widetilde{u}(x) = 0, \tag{4.26}$$

which can be viewed as the homogenised equation for the crystal, while the microscopic field is periodic and varies on the scale of s.

It is clear that the homogenised medium is very dispersive slightly above the critical frequency ω_*. Moreover, slightly below ω_*, the Bloch eigenmode oscillates highly (at the scale of the period of the crystal). It is worth emphasising that (4.26) describes only the behaviour of the low-frequency macroscopic field. The high-frequency homogenisation approach described here differs from the effective medium theory derived in Chapter 3, which is valid only for *dilute* systems of subwavelength resonators.

Finally, we perform some numerical experiments in the two-dimensional case, *i.e.*, for $d_l = d = 2$. We consider a square lattice with unit cell Y as depicted in Figure 4.2.

As D is a disk of radius R, we apply the multipole expansion method to compute the band structure (we refer to Appendix B.6 for details). We set $R = 0.25$ and $\delta = 0.001$. Figure 4.3 shows that the corresponding dispersion relation has a bandgap between the first and second bands, and the critical frequency ω_* of the first band is achieved at the symmetry point $\alpha^* = M = (\pi, \pi)$ in the Brillouin zone. We use $M = 3$ for the truncation order of cylindrical waves in the multipole expansion method.

Then we consider the dilute case. We set $R = 0.05$ and $\delta = 0.0002$. Figure 4.4 shows the computed band structure. It can be seen that a quite large subwavelength bandgap does exist. Here, we use $M = 7$ for the truncation order of cylindrical waves.

We compute the eigenmodes of the square crystal close to the critical frequency, namely $\omega = \omega_* - \varepsilon$. The small-scale behaviour of the eigenmodes, *i.e.*, the function

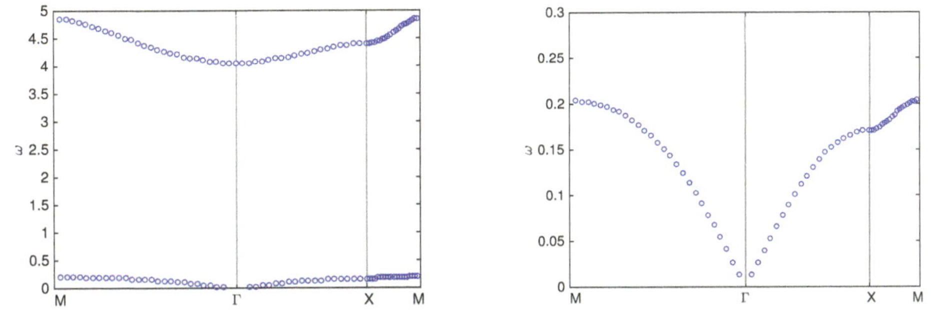

FIGURE 4.3. The band structure of a square array of circular resonators with radius $R = 0.25$ and contrast $\delta^{-1} = 1000$.

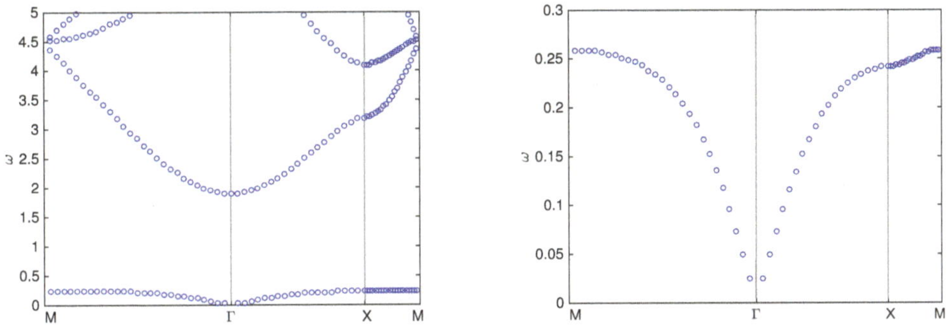

FIGURE 4.4. (Dilute case) The band structure of a square array of circular resonators with radius $R = 0.05$ and contrast $\delta^{-1} = 5000$.

$S(x/s)$, is shown in Figure 4.5. It can be seen that the function oscillates at the scale of the period of the crystal.

Next, the large-scale behaviour is considered. Figure 4.6 shows the Bloch eigenmode over many unit cells for $\widetilde{\alpha} = \begin{pmatrix} \widetilde{\alpha}_1 \\ 0 \end{pmatrix}$ and $\omega_* - \omega = 6 \cdot 10^{-5}$. The eigenmode varies at the large scale with a low-frequency macroscopic field $\widetilde{u}$.

Since D is a disk, the macroscopic equation (4.26) reduces to the isotropic form

$$\Delta \widetilde{u}(x) + \frac{\omega_*^2 - \omega^2}{\widetilde{\lambda}\delta} \widetilde{u}(x) = 0, \tag{4.27}$$

for some positive constant $\widetilde{\lambda}$. To illustrate the macroscpoic behaviour of (4.27), the spatial frequency f of the macroscopic field is computed for $\varepsilon \in [0, 0.01]$. Observe that due to the bandgap above ω_*, no eigenmodes exist for $\varepsilon < 0$. Figure 4.7 shows the spatial frequency for different ε. This figure shows that the spatial frequency scales like $\sqrt{\varepsilon}$ for small $\varepsilon > 0$, which is consistent with (4.27) since from (4.27) we expect a spatial frequency

$$f = \sqrt{\frac{\omega_*^2 - \omega^2}{\widetilde{\lambda}\delta}} = \sqrt{\varepsilon}\sqrt{\frac{\omega_* + \omega}{\widetilde{\lambda}\delta}}.$$

In Figures 4.6 and 4.7, the Bloch eigenmodes are computed using the multipole expansion method derived in Appendix B.6. Since the Bloch eigenmodes are

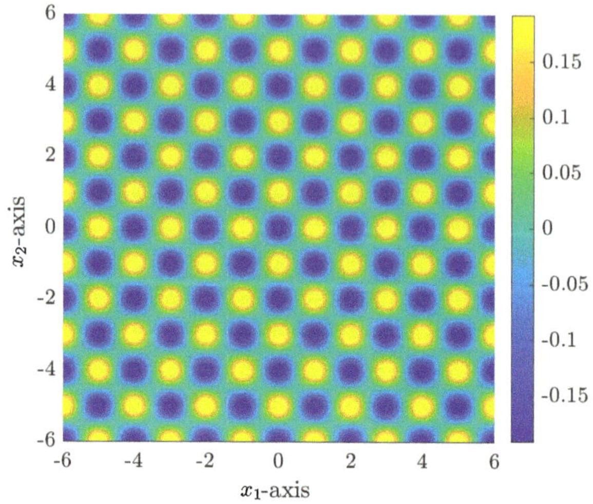

FIGURE 4.5. Real part of small-scale behaviour of the Bloch eigenmode of the square lattice at $\alpha = \alpha^*$ (imaginary part close to 0).

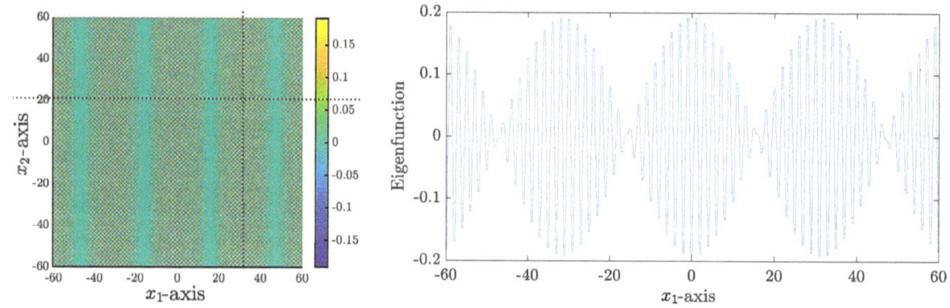

(A) Two-dimensional plot. (B) One-dimensional plot along the x_1-axis.

FIGURE 4.6. Real part of Bloch eigenmode of the square lattice shown over many unit cells.

asymptotically constant on D, we can accurately approximate functions by only the first term in the multipole expansion.

4.3. Honeycomb lattices

In this section, we consider a honeycomb lattice of subwavelength resonators and prove the existence of a Dirac dispersion cone in the subwavelength regime. We also show that near the Dirac points, honeycomb crystals of subwavelength resonators can be used as near-zero materials. To do so, we first derive an asymptotic formula for the quasiperiodic resonant frequencies close to the Dirac points in the Brillouin zone. Then, we obtain the linear dispersion relation of a Dirac cone. Finally, we show that near the Dirac points, the Bloch eigenmodes are the sum of two eigenmodes. Each eigenmode can be decomposed into two components: one which

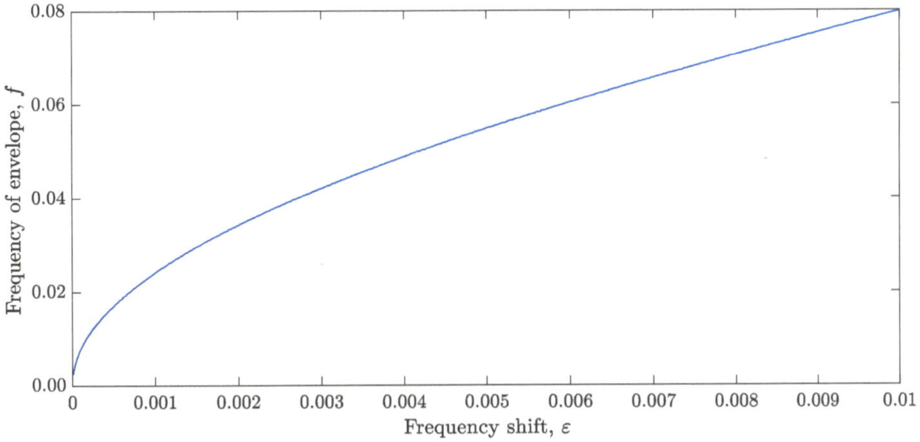

FIGURE 4.7. Spatial frequency of the envelope function $\widetilde{u}$ for different frequency shifts ε.

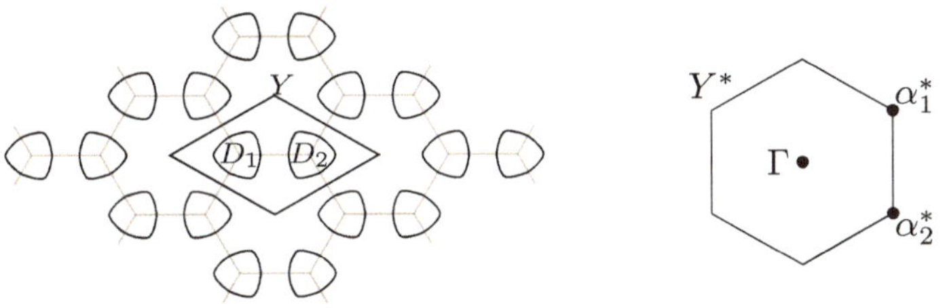

FIGURE 4.8. Honeycomb crystal and corresponding Brillouin zone Y^*.

is slowly varying and satisfies a homogenised equation, while the other is periodic across each elementary crystal cell and is highly oscillating. The slowly oscillating components (or the plane wave envelopes) of the eigenmodes satisfy a system of Dirac equations. The results presented in this section are from [**35, 45**].

4.3.1. Problem formulation. We consider a two-dimensional infinite honeycomb crystal in two dimensions depicted in Figure 4.8. We let the lattice Λ be generated by the lattice vectors

$$l_1 = L\left(\frac{\sqrt{3}}{2}, \frac{1}{2}\right), \ l_2 = L\left(\frac{\sqrt{3}}{2}, -\frac{1}{2}\right).$$

The dual lattice Λ^* is generated by α_1 and α_2 given by

$$\alpha_1 = \frac{2\pi}{L}\left(\frac{1}{\sqrt{3}}, 1\right), \ \alpha_2 = \frac{2\pi}{L}\left(\frac{1}{\sqrt{3}}, -1\right).$$

The points

$$\alpha_1^* = \frac{2\alpha_1 + \alpha_2}{3}, \quad \alpha_2^* = \frac{\alpha_1 + 2\alpha_2}{3}, \tag{4.28}$$

in the Brillouin zone Y^* are called *Dirac points.*

We assume that the unit cell $Y = \{ sl_1 + tl_2 : 0 \leq s, t \leq 1 \}$ contains two resonators, $D = D_1 \cup D_2$, where D_1 ad D_2 are disjoint, connected, smooth, and

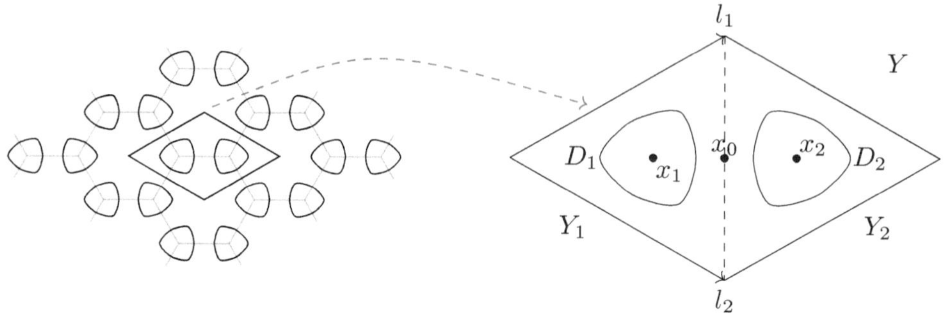

FIGURE 4.9. Illustration of quantities in the fundamental domain Y.

are such that each of them is invariant under rotation by $2\pi/3$ and so that D is invariant under rotation by π.

Let Y_1 and Y_2 be defined by

$$Y_1 = \{sl_1 + tl_2 : 0 \leq s, t, \text{ and } t + s \leq 1\}, \quad Y_2 = Y \setminus Y_1,$$

and denote the centres of Y, Y_1, and Y_2, respectively, by x_0, x_1 and x_2. See Figure 4.9.

Having defined the geometry, we now describe the wave propagation in the honeycomb crystal. We assume that the resonators D_1 and D_2 have the same material parameters ρ_r and v_r. We consider the following α-quasiperiodic wave propagation problem in Y:

$$(4.29) \qquad \begin{cases} \Delta u^\alpha + k^2 u^\alpha = 0 & \text{in } Y \backslash D, \\ \Delta u^\alpha + k_r^2 u^\alpha = 0 & \text{in } D, \\ u^\alpha|_+ - u^\alpha|_- = 0 & \text{on } \partial D, \\ \delta \frac{\partial u^\alpha}{\partial \nu}\Big|_+ - \frac{\partial u^\alpha}{\partial \nu}\Big|_- = 0 & \text{on } \partial D, \\ u^\alpha(x + l) = e^{\mathrm{i}\alpha \cdot l} u^\alpha(x) & \forall\, l \in \Lambda. \end{cases}$$

The solution to (4.29) can be represented by (4.6), where the single layer potentials $\mathcal{S}_D^{\alpha,k_r}$ and $\mathcal{S}_D^{\alpha,k}$ are associated to the honeycomb lattice Λ and the resonator system D. Moreover, with this definition of layer potentials, (4.29) is equivalent to (4.7).

For $\alpha \in Y^*, \alpha \neq 0$, we let $\psi_j^\alpha \in L^2(\partial D)$ be given by

$$(4.30) \qquad \mathcal{S}_D^{\alpha,0}[\psi_j^\alpha] = \chi_{\partial D_j} \quad \text{on } \partial D, \quad j = 1, 2.$$

Analogously to Lemma 4.3, the following result from [**35**] characterises the subwavelength Bloch resonant frequencies of a honeycomb crystal.

LEMMA 4.14. *For $\alpha \in Y^*, \alpha \neq 0$, there are precisely two Bloch resonant frequencies with positive real parts $\omega_j^\alpha = \omega_j^\alpha(\delta)$, $j = 1, 2$, such that $\omega_j^\alpha(0) = 0$, and ω_j^α depends on δ continuously.*

PROOF. In order to prove this result, we will apply the asymptotic Gohberg-Sigal theory in Appendix A.1. It is clear that $\omega = 0$ is a characteristic value of

$\mathcal{A}^\alpha(\omega, 0)$, defined by (4.9), because

$$\mathcal{A}^\alpha(0,0) = \begin{pmatrix} \mathcal{S}_D^{\alpha,0} & -\mathcal{S}_D^{\alpha,0} \\ -\frac{1}{2}I + (\mathcal{K}_D^{-\alpha,0})^* & 0 \end{pmatrix}$$

has a nontrivial kernel of two dimensions, which is generated by

$$\Psi_1^\alpha = \begin{pmatrix} \psi_1^\alpha \\ \psi_1^\alpha \end{pmatrix} \quad \text{and} \quad \Psi_2^\alpha = \begin{pmatrix} \psi_2^\alpha \\ \psi_2^\alpha \end{pmatrix}, \tag{4.31}$$

where ψ_1^α and ψ_2^α are defined in (4.30). From (B.98), we have that

$$\left(-\frac{1}{2}I + (\mathcal{K}_D^{-\alpha,k_r})^*\right)[\psi_i^\alpha](x) = \omega^2 h(x,\omega), \quad x \in \partial D, \quad i = 1,2, \tag{4.32}$$

for some function h that is holomorphic as a function of ω in a neighbourhood of 0. Since

$$\int_{D_i} \mathcal{S}_D^{\alpha,0}[\psi_i^\alpha](x)\ \mathrm{d}x \neq 0, \ i = 1,2,$$

it follows from (B.103) that $h(x,0)$ is not identically zero. Therefore, the rank of both Ψ_1^α and Ψ_2^α is 2. Since Ψ_1^α and Ψ_2^α are linearly independent, the multiplicity of the characteristic value $\omega = 0$ is 4.

Since the Neumann-Poincaré operator $(\mathcal{K}_D^{-\alpha,0})^* : L^2(\partial D) \to L^2(\partial D)$ is a compact operator, $-\frac{1}{2}I + (\mathcal{K}_D^{-\alpha,0})^*$ is Fredholm of index zero. On the other hand, since $\mathcal{S}_D^{\alpha,0}$ is invertible, also $\mathcal{A}^\alpha(0,0)$ is Fredholm of index zero. Since $(\mathcal{K}_D^{-\alpha,\omega})^*$ and $\mathcal{S}_D^{\alpha,\omega}$ are holomorphic as functions of ω in a neighbourhood of 0, it follows that $\mathcal{A}^\alpha(\omega,0)$ is of Fredholm type.

Let $V \subset \mathbb{C}$ be a disk around 0 with a small enough radius, chosen so that $\mathcal{A}^\alpha(\omega,0)$ is invertible on ∂V and $\omega = 0$ is the only characteristic value in V. From Lemma A.12, it follows that $\mathcal{A}^\alpha(\omega,0)$ is normal with respect to ∂V.

Now, we turn to the full operator $\mathcal{A}^\alpha(\omega,\delta)$. It is clear that

$$\begin{aligned} \mathcal{A}^\alpha(\omega,\delta) =& \mathcal{A}^\alpha(\omega,0) + \begin{pmatrix} 0 & 0 \\ 0 & -\delta\left(\frac{1}{2}I + (\mathcal{K}_D^{-\alpha,k})^*\right) \end{pmatrix} \\ :=& \mathcal{A}^\alpha(\omega,0) + A^{(1)}(\omega,\delta), \end{aligned}$$

where $A^{(1)}(\omega,\delta)$, as a function of ω, is holomorphic in V and continuous up to ∂V. For small enough δ, we have

$$\left\| (\mathcal{A}^\alpha(\omega,0))^{-1} A^{(1)}(\omega,\delta) \right\|_{\mathcal{L}\left((L^2(\partial D))^2,(L^2(\partial D))^2\right)} < 1, \quad \omega \in \partial V.$$

Hence, the generalised Rouché's theorem (Theorem A.16) shows that $\mathcal{A}^\alpha(\omega,\delta)$ has 4 characteristic values inside V, for small enough δ. Since the characteristic values are symmetric around the origin, it is clear that two of these, namely ω_1^α and ω_2^α, have positive real parts, while two characteristic values have negative real parts.

The fact that $\omega_1^\alpha(\delta)$ and $\omega_2^\alpha(\delta)$ are continuous in δ can be deduced in a similar way: If $U \subset \mathbb{C}$ is a neighbourhood of $\omega_i^\alpha(\delta_1)$, $i = 1,2$, we can write

$$\mathcal{A}^\alpha(\omega,\delta_2) = \mathcal{A}^\alpha(\omega,\delta_1) + (\delta_1 - \delta_2)\begin{pmatrix} 0 & 0 \\ 0 & \frac{1}{2}I + (\mathcal{K}_D^{-\alpha,k})^* \end{pmatrix},$$

and from the generalised Rouché's theorem it follows that $\omega_i^\alpha(\delta_2) \in U$ when $|\delta_1 - \delta_2|$ is small enough. □

4.3.2. Capacitance matrix formulation for the band structure. We first generalise the concept of quasiperiodic capacity introduced in (4.10) to the case where two resonators are in the unit cell Y.

DEFINITION 4.15 (Quasiperiodic capacitance matrix). *For $\alpha \in Y^*, \alpha \neq 0$, we define the capacitance coefficient matrix $C^\alpha = (C^\alpha_{ij})$ by*

$$C^\alpha_{ij} := -\int_{\partial D_j} \psi^\alpha_i \, \mathrm{d}\sigma, \quad i,j = 1,2, \tag{4.33}$$

where $\psi^\alpha_i, i = 1,2$ is given by (4.30). Equivalently, we can define C^α by

$$C^\alpha_{ij} := \int_{Y \setminus D} \overline{\nabla V^\alpha_i} \cdot \nabla V^\alpha_j \, \mathrm{d}x, \quad i,j = 1,2, \tag{4.34}$$

where V^α_i, $i = 1,2$, are the solutions to

$$\begin{cases} \Delta V^\alpha_i = 0 & \text{in } Y \setminus D, \\ V^\alpha_i = \delta_{ij} & \text{on } \partial D_j, \\ V^\alpha_i(x+l) = e^{\mathrm{i}\alpha \cdot l} V^\alpha_i(x) & \forall\, l \in \Lambda. \end{cases}$$

We assume that α is bounded away from zero. For some fixed $\alpha_0 > 0$, we denote by

$$Y^*_0 = \{\alpha \in Y^* : |\alpha| \geq \alpha_0\}.$$

The following characterisation of the subwavelength band functions holds.

THEOREM 4.16. *The subwavelength band functions $\omega^\alpha_j = \omega^\alpha_j(\delta)$, $j = 1,2$ of (4.29) can be approximated as*

$$\omega^\alpha_j = \sqrt{\frac{\delta \lambda^\alpha_j}{|D_1|}} v_r + \mathcal{O}(\delta),$$

uniformly for $\alpha \in Y^*_0$, where $|D_1|$ is the volume of one resonator and λ^α_j, $j = 1,2$, are the eigenvalues of the quasiperiodic capacitance matrix C^α.

PROOF. We seek solutions (ϕ, ψ) to the integral equation (4.7), normalised such that $\|\phi\|_{L^2(\partial D)} = 1$ and $\|\psi\|_{L^2(\partial D)} = 1$. Using the asymptotic expansions (B.98), we find that ϕ and ψ satisfy

$$\begin{aligned} \mathcal{S}^{\alpha,0}_D[\psi] - \mathcal{S}^{\alpha,0}_D[\phi] &= \mathcal{O}(\omega^2), \\ \left(-\frac{1}{2}I + (\mathcal{K}^{-\alpha,0}_D)^* + k_r^2 \mathcal{K}^{\alpha,0}_{D,1}\right)[\psi] - \delta\left(\frac{1}{2}I + (\mathcal{K}^{\alpha,0}_D)^*\right)[\phi] &= \mathcal{O}(\omega^4 + \delta^2), \end{aligned} \tag{4.35}$$

uniformly for $\alpha \in Y^*_0$, where the error terms are with respect to the norm in $L^2(\partial D)$. Observe that $\mathcal{S}^{\alpha,0}_D$ is invertible, and by the inverse mapping theorem together with the fact that Y^*_0 is a closed set we have that $\|(\mathcal{S}^{\alpha,0}_D)^{-1}\|_{\mathcal{L}(L^2(\partial D), H^1(\partial D))}$ is uniformly bounded for $\alpha \in Y^*_0$. Then, we get

$$\psi = \phi + \mathcal{O}(\omega^2),$$

uniformly for $\alpha \in Y^*_0$. Inserting the above approximation into (4.35), we obtain that

$$\left(-\frac{1}{2}I + (\mathcal{K}^{-\alpha,0}_D)^* + k_r^2 \mathcal{K}^{\alpha,0}_{D,1}\right)[\psi] - \delta\left(\frac{1}{2}I + (\mathcal{K}^{\alpha,0}_D)^*\right)[\psi] = \mathcal{O}(\omega^4 + \delta^2), \tag{4.36}$$

uniformly for $\alpha \in Y_0^*$. Recall that Ker $\left(-\frac{1}{2}I + (\mathcal{K}_D^{-\alpha,0})^*\right)$ is spanned by ψ_1^α and ψ_2^α, defined by (4.30). Then we write ψ as

$$\psi = a\psi_1^\alpha + b\psi_2^\alpha + \varphi, \tag{4.37}$$

where φ is orthogonal to span$\{\psi_1, \psi_2\}$ in $L^2(\partial D)$. We then have from (4.36) that

$$\left(-\frac{1}{2}I + (\mathcal{K}_D^{-\alpha,0})^*\right)[\varphi] = \mathcal{O}(\omega^2 + \delta),$$

uniformly for $\alpha \in Y_0^*$. Since $\left(-\frac{1}{2}I + (\mathcal{K}_D^{-\alpha,0})^*\right)$, restricted to the orthogonal complement of span$\{\psi_1^\alpha, \psi_2^\alpha\}$, is invertible with bounded inverse, it follows that, in the $L^2(\partial D)$-norm,

$$\varphi = \mathcal{O}(\omega^2 + \delta),$$

uniformly for $\alpha \in Y_0^*$. Moreover, we find that $|a| + |b| > 0$. Now, we substitute (4.37) into (4.36) and integrate over ∂D_i for $i = 1, 2$. Then, using (B.103), we get

$$-\frac{\omega^2|D_1|}{v_r^2}a + \delta(aC_{11}^\alpha + bC_{12}^\alpha) = \mathcal{O}(\omega^4 + \delta^2),$$
$$-\frac{\omega^2|D_1|}{v_r^2}b + \delta(aC_{21}^\alpha + bC_{22}^\alpha) = \mathcal{O}(\omega^4 + \delta^2),$$

uniformly for $\alpha \in Y_0^*$. Therefore, $\frac{\omega^2|D|}{\delta v_r^2}$ approximates the eigenvalues of the quasiperiodic capacitance matrix. This completes the proof. $\square$

4.3.3. Asymptotic band structure close to Dirac points. Theorem 4.16 gives an asymptotic formula of the band functions in terms of δ. In this subsection, we investigate the behaviour of this approximation for α close to the Dirac points.

Let R_1 and R_2 be the rotations by $-2\pi/3$ around x_1 and x_2. For $\alpha \in Y^*$, we define two transformations T_1^α and T_2 of α-quasiperiodic functions in Y by

$$(T_1^\alpha f)(x) := \begin{cases} e^{-\mathrm{i}\alpha\cdot l_1} f(R_1 x), & x \in Y_1, \\ e^{-2\mathrm{i}\alpha\cdot l_1} f(R_2 x), & x \in Y_2, \end{cases} \qquad (T_2 f)(x) := \overline{f(2x_0 - x)}.$$

Observe that $T_1^\alpha f$ is well defined on $\partial Y_1 \cap \partial Y_2$. For any $\alpha \in Y^*$ we have that $T_2 f$ is α-quasiperiodic, while at $\alpha = \alpha^*$ we have that $T_1^{\alpha^*} f$ is α^*-quasiperiodic. We will denote $T_1 := T_1^{\alpha^*}$ and define $\tau := e^{\mathrm{i}2\pi/3}$.

From (4.34), it is easy to see that the quasiperiodic capacitance matrix C^α is Hermitian for any α. At Dirac points, the following result holds [**35**].

LEMMA 4.17. *At the Dirac points, the quasiperiodic capacitance matrix C^α defined by* (4.33) *or equivalently by* (4.34) *is a constant multiple of the identity matrix.*

PROOF. Since $T_2 V_1^\alpha = V_2^\alpha$, we have

$$C_{11}^\alpha = \int_{Y\setminus D} \overline{\nabla V_1^\alpha} \cdot \nabla V_1^\alpha \, \mathrm{d}x = \int_{Y\setminus D} \overline{\nabla T_2 V_1^\alpha} \cdot \nabla T_2 V_1^\alpha \, \mathrm{d}x = C_{22}^\alpha.$$

Hence $C_{11}^\alpha = C_{22}^\alpha$ for any $\alpha \in Y_0^*$ and in particular $C_{11}^{\alpha^*} = C_{22}^{\alpha^*}$. We can also check that

$$T_1 V_1^{\alpha^*} = \tau V_1^{\alpha^*}, \; T_1 V_2^{\alpha^*} = \tau^2 V_2^{\alpha^*}.$$

Then, it follows that

$$C_{12}^{\alpha^*} = \int_{Y \setminus D} \overline{\nabla V_1^{\alpha^*}} \cdot \nabla V_2^{\alpha^*} \, \mathrm{d}x = \int_{Y \setminus D} \overline{\nabla T_1 V_1^{\alpha^*}} \cdot \nabla T_1 V_2^{\alpha^*} \, \mathrm{d}x = \tau C_{12}^{\alpha^*}.$$

Therefore, we have $C_{12}^{\alpha^*} = 0$ and so, $C_{21}^{\alpha^*} = 0$. □

Since C^α has a double eigenvalue at Dirac points, we have $\omega_1^{\alpha^*} = \omega_2^{\alpha^*} + \mathcal{O}(\delta)$. The following proposition shows that, in fact, $\omega_1^{\alpha^*} = \omega_2^{\alpha^*}$ and that this is a double characteristic value.

PROPOSITION 4.18. *At the Dirac point $\alpha = \alpha^*$ and for δ small enough, the first Bloch resonant frequency $\omega_* := \omega_1^{\alpha^*}$ is of multiplicity 2, i.e., $\mathcal{A}^\alpha(\omega_*, \delta)$, defined by* (4.8), *has a two-dimensional kernel.*

PROOF. From Lemma 4.14, we know that there are only two band functions (counted with multiplicity) converging to 0 as $\delta \to 0$. Hence, for small enough δ, the dimension of Ker $(\mathcal{A}^\alpha(\omega_*, \delta))$ is at most 2. Suppose that ω_* is of multiplicity 1. Suppose also that

$$\mathcal{A}^\alpha(\omega_*, \delta) \begin{pmatrix} \psi \\ \phi \end{pmatrix} = 0,$$

for a nontrivial pair (ϕ, ψ). Let

$$u(x) = \begin{cases} \mathcal{S}_D^{\alpha^*, k^*}[\phi](x), & x \in Y \setminus \overline{D}, \\ \mathcal{S}_D^{\alpha, k_r^*}[\psi](x), & x \in D, \end{cases} \tag{4.38}$$

where $k^* = \omega_*/v$ and $k_r^* = \omega_*/v_r$.

We can easily check that $T_1 u$ and $T_2 u$ also satisfy (4.29) for $k = k^*$ and $\alpha = \alpha^*$. Then u, $T_1 u$, and $T_2 u$ are linearly dependent and so

$$T_1 u = c_1 u, \; T_2 u = c_2 u,$$

for some nonzero constants c_1 and c_2. Here, we observe that

$$T_1^3 = I, \; T_2^2 = I, \; T_1 T_2 f = T_2 T_1 f.$$

Then, it follows that $c_1 \in \{1, \tau, \tau^2\}$, $c_2 = \pm 1$, and

$$c_1 c_2 u = T_1 T_2 u = T_2 T_1 u = \bar{c}_1 c_2 u.$$

Therefore, we get $c_1 = 1$, *i.e.*, $T_1 u = u$. However, taking the expression for ψ in (4.37), along with the representation (4.38), we find that u is constant on D_i, $i = 1, 2$, up to an error of order $\mathcal{O}(\delta)$. This contradicts $T_1 u = u$, which completes the proof. □

We define the curves Γ_1, Γ_2, and Γ_3 illustrated in Figure 4.10 as

$$\Gamma_1 := \partial Y_1 \cap \partial Y, \quad \Gamma_2 := \partial Y_2 \cap \partial Y, \quad \Gamma_3 := \partial Y_1 \cap \partial Y_2.$$

For $\alpha \in Y^*$, we write $\alpha = \begin{pmatrix} \alpha_1 \\ \alpha_2 \end{pmatrix}$. The following lemma from [**35**] holds.

LEMMA 4.19. *The quasiperiodic capacitance matrix coefficients C_{11}^α and C_{12}^α are differentiable with respect to α at $\alpha = \alpha^*$. Moreover,*

$$\nabla_\alpha C_{11}^\alpha \Big|_{\alpha = \alpha^*} = 0, \quad \nabla_\alpha C_{12}^\alpha \Big|_{\alpha = \alpha^*} = c \begin{pmatrix} 1 \\ -\mathrm{i} \end{pmatrix},$$

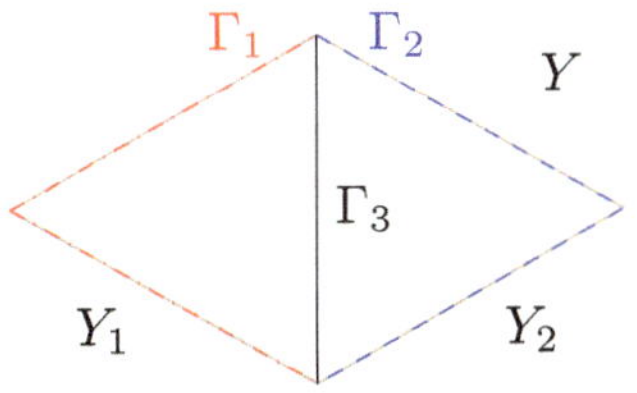

FIGURE 4.10. Fundamental domain Y and the curves Γ_1, Γ_2 and Γ_3 used in the proof of Lemma 4.19.

where

$$c := \frac{\partial C_{12}^\alpha}{\partial \alpha_1}\Big|_{\alpha=\alpha^*}. \tag{4.39}$$

Furthermore, if D is symmetric with respect to Γ_3, i.e., D satisfies

$$R_3 D = D,$$

where R_3 is the reflection across Γ_3, then we have $c \neq 0$.

PROOF. For a small $\varepsilon \in Y^*$, we have from (B.105)

$$\left(\mathcal{S}_D^{\alpha^*+\varepsilon,0}\right)^{-1} = \left(\mathcal{S}_D^{\alpha^*,0}\right)^{-1} + \left(\mathcal{S}_D^{\alpha^*,0}\right)^{-1}\left(\mathcal{S}_1^0 \cdot \varepsilon\right)\left(\mathcal{S}_D^{\alpha^*,0}\right)^{-1} + \mathcal{O}(|\varepsilon|^2),$$

where the error term is with respect to the operator norm. Therefore, from (4.34) it follows that the quasiperiodic capacitance coefficients are differentiable at $\alpha = \alpha^*$.

We have the relations

$$V_1^{R^2\alpha}(x) = \begin{cases} V_1^\alpha(R_1 x), & x \in Y_1, \\ e^{-\mathrm{i}\alpha\cdot l_1} V_1^\alpha(R_2 x), & x \in Y_2, \end{cases}$$

$$V_2^{R^2\alpha}(x) = \begin{cases} e^{\mathrm{i}\alpha\cdot l_1} V_2^\alpha(R_1 x), & x \in Y_1, \\ V_2^\alpha(R_2 x), & x \in Y_2, \end{cases}$$

from which it follows that

$$C_{11}^\alpha = C_{11}^{R\alpha} = C_{11}^{R^2\alpha}, \quad C_{12}^{R^2\alpha} = e^{\mathrm{i}\alpha\cdot l_1} C_{12}^\alpha.$$

Differentiating these expressions, and applying $R\alpha^* = \alpha^*$, we arrive at

$$\nabla_\alpha C_{11}^\alpha\Big|_{\alpha=\alpha^*} = 0, \quad \nabla_\alpha C_{12}^\alpha\Big|_{\alpha=\alpha^*} = c\begin{pmatrix} 1 \\ -\mathrm{i} \end{pmatrix},$$

where c is given by (4.39). It only remains to show that $c \neq 0$ if D is symmetric with respect to Γ_3. Let

$$\widehat{V}_j := \frac{\partial V_j^\alpha}{\partial \alpha_1}\Big|_{\alpha=\alpha^*},$$

which satisfies

$$\begin{cases} \Delta \widehat{V}_j = 0 & \text{in } Y \setminus \overline{D}, \\ \widehat{V}_j = 0 & \text{on } \partial D, \\ \widehat{V}_j(x+l) = e^{\mathrm{i}\alpha\cdot l}\widehat{V}_j(x) + \mathrm{i} l_1 e^{\mathrm{i}\alpha\cdot l} V_j^\alpha(x), & \forall\, l \in \Lambda, \end{cases}$$

where $l = \begin{pmatrix} l_1 \\ l_2 \end{pmatrix}$. Then, using the quasiperiodicity of V_j^α and $\widehat{V}_j$, we have

$$\begin{aligned} \frac{\partial C_{12}^\alpha}{\partial \alpha_1}\bigg|_{\alpha=\alpha^*} &= \int_{Y\setminus D} \left(\overline{\nabla V_1^{\alpha^*}} \cdot \nabla \widehat{V}_2 + \overline{\nabla \widehat{V}_1} \cdot \nabla V_2^{\alpha^*} \right) \, dx \\ &= \int_{\partial Y} \left(\overline{\frac{\partial V_1^{\alpha^*}}{\partial \nu}} \widehat{V}_2 + \overline{\widehat{V}_1} \frac{\partial V_2^{\alpha^*}}{\partial \nu} \right) \, d\sigma \\ &= \mathrm{i} \frac{\sqrt{3}a}{2} \int_{\Gamma_2} \left(\overline{\frac{\partial V_1^{\alpha^*}}{\partial \nu}} V_2^{\alpha^*} - \overline{V_1^{\alpha^*}} \frac{\partial V_2^{\alpha^*}}{\partial \nu} \right) \, d\sigma, \end{aligned} \tag{4.40}$$

where we have used the fact that $\sqrt{3}a/2$ is the l_1-component of the lattice vectors l_1 and l_2. Using (B.99), we find that

$$\int_{\Gamma_1} \overline{\frac{\partial V_1^{\alpha^*}}{\partial \nu}} V_2^{\alpha^*} \, d\sigma = - \int_{\Gamma_2} \overline{\frac{\partial V_1^{\alpha^*}}{\partial \nu}} V_2^{\alpha^*} \, d\sigma = - \int_{\Gamma_2} \overline{\frac{\partial (T_2 V_2^{\alpha^*})}{\partial \nu}} T_2 V_1^{\alpha^*} \, d\sigma,$$

and therefore,

$$\int_{\Gamma_1} \overline{\frac{\partial V_1^{\alpha^*}}{\partial \nu}} V_2^{\alpha^*} \, d\sigma = - \int_{\Gamma_1} \overline{V_1^{\alpha^*}} \frac{\partial V_2^{\alpha^*}}{\partial \nu} \, d\sigma. \tag{4.41}$$

Assume that $\frac{\partial C_{12}^\alpha}{\partial \alpha_1}\big|_{\alpha=\alpha^*} = 0$. Then, from (4.40) and (4.41), we obtain that

$$\int_{\Gamma_j} \overline{V_1^{\alpha^*}} \frac{\partial V_2^{\alpha^*}}{\partial \nu} \, d\sigma = \int_{\Gamma_j} \overline{\frac{\partial V_1^{\alpha^*}}{\partial \nu}} V_2^{\alpha^*} \, d\sigma = 0, \tag{4.42}$$

for $j = 1, 2$. Let $\nu_3 = \begin{pmatrix} 1 \\ 0 \end{pmatrix}$. Then it follows from (4.42) and $C_{12}^{\alpha^*} = 0$ that

$$\int_{\Gamma_3} \left(\overline{\frac{\partial V_1^{\alpha^*}}{\partial \nu_3}} V_2^{\alpha^*} - \overline{V_1^{\alpha^*}} \frac{\partial V_2^{\alpha^*}}{\partial \nu_3} \right) \, d\sigma = \int_{Y\setminus D} \overline{\nabla V_1^{\alpha^*}} \cdot \nabla V_2^{\alpha^*} \, dx = 0. \tag{4.43}$$

If we write $\alpha^* = \begin{pmatrix} \alpha_1^* \\ \alpha_2^* \end{pmatrix}$, then

$$\begin{pmatrix} -\alpha_1^* \\ \alpha_2^* \end{pmatrix} = \alpha^* \quad \text{in } Y^*.$$

If D is symmetric with respect to Γ_3, we observe that

$$V_1^{\alpha^*}(x) = V_2^{(-\alpha_1^*, \alpha_2^*)}(R_3 x) = V_2^{\alpha^*}(R_3 x).$$

From this, we have that

$$V_1^{\alpha^*} = V_2^{\alpha^*}, \quad \frac{\partial V_2^{\alpha^*}}{\partial \nu_3} = - \frac{\partial V_1^{\alpha^*}}{\partial \nu_3} \quad \text{on } \Gamma_3.$$

Together with (4.43), we get

$$2 \operatorname{Re} \int_{\Gamma_3} V_1^{\alpha^*} \overline{\frac{\partial V_1^{\alpha^*}}{\partial \nu_3}} \, d\sigma = \int_{\Gamma_3} \left(\overline{\frac{\partial V_1^{\alpha^*}}{\partial \nu_3}} V_2^{\alpha^*} - \overline{V_1^{\alpha^*}} \frac{\partial V_2^{\alpha^*}}{\partial \nu_3} \right) \, d\sigma = 0. \tag{4.44}$$

We recall that $T_1 V_1^{\alpha^*} = \tau V_1^{\alpha^*}$ and observe that $\Gamma_1 = \big(R_1 \Gamma_3\big) \cup \big((R_1)^2 \Gamma_3\big)$. Using these facts and (4.44), we get

$$\operatorname{Re} \int_{\Gamma_1} V_1^{\alpha^*} \overline{\frac{\partial V_1^{\alpha^*}}{\partial \nu}} \, d\sigma = 0,$$

so that

$$\int_{Y_2 \setminus D_2} |\nabla V_1^{\alpha^*}|^2 \, \mathrm{d}x = 0.$$

Combined with $V_1^{\alpha^*}\Big|_{\partial D_2} = 0$, this tells us that $V_1^{\alpha^*} = 0$ in $Y \setminus D$, which is a contradiction. Therefore, we can conclude that $\frac{\partial C_{12}^{\alpha}}{\partial \alpha_1}\Big|_{\alpha=\alpha^*} = c \neq 0$. □

Starting from the asymptotic formula for ω_j^α in terms of δ, we can now ascertain the asymptotic dependence for α close to α^*. From Lemma 4.19, we obtain that

$$C_{11}^{\alpha} = C_{11}^{\alpha^*} + \mathcal{O}(|\alpha - \alpha^*|^2), \quad |C_{12}^{\alpha}| = \left| \frac{\partial C_{12}^{\alpha}}{\partial \alpha_1}\Big|_{\alpha=\alpha^*} \right| |\alpha - \alpha^*| + \mathcal{O}(|\alpha - \alpha^*|^2).$$

Because C^α is Hermitian with identical diagonal elements, the eigenvalues are given by

$$\lambda_j^\alpha = C_{11}^\alpha + (-1)^j |C_{12}^\alpha|.$$

For α close to α^*, we find the following asymptotic behaviour:

$$\lambda_j^\alpha = C_{11}^{\alpha^*} + (-1)^j \left| \frac{\partial C_{12}^{\alpha}}{\partial \alpha_1}\Big|_{\alpha=\alpha^*} \right| |\alpha - \alpha^*| + \mathcal{O}(|\alpha - \alpha^*|^2).$$

Now, we conclude that $\lambda_1^{\alpha^*} = \lambda_2^{\alpha^*} = C_{11}^{\alpha^*}$ and

(4.45)

$$\omega_j^\alpha(\delta) = \sqrt{\frac{\delta C_{11}^{\alpha^*}}{|D|}} v_r \left(1 + (-1)^j \frac{\left| \frac{\partial C_{12}^{\alpha}}{\partial \alpha_1}\Big|_{\alpha=\alpha^*} \right|}{2 C_{11}^{\alpha^*}} |\alpha - \alpha^*| + \mathcal{O}(|\alpha - \alpha^*|^2) \right) + \mathcal{O}(\delta).$$

Equation (4.45) gives the asymptotic band structure for a small δ and suggests that the system has a Dirac cone. However, we do not know the behaviour of the error term $\mathcal{O}(\delta)$, so we cannot conclude the existence of a Dirac cone from (4.45) alone. This will be addressed in the next subsection.

REMARK 4.20. *Theorem 4.16 shows that* $(\omega_j^\alpha)^2$ *scales like* $\mathcal{O}(\delta)$ *for small* δ. *From Theorem 1.39, it can be seen that the subwavelength resonant frequency* ω_M *of a single resonator in free space scales according to* $\omega_M^2 \ln \omega_M = \mathcal{O}(\delta)$ *in two dimensions. Thus,* ω_j^α *has an asymptotic behaviour different from that of* ω_M. *The difference is explained by the fact that the quasiperiodic single layer potential does not exhibit a logarithmic singularity as* $\omega \to 0$.

4.3.4. Dirac cone dispersion in the band structure. Next, we will study the band functions and Bloch modes around the Dirac points. For simplicity, we only consider the analysis around the Dirac point $\alpha^* := \alpha_1^*$, the other point having a similar behaviour.

We recall that at $\alpha = \alpha^*$, the generalised capacitance matrix C^α has an eigenvalue of multiplicity 2: $\lambda_1^{\alpha^*} = \lambda_2^{\alpha^*}$. The next result shows that this asymptotic degeneracy is in fact an exact degeneracy, and moreover that the band functions intersect in a Dirac cone at this point (see Figure 4.11a). We refer to [**35**] for its proof.

THEOREM 4.21. *For* α *close to* α^* *and* δ *small enough, the first two band functions form a Dirac cone at* α^*, *i.e.,*

$$\begin{aligned} \omega_1^\alpha &= \omega_* - \mu |\alpha - \alpha^*| \big[1 + \mathcal{O}(|\alpha - \alpha^*|)\big], \\ \omega_2^\alpha &= \omega_* + \mu |\alpha - \alpha^*| \big[1 + \mathcal{O}(|\alpha - \alpha^*|)\big], \end{aligned} \tag{4.46}$$

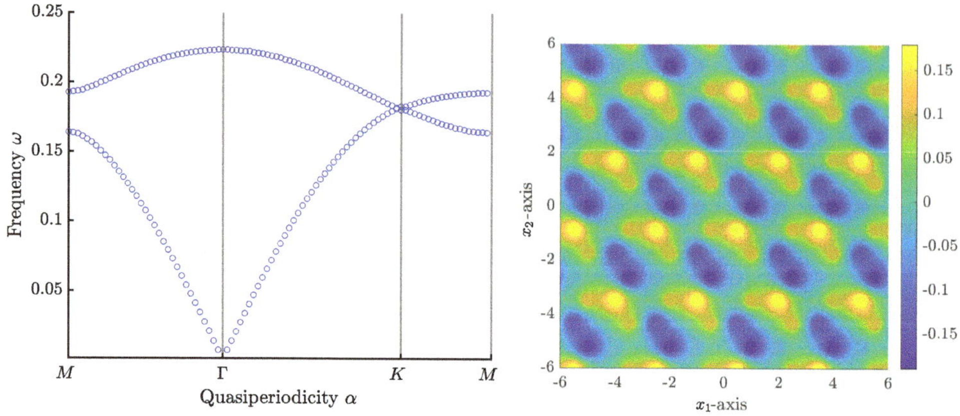

(A) Subwavelength band structure of a honeycomb structure, exhibiting a Dirac cone at $\alpha = \alpha^*$.

(B) Small-scale behaviour of the eigenmodes at the Dirac point.

FIGURE 4.11. The honeycomb structure exhibits a Dirac cone at the corner of the Brillouin zone. Corresponding eigenmodes are rapidly oscillating, and are periodic across one hexagon in the honeycomb structure.

where ω_* and μ are independent of α and satisfy, as $\delta \to 0$,

$$\omega_* = \sqrt{\lambda_1^{\alpha^*}} + \mathcal{O}(\delta) \quad \text{and} \quad \mu = |c|\sqrt{\delta}\mu_0 + \mathcal{O}(\delta) \tag{4.47}$$

with

$$\mu_0 = \frac{1}{2}\sqrt{\frac{v_{\mathrm{r}}^2}{|D_1|C_{11}^{\alpha^*}}} \quad \text{and} \quad c = \frac{\partial C_{12}^{\alpha}}{\partial \alpha_1}\bigg|_{\alpha=\alpha^*}.$$

Moreover, the error term $\mathcal{O}(|\alpha - \alpha^*|)$ in (4.46) is uniform in δ.

4.3.5. High-frequency homogenisation near the Dirac points. Analogously to the previous section, we expect the standard homogenisation to fail for the honeycomb crystal and instead seek a homogenised equation for the envelopes of the Bloch eigenmodes at the Dirac points. We will see that the near-zero property follows as a direct consequence of the Dirac cone. We consider a homogenisation setting close to the Dirac points. We rescale the unit cell by replacing Y with sY for some small $s > 0$. To fix the order of the resonant frequencies, *i.e.*, $\omega_i^{\alpha} = \mathcal{O}(1)$, we assume that $\delta = \mathcal{O}(s^2)$ as $s \to 0$. We then have the following result from [**45**], which is the analogue of Theorem 4.13 for the case of a honeycomb lattice.

THEOREM 4.22. *Let $S_j(x) = \mathcal{S}_D^{\alpha^*,0}[\psi_j^{\alpha^*}]$ for $j = 1, 2$, with $\psi_j^{\alpha^*}$ being defined by (4.30) for $\alpha = \alpha^*$. For frequencies ω close to the Dirac frequency ω_*, namely, $\omega - \omega_* = \beta\sqrt{\delta}$, the following asymptotic behaviour of the Bloch eigenmode $u_s^{\alpha^*/s+\widetilde{\alpha}}$ holds:*

$$u_s^{\alpha^*/s+\widetilde{\alpha}}(x) = \begin{pmatrix} Ae^{\mathrm{i}\widetilde{\alpha}\cdot x} \\ Be^{\mathrm{i}\widetilde{\alpha}\cdot x} \end{pmatrix} \cdot \begin{pmatrix} S_1(\frac{x}{s}) \\ S_2(\frac{x}{s}) \end{pmatrix} + \mathcal{O}(s),$$

where the macroscopic field

$$(\widetilde{u}_1, \widetilde{u}_2)^{\top} := (Ae^{\mathrm{i}\widetilde{\alpha}\cdot x}, Be^{\mathrm{i}\widetilde{\alpha}\cdot x})^{\top}$$

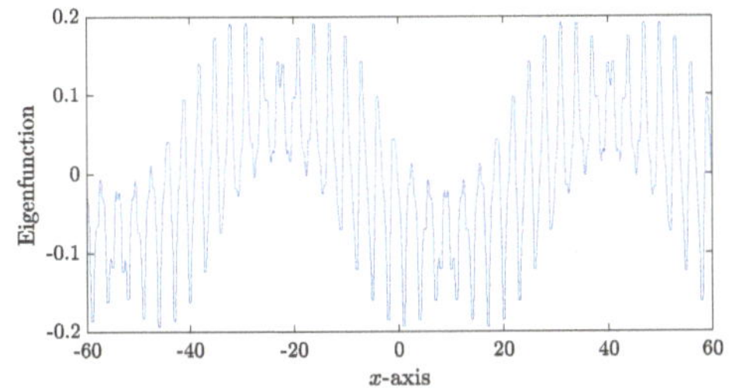

(A) Large-scale behaviour of the eigenmodes close to the Dirac point.

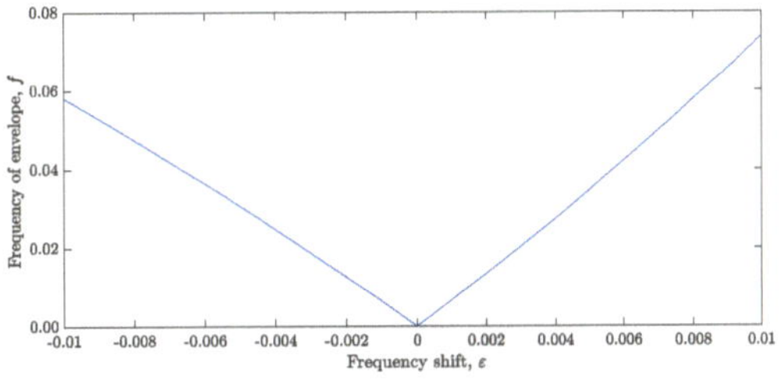

(B) Spatial frequency f of the envelopes as a function of the frequency shift $\varepsilon = \omega - \omega_*$.

FIGURE 4.12. For frequencies close to the Dirac frequency, the eigenmodes oscillate on two distinct scales. The spatial frequency of the large-scale oscillations vanish at the centre of the Dirac cone, corresponding to zero refractive index.

satisfies the two-dimensional Dirac equation

$$\mu_0 \begin{pmatrix} 0 & (-c\mathrm{i})(\partial_1 - \mathrm{i}\partial_2) \\ (-\bar{c}\mathrm{i})(\partial_1 + \mathrm{i}\partial_2) & 0 \end{pmatrix} \begin{pmatrix} \widetilde{u}_1 \\ \widetilde{u}_2 \end{pmatrix} = \frac{\omega - \omega_*}{\sqrt{\delta}} \begin{pmatrix} \widetilde{u}_1 \\ \widetilde{u}_2 \end{pmatrix}.$$

The system of Dirac equations can be considered as a homogenised equation for the honeycomb structure. Each $\widetilde{u}_j$ satisfies the Helmholtz equation

$$\Delta \widetilde{u}_j + \frac{(\omega - \omega_*)^2}{\delta \mu_0^2 |c|^2} \widetilde{u}_j = 0, \tag{4.48}$$

where μ_0 is defined in (4.47). In particular, at $\omega = \omega_*$ this effective equation reduces to the Laplace equation corresponding to the effective zero refractive index. Equation (4.48) describes the large-scale behaviour of the eigenmodes, illustrated in Figure 4.12. We emphasise that in addition to this large-scale behaviour, there will be small-scale oscillations described by the functions S_j as illustrated in Figure 4.11.

From (4.48), for $\omega = \omega_* + \varepsilon$ close to the Dirac frequency, each component $\widetilde{u}_j$, $j = 1, 2$, of the macroscopic field oscillates at a spatial frequency

$$f = \frac{|\varepsilon|}{\mu_0 |c| \sqrt{\delta}}, \tag{4.49}$$

which is linear in ε. This shows that close to the Dirac frequency the macroscopic behaviour of the honeycomb crystal can be described as a near-zero refractive index material.

To verify (4.49), we compute the large-scale spatial frequency of the eigenmodes for $\varepsilon \in [-0.01, 0.01]$, as shown in Figure 4.12. Note that the spatial frequencies of the macroscopic fields in the cases of a honeycomb lattice and a square lattice have different asymptotic behaviours close to the critical frequencies due to the fact that the square lattice cannot be mapped to a zero-index effective material.

REMARK 4.23. *The results in Theorems* 4.21 *and* 4.22 *are derived from first principles. They can be viewed as the classical wave analogues of those proved in* [**81**] *for Schrödinger operator. The time-evolution of wave-packets in honeycomb systems of subwavelength resonators, which are spectrally concentrated near conical points, can be studied in the same way as in* [**81, 169, 170**].

REMARK 4.24. *The results on the two-dimensional (double-degenerate) Dirac points in Theorems 4.21 and 4.22 can be extended to three dimensions and higher-order degeneracies* [**64, 99**].

4.4. Chains and screens of resonators

In this section, we consider a general lattice Λ and a dimension of periodicity $d_l < d$. By assuming that $d_l < d$, we focus on the study of chains and screens of subwavelength resonators. In particular, we are interested in describing the subwavelength resonances in the first radiation continuum. The corresponding results presented here are from [**26, 28**].

We let the quasiperiodic Green function $G^{\alpha,k}$ be the solution of

$$(\Delta + k^2)G^{\alpha,k}(x) = \Big(\sum_{m\in\Lambda} \delta_0((x_l,0) - m)e^{\mathrm{i}m\cdot\alpha}\Big)\delta_0(x_0) \quad \text{in } \mathbb{R}^d, \tag{4.50}$$

for $x = (x_l, x_0)$ and $\alpha \in Y^*$. Since the fundamental domain Y is unbounded in the x_0-direction, a radiation condition is needed to ensure the uniqueness of a solution to (4.50). We assume that $G^{\alpha,k}(x)$ for $x = (x_l, x_0)$ is bounded as $|x_0| \to +\infty$ and consists only on outgoing plane waves. Here, $x_l \in \mathbb{R}^{d_l}$ is the vector along the first d_l dimensions, $x_0 \in \mathbb{R}^{d-d_l}$, and the lattice Λ is aligned with the first d_l coordinate axes.

In the case $k \neq |\alpha + q|$ for all $q \in \Lambda^*$, we can also define the quasiperiodic Green function $G^{\alpha,k}(x)$ as the Floquet transform of $G^k(x)$ in the first d_l coordinate dimensions, *i.e.*,

$$G^{\alpha,k}(x) := \sum_{m\in\Lambda} G^k(x-m)e^{\mathrm{i}\alpha\cdot m}. \tag{4.51}$$

Here G^k is the usual Helmholtz Green function, defined in (1.7) for $d = 2$ or $d = 3$. The series in (4.51) converges uniformly for x and y in compact sets of $\mathbb{R}^d$, $x \neq y$, and $k \neq |\alpha + q|$ for all $q \in \Lambda^*$.

REMARK 4.25. *In order to exemplify the outgoing radiation condition, we consider the case of a screen, $d - d_l = 1$. Since $G^{\alpha,k}(x)e^{-\mathrm{i}\alpha\cdot x}$ for $\alpha \in Y^*$ is periodic in x_l, by using Fourier series expansion together with the Poisson summation formula*

$$\sum_{m\in\Lambda} \delta_0(x-m)e^{\mathrm{i}m\cdot\alpha} = \frac{1}{|Y_l|}\sum_{q\in\Lambda^*} e^{\mathrm{i}(\alpha+q)\cdot x},$$

it follows that

$$G^{\alpha,k}(x) = \frac{1}{|Y_l|}\sum_{q\in\Lambda^*} e^{\mathrm{i}(\alpha+q)\cdot x}\varphi_q^\alpha(x_0), \tag{4.52}$$

where φ_q^α satisfies the differential equation

$$\frac{d^2}{dx_0^2}\varphi_q^\alpha + (k^2 - |q+\alpha|^2)\varphi_q^\alpha = \delta_0 \quad \textit{on } \mathbb{R}.$$

The outgoing radiation condition imposed on $G^{\alpha,k}(x)$ shows that

$$\varphi_q^\alpha(x_0) = \begin{cases} \frac{1}{2}|x_0| & \textit{if } k = |q+\alpha|, \\ \frac{1}{2\mathrm{i}\sqrt{k^2-|q+\alpha|^2}}e^{\mathrm{i}\sqrt{k^2-|q+\alpha|^2}|x_0|} & \textit{if } k > |q+\alpha|, \\ -\frac{1}{2\sqrt{|q+\alpha|^2-k^2}}e^{-\sqrt{|q+\alpha|^2-k^2}|x_0|} & \textit{if } k < |q+\alpha|. \end{cases} \tag{4.53}$$

Therefore, assuming that $k \neq |\alpha+q|$ for all $q \in \Lambda^*$, the quasiperiodic Green function admits the following spectral representation:

$$
\begin{aligned}
G^{\alpha,k}(x) = & \sum_{q\in\Lambda^*, |\alpha+q|<k} \underbrace{\frac{e^{\mathrm{i}(\alpha+q)\cdot x} e^{\mathrm{i}\sqrt{k^2-|\alpha+q|^2}|x_0|}}{2\mathrm{i}|Y_l|\sqrt{k^2-|\alpha+q|^2}}}_{\textit{outgoing modes}} \\
& - \sum_{q\in\Lambda^*, |\alpha+q|>k} \underbrace{\frac{e^{\mathrm{i}(\alpha+q)\cdot x} e^{-\sqrt{|\alpha+q|^2-k^2}|x_0|}}{2|Y_l|\sqrt{|\alpha+q|^2-k^2}}}_{\textit{evanescent modes}},
\end{aligned} \tag{4.54}
$$

where $x = (x_l, x_0)$. The series in (4.61) *converges uniformly for x in compact sets of $\mathbb{R}^d$, $x \neq 0$; see,* e.g., [**37**, Section 2.12].

Using the quasiperiodic Green function $G^{\alpha,k}$ we can define analogous quantities and get analogous results as in the finite case. The quasiperiodic single layer potential $\mathcal{S}_D^{\alpha,k}$ is then defined as

$$
\mathcal{S}_D^{\alpha,k}[\varphi](x) := \int_{\partial D} G^{\alpha,k}(x-y)\varphi(y)\, \mathrm{d}\sigma(y), \quad x \in \mathbb{R}^d.
$$

The quasiperiodic single layer potential satisfies many conceptually similar properties as the "regular" single layer potential (which is a consequence of the fact that the singularity of corresponding Green's functions are the same, *i.e.*, $G^{\alpha,k} - G^k$ is a smooth function of x around the origin). For example, the quasiperiodic single layer potential satisfies the following jump relations on ∂D:

$$
\mathcal{S}_D^{\alpha,k}[\varphi]\big|_+ = \mathcal{S}_D^{\alpha,k}[\varphi]\big|_- \quad \text{on } \partial D, \tag{4.55}
$$

and

$$
\frac{\partial}{\partial \nu}\mathcal{S}_D^{\alpha,k}[\varphi]\Big|_{\pm} = \left(\pm\frac{1}{2}I + (\mathcal{K}_D^{-\alpha,k})^*\right)[\varphi] \quad \text{on } \partial D, \tag{4.56}
$$

where $(\mathcal{K}_D^{-\alpha,k})^*$ is the quasiperiodic Neumann-Poincaré operator, given by

$$
(\mathcal{K}_D^{-\alpha,k})^*[\varphi](x) := \int_{\partial D} \frac{\partial}{\partial \nu(x)} G^{\alpha,k}(x-y)\varphi(y)\, \mathrm{d}\sigma(y).
$$

LEMMA 4.26. *The quasiperiodic single layer potential $\mathcal{S}_D^{\alpha,k} : L^2(\partial D) \to H^1(\partial D)$ is invertible if k is small enough and $k \neq |\alpha+q|$ for all $q \in \Lambda^*$.*

PROOF. If $\varphi \in L^2(\partial D)$ satisfies $\mathcal{S}_D^{\alpha,k}[\varphi] = 0$ on ∂D, then $u := \mathcal{S}_D^{\alpha,k}[\varphi]$ satisfies $\Delta u + k^2 u = 0$ in $Y \setminus \partial D$. Since 0 is not a Dirichlet eigenvalue of $-\Delta$ in D, and neither a Dirichlet eigenvalue of $-\Delta$ on $Y \setminus D$ with quasiperiodic conditions on ∂Y, it follows that $u = 0$ for small enough k. Then, from the jump condition (4.56) we have $\varphi = \partial u/\partial\nu|_+ - \partial u/\partial\nu|_- = 0$, which proves the claim. □

Above, we assumed that $k \neq |\alpha+q|$ for all $q \in \Lambda^*$. When k is small and lies in the subwavelength regime, this condition separates into two cases:

- $k < \inf_{q\in\Lambda^*} |\alpha+q|$. Waves in this regime are exponentially decaying away from the structure. Such waves, which vanish in the far field, are known as *evanescent* waves;

- $|\alpha| < k < \inf_{q\in\Lambda^*\setminus\{0\}} |\alpha+q|$. Waves in this regime are propagating far away from the structure, and this regime is known as the *first radiation continuum*.

Consider the periodic array of resonators $\mathcal{D}$ given by (4.1). If we assume that v_i and δ_i are real, then we can interpret the two regimes as follows. When $k < \inf_{q\in\Lambda^*} |\alpha+q|$, the problem (4.4) can be viewed as a spectral problem for a self-adjoint operator, and the resonant frequencies are real. When $|\alpha| < k < \inf_{q\in\Lambda^*\setminus\{0\}} |\alpha+q|$, due to the radiation condition, this equation no longer corresponds to a self-adjoint operator. Therefore, the resonators attain a small but nonzero imaginary part corresponding to the coupling with the far field. The transition between these two regimes occurs when $k = |\alpha|$ for some q (known as a *Rayleigh-Wood anomaly*), which are the points where the spectrum becomes real and the modes become localised to the structure.

In the two regimes mentioned above, we have the following integral representation (analogously to Lemma 1.2).

LEMMA 4.27. *Let $d \in \{2,3\}$ and $0 < d_l < d$. Assume that $k \neq |\alpha+q|$ for all $q\in\Lambda^*$. Then the Helmholtz problem (4.4) is equivalent to finding $\phi,\psi \in L^2(\partial D)$ such that*

$$\mathcal{A}^\alpha(\omega,\delta)\begin{pmatrix}\psi\\ \phi\end{pmatrix} = \begin{pmatrix}0\\0\end{pmatrix}, \tag{4.57}$$

where the operator $\mathcal{A}^\alpha(\omega,\delta) : L^2(\partial D)\times L^2(\partial D) \to H^1(\partial D)\times L^2(\partial D)$ is defined as

$$\mathcal{A}^\alpha(\omega,\delta) = \begin{pmatrix} \widetilde{\mathcal{S}}_D^\omega & -\mathcal{S}_D^{\alpha,k} \\ -\frac{1}{2}I + \widetilde{\mathcal{K}}_D^{\omega,*} & -\widetilde{\delta}\left(\frac{1}{2}I + (\mathcal{K}_D^{-\alpha,k})^*\right)\end{pmatrix}, \tag{4.58}$$

with $\widetilde{\delta}$, $\widetilde{\mathcal{S}}_D^\omega$ and $\widetilde{\mathcal{K}}_D^{\omega,*}$ being defined in (1.4), (1.10) and (1.12), respectively.

REMARK 4.28. *Note that the formulation in Lemma 4.27 is slightly different from the one previously used in (4.7)–(4.8), where quasiperiodic layer potentials are used inside the resonators. Both formulations characterise the quasiperiodic subwavelength resonant frequencies. Nevertheless, computations may be easier using one of these formulations than using the other.*

4.4.1. Evanescent-mode resonances. If we assume $|\alpha| > c > 0$ for some constant c independent of ω and δ, the quasiperiodic Helmholtz single layer potential is well approximated by the corresponding Laplace single layer potential in the sense that $\mathcal{S}_D^{\alpha,\omega}$ converges to $\mathcal{S}_D^{\alpha,0}$ in the operator norm as $\omega \to 0$. Due to this result, which holds even in the case $d = 2$, we can use methods similar to those outlined in Chapter 1, and obtain similar results.

LEMMA 4.29. *Assume $|\alpha| > c > 0$ for some constant c independent of ω and δ, and consider a fundamental cell Y containing N subwavelength resonators $D = D_1 \cup \ldots D_N$. Then, it holds that $\operatorname{Ker}(-\frac{1}{2}I + (\mathcal{K}_D^{-\alpha,0})^*) = \operatorname{span}\{\psi_1^\alpha, \psi_2^\alpha, \ldots, \psi_N^\alpha\}$, where $\psi_i^\alpha := (\mathcal{S}_D^{\alpha,0})^{-1}[\chi_{\partial D_i}]$.*

THEOREM 4.30. *Assume $|\alpha| > c > 0$ for some constant c independent of ω and δ, and consider a fundamental cell Y containing N subwavelength resonators $D = D_1 \cup \ldots D_N$. For sufficiently small $\delta > 0$, there exist N subwavelength resonant frequencies $\omega_1^\alpha(\delta), \ldots, \omega_N^\alpha(\delta)$ with non-negative real parts.*

REMARK 4.31. *The above theorem describes only the* subwavelength *part of the band structure. For small enough δ, there will be a bandgap between the first N*

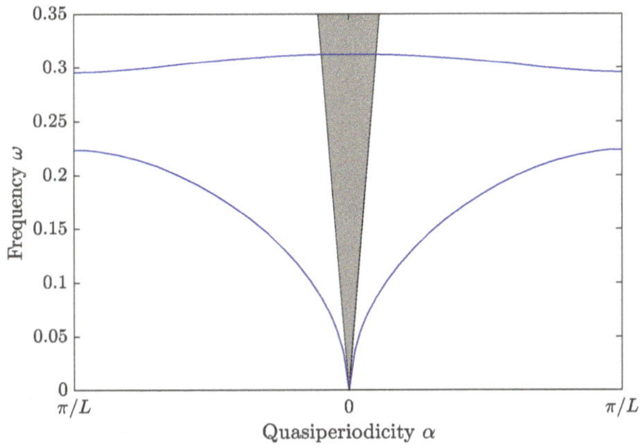

FIGURE 4.13. Example of the subwavelength band structure of a resonator array with two resonators in the unit cell. The shaded region is the first radiation continuum, defined by $|\alpha| < \omega/v < \inf_{q\in\Lambda^*\setminus\{0\}} |\alpha+q|$, while the unshaded region corresponds to evanescent modes. Here we see an example of bandgap opening: there is an uncovered interval between the first and the second bands. Moreover, between the subwavelength bands and the higher (non-subwavelength) bands there will always be a bandgap.

bands (which are in the subwavelength regime) and the higher bands (which are not close to 0 *for small* δ*; see Figure* 4.13*).*

Analogously to Definition 4.15, we can define the quasiperiodic capacitance matrix for a chain or a screen of resonators as follows.

DEFINITION 4.32 (Quasiperiodic capacitance matrix). *Assume* $\alpha \neq 0$. *For a system of* $N \in \mathbb{N}$ *resonators* $D_1, \ldots, D_N$ *in* Y*, the quasiperiodic capacitance matrix* $C^\alpha = (C_{ij}^\alpha) \in \mathbb{C}^{N\times N}$ *is the square matrix given by*

$$C_{ij}^\alpha := -\int_{\partial D_i} (\mathcal{S}_D^{\alpha,0})^{-1}[\chi_{\partial D_j}]\, \mathrm{d}\sigma, \quad i,j = 1,\ldots,N,$$

or equivalently, by

$$C_{ij}^\alpha := \int_{Y\setminus D} \overline{\nabla V_i^\alpha} \cdot \nabla V_j^\alpha \, \mathrm{d}x, \quad i,j = 1,\ldots,N, \tag{4.59}$$

where V_i^α, $i = 1,\ldots,N$, *are the solutions to*

$$\begin{cases} \Delta V_i^\alpha = 0 & \text{in } Y \setminus D, \\ V_i^\alpha = \delta_{ij} & \text{on } \partial D_j, \\ V_i^\alpha(x+l) = e^{\mathrm{i}\alpha\cdot l} V_i^\alpha(x) & \forall\, l \in \Lambda, \\ V_i^\alpha(x) \to 0 & \text{as } |x_0| \to +\infty, \end{cases} \tag{4.60}$$

with $x = (x_l, x_0)$.

The following result follows immediately from the characterisation (4.59) of C^α.

LEMMA 4.33. *The quasiperiodic capacitance matrix is a Hermitian matrix.*

DEFINITION 4.34 (Generalised quasiperiodic capacitance matrix). *Assume $\alpha \neq 0$. For a system of $N \in \mathbb{N}$ resonators $D_1, \ldots, D_N$ in Y, we can define the generalised quasiperiodic capacitance matrix, denoted by $\mathcal{C}^\alpha = (\mathcal{C}_{ij}^\alpha) \in \mathbb{C}^{N\times N}$, to be the square matrix given by*

$$\mathcal{C}_{ij}^\alpha = \frac{\delta_i v_i^2}{|D_i|} C_{ij}^\alpha, \quad i,j = 1,\ldots,N.$$

The next result characterises the first N quasiperiodic resonances of the periodic structure and shows that they are in the subwavelength regime.

THEOREM 4.35. *Consider a system of N subwavelength resonators in Y, and assume that $|\alpha| > c > 0$ for some constant c independent of ω and δ. As $\delta \to 0$, the N subwavelength quasiperiodic resonant frequencies satisfy the asymptotic formula*

$$\omega_n^\alpha = \sqrt{\lambda_n^\alpha} + \mathcal{O}(\delta^{3/2}), \quad n = 1,\ldots,N,$$

where $\{\lambda_n^\alpha : n = 1,\ldots,N\}$ are the eigenvalues of the generalised quasiperiodic capacitance matrix $\mathcal{C}^\alpha \in \mathbb{C}^{N\times N}$, which satisfy $\lambda_n^\alpha = \mathcal{O}(\delta)$ as $\delta \to 0$.

REMARK 4.36. *The error term $\mathcal{O}(\delta^{3/2})$ has higher-order compared to the error term $\mathcal{O}(\delta)$ in Theorem 1.9. This is a consequence of the fact that the $\mathcal{O}(\omega)$-term in the expansion of $\mathcal{S}_D^{\alpha,\omega}$ with respect to ω vanishes; see* (B.98).

COROLLARY 4.37. *Assume that $|\alpha| > c > 0$ for some constant c independent of ω and δ. Let $\mathbf{v}_n^\alpha$ be the eigenvector of $\mathcal{C}^\alpha$ associated with the eigenvalue λ_n^α. Then the resonant mode u_n^α associated to the resonant frequency ω_n^α is given, as $\delta \to 0$, by*

$$u_n^\alpha(x) = \begin{cases} \mathbf{v}_n^\alpha \cdot \mathbf{S}_D^{\alpha,\omega_n^\alpha/v}(x) + \mathcal{O}(\delta^{1/2}), & x \in \mathbb{R}^d \setminus \overline{\mathcal{D}}, \\ \mathbf{v}_n^\alpha \cdot \mathbf{S}_D^{\alpha,\omega_n^\alpha/v_i}(x) + \mathcal{O}(\delta^{1/2}), & x \in \mathcal{D}_i, \end{cases}$$

where $\mathbf{S}_D^{\alpha,\omega_n^\alpha/v} : \mathbb{R}^d \to \mathbb{C}^N$ is the vector-valued function given by

$$\mathbf{S}_D^{\alpha,\omega_n^\alpha/v}(x) = \begin{pmatrix} \mathcal{S}_D^{\alpha,\omega_n^\alpha/v}[\psi_1^\alpha](x) \\ \vdots \\ \mathcal{S}_D^{\alpha,\omega_n^\alpha/v}[\psi_N^\alpha](x) \end{pmatrix},$$

with $\psi_i^\alpha := (\mathcal{S}_D^{\alpha,0})^{-1}[\chi_{\partial D_i}]$, and $\mathbf{S}_D^{\alpha,\omega_n^\alpha/v_i}$ is defined analogously.

REMARK 4.38. *As shown in Section 1.8, the discrete formulations for approximating subwavelength resonant frequencies for finite systems in two and three dimensions are slightly different. In contrast with the finite system setting, for infinitely periodic systems, exactly the same capacitance matrix formulation holds in both two and three dimensions. As said in Remark 4.20, this is due to the fact that the single layer potential associated with the Helmholtz equation in free space has a logarithmic singularity in ω near zero (see* (1.75)*) while the α-quasiperiodic single layer potentials are smooth functions of ω around zero when $\alpha \neq 0$.*

4.4.2. Resonances in the first radiation continuum. Here, we study the regime $|\alpha| < k < \inf_{q\in\Lambda^*\setminus\{0\}} |\alpha + q|$. Since the analysis depends on the dimensionalities, we will exemplify it in the case of a screen, $d - d_l = 1$.

Recall that $k = \omega/v$. In the current regime, we must have $\alpha \to 0$ as $\omega \to 0$. Therefore, we assume that

$$\alpha = \omega\alpha_0 \in Y^*,$$

for some $\alpha_0 \neq 0$, independent of ω and such that $|\alpha_0| < 1/v$. In scattering problems, this limit corresponds to incident waves with a fixed direction of incidence (specified by α_0) and a frequency ω in the subwavelength regime.

4.4.2.1. *Green's function and capacitance matrix formulation.* If $|\alpha| < k = \omega/v < \inf_{q\in\Lambda^*\setminus\{0\}} |\alpha + q|$, then from (4.54) it follows that

$$G^{\alpha,k}(x) = \frac{e^{\mathrm{i}\alpha\cdot x}e^{\mathrm{i}k_0|x_0|}}{2\mathrm{i}k_0|Y_l|} - \sum_{q\in\Lambda^*\setminus\{0\}} \frac{e^{\mathrm{i}(\alpha+q)\cdot x}e^{-\sqrt{|\alpha+q|^2-k^2}|x_0|}}{2|Y_l|\sqrt{|\alpha+q|^2-k^2}}, \tag{4.61}$$

where $x = (x_l, x_0)$ and $k_0 = \sqrt{k^2 - |\alpha|^2}$.

In the case where $k = \alpha = 0$, we deduce from (4.52) and (4.53) that *periodic* Green's function $G^{0,0}$ is given by

$$G^{0,0}(x) = \frac{|x_0|}{2|Y_l|} - \sum_{q\in\Lambda^*\setminus\{0\}} \frac{e^{\mathrm{i}q\cdot x}e^{-|q||x_0|}}{2|Y_l||q|}. \tag{4.62}$$

Here, $G^{0,0}$ is periodic in the x_l-variable. We let $\mathcal{S}_D^{0,0}$ be the periodic single layer potential associated with $G^{0,0}$.

When $\omega \to 0$, we then have the asymptotic expansion

$$G^{\omega\alpha_0,k}(x) = \frac{1}{2\mathrm{i}k_0|Y_l|} + G^{0,0}(x) + \frac{\alpha\cdot x}{2k_0|Y_l|} + \mathcal{O}(\omega). \tag{4.63}$$

In particular, since $k_0 = \omega\sqrt{1/v^2 - |\alpha_0|^2}$, the Green function has a singularity when $\omega \to 0$. In fact, this will make the analysis conceptually similar to the case in Section 1.8. We define the operator $\widehat{\mathcal{S}}_D^{\alpha,k} : L^2(\partial D) \to H^1(\partial D)$ as

$$\widehat{\mathcal{S}}_D^{\alpha,k}[\varphi](x) = \mathcal{S}_D^{0,0}[\varphi](x) - \frac{\mathrm{i} - \alpha\cdot x}{2k_0|Y_l|}\int_{\partial D}\varphi(y)\,\mathrm{d}\sigma(y) - \int_{\partial D}\frac{\alpha\cdot y}{2k_0|Y_l|}\varphi(y)\,\mathrm{d}\sigma(y). \tag{4.64}$$

As $\omega \to 0$, using (4.61) we then have the asymptotic expansion

$$\mathcal{S}_D^{\omega\alpha_0,k} = \widehat{\mathcal{S}}_D^{\omega\alpha_0,k} + \omega\mathcal{S}_1^{\alpha_0} + \mathcal{O}(\omega^2), \tag{4.65}$$

with respect to the operator norm, for some operator $\mathcal{S}_1^{\alpha_0}$ that is independent of ω.

The periodic single layer potential $\mathcal{S}_D^{0,0}$ could fail to be invertible.

LEMMA 4.39. *The dimension of* $\operatorname{Ker}\mathcal{S}_D^{0,0}$ *is at most one. Moreover,* $\mathcal{S}_D^{0,0}$ *is invertible from the mean zero space* $L_0^2(\partial D)$ *onto its image.*

PROOF. For small but nonzero ω, we know from Lemma 4.26 that $\mathcal{S}_D^{0,k}$ is invertible, and therefore, by (4.65), $\widehat{\mathcal{S}}_D^{0,k}$ is also invertible for small ω. We can write $\mathcal{S}_D^{0,0}$ as

$$\mathcal{S}_D^{0,0}[\varphi] = \widehat{\mathcal{S}}_D^{0,k}[\varphi] + \frac{\mathrm{i}}{2k_0|Y_l|}\int_{\partial D}\varphi\,\mathrm{d}\sigma,$$

or, in other words, $\mathcal{S}_D^{0,0}$ is a rank-one perturbation of the invertible operator $\widehat{\mathcal{S}}_D^{0,k}$. This shows that $\dim\operatorname{Ker}\mathcal{S}_D^{0,0} \leq 1$ and, moreover, that any nonzero $\varphi \in \operatorname{Ker}\mathcal{S}_D^{0,0}$ satisfies $\int_{\partial D}\varphi\,\mathrm{d}\sigma \neq 0$ and then since $\varphi \in L_0^2(\partial D)$, φ has to be zero. □

Similarly to Lemma B.27 for the crystal case, the following result holds.

LEMMA 4.40. *If* $\mathcal{S}_D^{0,0}[\varphi] = K$ *on* ∂D *for some constant* K *and some* $\varphi \in L^2(\partial D)$ *that satisfy* $\int_{\partial D}\varphi\,\mathrm{d}\sigma = 0$*, then* $\varphi = 0$.

PROOF. For $x \in \mathbb{R}^d \setminus \mathcal{D}$, define $V(x) := \mathcal{S}_D^{0,0}[\varphi](x)$. Then, from $\int_{\partial D} \varphi \, \mathrm{d}\sigma = 0$, V solves the following differential problem:

$$
\begin{cases}
\Delta V = 0 & \text{in } \mathbb{R}^d \setminus \mathcal{D}, \\
V|_+ = K & \text{on } \partial \mathcal{D}, \\
V(x+l) = V(x) & \text{for all } l \in \Lambda, \\
V(x) \to \pm V_\infty & \text{as } x_0 \to \pm\infty,
\end{cases}
\tag{4.66}
$$

where the constant

$$
V_\infty = -\frac{1}{2|Y_l|} \int_{\partial D} y_0 \varphi(y) \, \mathrm{d}\sigma(y).
$$

Moreover, using the jump relations satisfied by the periodic single layer potential $\mathcal{S}_D^{0,0}$ and integration by parts, we have that

$$
\int_{\partial D} \varphi \, \mathrm{d}\sigma = K \int_{Y \setminus D} |\nabla V|^2 \, \mathrm{d}x = 0.
$$

If $K \neq 0$, from (4.66) it follows that $\int_{Y \setminus D} |\nabla V|^2 \, \mathrm{d}x \neq 0$. In other words, we must have $K = 0$ and, since $\int_{\partial D} \varphi \, \mathrm{d}\sigma = 0$, it follows from Lemma 4.39 that $\varphi = 0$. □

LEMMA 4.41. *For any $\alpha_0 \in Y^*$ with $0 < |\alpha_0| < 1/v$, $\left(\mathcal{S}_D^{\omega\alpha_0,k}\right)^{-1}$ is a holomorphic operator-valued function of ω in a neighbourhood of $\omega = 0$.*

PROOF. The assumption that $|\alpha_0| < 1/v$ is required for $\widehat{\mathcal{S}}_D^{\omega\alpha_0,k}$ to be well-defined. We know that $\widehat{\mathcal{S}}_D^{\omega\alpha_0,k}$ is a meromorphic operator-valued function of ω with a pole at $\omega = 0$. From Corollary A.11, we find that $\left(\widehat{\mathcal{S}}_D^{\omega\alpha_0,k}\right)^{-1}$ is also meromorphic for ω in a neighbourhood of 0:

$$
\left(\widehat{\mathcal{S}}_D^{\omega\alpha_0,k}\right)^{-1} = \sum_{i \geq -n} \omega^i S_i,
$$

for operators $S_i : H^1(\partial D) \to L^2(\partial D)$ independent of ω. It remains to show that the principal part vanishes. To reach a contradiction, we assume that $S_{-n} \neq 0$ for some $n \geq 1$. Pick $\psi \in H^1(\partial D) \setminus \operatorname{Ker} S_{-n}$, independent of ω. We can then write

$$
\left(\widehat{\mathcal{S}}_D^{\omega\alpha_0,k}\right)^{-1}[\psi] := \omega^{-n} \phi
$$

for some $\phi \in L^2(\partial D)$, depending on ω, such that $\|\phi\|_{L^2(\partial D)} = \mathcal{O}(1)$. It follows that

$$
\widehat{\mathcal{S}}_D^{\omega\alpha_0,k}[\phi] = \omega^n \psi = \mathcal{O}(\omega^n).
\tag{4.67}
$$

We can rewrite ϕ as $\phi = \phi_0 + \omega\phi_1$, where ϕ_0 is nonzero and independent of ω, while ϕ_1 depends on ω and satisfies $\phi_1 = \mathcal{O}(1)$. Then the singular part of $\widehat{\mathcal{S}}_D^{\omega\alpha_0,k}$ must vanish on ϕ_0, *i.e.*,

$$
\int_{\partial D} \phi_0 \, \mathrm{d}\sigma = 0.
$$

Substituting this into (4.64) we find, as $\omega \to 0$,

$$
\begin{aligned}
\widehat{\mathcal{S}}_D^{\omega\alpha_0,k}[\phi](x) = \mathcal{S}_D^{0,0}[\phi_0](x) - \int_{\partial D} \frac{\alpha_0 \cdot y}{2w_0|Y_l|} \phi_0(y) \, \mathrm{d}\sigma(y) \\
- \frac{\mathrm{i}}{2w_0|Y_l|} \int_{\partial D} \phi_1 \, \mathrm{d}\sigma + \mathcal{O}(\omega),
\end{aligned}
\tag{4.68}
$$

where $w_0 := \sqrt{1/v^2 - |\alpha_0|^2}$. In view of (4.67), the term in (4.68) of order $\mathcal{O}(1)$ must vanish. Additionally, we observe that the latter two terms on the right-hand side of (4.68) are constant as functions of x. Thus, for some constant K, we have $\mathcal{S}_D^{0,0}[\phi_0] = K$ on ∂D. It then follows from Lemma 4.40 that $\phi_0 = 0$, which contradicts the fact that $\|\phi\| = \mathcal{O}(1)$. $\square$

Crucially, Lemma 4.41 shows that $(\mathcal{S}_D^{\omega\alpha_0,k})^{-1}$ does not have ω^{-1}-singularity around $\omega = 0$. This will allow us to define capacitance coefficients in the current setting. Since $(\mathcal{S}_D^{\omega\alpha_0,k})^{-1}$ is a holomorphic function of ω, we have

$$(4.69)\qquad \left(\mathcal{S}_D^{\omega\alpha_0,k}\right)^{-1} = \mathcal{S}_0^{\alpha_0} + \omega\mathcal{S}_{-1}^{\alpha_0} + \mathcal{O}(\omega^2) \quad \text{as } \omega \to 0,$$

for some operators $\mathcal{S}_0^{\alpha_0}$ and $\mathcal{S}_{-1}^{\alpha_0}$ which are independent of ω.

For $\alpha \in Y^*$, we let

$$\psi_i^{\alpha,k} := \left(\widehat{\mathcal{S}}_D^{\alpha,k}\right)^{-1}[\chi_{\partial D_i}].$$

Then, if $\alpha = \omega\alpha_0$ for some fixed α_0 with $|\alpha_0| < 1/v$, we obtain from Lemma 4.41 the following expansion:

$$(4.70)\qquad \left(\mathcal{S}_D^{\omega\alpha_0,k}\right)^{-1}[\chi_{\partial D_i}] = \psi_i^0 + \omega\psi_i^{1,\alpha_0} + \mathcal{O}(\omega^2),$$

as $\omega \to 0$, for some $\psi_i^0, \psi_i^{1,\alpha_0} \in L^2(\partial D)$ independent of ω. We can also easily see that $\psi_i^0 = \mathcal{S}_0^{\alpha_0}[\chi_{\partial D_i}]$ and $\psi_i^{1,\alpha_0} = \mathcal{S}_{-1}^{\alpha_0}[\chi_{\partial D_i}]$. On the other hand, from (4.65) and using the Neumann series, we have

$$(4.71)\qquad \left(\widehat{\mathcal{S}}_D^{\omega\alpha_0,k}\right)^{-1}[\chi_{\partial D_i}] = \psi_i^{\omega\alpha_0,k} = \psi_i^0 + \omega\widehat{\psi}_i^{1,\alpha_0} + \mathcal{O}(\omega^2),$$

as $\omega \to 0$, where $\widehat{\psi}_i^{1,\alpha_0} = \psi_i^{1,\alpha_0} + \mathcal{S}_0^{\alpha_0}\mathcal{S}_1^{\alpha_0}[\psi_i^0]$. Furthermore, since $\widehat{\mathcal{S}}_D^{\omega\alpha_0,k}[\psi_i^{\omega\alpha_0,k}]$ is bounded as $\omega \to 0$, the singular part of $\widehat{\mathcal{S}}_D^{\omega\alpha_0,k}$ must vanish on ψ_i^0, or in other words

$$(4.72)\qquad \int_{\partial D} \psi_i^0 \, d\sigma = 0.$$

Definition 4.42 (Periodic capacitance matrix). *For α_0 with $|\alpha_0| < 1/v$, and for a system of $N \in \mathbb{N}$ resonators $D_1, \ldots, D_N$ in Y we can define the periodic capacitance matrix $C^0 = (C_{ij}^0) \in \mathbb{R}^{N\times N}$ to be the square matrix given by*

$$(4.73)\qquad C_{ij}^0 = -\int_{\partial D_i} \underbrace{\mathcal{S}_0^{\alpha_0}[\chi_{\partial D_j}]}_{=\psi_j^0} \, d\sigma, \quad i,j = 1,\ldots,N.$$

Remark 4.43. *The periodic capacitance matrix C^0 may* a priori *depend on α_0, but we will see in Proposition 4.78 that, for a dimer under inversion symmetry, ψ_i^0 and consequently C^0 are independent of α_0.*

Similarly to Lemma 1.13, the periodic capacitance matrix can alternatively be written as

$$(4.74)\qquad C_{ij}^0 = \int_{Y\setminus D} \nabla V_i^0 \cdot \nabla V_j^0 \, dx,$$

where $V_i^0 := \mathcal{S}_D^{0,0}[\psi_i^0]$. Notice that, analogously to (4.66), V_i^0 is the unique solution to the following problem:

$$\begin{cases} \Delta V_i^0 = 0 & \text{in } Y \setminus D, \\ V_i^0 = \delta_{ij} & \text{on } \partial D_j, \\ V_i^0(x_l, x_0) & \text{is } \Lambda\text{-periodic in } x_l, \\ V_i^0(x_l, x_0) \to \pm V_\infty^i & \text{as } x_0 \to \pm\infty, \end{cases}$$

where the constant

$$V_\infty^i = -\frac{1}{2|Y_l|}\int_{\partial D} y_0 \psi_i^0(y)\, \mathrm{d}\sigma(y), i = 1, \ldots, N,$$

and may depend on α_0.

Because of (4.74), C^0 shares many of the properties of the capacitance matrices in previous settings.

LEMMA 4.44. *The periodic capacitance matrix C^0 is a real, symmetric, positive semidefinite matrix with one vanishing eigenvalue.*

PROOF. From (4.74), it follows that C^0 is real, symmetric and positive semidefinite. Furthermore, (4.72) yields

$$\sum_{j=1}^{N} C_{ij}^0 = 0 \quad \text{for all } i. \tag{4.75}$$

Thus 0 is an eigenvalue of C^0 associated with the eigenvector $(1, \ldots, 1)^\top$. Since $\operatorname{Ker} \mathcal{S}_D^{0,0}$ is of dimension one, 0 is a simple eigenvalue. □

The name *periodic* comes from the fact that an α-quasiperiodic function is, in the case $\alpha = 0$, a periodic function. Since $\alpha = \omega\alpha_0$ and ω is small, we are working close to the periodic case. When C^0 is independent of α_0, all the relations involving C^0 are equal to the periodic case $\alpha_0 = 0$.

DEFINITION 4.45 (Generalised periodic capacitance matrix). *For a system of $N \in \mathbb{N}$ resonators $D_1, \ldots, D_N$ in Y, we can define the generalised periodic capacitance matrix, denoted by $\mathcal{C}^0 = (\mathcal{C}_{ij}^0) \in \mathbb{C}^{N\times N}$, to be the square matrix given by*

$$\mathcal{C}_{ij}^0 = \frac{\delta_i v_i^2}{|D_i|} C_{ij}^0, \quad i, j = 1, \ldots, N.$$

Since $\mathcal{S}_D^{\omega\alpha_0,k}$ has a singularity of order ω^{-1} as $\omega \to 0$, we cannot apply the asymptotic perturbation theory of Gohberg and Sigal to study the characteristic value perturbation of $\mathcal{A}^\alpha(\omega, \delta)$, defined in Lemma 4.27, as $\delta \to 0$. Instead, we rephrase the system in the following result [**28**].

LEMMA 4.46. *Let $k = \omega/v$ and assume that $\alpha = \omega\alpha_0$ for some α_0 independent of ω and δ such that $|\alpha_0| < 1/v$. Then the Helmholtz problem (4.2) is equivalent to finding $\eta \in H^1(\partial D)$ such that*

$$\widehat{\mathcal{A}}^\alpha(\omega, \delta)[\eta] = 0, \tag{4.76}$$

where the operator $\widehat{\mathcal{A}}^\alpha(\omega, \delta) : H^1(\partial D) \to L^2(\partial D)$ is defined as

$$\widehat{\mathcal{A}}^\alpha(\omega, \delta) = \left(-\frac{1}{2}I + \widetilde{\mathcal{K}}_D^{\omega,*}\right)\left(\widetilde{\mathcal{S}}_D^{\omega}\right)^{-1} - \widetilde{\delta}\left(\frac{1}{2}I + (\mathcal{K}_D^{-\alpha,k})^*\right)\left(\mathcal{S}_D^{\omega\alpha_0,k}\right)^{-1}, \tag{4.77}$$

with $\widetilde{\mathcal{K}}_D^{\omega,*}$, $\widetilde{\mathcal{S}}_D^{\omega}$, *and* $\widetilde{\delta}$ *being given by* (1.12), (1.10), *and* (1.4).

Now, we can apply the functional analytic approach, outlined in Section 1.3, to derive the following result on the subwavelength resonant frequencies in the current setting [**26, 28**].

THEOREM 4.47. *Let* $d \in \{2,3\}$ *and* $d - d_l = 1$. *Let* $k = \omega/v$ *and assume that* $\alpha = \omega\alpha_0$ *for some* α_0 *independent of* ω *and* δ *such that* $|\alpha_0| < 1/v$. *As* $\delta \to 0$, *there are* N *subwavelength resonant frequencies which satisfy the asymptotic formula*

$$\omega_n^0 = \sqrt{\lambda_n^0} + \mathcal{O}(\delta), \qquad n = 1, \ldots, N,$$

where $\{\lambda_n^0 : n = 1, \ldots, N\}$ *are the eigenvalues of the generalised periodic capacitance matrix* $\mathcal{C}^0 \in \mathbb{C}^{N\times N}$, *which satisfy* $\lambda_n^0 = \mathcal{O}(\delta)$ *as* $\delta \to 0$.

4.4.2.2. *Higher-order approximations.* In the case $\delta_i v_i^2 \in \mathbb{R}$ for all i, the leading-order parts of the resonances, given in Theorem 4.47, are real. It is also interesting to compute the corresponding imaginary parts, which will specify the bandwidth of the resonant behaviour. Then, in view of (4.69), we can define the "higher-order" matrix C^{1,α_0} as

$$C_{ij}^{1,\alpha_0} = -\int_{\partial D_i} \underbrace{\mathcal{S}_{-1}^{\alpha_0}[\chi_{\partial D_j}]}_{=\psi_j^{1,\alpha_0}} \, \mathrm{d}\sigma, \tag{4.78}$$

and corresponding generalised matrix $\mathcal{C}^{1,\alpha_0}$ as

$$\mathcal{C}_{ij}^{1,\alpha_0} = \frac{\delta_i v_i^2}{|D_i|} C_{ij}^{1,\alpha_0}.$$

Therefore, we have the following result, which gives the resonances ω_i^α to a higher-order [**28**].

THEOREM 4.48. *Let* $d \in \{2,3\}$ *and* $d - d_l = 1$. *Let* $k = \omega/v$ *and assume that* $\alpha = \omega\alpha_0$ *for some* α_0 *independent of* ω *and* δ *such that* $|\alpha_0| < 1/v$. *As* $\delta \to 0$, *the* N *subwavelength resonant frequencies satisfy* $\omega_n^\alpha = \widehat{\omega}_n^\alpha + \mathcal{O}(\delta^{3/2})$, *where* $\widehat{\omega}_n^\alpha$, *for* $n = 1, \ldots, N$, *are the roots* $\omega = \widehat{\omega}_n^\alpha$ *of the equation*

$$\det\left(\mathcal{C}^0 + \omega\mathcal{C}^{1,\alpha_0} - \omega^2 I\right) = 0.$$

We define $\mathbf{c}_n \in \mathbb{C}^d$ by

$$\mathbf{c}_n := \int_{\partial D} y\psi_n^0(y) \, \mathrm{d}\sigma(y) = \int_{\partial D} y\mathcal{S}_0^{\alpha_0}[\chi_{\partial D_n}](y) \, \mathrm{d}\sigma(y), \quad n = 1, \ldots, N. \tag{4.79}$$

While the capacitance coefficients can be thought of as total charge (or "mass"), $\mathbf{c}_n$ is the centre of mass (up to rescaling). Briefly put, we can compute $\mathcal{C}^{1,\alpha_0}$ in terms of these coefficients, which allows us to compute explicit expressions of $\widehat{\omega}_n^\alpha$.

4.4.2.3. *Modal decomposition.* The "higher-order" term $\mathcal{S}_{-1}^{\alpha_0}$ in the expansion of the single layer potential enters the expression for the resonant modes in this case. From the arguments used to derive Theorem 4.48, we have the following result on the resonant modes [**28**].

THEOREM 4.49. *Let* $d \in \{2,3\}$ *and* $d - d_l = 1$. *Let* $k = \omega/v$ *and assume that* $\alpha = \omega\alpha_0$ *for some* α_0 *independent of* ω *and* δ *such that* $|\alpha_0| < 1/v$. *Let* $\mathbf{v}_n^0$ *be the* *eigenvector of $\mathcal{C}^0$ associated with the eigenvalue λ_n^0. Then the resonant mode u_n^α associated with the resonant frequency ω_n^α is given, as $\delta \to 0$, by*

$$u_n^\alpha(x) = \begin{cases} \mathbf{v}_n^0 \cdot \mathbf{S}_D^{\alpha,\omega_n^\alpha/v}(x) + \mathcal{O}(\delta^{1/2}), & x \in \mathbb{R}^d \setminus \overline{\mathcal{D}}, \\ \mathbf{v}_n^0 \cdot \mathbf{S}_D^{\alpha,\omega_n^\alpha/v_i}(x) + \mathcal{O}(\delta^{1/2}), & x \in \mathcal{D}_i, \end{cases}$$

where $\mathbf{S}_D^{\alpha,\omega_n^\alpha/v} : \mathbb{R}^d \to \mathbb{C}^N$ is the vector-valued function given by

$$\mathbf{S}_D^{\alpha,\omega_n^\alpha/v}(x) = \begin{pmatrix} \mathcal{S}_D^{\alpha,\omega_n^\alpha/v}[\psi_1^0 + \frac{\omega_n^\alpha}{v}\psi_1^{1,\alpha_0}](x) \\ \vdots \\ \mathcal{S}_D^{\alpha,\omega_n^\alpha/v}[\psi_N^0 + \frac{\omega_n^\alpha}{v}\psi_N^{1,\alpha_0}](x) \end{pmatrix},$$

with $\psi_i^0 := \mathcal{S}_0^{\alpha_0}[\chi_{\partial D_i}]$ and $\psi_i^{1,\alpha_0} := \mathcal{S}_{-1}^{\alpha_0}[\chi_{\partial D_i}]$. The vector-valued function $\mathbf{S}_D^{\alpha,\omega_n^\alpha/v_i}$ is defined analogously.

With these resonant modes at hand, we can easily solve the associated scattering problem

$$\begin{cases} \Delta u^\alpha + k^2 u^\alpha = 0 & \text{in } \mathbb{R}^d \setminus \mathcal{D}, \\ \Delta u^\alpha + k_i^2 u^\alpha = 0 & \text{in } \mathcal{D}_i,\ i = 1,\dots,N, \\ u^\alpha|_+ - u^\alpha|_- = 0 & \text{on } \partial\mathcal{D}, \\ \delta_i \dfrac{\partial u^\alpha}{\partial\nu}\Big|_+ - \dfrac{\partial u^\alpha}{\partial\nu}\Big|_- = 0 & \text{on } \partial\mathcal{D}_i,\ i = 1,\dots,N, \\ u^\alpha(x_l, x_0) \text{ is } \alpha\text{-quasiperiodic in } x_l, \\ u^\alpha - u_{\text{in}} \text{ satisfies } \alpha\text{-quasiperiodic radiation condition as } |x_0| \to +\infty. \end{cases} \tag{4.80}$$

Here, we assume that u_{in} is a plane wave with some wave vector $\mathbf{k}$; $u_{\text{in}}(x) = e^{\mathrm{i}\mathbf{k}\cdot x}$. Furthermore, α is now specified in terms of $\mathbf{k}$ as $\alpha = P_l\mathbf{k}$, where (as before) P_l is the projection to the first d_l coordinates. Then, for small ω we assume that $\mathbf{k} = \omega\mathbf{w}$ for some fixed vector $\mathbf{w}$.

THEOREM 4.50. *Let $d \in \{2,3\}$ and $d - d_l = 1$. Assume that $u_{\text{in}} = e^{\mathrm{i}\mathbf{k}\cdot x}$, where $\mathbf{k} = \omega\mathbf{w}$. Also, assume that $|\omega - \omega_n^0| > K\sqrt{\delta}$ for $n = 1, \dots, N$, for some constant $K > 0$. Then, as $\delta \to 0$, the solution to the scattering problem (4.80) can be written uniformly for x in compact subsets of $\mathbb{R}^d$ as*

$$(u^\alpha - u_{\text{in}})(x) = \sum_{n=1}^N a_n u_n^\alpha(x) - \mathcal{S}_D^{\alpha,k}\left(\mathcal{S}_D^{\alpha,k}\right)^{-1}[u_{\text{in}}](x) + \mathcal{O}(\sqrt{\delta}), \tag{4.81}$$

where V is the matrix of eigenvectors of $\mathcal{C}^0$ and the coefficients $a_n = a_n(\omega)$ satisfy

$$V\begin{pmatrix} \omega^2-(\omega_1^0)^2 & & \\ & \ddots & \\ & & \omega^2-(\omega_N^0)^2 \end{pmatrix}\begin{pmatrix} a_1 \\ \vdots \\ a_N \end{pmatrix} = \begin{pmatrix} \frac{\delta_1 v_1^2}{|D_1|}\int_{\partial D_1}\left(\mathcal{S}_D^{\alpha,k}\right)^{-1}[u_{\text{in}}]\,\mathrm{d}\sigma \\ \vdots \\ \frac{\delta_N v_N^2}{|D_N|}\int_{\partial D_N}\left(\mathcal{S}_D^{\alpha,k}\right)^{-1}[u_{\text{in}}]\,\mathrm{d}\sigma \end{pmatrix}.$$

4.4.2.4. *Connection to quasiperiodic capacitance matrix.* In the previous subsections, we have used the notation C^0 for the periodic capacitance matrix, and C^α for the quasiperiodic capacitance matrix, defined for $\alpha \in Y^* \setminus \{0\}$. This choice of suggestive notation is deliberate, and in this subsection we will show that these capacitance matrices can be combined into a continuous function of $\alpha \in Y^*$, in the sense that $C^0 = \lim C^\alpha$, as $\alpha \to 0$ with $\alpha/|\alpha|$ being fixed.

For $\alpha \in Y^* \setminus \{0\}$, the Green function $G^{\alpha,0}$ satisfies

$$G^{\alpha,0}(x) = -\frac{1}{2|Y_l|} \sum_{q\in\Lambda^*} \frac{e^{\mathrm{i}(\alpha+q)\cdot x} e^{-|\alpha+q||x_0|}}{|\alpha+q|}, \tag{4.82}$$

so that for small α we have

$$G^{\alpha,0}(x) = -\frac{1}{2|\alpha||Y_l|} - \frac{\mathrm{i}\alpha\cdot x}{2|\alpha||Y_l|} + G^{0,0}(x) + \mathcal{O}(|\alpha|). \tag{4.83}$$

In this setting, we define the operator $\widehat{\mathcal{S}}_D^{\alpha,0} : L^2(\partial D) \to H^1(\partial D)$ as

$$\widehat{\mathcal{S}}_D^{\alpha,0}[\varphi](x) = \mathcal{S}_D^{0,0}[\varphi](x) - \frac{1+\mathrm{i}\alpha\cdot x}{2|\alpha||Y_l|}\int_{\partial D}\varphi \,\mathrm{d}\sigma + \int_{\partial D} \frac{\mathrm{i}\alpha\cdot y}{2|\alpha||Y_l|}\varphi(y)\,\mathrm{d}\sigma(y). \tag{4.84}$$

The structure of this operator is entirely analogous to $\widehat{\mathcal{S}}_D^{\alpha,k}$ defined in (4.64) (where, conceptually, the limit $\omega \to 0$ now corresponds to $|\alpha| \to 0$ with $\alpha/|\alpha| = \alpha_0$, for a fixed direction α_0). Therefore, we can apply the same method to obtain the following result [**26**].

LEMMA 4.51. *The periodic capacitance matrix C^0 and the quasiperiodic capacitance matrix C^α, for $\alpha \in Y^* \setminus \{0\}$, satisfy*

$$C^0 = \lim_{\alpha\to 0, \alpha/|\alpha|=\alpha_0} C^\alpha.$$

As will be shown in Proposition 4.78, the limit C^0 is independent from the direction α_0 under the inversion symmetry.

4.5. Subwavelength guiding of waves

The study of point and line defects in bandgap materials has had an immense impact on technological applications. The capability of subwavelength devices to trap and guide waves significantly below the diffraction limit is considered to be the key to the development of the next generation of communication, biomedical, quantum, and sensing technologies [**65**].

As we have seen, bandgaps can be found in any type of wave propagation systems. In this section, we show that defects in a subwavelength bandgap material can be used to enable trapped or guided waves on very small length scales. The results presented in this section are from [**36, 46**].

To fix the setting, we consider the Helmholtz resonance problem in the fully periodic case $d = d_l = 2$. We study the equation (4.2), where D consists of a single circular resonator inside Y. However, instead of considering the periodic crystal $\mathcal{D}$, we detune the size of some resonators, thereby creating a defect. We will consider the two defects illustrated in Figure 4.14, where a single resonator or a line of resonators is detuned. We choose a square lattice with unit cell Y

$$Y = \left[-\frac{1}{2}, \frac{1}{2}\right] \times \left[-\frac{1}{2}, \frac{1}{2}\right],$$

and let D be a circle of radius R and D_d a circle of radius $R+\varepsilon$ for some $-R < \varepsilon < 1-R$. Then, we define the defect crystals

$$\mathcal{D}_{\mathrm{pt}} = \left(\bigcup_{m\in\mathbb{Z}^2\setminus\{(0,0)\}} D+m\right) \cup D_d,$$

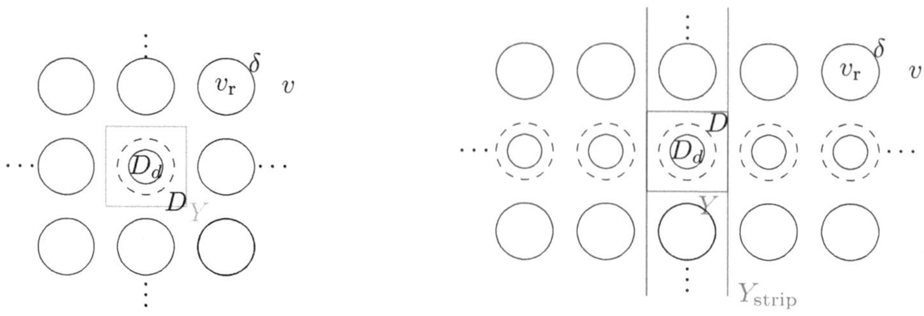

(A) Point defect crystal. (B) Line defect crystal.

FIGURE 4.14. Illustration of the two different defect crystals, and the material parameters, in the case of smaller defect resonator.

and

$$\mathcal{D}_{\text{ln}} = \left(\bigcup_{m_1 \in \mathbb{Z}, m_2 \in \mathbb{Z} \setminus \{0\}} D + (m_1, m_2) \right) \cup \left(\bigcup_{m \in \mathbb{Z} \times \{0\}} D_d + m \right).$$

We will always consider defect structures in relation to a corresponding unperturbed periodic structure, in this case $\mathcal{D}$.

DEFINITION 4.52 (Subwavelength bandgap frequency). *A subwavelength resonant frequency of a resonator structure with a defect is called a subwavelength bandgap frequency if it lies inside a subwavelength bandgap of the unperturbed structure.*

We remark that since we consider the fully periodic case, both the subwavelength bands and the subwavelength bandgap frequencies are real.

4.5.1. Fictitious source method for point defects. We wish to solve (4.2) where the periodic crystal $\mathcal{D}$ is replaced by $\mathcal{D}_{\text{pt}}$ or $\mathcal{D}_{\text{ln}}$. We first observe that $\mathcal{D}_{\text{ln}}$ is periodic in the x_1-direction, so the corresponding problem can be reduced to the quasiperiodic problem in the strip $Y_{\text{strip}} = \left[-\frac{1}{2}, \frac{1}{2}\right] \times \mathbb{R}$ by Floquet-Bloch theory. This makes the analysis of the two problems ($\mathcal{D}_{\text{pt}}$ on $\mathbb{R}^2$ or $\mathcal{D}_{\text{ln}}$ on Y_{strip}) conceptually very similar. We will outline the method of *fictitious sources* in the case of a point defect. For $d = d_l = 2$, we are studying the problem

$$\begin{cases} \Delta u + k^2 u = 0 & \text{in } \mathbb{R}^2 \setminus \mathcal{D}_{\text{pt}}, \\ \Delta u + k_r^2 u = 0 & \text{in } \mathcal{D}_{\text{pt}}, \\ u|_+ - u|_- = 0 & \text{on } \partial \mathcal{D}_{\text{pt}}, \\ \delta \left.\dfrac{\partial u}{\partial \nu}\right|_+ - \left.\dfrac{\partial u}{\partial \nu}\right|_- = 0 & \text{on } \partial \mathcal{D}_{\text{pt}}, \end{cases} \tag{4.85}$$

where $k_r = \omega / v_{\text{r}}$ and v_{r} is the wave speed inside the resonators. The idea is to replace the defected resonator with an unperturbed resonator, along with fictitious sources which are designed to make the new problem equivalent to the original.

More precisely, we consider the problem

$$(4.86)\qquad \begin{cases} \Delta\widetilde{u} + k^2\widetilde{u} = 0 & \text{in } \mathbb{R}^2\backslash\mathcal{D}, \\ \Delta\widetilde{u} + k_r^2\widetilde{u} = 0 & \text{in } \mathcal{D}, \\ \widetilde{u}|_+ - \widetilde{u}|_- = f\chi_{\partial D} & \text{on } \partial\mathcal{D}, \\ \delta\dfrac{\partial\widetilde{u}}{\partial\nu}\Big|_+ - \dfrac{\partial\widetilde{u}}{\partial\nu}\Big|_- = g\chi_{\partial D} & \text{on } \partial\mathcal{D}, \end{cases}$$

where f, g are the source terms and $\chi_{\partial D}$ is, as usual, the characteristic function of ∂D. Note that (4.86) is posed on the periodic geometry $\mathcal{D}$, and that the nonzero sources f and g are present only on the boundary of the central resonator D. Note also that since we are in the fully periodic case ($d = d_l$), we do not assume any radiation condition in (4.85) and (4.86). We seek solutions u of (4.85) and $\widetilde{u}$ of (4.86) for ω inside the subwavelength bandgap of the periodic problem. Inside the unit cell Y, we can represent the solution $\widetilde{u}$ as

$$(4.87)\qquad \widetilde{u} = \begin{cases} H + \mathcal{S}_D^k[\phi] & \text{in } Y \setminus \overline{D}, \\ \mathcal{S}_D^{k_r}[\psi] & \text{in } D, \end{cases}$$

where H satisfies the homogeneous equation $\Delta H + k^2 H = 0$ in Y. Using this approach, we can explicitly compute f, g in terms of ϕ, ψ so that $\widetilde{u}$ coincides with u in $Y \setminus (D \cap D_d)$.

LEMMA 4.53. *The density pair* (ϕ, ψ) *and the effective sources* (f, g) *satisfy the relation*

$$(4.88)\qquad (\mathcal{A}^\varepsilon - \mathcal{A})\begin{pmatrix}\psi\\ \phi\end{pmatrix} = \begin{pmatrix} f \\ g\end{pmatrix},$$

where $\mathcal{A}^\varepsilon$ *is defined as*

$$(4.89)\qquad \mathcal{A}^\varepsilon := (\mathcal{P}_2)^{-1}\mathcal{A}_{D_d}\mathcal{P}_1.$$

Here, $\mathcal{A}$ *and* $\mathcal{A}_{D_d}$ *are the operators defined in Lemma 1.2 for the domains* D *and* D_d*, respectively, and the operators* $\mathcal{P}_1 : L^2(\partial D) \times L^2(\partial D) \to L^2(\partial D_d) \times L^2(\partial D_d)$ *and* $\mathcal{P}_2 : L^2(\partial D) \times L^2(\partial D) \to L^2(\partial D_d) \times L^2(\partial D_d)$ *are defined by*

$$\mathcal{P}_1\begin{pmatrix} e^{in\theta} \\ e^{im\theta}\end{pmatrix} = \delta_{mn}\frac{R}{R_d}\begin{pmatrix} \dfrac{H_n^{(1)}(k_r R)}{H_n^{(1)}(k_r R_d)} e^{in\theta} \\ \dfrac{J_n(kR)}{J_n(kR_d)} e^{in\theta} \end{pmatrix}, \quad \mathcal{P}_2\begin{pmatrix} e^{in\theta} \\ e^{im\theta}\end{pmatrix} = \delta_{mn}\begin{pmatrix} \dfrac{J_n(kR_d)}{J_n(kR)} e^{in\theta} \\ \dfrac{J_n'(kR_d)}{J_n'(kR)} e^{in\theta} \end{pmatrix}.$$

Here, the Bessel function J_n and the Hankel function $H_n^{(1)}$ are defined in Appendix B.2.1.

PROOF. We first observe that the densities ϕ, ψ in (4.87) satisfy

$$(4.90)\qquad \mathcal{A}\begin{pmatrix}\psi\\ \phi\end{pmatrix} := \begin{pmatrix} \mathcal{S}_D^{k_r} & -\mathcal{S}_D^k \\ \dfrac{\partial \mathcal{S}_D^{k_r}}{\partial\nu}\Big|_- & -\delta\dfrac{\partial\mathcal{S}_D^k}{\partial\nu}\Big|_+ \end{pmatrix}\begin{pmatrix}\psi\\ \phi\end{pmatrix} = \begin{pmatrix} H|_{\partial D} - f \\ \dfrac{\partial H}{\partial\nu}\Big|_{\partial D} - g\end{pmatrix}.$$

Then, analogously to (4.87), we represent the solution u of (4.86) inside the unit cell Y as

$$u = \begin{cases} H + \mathcal{S}_D^k[\phi_d] & \text{in } Y \setminus \overline{D_d}, \\ \mathcal{S}_D^{k_r}[\psi_d] & \text{in } D_d, \end{cases}$$

where the densities ϕ_d, ψ_d satisfy

$$\mathcal{A}_{D_d}\begin{pmatrix}\psi_d\\ \phi_d\end{pmatrix} := \begin{pmatrix} \mathcal{S}_{D_d}^{k_r} & -\mathcal{S}_{D_d}^{k} \\ \left.\frac{\partial \mathcal{S}_{D_d}^{k_r}}{\partial \nu}\right|_- & -\delta \left.\frac{\partial \mathcal{S}_{D_d}^{k}}{\partial \nu}\right|_+ \end{pmatrix}\begin{pmatrix}\psi_d\\ \phi_d\end{pmatrix} = \begin{pmatrix} H|_{\partial D_d} \\ \left.\frac{\partial H}{\partial \nu}\right|_{\partial D_d}\end{pmatrix}. \tag{4.91}$$

Introduce the notation

$$D_l = \begin{cases} D_d & \text{if } \varepsilon > 0, \\ D & \text{if } \varepsilon < 0, \end{cases} \qquad D_s = \begin{cases} D & \text{if } \varepsilon > 0, \\ D_d & \text{if } \varepsilon < 0. \end{cases}$$

Thus, D_l and D_s denote the largest and smallest of D, D_d, respectively. The condition $\widetilde{u} = u$ in $Y \setminus (D \cap D_d)$ implies that

$$\mathcal{S}_{D_d}^{k_r}[\psi_d] \equiv \mathcal{S}_D^{k_r}[\psi] \quad \text{in } D_s, \tag{4.92}$$

and

$$\mathcal{S}_{D_d}^{k}[\phi_d] \equiv \mathcal{S}_D^{k}[\phi] \quad \text{in } Y \setminus \overline{D_l}. \tag{4.93}$$

Since D and D_d are circular disks, we can use a Fourier basis for functions on ∂D or ∂D_d in polar coordinates (r, θ) to make (4.90)–(4.93) more explicit. Let us write ϕ and ψ in the form of a Fourier series:

$$\phi(\theta) = \sum_{n\in\mathbb{Z}} \phi_n e^{\mathrm{i}n\theta}, \quad \psi(\theta) = \sum_{n\in\mathbb{Z}} \psi_n e^{\mathrm{i}n\theta}.$$

Similarly, we also write ϕ_d and ψ_d as

$$\phi_d(\theta) = \sum_{n\in\mathbb{Z}} \phi_{d,n} e^{\mathrm{i}n\theta}, \quad \psi_d(\theta) = \sum_{n\in\mathbb{Z}} \psi_{d,n} e^{\mathrm{i}n\theta}.$$

We define the subspace V_{mn} of $L^2(\partial D) \times L^2(\partial D)$ as

$$V_{mn} := \mathrm{span}\{e^{\mathrm{i}m\theta}\} \times \mathrm{span}\{e^{\mathrm{i}n\theta}\}, \quad m, n \in \mathbb{Z}.$$

Similarly, let V_{mn}^d be the subspace of $L^2(\partial D_d) \times L^2(\partial D_d)$ with the same Fourier basis.

From (B.133), it follows that the operator $\mathcal{A}$ in (4.90) has the following matrix representation as an operator from V_{mn} to $V_{m'n'}$:

$$(\mathcal{A})_{V_{mn}\to V_{m'n'}} = \delta_{mn}\delta_{m'n'}\frac{-\mathrm{i}\pi R}{2}\begin{pmatrix} J_n(k_rR)H_n^{(1)}(k_rR) & -J_n(kR)H_n^{(1)}(kR) \\ k_rJ_n'(k_rR)H_n^{(1)}(k_rR) & -\delta kJ_n(kR)\big(H_n^{(1)}\big)'(kR)\end{pmatrix}. \tag{4.94}$$

Similarly, the operator $\mathcal{A}_{D_d}$ in (4.91), as a mapping from V_{mn}^d to $V_{m'n'}^d$, is represented as follows:

$$\begin{aligned}(\mathcal{A}_{D_d})_{V_{mn}^d\to V_{m'n'}^d} &= \delta_{mn}\delta_{m'n'}\frac{-\mathrm{i}\pi R_d}{2} \\ &\times \begin{pmatrix} J_n(k_rR_d)H_n^{(1)}(k_rR_d) & -J_n(kR_d)H_n^{(1)}(kR_d) \\ k_rJ_n'(k_rR_d)H_n^{(1)}(k_rR_d) & -\delta kJ_n(kR_d)\big(H_n^{(1)}\big)'(kR_d)\end{pmatrix}.\end{aligned} \tag{4.95}$$

Now, we consider (4.92) and (4.93). We have from (B.133) that, inside D_s,

$$\begin{aligned}
\mathcal{S}_D^{k_r}[\psi] &= \sum_{n\in\mathbb{Z}} \psi_n \mathcal{S}_D^{k_r}[e^{\mathrm{i}n\theta}] = \frac{-\mathrm{i}\pi R}{2}\sum_{n\in\mathbb{Z}} \psi_n H_n^{(1)}(k_r R) J_n(k_r r) e^{\mathrm{i}n\theta} \\
&= \frac{-\mathrm{i}\pi R}{2}\sum_{n\in\mathbb{Z}} \psi_n \frac{H_n^{(1)}(k_r R)}{H_n^{(1)}(k_r R_d)} H_n^{(1)}(k_r R) J_n(k_r r) e^{\mathrm{i}n\theta} \\
&= \mathcal{S}_{D_d}^{k_r}\left[\sum_{n\in\mathbb{Z}} \psi_n \frac{R}{R_d}\frac{H_n^{(1)}(k_r R)}{H_n^{(1)}(k_r R_d)} e^{\mathrm{i}n\theta}\right].
\end{aligned}$$

Similarly, outside D_l,

$$\mathcal{S}_D^k[\phi] = \mathcal{S}_{D_d}^k\left[\sum_{n\in\mathbb{Z}} \phi_n \frac{R}{R_d}\frac{J_n(kR)}{J_n(kR_d)} e^{\mathrm{i}n\theta}\right].$$

So, from (4.92) and (4.93), we see that

$$\begin{pmatrix}\psi_d\\ \phi_d\end{pmatrix} = \mathcal{P}_1\begin{pmatrix}\psi\\ \phi\end{pmatrix},$$

where the operator $\mathcal{P}_1 : L^2(\partial D)\times L^2(\partial D) \to L^2(\partial D_d)\times L^2(\partial D_d)$ is given by

$$(\mathcal{P}_1)_{V_{mn}\to V^d_{m'n'}} = \delta_{mn}\delta_{m'n'}\frac{R}{R_d}\begin{pmatrix}\dfrac{H_n^{(1)}(k_r R)}{H_n^{(1)}(k_r R_d)} & 0\\ 0 & \dfrac{J_n(kR)}{J_n(kR_d)}\end{pmatrix}.$$

In the same way, we can obtain that

$$\begin{pmatrix} H|_{\partial D_d}\\ \frac{\partial H}{\partial \nu}\big|_{\partial D_d}\end{pmatrix} = \mathcal{P}_2\begin{pmatrix} H|_{\partial D}\\ \frac{\partial H}{\partial \nu}\big|_{\partial D}\end{pmatrix},$$

where the operator $\mathcal{P}_2 : L^2(\partial D)\times L^2(\partial D) \to L^2(\partial D_d)\times L^2(\partial D_d)$ is given by

$$(\mathcal{P}_2)_{V_n\to V^d_m} = \delta_{mn}\begin{pmatrix}\frac{J_n(kR_d)}{J_n(kR)} & 0\\ 0 & \dfrac{J_n'(kR_d)}{J_n'(kR)}\end{pmatrix}.$$

Therefore, using (4.91), we arrive as desired at (4.89), which yields (4.88). □

As in previous settings, we can use an integral equation formulation of the problem. However, in the defect crystal setting, the formulation is slightly different from that previously.

LEMMA 4.54. *The subwavelength bandgap frequencies of* (4.85) *are precisely the characteristic values* $\omega = \omega^\varepsilon(\delta)$ *of the operator-valued function*

$$\omega \mapsto \mathcal{M}^\varepsilon(\omega,\delta) := I + \frac{1}{(2\pi)^2}\big(\mathcal{A}^\varepsilon(\omega,\delta) - \mathcal{A}(\omega,\delta)\big)\int_{Y^*}\mathcal{A}^\alpha(\omega,\delta)^{-1}\,\mathrm{d}\alpha \tag{4.96}$$

inside the subwavelength bandgap of $\mathcal{D}$, *such that* $\omega^\varepsilon \to 0$ *as* $\delta\to 0$. *Here,* $\mathcal{A}^\alpha$ *is the operator defined in Lemma 4.27, which is invertible for small enough* δ *and for* ω *inside the bandgap.*

PROOF. Applying the Floquet transform to the solution $\widetilde{u}$ of (4.86) shows that

$$\widetilde{u}^\alpha := \sum_{m\in\mathbb{Z}^2} \widetilde{u}(x-m)e^{\mathrm{i}\alpha\cdot m}$$

satisfies

$$\begin{cases} \Delta\widetilde{u}^\alpha + k^2\widetilde{u}^\alpha = 0 & \text{in } Y\backslash\overline{D}, \\ \Delta\widetilde{u}^\alpha + k_r^2\widetilde{u}^\alpha = 0 & \text{in } D, \\ \widetilde{u}^\alpha|_+ - \widetilde{u}^\alpha|_- = f & \text{on } \partial D, \\ \delta \dfrac{\partial \widetilde{u}^\alpha}{\partial \nu}\bigg|_+ - \dfrac{\partial \widetilde{u}^\alpha}{\partial \nu}\bigg|_- = g & \text{on } \partial D, \\ e^{-\mathrm{i}\alpha\cdot x}\widetilde{u}^\alpha(x) & \text{is periodic in } x. \end{cases} \tag{4.97}$$

Then $\widetilde{u}^\alpha$ can be represented using quasiperiodic layer potentials as

$$\widetilde{u}^\alpha = \begin{cases} \mathcal{S}_D^{\alpha,k}[\phi^\alpha] & \text{in } Y\setminus\overline{D}, \\ \mathcal{S}_D^{k_r}[\psi^\alpha] & \text{in } D, \end{cases} \tag{4.98}$$

where the pair $(\phi^\alpha, \psi^\alpha) \in L^2(\partial D)\times L^2(\partial D)$ is the solution to

$$\mathcal{A}^\alpha(\omega,\delta)\begin{pmatrix}\psi^\alpha\\ \phi^\alpha\end{pmatrix} = -\begin{pmatrix} f \\ g\end{pmatrix}.$$

Since $\mathcal{A}^\alpha$ is invertible for small enough δ and for ω inside the subwavelength bandgap, we have

$$\begin{pmatrix}\psi^\alpha\\ \phi^\alpha\end{pmatrix} = -\mathcal{A}^\alpha(\omega,\delta)^{-1}\begin{pmatrix} f \\ g\end{pmatrix}.$$

The solution $\widetilde{u}$ of (4.86) can be recovered using the inversion formula

$$\widetilde{u}(x) = \frac{1}{(2\pi)^2}\int_{Y^*} \widetilde{u}^\alpha(x)\, \mathrm{d}\alpha.$$

Then, for $x\in D$, it follows that

$$\widetilde{u}(x) = \mathcal{S}_D^{k_r}\left[\frac{1}{(2\pi)^2}\int_{Y^*}\psi^\alpha(x)\,\mathrm{d}\alpha\right],$$

and for $x \in Y\setminus\overline{D}$,

$$\begin{aligned} \widetilde{u}(x) &= \mathcal{S}_D^{k_r}\left[\frac{1}{(2\pi)^2}\int_{Y^*}\phi^\alpha(x)\,\mathrm{d}\alpha\right] \\ &\quad + \frac{1}{(2\pi)^2}\int_{Y^*}\sum_{m\in\mathbb{Z}^2, m\neq 0} \mathcal{S}_D^k[\phi^\alpha](x-m)e^{\mathrm{i}\alpha\cdot m}\,\mathrm{d}\alpha. \end{aligned} \tag{4.99}$$

Note that the second term in the right-hand side of (4.99) satisfies the homogeneous Helmholtz equation with wave number k in Y. So, in view of (4.87), we can identify ϕ, ψ, and H as follows:

$$\phi(x) = \frac{1}{(2\pi)^2}\int_{Y^*}\phi^\alpha(x)\,\mathrm{d}\alpha, \quad \psi(x) = \frac{1}{(2\pi)^2}\int_{Y^*}\psi^\alpha(x)\,\mathrm{d}\alpha,$$

and

$$H(x) = \frac{1}{(2\pi)^2}\int_{Y^*}\sum_{m\in\mathbb{Z}^2, m\neq 0}\mathcal{S}_D^k[\phi^\alpha](x-m)e^{\mathrm{i}\alpha\cdot m}\,\mathrm{d}\alpha.$$

Therefore, from (4.88) it follows that

$$\begin{pmatrix}\psi\\ \phi\end{pmatrix} = -\frac{1}{(2\pi)^2}\left(\mathcal{A}^\varepsilon(\omega,\delta) - \mathcal{A}(\omega,\delta)\right)\left(\int_{Y^*} \mathcal{A}^\alpha(\omega,\delta)^{-1}\,\mathrm{d}\alpha\right)\begin{pmatrix}\psi\\ \phi\end{pmatrix}.$$

□

REMARK 4.55. *The formulation in Lemma 4.54 is in terms of the densities ϕ and ψ. An equivalent formulation in terms of the effective sources f and g instead of the densities ϕ and ψ yields a characterisation of the subwavelength bandgap frequencies as characteristic values of the operator-valued function*

$$\omega \mapsto I + \frac{1}{(2\pi)^2}\left(\int_{Y^*} \mathcal{A}^\alpha(\omega,\delta)^{-1}\,\mathrm{d}\alpha\right)\left(\mathcal{A}^\varepsilon(\omega,\delta) - \mathcal{A}(\omega,\delta)\right).$$

We let $\omega_1^* = \omega_1^*(\delta)$ be the maximum of the first band:

$$\omega_1^*(\delta) = \max_{\alpha\in Y^*} \omega_1^\alpha(\delta).$$

The following theorem from [**36**] demonstrates the existence of bandgap frequencies in the point defect crystal. We refer to Appendix C.4 for its proof.

THEOREM 4.56. *Assume that δ and ε are small enough and the pair (R, ε) satisfies either*

(i) *R is small enough and $\varepsilon < 0$ (dilute regime); or*
(ii) *R is close enough to $1/2$ and $\varepsilon > 0$ (nondilute regime).*

Then there exists a subwavelength bandgap frequency ω^ε of (4.85). *In each case, we have the asymptotic expansion*

$$\omega^\varepsilon - \omega_1^* = \exp\left(-\frac{\mu}{\delta\varepsilon} + \mathcal{O}\left(\frac{1}{|\varepsilon \ln \delta|} + 1\right)\right),$$

when ε and δ go to zero, for some constant μ. In both cases, $\mu/\varepsilon > 0$.

REMARK 4.57. *An explicit expression for μ is given in Appendix* C.4. *Figure* C.1 *illustrates the dilute and nondilute regimes.*

REMARK 4.58. *In the reverse cases, i.e., if R is close to $1/2$ and $\varepsilon < 0$ or if R is small and $\varepsilon > 0$, there will be no bandgap frequencies. In these cases, the resonance frequency of the defected resonator is shifted downwards and will therefore lie outside of the subwavelength bandgap.*

In Figure 4.15, we provide a numerical example showing the existence of a subwavelength localised mode. We show that the subwavelength bandgap frequency ω^ε is above the maximum frequency ω_1^*, which is attained at the symmetry point M.

4.5.2. Guided waves in line defects. We let $\mathcal{D}_{\mathrm{ln}}$ be the line defect crystal (see Figure 4.14b) and study the problem

$$\begin{cases} \Delta u + k^2 u = 0 & \text{in } \mathbb{R}^2\backslash\mathcal{D}_{\mathrm{ln}},\\ \Delta u + k_r^2 u = 0 & \text{in } \mathcal{D}_{\mathrm{ln}},\\ u|_+ - u|_- = 0 & \text{on } \partial\mathcal{D}_{\mathrm{ln}},\\ \delta \left.\frac{\partial u}{\partial \nu}\right|_+ - \left.\frac{\partial u}{\partial \nu}\right|_- = 0 & \text{on } \partial\mathcal{D}_{\mathrm{ln}},\\ u(x_1,x_2)e^{-\mathrm{i}\alpha_1 x_1} & \text{is periodic in } x_1, \end{cases} \tag{4.100}$$

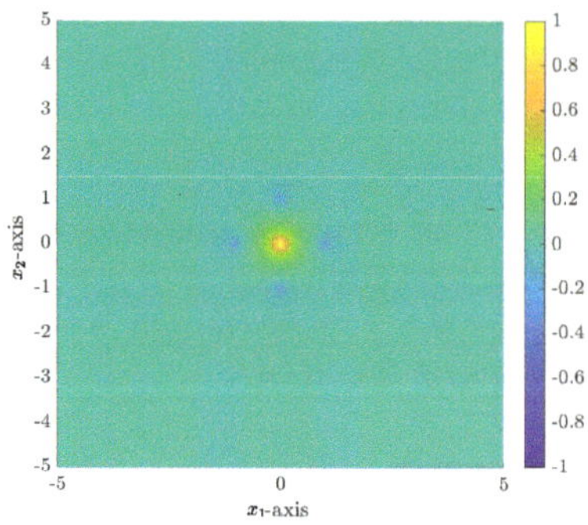

(A) Localised mode.

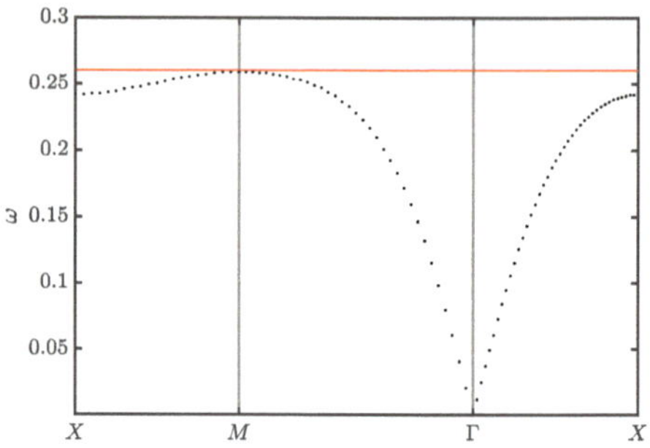

(B) Subwavelength bulk band (dotted black) and bandgap frequency (solid red).

FIGURE 4.15. Localised mode on subwavelength scales (a) and corresponding bandgap frequency (b). These point defect structures can provide a good degree of localisation. However, as is apparent from (b), the bandgap frequency is exponentially close to the edge of the bulk band and is therefore sensitive to imperfections of the structure.

where $\alpha = (\alpha_1, \alpha_2)$. In this case, since the defect is periodic in the x_1-direction, we can reduce (4.100) to a Helmholtz resonance problem posed on the infinite strip Y_{strip} shown in Figure 4.14b and have a very similar characterisation of the bandgap frequencies to the one given in Lemma 4.54.

The following lemma holds by exactly the same arguments as those for proving Lemma 4.54.

LEMMA 4.59. *The subwavelength bandgap frequencies of* (4.85) *are precisely the characteristic values* $\omega = \omega^\varepsilon(\delta, \alpha_1)$ *of the operator-valued function*

(4.101)
$$\omega \mapsto \mathcal{M}^{\varepsilon,\alpha_1}(\omega,\delta) := I + \frac{1}{2\pi}\big(\mathcal{A}^\varepsilon(\omega,\delta) - \mathcal{A}(\omega,\delta)\big)\left(\int_{-\pi}^{\pi} \mathcal{A}^{(\alpha_1,\alpha_2)}(\omega,\delta)^{-1}\, \mathrm{d}\alpha_2\right)$$

inside the bandgap of $\mathcal{D}$, *such that* $\omega^\varepsilon \to 0$ *as* $\delta \to 0$.

We now let $\omega_1^{\alpha_1,*}$ be the maximum of the first band over α_2:

$$\omega_1^{\alpha_1,*}(\delta) = \max_{\alpha_2 \in [-\pi,\pi]} \omega_1^{(\alpha_1,\alpha_2)}(\delta).$$

Then we have the following result from [**46**].

THEOREM 4.60. *Assume that* δ *and* ε *are small enough and the pair* (R, ε) *satisfies either*

(i) R *is small enough and* $\varepsilon < 0$ *(dilute regime); or*
(ii) R *is close enough to* $1/2$ *and* $\varepsilon > 0$ *(nondilute regime).*

Then there exists a subwavelength resonant frequency ω^ε *of* (4.100) *satisfying*

$$\omega^\varepsilon > \omega_1^{\alpha_1,*}.$$

Moreover, as $\delta, \varepsilon \to 0$, *we have*

$$\omega^\varepsilon(\delta,\alpha_1) = \omega_1^{\alpha_1,*}(\delta) + \mu(\alpha_1)\sqrt{\delta}\varepsilon^2 + \mathcal{O}\left(\varepsilon^2\sqrt{\delta}\left(\frac{1}{|\ln\delta|} + |\varepsilon|\right)\right) \quad (4.102)$$

for some $\mu = \mu(\alpha_1) > 0$ which is independent of ε and δ.

We call ω^ε, viewed as a function of α_1, a *defect band.* We emphasise that parts of the defect band might not correspond to bandgap frequencies in the sense of Definition 4.52. From Theorem 4.60 we know that $\omega^\varepsilon(\alpha_1) > \omega_1^{\alpha_1,*}$ while in order for ω^ε to lie in the bandgap of $\mathcal{D}$ we need $\omega^\varepsilon(\alpha_1) > \omega_1^*$ where, as before,

$$\omega_1^* = \max_{\alpha_1 \in [-\pi,\pi]} \omega_1^{\alpha_1,*}.$$

In order for the whole defect band to lie in the subwavelength bandgap, we need a sufficiently large ε. Since Theorem 4.60 is based on asymptotically expanding $\mathcal{M}^{\varepsilon,\alpha_1}$ for small ε, a different analysis is needed to handle this case. Instead of the formulation given in Lemma 4.54 which is useful in the case of small ε, we use the one given in Remark 4.55. The strength of Remark 4.55 is that it is valid for fixed ε but R small.

The following result is based on asymptotics in the dilute regime (*i.e.*, $R \to 0$) and is valid even for ε with a large magnitude [**46**]. Its proof uses an approach similar to that for showing Theorem 4.56, which is based on the asymptotic Gohberg-Sigal theory.

THEOREM 4.61. *For δ and R small enough, and for fixed $\varepsilon \in (-R, 0)$, there exists a unique subwavelength resonant frequency ω^ε of (4.100) satisfying $\omega^\varepsilon > \omega_1^{\alpha_1,*}$. For $\alpha_1 \neq 0$,*

$$\omega^\varepsilon(\alpha_1) = \widehat{\omega} + \mathcal{O}\left(R^2 + \delta\right),$$

where $\widehat{\omega}$ is the root of the following equation:

$$1 + \frac{1}{2\pi}\left(\frac{\widehat{\omega}^2 R^2}{2\delta}\ln\frac{R}{R_d} + \left(1 - \frac{R^2}{R_d^2}\right)\right)\int_{-\pi}^{\pi} \frac{(\omega^\alpha)^2}{\widehat{\omega}^2 - (\omega^\alpha)^2}\, d\alpha_2 = 0. \tag{4.103}$$

For δ and R small enough, and for fixed $\varepsilon \in [0, 1-R)$, there are no resonant frequencies satisfying $\omega^\varepsilon > \omega_1^{\alpha_1,*}$.

PROPOSITION 4.62. *For R and δ small enough, there exists an $\varepsilon_0 > 0$ such that for any $\varepsilon \in (-R, -\varepsilon_0)$ we have*

$$\omega^\varepsilon(\alpha_1) > \omega_1^*$$

for all $\alpha_1 \in [-\pi, \pi]$.

PROOF. For $\alpha = (\alpha_1, \alpha_2) \in Y^*$, let

$$I(\omega, \alpha_1) := \frac{1}{2\pi}\int_{-\pi}^{\pi} \frac{(\omega_1^\alpha)^2}{\omega^2 - (\omega_1^\alpha)^2}\, d\alpha_2.$$

From Theorem 4.7, we know that

$$\omega_1^* := \max_{\alpha \in Y^*} \omega_1^\alpha$$

is attained at $\alpha = \alpha^* = (\pi, \pi)$. Using Theorem 4.61, we conclude that if $\omega^\varepsilon(\alpha_1)$ is equal to ω_1^* for some $\alpha_1 \neq 0$, this α_1 shall satisfy the equation

$$\left(\frac{R^2}{R_d^2} - \frac{\ln R_d}{\ln R}\right) = \frac{1}{I(\omega_1^*, \alpha_1)} + \mathcal{O}\left(\sqrt{\delta} + R\right).$$

For small enough R and δ, the right-hand side is positive, while the left-hand side ranges from 0 to $+\infty$ for $\varepsilon \in (-R, 0)$. Moreover, $I(\omega, \alpha_1)$ is continuous for α_1 in a neighbourhood of 0. Hence, for each $\alpha_1 \in [-\pi, \pi]$, we can find a solution $\widetilde{\varepsilon}(\alpha_1)$ to this equation. Define

$$\varepsilon_0 = \max_{\alpha_1 \in [-\pi,\pi]} \widetilde{\varepsilon}(\alpha_1).$$

With the choice $\varepsilon = \varepsilon_0$, for every fixed $\alpha_1 \neq 0$, $\omega^\varepsilon(\alpha_1)$ is a bandgap frequency for R and δ small enough. Moreover, since $\omega^\varepsilon(\alpha_1)$ is continuous, by choosing $\varepsilon > \varepsilon_0$ the statement holds for any $\alpha_1 \in [-\pi, \pi]$. □

In order for the line defect crystal to be useful as a waveguide, we need the localised modes to propagate along the defect line. In other words, we must exclude the case of *bound modes*, which are modes that are localised along the direction of the line. If there is such a mode u, corresponding to a frequency ω, we can apply the Floquet transform so that $u^\alpha(x) := \mathcal{U}[u](x, \alpha)$ with $\mathcal{U}$ being the Floquet transform solves (4.85) for any α. The corresponding band function attains the same value ω for any α, so we conclude that *bound modes are associated to flat band functions.* The next result from [**46**] shows that the defect modes in our case are not bound along the defect line.

PROPOSITION 4.63. *For δ and R small enough, and for $\alpha_1 \notin \{0, \pi\}$, the subwavelength resonant frequency $\omega^\varepsilon = \omega^\varepsilon(\alpha_1)$ satisfies*

$$\frac{\partial \omega^\varepsilon}{\partial \alpha_1} \neq 0.$$

PROOF. To simplify the computations, we introduce the following notation:

$$a = \frac{R^2}{4\pi\delta} \ln \frac{R}{R_d}, \qquad b = \frac{1}{2\pi}\left(1 - \frac{R^2}{R_d^2}\right),$$
$$x = x(\alpha_1) = \hat{\omega}^2, \qquad y = y(\alpha_1, \alpha_2) = (\omega^\alpha)^2.$$

Then equation (4.103) reads

$$(ax + b) \int_{-\pi}^{\pi} \frac{y}{x - y} \, \mathrm{d}\alpha_2 = 1.$$

Denote $x' = \frac{\partial x}{\partial \alpha_1}$ and $y' = \frac{\partial y}{\partial \alpha_1}$. Then

$$ax' \int_{-\pi}^{\pi} \frac{y}{x - y} \, \mathrm{d}\alpha_2 - (ax + b) \int_{-\pi}^{\pi} \frac{x'y - xy'}{(x - y)^2} \, \mathrm{d}\alpha_2 = 0,$$

or equivalently,

$$x'A + B = 0,$$

where

$$A = a \int_{-\pi}^{\pi} \frac{y}{x - y} \, \mathrm{d}\alpha_2 - (ax + b) \int_{-\pi}^{\pi} \frac{y}{(x - y)^2} \, \mathrm{d}\alpha_2,$$
$$B = (ax + b)x \int_{-\pi}^{\pi} \frac{y'}{(x - y)^2} \, \mathrm{d}\alpha_2.$$

First, we show that $A \neq 0$, which implies that the zeros of x' coincide with the zeros of B. We have

$$\begin{aligned} A &= a \int_{-\pi}^{\pi} \frac{y}{x-y} \, d\alpha_2 - (ax+b) \int_{-\pi}^{\pi} \frac{y}{(x-y)^2} \, d\alpha_2 \\ &= \int_{-\pi}^{\pi} \frac{ay(x-y) - (ax+b)y}{(x-y)^2} \, d\alpha_2, \\ &= - \int_{-\pi}^{\pi} \frac{y(ay+b)}{(x-y)^2} \, d\alpha_2 < 0, \end{aligned}$$

since $y(ay+b) > 0$ for all $(\alpha_1, \alpha_2) \in Y^*$.

Next, we show that the leading-order of B vanishes exactly at the points $\alpha_1 = 0$ and $\alpha = \pi$. Using (4.11) and (B.119), we have

$$\begin{aligned} y' &= \frac{\partial}{\partial \alpha_1} (\omega^\alpha)^2 \\ &= \frac{\partial}{\partial \alpha_1} \left(\frac{-2\delta}{R^2 \ln R + 2\pi R^3 \mathcal{R}_\alpha(0)} \right) + \mathcal{O}\left(\frac{R^3}{\ln R} + \delta^2 \right) \\ &= \frac{4\pi R^3 \delta}{(R^2 \ln R + 2\pi R^3 \mathcal{R}_\alpha(0))^2} \frac{\partial}{\partial \alpha_1} \mathcal{R}_\alpha(0) + \mathcal{O}\left(\frac{R^3}{\ln R} + \delta^2 \right), \end{aligned}$$

where $\mathcal{R}_\alpha := G^{\alpha,0} - G^0$. Using Lemma B.30 we conclude that for δ and R small enough and $\alpha_1 \neq 0, \pi$, y' is nonzero for any α_2. Hence, B is nonzero, which concludes the proof. □

The integral representation for the localised and guided waves (Lemmas 4.54 and 4.59) can be discretised using the multipole expansion method, which provides an efficient method to compute the bandgap frequency and the corresponding localised mode (see Figures 4.15 and 4.16).

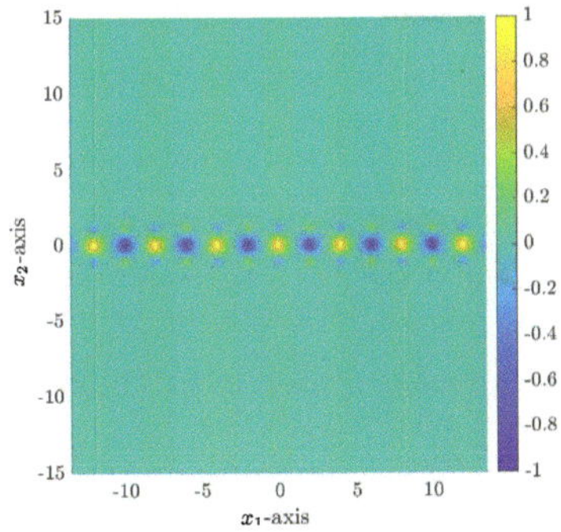

(A) Guided mode.

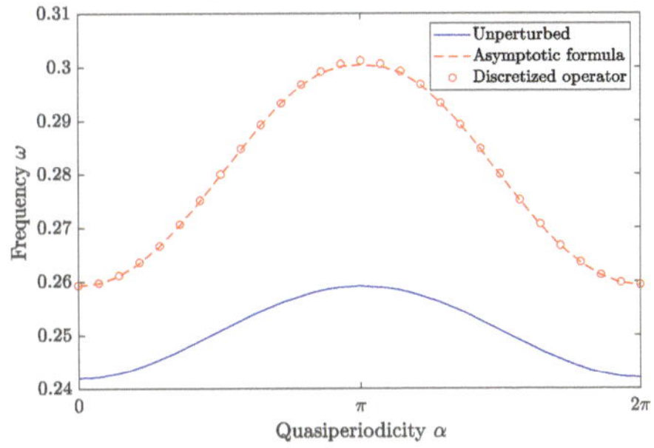

(B) Subwavelength bulk band (solid blue) and defect band (dashed red).

FIGURE 4.16. Guided mode at subwavelength scales (a) and corresponding bandgap frequency (b). Here $\varepsilon < -\varepsilon_0$, which means that the whole defect band will be above the bulk band. Moreover, we see that the defect band is not flat, and corresponding modes are propagating along the line defect. The defect band was computed using the asymptotic formula (4.103) (red dashed) and by discretising the operator $\mathcal{M}^{\varepsilon,\alpha_1}(\omega, \delta)$ (defined in (4.101)) using the multipole expansion method (red circles).

4.6. Robust guiding at subwavelength scales

There is a fundamental restriction of the practical applicability of the localised modes studied in Section 4.5. Taking the point defect as an example, we have from Theorem 4.56 that the bandgap frequency is exponentially close to the edge of the bulk bands (see Figure 4.15). When fabricating such structures, there is a large risk that small imperfections will cause the bandgap frequency to be lost inside the bulk bands, so that the desirable localisation property disappears as any incident radiation or source will be able to couple with the propagating Bloch modes.

For structures to have physically achievable and useful localisation properties, such properties must be robust against imperfections in their design. To accomplish this, we take inspiration from the field of topological insulators. These are materials that can conduct current along the edges of the material, while they are insulating in the bulk. The striking property of topological insulators is that these conducting edge modes originate from the structure of the bulk, rather than the edge itself. Moreover, they are localised to the edges and exhibit a remarkable robustness against perturbations of the system. These concepts have recently been studied mathematically in a variety of settings; see, for instance, [**2, 3, 9, 67, 70, 72, 79, 93, 94, 112, 113, 118, 137, 141, 153**].

In this section, we study a high-contrast resonator analogue of the Su-Schrieffer-Heeger model [**150**]. We consider the case of a chain of resonators, corresponding to $d = 3$ and $d_l = 1$, where

$$Y = \left[-\frac{L}{2}, \frac{L}{2}\right] \times \mathbb{R}^2, \quad \Lambda = L\mathbb{Z}, \quad Y^* = \left[-\frac{\pi}{L}, \frac{\pi}{L}\right], \quad \Lambda^* = \frac{2\pi}{L}\mathbb{Z}.$$

Moreover, we consider the case of a dimer of resonators $D = D_1 \cup D_2$. We need two assumptions of symmetry for the analysis that follows. The first assumption is that each individual resonator is symmetric in the sense that there exists some $x_1 \in \mathbb{R}$ such that

$$R_1 D_1 = D_1, \quad R_2 D_2 = D_2, \tag{4.104}$$

where R_1 and R_2 are the reflections in the planes

$$p_1 = \{-x_1\} \times \mathbb{R}^2 \quad \text{and} \quad p_2 = \{x_1\} \times \mathbb{R}^2,$$

respectively. We also assume, as in (2.4), that the dimer is symmetric in the sense that

$$\mathcal{P} D_1 = D_2, \tag{4.105}$$

where $\mathcal{P}(x) = -x$. We denote the resonator separation within the cell as l, and between the cells as l', that is,

$$l = 2x_1, \qquad l' = L - l,$$

see Figure 4.17.

For simplicity, we choose $v = v_1 = v_2 = 1$, and study effects originating from the geometry of the structure. As we shall see, a topological phase transition occurs when l changes across the symmetry point $l = L/2$ (corresponding to $l = l'$). We

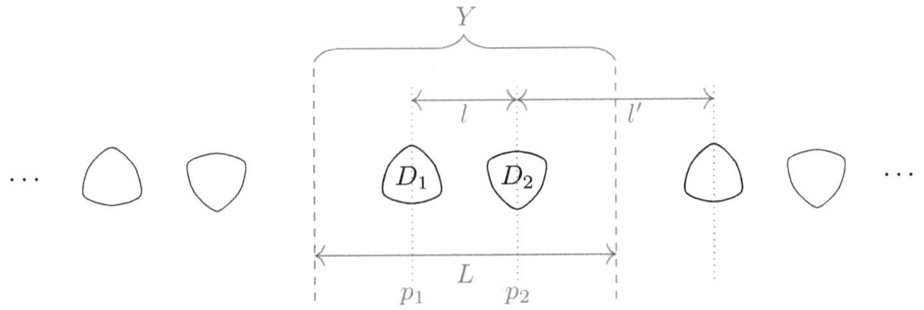

FIGURE 4.17. Example of the array, drawn to illustrate the symmetry assumptions.

begin by considering the periodic equation

$$(4.106)\qquad \begin{cases} \Delta u^\alpha + \omega^2 u^\alpha = 0 & \text{in } \mathbb{R}^3 \setminus \partial\mathcal{D}, \\ u^\alpha|_+ - u^\alpha|_- = 0 & \text{on } \partial\mathcal{D}, \\ \delta \frac{\partial u^\alpha}{\partial \nu}\Big|_+ - \frac{\partial u^\alpha}{\partial \nu}\Big|_- = 0 & \text{on } \partial\mathcal{D}, \\ e^{-\mathrm{i}\alpha_1 x_1} u^\alpha(x_1, x_2, x_3) \ \text{ is periodic in } x_1, \\ u^\alpha(x_1,x_2,x_3) \text{ satisfies the } \alpha\text{-quasiperiodic outgoing} \\ \ \text{radiation condition as } \sqrt{x_2^2 + x_3^2} \to +\infty, \end{cases}$$

where $\alpha = (\alpha_1, 0, 0)$ and the α-quasiperiodic outgoing radiation condition says that the wave is the sum of a finite number of outgoing propagating plane waves and an infinite number of evanescent waves (*i.e.*, exponentially decaying in the directions x_2 and x_3) as in (4.54).

LEMMA 4.64. *Under the symmetry assumption* (4.105)*, the quasiperiodic capacitance matrix* C^α *introduced in Definition* 4.32 *is Hermitian with constant diagonal, that is,*

$$C_{11}^\alpha = C_{22}^\alpha \in \mathbb{R}, \quad C_{12}^\alpha = \overline{C_{21}^\alpha} \in \mathbb{C}.$$

PROOF. From Lemma 4.33, C^α is Hermitian. To show that $C_{11}^\alpha = C_{22}^\alpha$, we define the mapping P by

$$(4.107)\qquad (Pf)(x) := f(-x).$$

Then, thanks to the assumed symmetry of the dimer in (4.105), it holds that $PV_1^\alpha = \overline{V_2^\alpha}$ and $PV_2^\alpha = \overline{V_1^\alpha}$. Therefore, it follows that

$$\begin{aligned} C_{11}^\alpha &= \int_{Y\setminus D} \overline{\nabla V_1^\alpha} \cdot \nabla V_1^\alpha \, dx \\ &= \int_{Y\setminus D} \overline{\nabla P V_1^\alpha} \cdot \nabla P V_1^\alpha \, dx \\ &= \int_{Y\setminus D} \nabla V_2^\alpha \cdot \overline{\nabla V_2^\alpha} \, dx \\ &= C_{22}^\alpha. \end{aligned}$$

□

The following lemma follows directly.

LEMMA 4.65. *The eigenvalues and corresponding eigenvectors of the quasiperiodic capacitance matrix* C^α *are given by*

$$\lambda_1^\alpha = C_{11}^\alpha - |C_{12}^\alpha|, \qquad \begin{pmatrix} a_1 \\ b_1 \end{pmatrix} = \frac{1}{\sqrt{2}} \begin{pmatrix} -e^{\mathrm{i}\theta_\alpha} \\ 1 \end{pmatrix},$$
$$\lambda_2^\alpha = C_{11}^\alpha + |C_{12}^\alpha|, \qquad \begin{pmatrix} a_2 \\ b_2 \end{pmatrix} = \frac{1}{\sqrt{2}} \begin{pmatrix} e^{\mathrm{i}\theta_\alpha} \\ 1 \end{pmatrix},$$

where, for α *such that* $C_{12}^\alpha \neq 0$, $\theta_\alpha \in [0, 2\pi)$ *is defined to be such that*

$$e^{\mathrm{i}\theta_\alpha} = \frac{C_{12}^\alpha}{|C_{12}^\alpha|}. \tag{4.108}$$

To enable explicit computations of asymptotic expansions for the band structure and its topological properties, we will assume that the resonators are dilute, in the sense that they occupy a small volume compared to the surrounding medium. As in Chapter 2, we assume that the resonators can be obtained by rescaling fixed domains B_1, B_2 as follows:

$$D_1 = \varepsilon B_1 - \left(\frac{l}{2}, 0, 0\right), \quad D_2 = \varepsilon B_2 + \left(\frac{l}{2}, 0, 0\right), \tag{4.109}$$

for some small parameter $\varepsilon > 0$.

Analogously to Lemma 1.19 for dilute finite systems of resonators, the following asymptotic expansion of C^α in terms of ε holds. We refer to Appendix C.5 for its proof.

LEMMA 4.66. *We assume that the resonators are in the dilute regime specified by* (4.109). *We let* $B = B_i$ *for* $i = 1$ *or* $i = 2$. *We also assume that* $\alpha \in Y^*, \alpha \neq 0$, *is fixed. Then we have the following asymptotics of the capacitance matrix* C_{ij}^α *as* $\varepsilon \to 0$:

$$C_{11}^\alpha = \varepsilon \mathrm{Cap}_B - \frac{(\varepsilon \mathrm{Cap}_B)^2}{4\pi} \sum_{m \neq 0} \frac{e^{\mathrm{i}m\alpha L}}{|mL|} + \mathcal{O}(\varepsilon^3), \tag{4.110}$$

$$C_{12}^\alpha = -\frac{(\varepsilon \mathrm{Cap}_B)^2}{4\pi} \sum_{m=-\infty}^{+\infty} \frac{e^{\mathrm{i}m\alpha L}}{|mL + l|} + \mathcal{O}(\varepsilon^3), \tag{4.111}$$

where Cap_B *is defined by* (1.64) *for* $D = B$. *Taking the imaginary part of* (4.111), *the corresponding asymptotic formula holds uniformly in* $\alpha \in Y^*$.

By observing that the following sum has a logarithmic singularity as $\alpha \to 0$:

$$\sum_{m \neq 0} \frac{e^{\mathrm{i}m\alpha L}}{|m|} = -\ln\big(2 - 2\cos(\alpha L)\big),$$

we can easily deduce from Lemma 4.66 that, for every $\varepsilon_0 > 0$ and $p \in \mathbb{N}$, there exists a constant A_p such that (4.110) and (4.111) for $\varepsilon < \varepsilon_0$ (after replacing $\mathcal{O}(\varepsilon^3)$ with $o(\varepsilon^2)$) hold uniformly in α for $|\alpha| \geq A_p \varepsilon^p$.

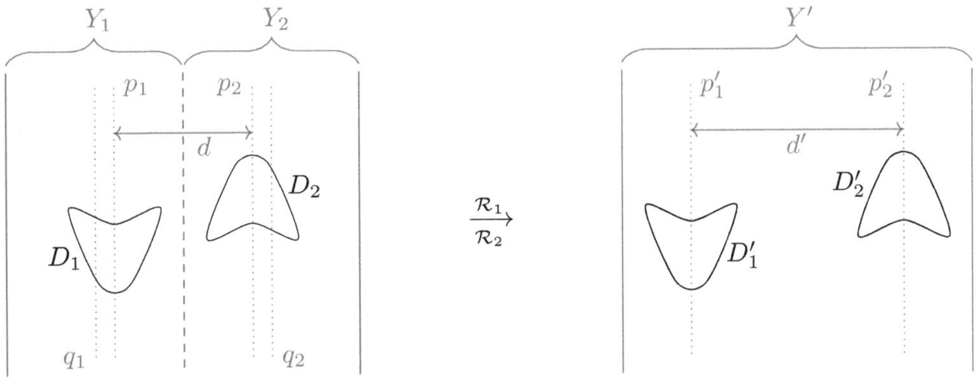

FIGURE 4.18. Reflections taking D to D'.

Define normalised extensions of V_j^α given by (4.60) as

$$S_j^\alpha(x) := \begin{cases} \frac{1}{\sqrt{|D_1|}}\delta_{ij} & x \in D_i,\ i = 1, 2, \\ \frac{1}{\sqrt{|D_1|}} V_j^\alpha(x) & x \in Y \setminus D, \end{cases} \tag{4.112}$$

where $|D_1|$ is the volume of one of the resonators. Note that $|D_1| = |D_2|$ thanks to the dimer's symmetry (4.105). Using arguments similar to those given in Theorem 4.35 and Corollary 4.37, the following approximation results can be proved.

THEOREM 4.67. *The subwavelength quasiperiodic resonant frequencies* $\omega_j^\alpha = \omega_j^\alpha(\delta)$, *for* $j = 1, 2$, *can be approximated as*

$$\omega_j^\alpha = \sqrt{\frac{\delta \lambda_j^\alpha}{|D_1|}} + \mathcal{O}(\delta),$$

where λ_j^α, $j = 1, 2$, *are the eigenvalues of the capacitance matrix* C^α. *Moreover, the Bloch eigenmodes* u_j^α, $j = 1, 2$, *corresponding to the resonances* ω_j^α, *can be approximated as*

$$u_j^\alpha(x) = a_j S_1^\alpha(x) + b_j S_2^\alpha(x) + \mathcal{O}(\delta),$$

where $\begin{pmatrix} a_j \\ b_j \end{pmatrix}$, $j = 1, 2$, *are the eigenvectors of the quasiperiodic capacitance matrix* C^α, *as given by Lemma* 4.65.

We now introduce a notation that, due to the assumed symmetry of the resonators, will allow us to prove topological properties of the chain. Divide Y into two subsets $Y = Y_1 \cup Y_2$, where $Y_1 := [-\frac{L}{2}, 0] \times \mathbb{R}^2$ and let $Y_2 := [0, \frac{L}{2}] \times \mathbb{R}^2$, as depicted in Figure 4.18. Define q_1 and q_2 as the central planes of Y_1 and Y_2, that is, the planes $q_1 := \{-\frac{L}{4}\} \times \mathbb{R}^2$ and $q_2 := \{\frac{L}{4}\} \times \mathbb{R}^2$. Let $\mathcal{R}_1$ and $\mathcal{R}_2$ be reflections in the respective planes. Observe that, due to the assumed symmetry of each resonator (4.104), the "complementary" dimer $D' = D_1' \cup D_2'$, given by swapping l and l', satisfies $D_i' = \mathcal{R}_i D_i$ for $i = 1, 2$.

Define the operator T_α on the set of α-quasiperiodic functions f on Y as

$$T_\alpha f(x) := \begin{cases} e^{-\mathrm{i}\alpha L}\overline{f(\mathcal{R}_1 x)}, & x \in Y_1, \\ \overline{f(\mathcal{R}_2 x)}, & x \in Y_2, \end{cases}$$

where the factor $e^{-\mathrm{i}\alpha L}$ is chosen so that the image of a continuous (α-quasiperiodic) function is continuous.

We will now proceed to use T_α to analyse the different topological properties of the two dimer configurations. Define the quantity ${C_{12}^\alpha}'$ analogously to C_{12}^α but on the dimer D', that is, to be the top-right element of the corresponding quasiperiodic capacitance matrix.

LEMMA 4.68. *We have*

$$ {C_{12}^\alpha}' = e^{-\mathrm{i}\alpha L}\overline{C_{12}^\alpha}. $$

Consequently, if $l = l' = L/2$ *then* $C_{12}^{\pi/L} = 0$.

PROOF. Define ${V_1^\alpha}', {V_2^\alpha}'$ by (4.60) but on D' instead of D. Observe that $T_\alpha {V_1^\alpha}' = e^{-\mathrm{i}\alpha L}\overline{V_1^\alpha}$ and $T_\alpha {V_2^\alpha}' = \overline{V_1^\alpha}$. Then, we find that

$$ \begin{aligned} {C_{12}^\alpha}' &= \int_{Y\setminus D'} \overline{\nabla {V_1^\alpha}'} \cdot \nabla {V_2^\alpha}' \, \mathrm{d}x \\ &= \int_{Y\setminus D} \overline{\nabla T_\alpha {V_1^\alpha}'} \cdot \nabla T_\alpha {V_2^\alpha}' \, \mathrm{d}x \\ &= e^{-\mathrm{i}\alpha L} \int_{Y\setminus D} \nabla V_1^\alpha \cdot \overline{\nabla V_2^\alpha} \, \mathrm{d}x \\ &= e^{-\mathrm{i}\alpha L}\overline{C_{12}^\alpha}. \end{aligned} $$

At $\alpha = \pi/L$, we have ${C_{12}^{\pi/L}}' = -\overline{C_{12}^{\pi/L}}$. Moreover, if $l = l'$, the symmetry of the structure means that ${C_{12}^{\pi/L}}' = C_{12}^{\pi/L}$, so it must be the case that $C_{12}^{\pi/L} = 0$. □

The following results will be useful.

LEMMA 4.69. *We assume that* D *is in the dilute regime specified by* (1.32). *Then, for* ε *small enough,*

(i) $\operatorname{Im} C_{12}^\alpha > 0$ *for* $0 < \alpha < \pi/L$ *and* $\operatorname{Im} C_{12}^\alpha < 0$ *for* $-\pi/L < \alpha < 0$. *In particular,* $\operatorname{Im} C_{12}^\alpha$ *is zero if and only if* $\alpha \in \{0, \pi/L\}$*;*
(ii) C_{12}^α *is zero if and only if both* $l = l'$ *and* $\alpha = \pi/L$*;*
(iii) $C_{12}^{\pi/L} < 0$ *when* $d < d'$ *and* $C_{12}^{\pi/L} > 0$ *when* $l > l'$. *In both cases, we have* $C_{12}^0 < 0$.

The proof of Lemma 4.69 is given in Appendix C.6. This lemma describes the crucial properties of the behaviour of the curve $\{C_{12}^\alpha : \alpha \in Y^*\}$ in the complex plane. The periodic nature of Y^* means that this is a closed curve. Part (i) tells us that this curve crosses the real axis in precisely two points. Taken together with (iii), we know that this curve winds around the origin in the case $l > l'$, but not in the case $l < l'$.

LEMMA 4.70. *Let* λ_1^α *and* λ_2^α *be defined as in Lemma* 4.65. *Then, in the dilute regime, we have*

$$ \max_{\alpha\in Y^*} \lambda_1^\alpha = \lambda_1^{\pi/L}, \qquad \min_{\alpha\in Y^*} \lambda_2^\alpha = \lambda_2^{\pi/L}, $$

for ε *small enough.*

PROOF. We have

$$\lambda_1^\alpha = C_{11}^\alpha - |C_{12}^\alpha| \\ \leq \underbrace{C_{11}^\alpha + \mathrm{Re}\,(C_{12}^\alpha)}_{=\frac{1}{2}\mathrm{Cap}_{\alpha,D}},$$

where $\mathrm{Cap}_{\alpha,D}$ is the α-quasiperiodic capacity of D defined as in (B.115). Using the variational characterisation of $\mathrm{Cap}_{\alpha,D}$, in the same way as in Theorem 4.7, it follows that the maximum of $\mathrm{Cap}_{\alpha,D}$ is attained at $\alpha = \pi/L$. Moreover, in the dilute regime, from Lemma 4.69 we know that $C_{12}^{\pi/L}$ is a non-positive real number. We therefore have

$$\lambda_1^{\pi/L} = \frac{1}{2}\mathrm{Cap}_{\pi/L,D},$$

so the maximum of λ_1^α is attained at $\alpha = \pi/L$.

We now turn to the second eigenvalue λ_2^α. Similarly, we have

$$\lambda_2^\alpha = C_{11}^\alpha + |C_{12}^\alpha| \\ \geq C_{11}^\alpha - \mathrm{Re}\,(C_{12}^\alpha)\,. \tag{4.113}$$

We can formulate a variational characterisation for $C_{11}^\alpha - \mathrm{Re}\,(C_{12}^\alpha)$ in terms of the Dirichlet energy. Let C_α^∞ be the set of functions in $C^\infty(Y)$ that can be extended to α-quasiperiodic functions in $C^\infty(\mathbb{R}^3)$ decaying as

$$\mathcal{O}\left(1/\sqrt{x_2^2 + x_3^2}\right) \text{ as } \sqrt{x_2^2 + x_3^2} \to +\infty.$$

Let

$$\mathcal{H} = \left\{v \in H^1_{\mathrm{loc}}(Y) \ : \ v(x_1, x_2, x_3) = \mathcal{O}\left(\frac{1}{\sqrt{x_2^2 + x_3^2}}\right) \text{ as } \sqrt{x_2^2 + x_3^2} \to +\infty\right\}$$

and let $\mathcal{H}_\alpha$ be the closure of C_α^∞ in $\mathcal{H}$. Then define

$$\mathcal{V}_\alpha = \left\{v \in \mathcal{H}_\alpha \ : \ v = -\frac{1}{\sqrt{2}} \text{ on } \partial D_1, v = \frac{1}{\sqrt{2}} \text{ on } \partial D_2\right\}.$$

We then have the variational characterisation

$$C_{11}^\alpha - \mathrm{Re}\,(C_{12}^\alpha) = \min_{v \in \mathcal{V}_\alpha} \int_{Y \setminus D} |\nabla v|^2 \, dx. \tag{4.114}$$

Indeed, the minimiser v_0 satisfies $\Delta v_0 = 0$ in $Y \setminus D$ and therefore

$$v_0 = \frac{1}{\sqrt{2}}\left(-V_1^\alpha + V_2^\alpha\right).$$

Equation (4.114) then follows by expanding the integral.

Define $\mathcal{V} = \cup_{\alpha \in Y^*} \mathcal{V}_\alpha$. From (4.114), we find

$$\min_{\alpha \in Y^*} \left[C_{11}^\alpha - \mathrm{Re}\,(C_{12}^\alpha)\right] = \min_{v \in \mathcal{V}} \int_{Y \setminus D} |\nabla v|^2 \, dx.$$

Due to the symmetry of D, the corresponding minimiser v_1 is an odd function in x_1. In other words, v_1 is a π/L-quasiperiodic function, so

$$\min_{\alpha \in Y^*} \left[C_{11}^\alpha - \mathrm{Re}\,(C_{12}^\alpha)\right] = C_{11}^{\pi/L} - \mathrm{Re}\left(C_{12}^{\pi/L}\right). \tag{4.115}$$

At $\alpha = \pi/L$, (4.113) is an equality. This, together with (4.115), proves that the minimum of λ_2^α is attained at $\alpha = \pi/L$. □

The next result shows that, in addition to the bandgap above the second band, there is also a bandgap between the first two bands in the case $l \neq L/2$ [**22**].

THEOREM 4.71. *In the dilute regime and with δ sufficiently small, there exists a subwavelength bandgap between the first two Bloch band functions $\alpha \mapsto \omega_j^\alpha$, $j = 1,2$, if $l \neq L/2$, i.e.,*

$$\max_{\alpha \in Y^*} \omega_1^\alpha < \min_{\alpha \in Y^*} \omega_2^\alpha,$$

for ε and δ small enough.

The proof of Theorem 4.71 is given in Appendix C.7. The argument is based on representing the first and second resonant frequencies as

$$\omega_1^\alpha = \sqrt{\frac{\delta\left(C_{11}^\alpha - |C_{12}^\alpha|\right)}{|D_1|}} + \mathcal{O}(\delta), \qquad \omega_2^\alpha = \sqrt{\frac{\delta\left(C_{11}^\alpha + |C_{12}^\alpha|\right)}{|D_1|}} + \mathcal{O}(\delta),$$

and using the fact that, in the dilute regime and for fixed $\alpha \neq 0$, the capacitance coefficients can be expanded using Lemma 4.66.

In the case $l = l' = L/2$, the first and the second bands will have a degeneracy at $\alpha = \pi/L$. Moreover, the two bands will intersect linearly in a so-called *Dirac cone*. By altering l, this Dirac cone will open a bandgap. As we shall see, the nature of this bandgap is fundamentally different in the two cases $l < l'$ and $l > l'$.

4.6.1. Topological indices and band inversion. We will now define an index which quantifies the topological properties of the band structure of our chain of resonators. Such index is the one-dimensional analogue of Chern numbers in higher periodicity dimensions; see, for instance, [**59, 97**].

DEFINITION 4.72 (Zak phase). *For a nondegenerate band ω_j^α, we let u_j^α be a family of normalised eigenmodes that continuously depend on α. We then define the Zak phase φ_j^{zak} as*

$$\varphi_j^{\mathrm{zak}} := \mathrm{i} \int_{Y^*} \langle u_j^\alpha, \frac{\partial}{\partial \alpha} u_j^\alpha \rangle \,\mathrm{d}\alpha,$$

where $\langle \cdot, \cdot \rangle$ denotes the $L^2(D)$-inner product.

THEOREM 4.73. *We assume that D is in the dilute regime specified by (4.109). Then the Zak phase satisfies*

$$\varphi_j^z = \begin{cases} 0 & \text{if } l < l', \\ \pi & \text{if } l > l', \end{cases} \tag{4.116}$$

for ε and δ small enough.

PROOF. We compute the Zak phase φ_j^z, $j = 1,2$, of the first and second band, respectively. Let $S_j^\alpha, j = 1,2$ be defined by (4.112). Observe that

$$\langle S_1^\alpha, S_1^\alpha \rangle = 1, \qquad \langle S_2^\alpha, S_2^\alpha \rangle = 1, \qquad \langle S_1^\alpha, S_2^\alpha \rangle = 0$$

and in D we have

$$\frac{\partial}{\partial \alpha} S_1^\alpha \equiv 0, \qquad \frac{\partial}{\partial \alpha} S_2^\alpha \equiv 0,$$

for all $\alpha \in Y^*$. By Theorem 4.67 and Lemma 4.65, it follows that

$$\langle u_j^\alpha, \frac{\partial}{\partial \alpha} u_j^\alpha \rangle = \frac{\mathrm{i}}{2} \frac{\partial \theta_\alpha}{\partial \alpha} + \mathcal{O}(\delta),$$

so the Zak phase is given by

$$\varphi_j^z = -\frac{1}{2}\left[\theta_\alpha\right]_{Y^*} + \mathcal{O}(\delta).$$

Since we know that φ_j^z is an integer multiple of π, we have for small enough δ that

$$\varphi_j^z = -\frac{1}{2}\left[\theta_\alpha\right]_{Y^*}. \tag{4.117}$$

This representation of φ_j^z shows that the Zak phase is given by the change in the argument of C_{12}^α as α varies over the Brillouin zone Y^*. We can see from parts (i) and (iii) of Lemma 4.69 that the winding number of θ_α around the origin depends on whether $l < l'$ or $l > l'$. In the two cases, we have, respectively, $[\theta_\alpha]_{Y^*} = 0$ and $[\theta_\alpha]_{Y^*} = -2\pi$. Therefore, if δ is small enough, then (4.116) holds. □

Qualitatively, a nonzero Zak phase means that the crystal has undergone *band inversion*, meaning that at some point in the Brillouin zone the monopole/dipole nature of the first/second Bloch eigenmodes has swapped. In the current setting, a monopole mode is a mode with even parity, while a dipole mode is a mode with odd parity. For the two-resonator chain studied here, the first mode will always be of monopole nature at the origin $\alpha = 0$ (which is a consequence of the fact that the eigenvectors of the periodic capacitance matrix C^0 are, respectively, $\begin{pmatrix}1\\1\end{pmatrix}$ and $\begin{pmatrix}1\\-1\end{pmatrix}$). At $\alpha = \pi/L$, the modes will also be of monopole/dipole nature, but which of these corresponds to the first or second mode depends on the geometry of the structure [**29**].

The Zak phase predicts band inversion in the case $l > l'$, which is the statement of the next result.

PROPOSITION 4.74. *For ε and δ small enough, the crystal has undergone band inversion in the case $l > l'$, but not if $l < l'$. In other words, the eigenmodes associated with the first and second bands at $\alpha = \pi/L$ satisfy*

$$u_1^{\pi/L}(-x) = u_1^{\pi/L}(x), \qquad u_2^{\pi/L}(-x) = -u_2^{\pi/L}(x), \qquad \textit{when } l < l',$$

$$u_1^{\pi/L}(-x) = -u_1^{\pi/L}(x), \qquad u_2^{\pi/L}(-x) = u_2^{\pi/L}(x), \qquad \textit{when } l > l'.$$

As $\delta \to 0$, these eigenmodes are given, respectively, by

$$u_1^{\pi/L}(x) = \left(\begin{smallmatrix}1\\1\end{smallmatrix}\right)\cdot \mathbf{S}_D^{\pi/L,\omega}(x) + \mathcal{O}(\delta), \quad u_2^{\pi/L}(x) = \left(\begin{smallmatrix}1\\-1\end{smallmatrix}\right)\cdot \mathbf{S}_D^{\pi/L,\omega}(x) + \mathcal{O}(\delta),$$

when $l < l'$, and

$$u_1^{\pi/L}(x) = \left(\begin{smallmatrix}1\\-1\end{smallmatrix}\right)\cdot \mathbf{S}_D^{\pi/L,\omega}(x) + \mathcal{O}(\delta), \quad u_2^{\pi/L}(x) = \left(\begin{smallmatrix}1\\1\end{smallmatrix}\right)\cdot \mathbf{S}_D^{\pi/L,\omega}(x) + \mathcal{O}(\delta),$$

when $l > l'$. Here, $\mathbf{S}_D^{\alpha,\omega}(x)$ is the function defined in Corollary 4.37.

4.6.2. Robustness of interface modes. We now study a finite chain of resonators that supports topologically protected interface modes. Specifically, we assume that D has the form

(4.118)

$$D = \left(\bigcup_{n=-M}^{M} D_0 + n(l+l',0,0)\right) \bigcup \left(\bigcup_{n=-M+1}^{M} D_0 + n(l+l',0,0) - (l',0,0)\right),$$

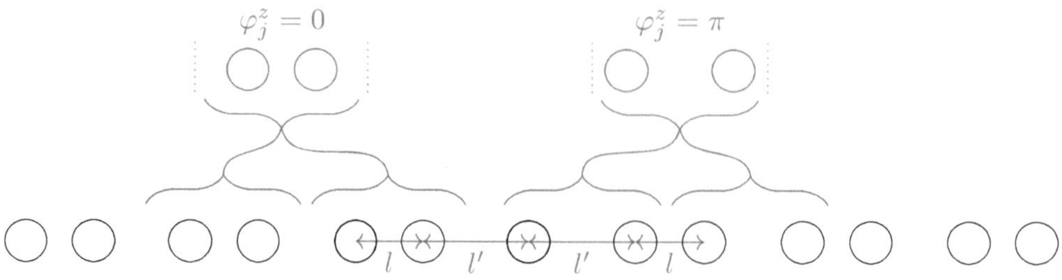

FIGURE 4.19. Two-dimensional cross-section of a finite dimer chain with 13 resonators, heuristically showing how to identify unit cells with different Zak phases on either side of the interface.

where D_0 is a single repeating resonator. In other words, D consists of an odd number N of identical resonators ($N = 4M + 1$) with alternating distances l and l' that are swapped at the middle resonator. An example of such a configuration is depicted in Figure 4.19. In this figure, it is shown how to associate different Zak phases on either side of the central resonator (which constitutes the interface). Based on the principle of *bulk-boundary correspondence*, we thereby expect robust localised modes around this interface.

Since $v = v_1 = v_2 = 1$ and $\delta_1 = \delta_2 = \delta$, we model wave propagation in the crystal $\mathcal{D}$ by the Helmholtz problem

$$
\begin{cases}
\Delta u + \omega^2 u = 0 & \text{in } \mathbb{R}^3 \setminus \partial\mathcal{D}, \\
u|_+ - u|_- = 0 & \text{on } \partial\mathcal{D}, \\
\delta \left.\frac{\partial u}{\partial \nu}\right|_+ - \left.\frac{\partial u}{\partial \nu}\right|_- = 0 & \text{on } \partial\mathcal{D}, \\
|x|\left(\frac{\partial}{\partial|x|} - \mathrm{i}\omega\right)u \to 0 & \text{as } |x| \to +\infty.
\end{cases}
\tag{4.119}
$$

Figure 4.20 shows a comparison between the bandgap frequencies of a chain with a topological defect (as in Figure 4.19), compared against a chain with a point defect (analogously as in Section 4.5). Not only is the topological bandgap frequency further away from the edges of the bandgap, but it also exhibits a lower variance when random Gaussian errors are imposed on the resonator locations. For large error, we see that the bandgap frequency of the topological defect chain (Figure 4.20a) is much more robust than the bandgap frequency of the point defect chain (and, similarly, more robust than any typical eigenvalue of a subwavelength resonant array [**69**]).

In Section 1.10 we have observed that, under the dilute assumption, the capacitance formulation is analogous to the tight-binding model . The tight-binding model is often coupled with a nearest-neighbour approximation, whereby long-range interactions are neglected. Figure 4.21b shows the bulk and bandgap frequencies computed using such nearest-neighbour approximation. Compared to Figure 4.21a, where all interactions are taken into account, we observe that the nearest-neighbour approximation is not accurate. This discrepancy shows a fundamental difference between topological interface modes in the setting of classical waves and quantum-mechanical waves. Moreover, we see that the bandgap frequency in Figure 4.21b is not affected by the error. This can be seen as a consequence of the chiral symmetry, which is not present without the nearest-neighbour approximation.

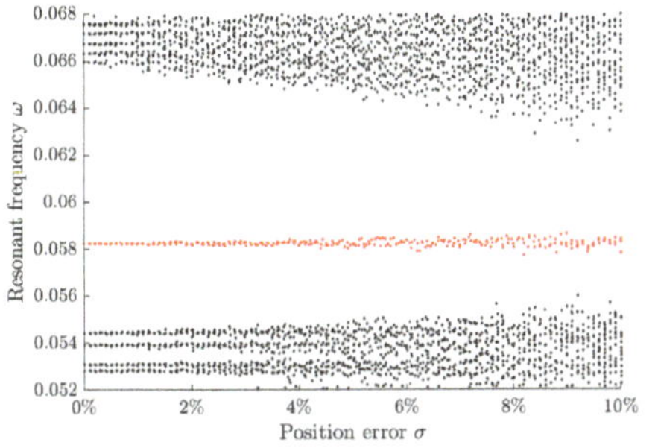

(A) Dimer chain with 41 resonators, separation distances $d = 3, d' = 6$.

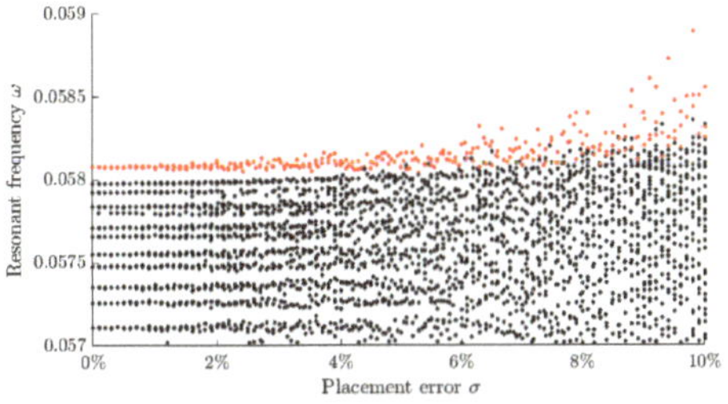

(B) Point defect chain with 41 resonators, separation distance $l = 12$ and defect radius $R_d = 0.99$.

FIGURE 4.20. Simulation of bandgap frequency (red) and bulk frequencies (black) of a topological defect (a) and point defect (b) chain, with Gaussian $\mathcal{N}(0, \sigma^2)$ errors added to the resonator positions. The standard deviation σ is expressed as a percentage of the average resonator separation.

Chiral symmetry is a geometric property that a matrix $\widetilde{C}$ is said to possess if there is a unitary matrix Σ with $\Sigma^2 = I$ such that $\widetilde{C}$ satisfies

$$\Sigma\widetilde{C}\Sigma = -\widetilde{C}.$$

The significance of this property is that a chirally symmetric matrix will have a symmetric spectrum. This can be easily seen from the fact that if (λ, v) is an eigenpair for $\widetilde{C}$ then so is $(-\lambda, \Sigma v)$.

Finite chains that have an odd number of resonators (such as the example studied here) will have an odd number of resonant frequencies, hence there must be a middle frequency. Thus, if one can design a chain which has a bandgap and is chirally symmetric, there must be a midgap frequency. Moreover, this midgap frequency is stable with respect to perturbations which preserve the chiral symmetry.

Let the matrix $\widetilde{C}$ obtained by subtracting the constant diagonal elements from the capacitance matrix C and then using a nearest-neighbour approximation. Then $\widetilde{C}$ is a bisymmetric tridiagonal matrix with odd size and zero diagonal. Such a matrix is chirally symmetric, and therefore has a zero eigenvalue. Moreover, $\widetilde{C}$ will retain its chiral symmetry when errors are made in the position of the resonators. This is because such errors will not affect the diagonal entries of $\widetilde{C}$ and, away from the diagonals, $\widetilde{C}_{ij}$ and $\widetilde{C}_{ji}$ will experience the same effects. Therefore, the midgap frequencies of $\widetilde{C}$ are stable with respect to errors in the resonator position.

4.6.3. Edge modes in a dislocated chain. In Section 4.6.1 we saw that the structure under consideration has different topological properties in the different cases $l > l'$ and $l < l'$. These two regimes only differ by the choice of unit cell: if we shift the unit cell Y by $L/2$ the regimes will be swapped. Correspondingly, edges between the different topological regimes are created by introducing a shift between the unit cells on either side of the edge. This is exemplified in Figure 4.19. Another example of this principle is to translate, or *dislocate*, half of the chain by some distance d as illustrated in Figure 4.22. In this case, we consider a chain of

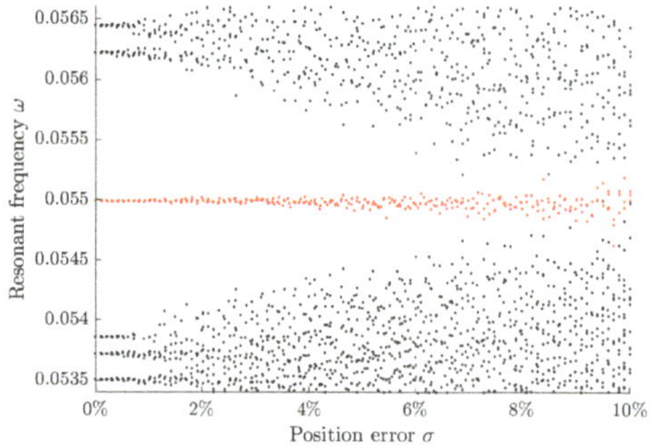

(A) Dilute dimer chain with 41 resonators, separation distances $l = 12, l' = 42$.

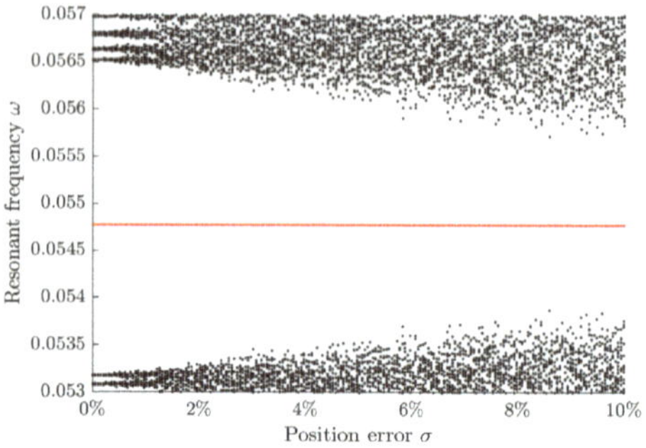

(B) Nearest-neighbour approximation for the dilute dimer chain from (a).

FIGURE 4.21. Similar simulations as in Figure 4.20, but for a dilute chain. (a): fully coupled simulations, where all interactions are taken into account. (b): nearest-neighbour approximation, where long-range interactions are neglected. The nearest-neighbour approximation provides a chirally symmetric problem, in which the centre frequency is preserved. This approximation is not accurate, and the centre frequency will have a small but nonzero variance corresponding to small but nonzero long-range interactions.

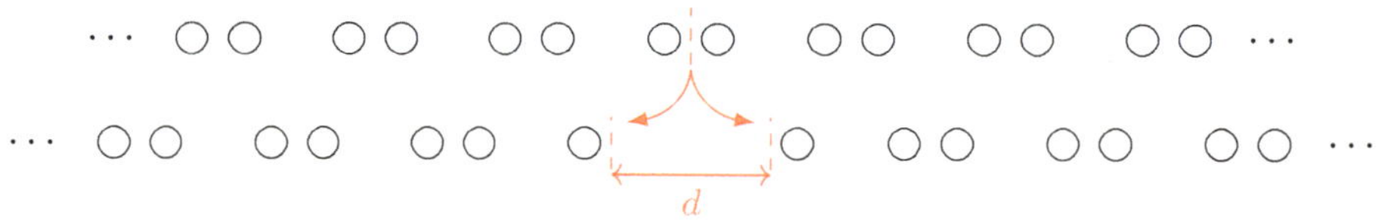

FIGURE 4.22. We start with an array of pairs of subwavelength resonators, known to have a subwavelength bandgap. A dislocation (with size $d > 0$) is introduced to create bandgap frequencies.

resonators given by

$$\mathcal{D}_{\text{disloc}} = \left(\bigcup_{m=-\infty}^{-1} D + (mL, 0, 0) \right) \cup \left(\bigcup_{m=0}^{+\infty} D + (mL + d, 0, 0) \right),$$

where D is the resonator dimer defined as in the beginning of this section.

Qualitatively, it is straightforward to understand how the dislocation will affect the existence and behaviour of bandgap frequencies. As d increases from 0, we will detune the dimer coupling, which is responsible for the bandgap, and bandgap frequencies will therefore appear from each edge of the bandgap. On the other hand, when d is very large, the two half-space chains will decouple, and the bulk-boundary correspondence suggests that each of these half-chains will support a single bandgap frequency. We therefore expect only a single bandgap frequency in this limit. As d varies between 0 and $+\infty$ these two frequencies hybridise, and will together cover the whole bandgap. This is sketched in Figure 4.23.

We now set out to prove that the picture outlined above is correct. Similarly to Section 4.5, we will apply a fictitious source method to model the dislocated structure $\mathcal{D}_{\text{disloc}}$ in terms of the original structure $\mathcal{D}$ along with fictitious sources on the dislocated resonators. The analysis is divided into three parts, depending on the value of d:

- $d \ll 1$. In this case, we can use asymptotic expansions in terms of d to prove that there is a bandgap frequency emerging from each edge of the bandgap;
- $d = mL$ for some $m > 0$. This particular dislocation is equivalent to removing m dimers from $\mathcal{D}$. This observation simplifies the problem and allows explicit computations of the bandgap frequencies in terms of the eigenvalue problem of certain Toeplitz matrices; see Appendix C.8;
- $d > d_0$, where d_0 is the width of one resonator. For technical reasons, we restrict the remaining values of d, so that the dislocated and original resonators do not overlap.

We introduce the notation $l_0 = l/L$, *i.e.*, l_0 is the ratio of the separation of the resonators to the unit cell length. There are two fundamentally different cases: $l_0 < 1/2$ and $l_0 > 1/2$. In the first case, the dislocation occurs between dimers of resonators, keeping each pair of resonators intact. The second case corresponds to the dislocation that occurs within a dimer, splitting one pair of resonators into two "edge" resonators.

With these ideas at hand, the following two theorems were proved in [**22**]. The first result, valid for small d, shows the emergence of a bandgap frequency from each edge of the bandgap.

THEOREM 4.75. *Assume that D_1 and D_2 are strictly convex. For small enough d and δ, and in the case $l_0 > 1/2$, there are two bandgap frequencies $\omega_1(d), \omega_2(d)$ such that $\omega_j(d) \to \omega_j^\diamond, j = 1, 2$ as $d \to 0$. In the case $l_0 < 1/2$, there are no bandgap frequencies as $d, \delta \to 0$.*

THEOREM 4.76. *Assume that the resonators are in the dilute regime specified by (4.109) and that $l_0 > 1/2$. Then, for small enough δ and ε, there exists some $d_0 = \mathcal{O}(\varepsilon)$ such that there are two bandgap frequencies $\omega_1(d)$ and $\omega_2(d)$ for all $d \in [d_0, +\infty)$, both of which converge to the same value ω_∞ as $d \to +\infty$.*

In particular, Theorem 4.76 states that the bandgap frequencies will cover an interval

$$\mathcal{I} := [\omega_1(d_0), \omega_2(d_0)]$$

inside the bandgap, which allows us to fine-tune the system to achieve optimal robustness.

In Figure 4.24, we see numerically computed eigenmodes, clearly showing how the edge modes of the "half system" hybridise to create the two bandgap frequencies.

4.7. Bound states in the continuum and Fano resonances

Sometimes, localised modes can exist in periodic structures without a defect. This typically happens when the structure has certain symmetries, resulting in resonant modes in the radiation continuum whose far field radiation vanishes. Such states are known as *bound states in the continuum* (BICs).

In this section, we prove the existence of such modes in systems of subwavelength resonators. When the symmetry is slightly broken, these resonant modes can be excited from the far field. Remarkably, this may create asymmetric (Fano-type)

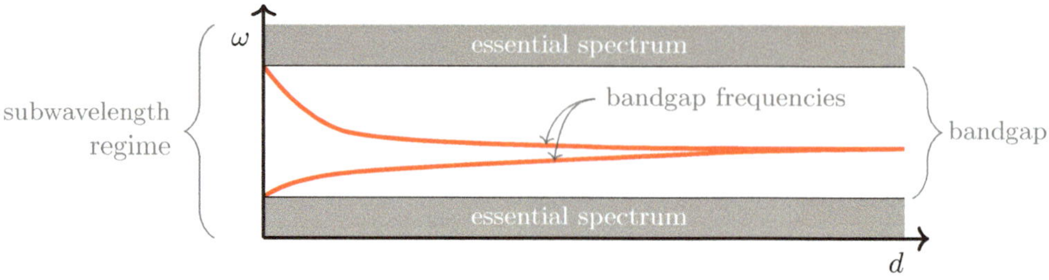

FIGURE 4.23. As the dislocation size d increases from zero, a bandgap frequency appears from each edge of the subwavelength bandgap. These two frequencies converge to a single value within the subwavelength bandgap as $d \to +\infty$.

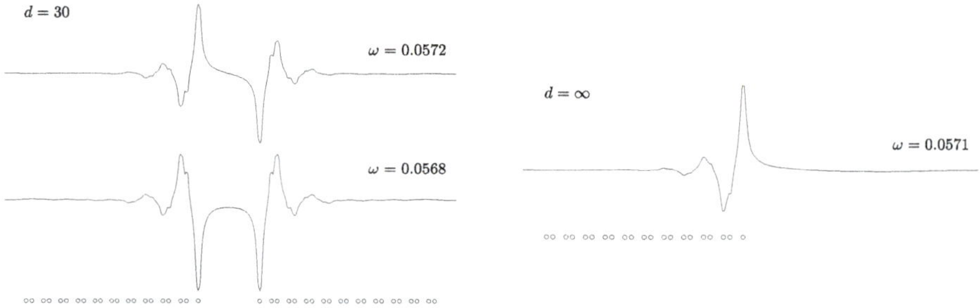

FIGURE 4.24. *Left:* The two edge modes for an array of 42 spherical resonators of radius 1. Here, we simulate an array with parameters $L = 9$, $l = 6$, $d = 30$ and $\delta = 1/7000$. *Right:* For comparison, the edge mode of the corresponding 'half system' is shown, which can be thought of as the $d = +\infty$ case. In both cases, the eigenmodes are shown directly above the corresponding system of resonators.

scattering behavior where the transmission is fundamentally different for frequencies on either side of the resonant frequency. Using asymptotic analysis, we compute the scattering matrix of the system explicitly, thereby characterising this Fano-type transmission anomaly.

We start by introducing the following definition.

DEFINITION 4.77. *A resonant mode u_n^α of (4.4) is said to be a bound state in the continuum if the corresponding resonant frequency ω_n^α is real, satisfies $|\alpha| < \omega_n^\alpha / v$ and the mode satisfies*

$$u_n^\alpha(x_l, x_0) = \mathcal{O}(e^{-K|x_0|}), \quad |x_0| \to +\infty, \quad K > 0.$$

As we shall see, there are two symmetry conditions required to achieve bound states in the continuum in the subwavelength resonator arrays. The first condition is a symmetry condition of the structure, while the second condition is $\alpha = 0$, which corresponds to modes that radiate perpendicularly to the structure.

Assume that $d - d_l = 1$, $D = D_1 \cup D_2 \Subset Y$ satisfies the symmetry conditions

$$\mathcal{P} D_1 = D_2, \quad \mathcal{P}_0 D_1 = D_1, \quad \mathcal{P}_0 D_2 = D_2, \tag{4.120}$$

and

$$\delta_1 v_2^2 = \delta_2 v_2^2 \in \mathbb{R}, \tag{4.121}$$

where $\mathcal{P}, \mathcal{P}_0 : \mathbb{R}^d \to \mathbb{R}^d$ are the parity operators

$$\mathcal{P}(x) = -x \quad \text{and} \quad \mathcal{P}_0(x_l, x_0) = (x_l, -x_0). \tag{4.122}$$

Moreover, assume that $\alpha = 0$. Then, for small enough δ, one can prove that the second resonant mode u_2^0 is a bound state in the continuum [**26**]. Interestingly, one can never have the first resonant mode u_1^0 as a bound state in the continuum. This is due to the fact that the first mode corresponds to the broad low-frequency response of the metascreen which is not due to the local resonance of the structure.

An interesting problem now is to describe the behaviour when the symmetry conditions are no longer satisfied. In particular, when $\mathcal{P}_0 D_i \neq D_i$ for $i = 1, 2$, we can use Theorem 4.48 to conclude that ω_2^α has a small but nonzero imaginary part. As we shall see next, this causes an interesting *Fano-type* transmission scattering. This phenomenon emerges from the interference between the two coupled subwavelength resonant frequencies of the pair D of resonators, which have significantly different imaginary parts. The first resonant frequency corresponds to the "continuum" of states while the second resonant frequency originates from the resonant behaviour of the periodic structure and has a comparatively sharp response, corresponding to a "discrete state". The interaction between these two states leads to the creation of a Fano-type asymmetric transmission anomaly [**26, 117**].

Given a unit vector $\mathbf{w} = (w_l, w_0) \in \mathbb{R}^d$ with $w_0 > 0$, we define the wave vectors

$$\mathbf{k}_+ = \frac{\omega}{v}\begin{pmatrix} w_l \\ w_0 \end{pmatrix}, \qquad \mathbf{k}_- = \frac{\omega}{v}\begin{pmatrix} w_l \\ -w_0 \end{pmatrix}. \tag{4.123}$$

We also let $\alpha = \frac{\omega}{v}\begin{pmatrix} w_l \\ 0 \end{pmatrix}$. We now assume that

$$u_{\text{in}}(x) = c_1 e^{\mathrm{i}\mathbf{k}_- \cdot x} + c_2 e^{\mathrm{i}\mathbf{k}_+ \cdot x}, \tag{4.124}$$

and seek the behaviour of the solution u of the scattering problem (4.80). In the first radiation continuum, the scattered field $u - u_{\text{in}}$ consists of a single propagating mode as $|x_0| \to +\infty$. We will write $f \sim g$ to denote that two functions f, g are equal up to exponentially decaying factors, in the sense that there is some constant $K > 0$ such that

$$|f(x_l, x_0) - g(x_l, x_0)| = \mathcal{O}(e^{-K|x_0|}) \text{ as } |x_0| \to +\infty.$$

We therefore have

$$u \sim \begin{cases} c_1 e^{\mathrm{i}\mathbf{k}_- \cdot x} + d_1 e^{\mathrm{i}\mathbf{k}_+ \cdot x}, & x_0 \to +\infty, \\ c_2 e^{\mathrm{i}\mathbf{k}_+ \cdot x} + d_2 e^{\mathrm{i}\mathbf{k}_- \cdot x}, & x_0 \to -\infty, \end{cases} \tag{4.125}$$

where

$$\begin{pmatrix} d_1 \\ d_2 \end{pmatrix} = S \begin{pmatrix} c_1 \\ c_2 \end{pmatrix}, \qquad S = \begin{pmatrix} r_+ & t_- \\ t_+ & r_- \end{pmatrix}. \tag{4.126}$$

Here, S is known as the scattering matrix. The reflection and transmission coefficients r_+, t_+ are the coefficients of the outgoing part of the field in the case $u_{\text{in}}(x) = e^{\mathrm{i}\mathbf{k}_- \cdot x}$, *i.e.*, when the incident field is a plane wave from the positive x_0-direction (and reversely for r_-, t_-). In the low-frequency regime, the scattering matrix S contains two transmission peaks originating from the resonances ω_1^α and

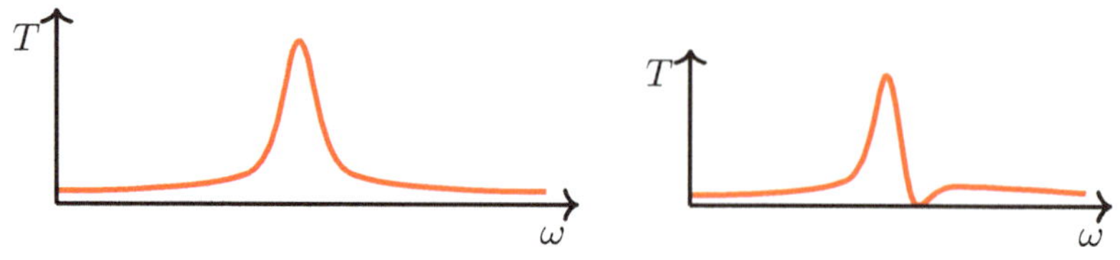

(A) Lorentzian line shape. (B) Fano-type line shape.

FIGURE 4.25. When the frequency of incident waves varies, the transmittance T (*i.e.*, the intensity of the transmitted field) will have peaks at certain frequencies. A Lorentzian line shape, which is a symmetric peak, is typically found in scattering problems. In this section, the setting exhibits a Fano-type line shape, which is asymmetric and rapidly drops from 1 to 0.

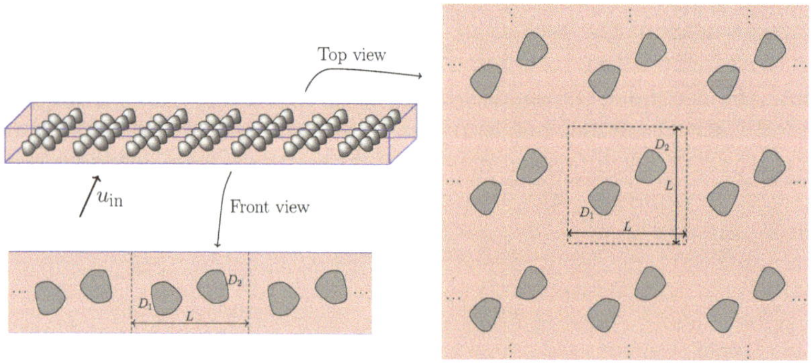

FIGURE 4.26. A metascreen is composed of a $\mathcal{P}$-symmetric resonator dimer $D = D_1 \cup D_2$ repeated periodically in a planar configuration with an incident plane wave u_{in}.

ω_2^α. Since the imaginary part of ω_2^α is very small, this corresponds to a sharp peak which will interfere with the broader peak associated to ω_1^α to create an asymmetric, Fano-type, transmission peak, as schematised in Figure 4.25.

To illustrate the results above, we consider dimers in a two-dimensional square lattice with period L in $\mathbb{R}^3$. The dual lattice Λ^* of Λ is defined as $\Lambda^* = (2\pi/L)\Lambda$. The unit cell is $Y = [-L/2, L/2] \times [-L/2, L/2] \times \mathbb{R}$ and the Brillouin zone is $Y^* = \mathbb{R}^2 \times \{0\}/\Lambda^*$. We assume that the dimer $D = D_1 \cup D_2$ satisfies the symmetry conditions (4.120) and, for simplicity, that $v = v_1 = v_2 = 1$ and $\delta_1 = \delta_2 = \delta$. The structure, called a metascreen, is depicted in Figure 4.26.

In order to investigate the subwavelength resonant frequencies of the metascreen, we first note that the corresponding periodic capacitance matrix C^0 introduced in Definition 4.42 satisfies the following properties.

PROPOSITION 4.78. *Under the assumption of inversion symmetry*

$$\mathcal{P}D_1 = D_2 \tag{4.127}$$

with $\mathcal{P}$ being the parity operator $\mathcal{P}(x) = -x$, the periodic capacitance matrix C^0 is given by

$$C^0 = C_{11}^0 \begin{pmatrix} 1 & -1 \\ -1 & 1 \end{pmatrix}.$$

Moreover, ψ_i^0 in (4.73) is independent of α_0, and consequently C^0 and $\mathbf{c}_i$, $i = 1, 2$, defined by (4.79), are independent of α_0.

PROOF. Since C^0 is symmetric, $C_{11}^0 = C_{22}^0$ by $\mathcal{P}$-symmetry and $C_{11}^0 + C_{12}^0 = 0$ by (4.75). The independence of ψ_i^0, $i = 1, 2$, from α_0 is shown in Appendix C.9. □

Again, from Appendix C.9, we have $\mathbf{c}_1 = -\mathbf{c}_2$. We write the vector $\mathbf{c}_1$ as

$$\mathbf{c}_1 = \begin{pmatrix} c_1 \\ c_2 \\ c_0 \end{pmatrix}. \tag{4.128}$$

Moreover, we also note that the following result on C^{1,α_0} holds.

PROPOSITION 4.79. *For α_0 satisfying $|\alpha_0| < 1$, the "higher-order" coefficients introduced in (4.78) are given, for $i, j = 1, 2$, by*

$$C_{ij}^{1,\alpha_0} = -\frac{\mathrm{i} w_0 L^2}{2} \begin{pmatrix} 1 & 1 \\ 1 & 1 \end{pmatrix} + \mathrm{i}\alpha_0 \cdot \mathbf{c}_1 \begin{pmatrix} 0 & 1 \\ -1 & 0 \end{pmatrix} - \frac{\mathrm{i} w_0 c_0^2}{2L^2} \begin{pmatrix} 1 & -1 \\ -1 & 1 \end{pmatrix}, \tag{4.129}$$

where w_0, first defined in (4.123), is given by

$$w_0 = \sqrt{1 - |\alpha_0|^2}. \tag{4.130}$$

Then by analogous steps as those in the proof of Lemma 4.14, we obtain that for $\alpha = \omega\alpha_0$, there are precisely two quasiperiodic subwavelength resonant frequencies $\omega_1(\delta)$ and $\omega_2(\delta)$ which continuously depend on δ and are such that $\omega_i(0) = 0$ for $i = 1, 2$. Moreover, ω_1 and ω_2 are characterised as the characteristic values of the operator-valued function

$$\mathcal{A}^\omega := \left(-\frac{1}{2}I + \mathcal{K}_D^{\omega,*}\right)(\mathcal{S}_D^\omega)^{-1} - \delta\left(\frac{1}{2}I + (\mathcal{K}_D^{-\omega\alpha_0,\omega})^*\right)(\mathcal{S}_D^{\omega\alpha_0,\omega})^{-1}, \tag{4.131}$$

which are near zero. Note that in our setting here (*i.e.*, with $\alpha = \omega\alpha_0$, $v = v_1 = v_2 = 1$, and $\delta = \delta_1 = \delta_2$), the operator introduced in (4.77) reduces to $\mathcal{A}^\omega$. Given the existence of ω_1 and ω_2, we may derive from Theorem 4.48 their asymptotic behaviours as $\delta \to 0$, which will be used to describe the Fano-type resonance of the metascreen.

THEOREM 4.80. *Let $\alpha = \omega\alpha_0$ for some α_0 that is independent of δ and satisfies $|\alpha_0| < 1$. Let c_0 and w_0 be, respectively, defined by (4.128) and (4.130). Then, as $\delta \to 0$,*

$$\omega_1^\alpha = -\frac{\mathrm{i}\delta w_0 L^2}{|D_1|} + \mathcal{O}(\delta^2),$$

$$\omega_2^\alpha = \sqrt{\frac{2\delta C_{11}^0}{|D_1|}} - \frac{\mathrm{i}\delta w_0 c_0^2}{2|D_1|L^2} + \mathcal{O}(\delta^{3/2}).$$

In addition to the symmetry conditions (4.120) (satisfied, for example, in Figure 4.27 with $\theta = 0$), we will assume that $\alpha_0 = 0$, which corresponds to the case where the incident wave is perpendicular to the metascreen.

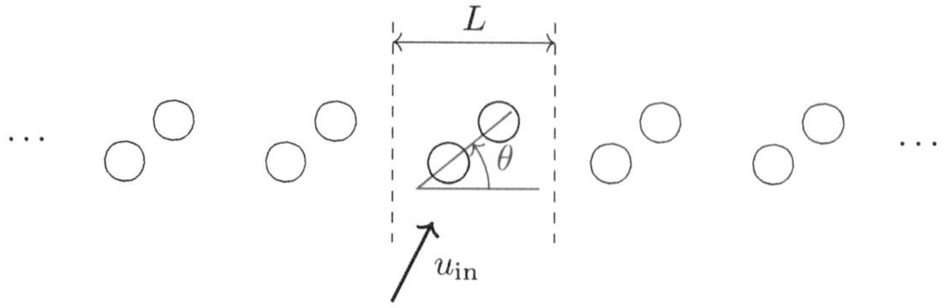

FIGURE 4.27. The front view of a symmetric metascreen with an incident plane wave u_{in}. In this case, we have circular resonators arranged in a $\mathcal{P}$-symmetric dimer that is inclined at an angle of θ to the plane of the metascreen.

From $\mathcal{P}_0 D_i = D_i$ for $i = 1, 2$, by symmetry arguments, it follows that $c_0 = 0$. Therefore, Theorem 4.80 shows that the $\mathcal{O}(\delta)$-imaginary part of ω_2 vanishes when the structure is symmetric. In fact, we will prove that the second resonance ω_2^α becomes exactly real under the additional assumption of perpendicular incidence.

PROPOSITION 4.81. *Assume that $\mathcal{P}_0 D_i = D_i$ for $i = 1, 2$, and that $\alpha_0 = 0$. Then ω_2 is real.*

PROOF. We introduce $\mathcal{B} : L^2(\partial D) \to L^2(\partial D)$ as

$$\mathcal{B}[\varphi] = -\chi_{\partial D_1} \int_{\partial D_1} \mathcal{S}_0^{\alpha_0}[\varphi] \, \mathrm{d}\sigma - \chi_{\partial D_2} \int_{\partial D_2} \mathcal{S}_0^{\alpha_0}[\varphi] \, \mathrm{d}\sigma,$$

and then we define

$$\mathcal{A}_0^\omega = \left(-\frac{1}{2}I + \mathcal{K}_D^{0,*}\right)(\mathcal{S}_D^0)^{-1} - \omega^2|D_1|I + \delta\mathcal{B}. \tag{4.132}$$

It is clear that $\mathcal{A}_0^\omega$ has a characteristic value given by

$$\omega_0 = \sqrt{\frac{2\delta C_{11}^0}{|D_1|}},$$

and that $\eta_0 = \frac{1}{\sqrt{2|D_1|}}(\chi_{\partial D_1} - \chi_{\partial D_2})$ spans the kernels of $\mathcal{A}_0^{\omega_0}$ and its adjoint $(\mathcal{A}_0^{\omega_0})^*$:

$$\mathrm{Ker}\mathcal{A}_0^{\omega_0} = \mathrm{span}\{\eta_0\} \quad \text{and} \quad \mathrm{Ker}(\mathcal{A}_0^{\omega_0})^* = \mathrm{span}\{\eta_0\}.$$

From (4.132), we have

$$\left\langle \frac{\mathrm{d}}{\mathrm{d}\omega}\mathcal{A}_0^\omega\big|_{\omega=\omega_0}[\eta_0], \eta_0 \right\rangle = -2\omega_0|D_1|.$$

Then, by the same arguments as those in Appendix C.1, we have the following pole-pencil decomposition for ω close to ω_0:

$$(\mathcal{A}_0^\omega)^{-1} = -\frac{\langle \cdot, \eta_0 \rangle \, \eta_0}{2\omega_0|D_1|(\omega - \omega_0)} + R(\omega),$$

where $\langle \cdot, \cdot \rangle$ denotes the $L^2(\partial D)$-inner product and $R(\omega)$ is a holomorphic function of ω in a neighbourhood of ω_0 satisfying

$$R(\omega)[\eta_0] = r(\omega)\eta_0,$$

for some function $r(\omega)$ that is real-valued for real ω.

By the characteristic value perturbation theory, as in the proof of Theorem C.2, we have

$$\omega_2 - \omega_0 = \frac{\mathrm{tr}}{2\pi \mathrm{i}} \sum_{p=1}^{+\infty} \frac{1}{p} \int_{\partial V} \left((\mathcal{A}_0^\omega)^{-1} (\mathcal{A}_0^\omega - \mathcal{A}^\omega) \right)^p \, \mathrm{d}\omega,$$

where $\mathcal{A}^\omega$ is defined by (4.131) and V is a small neighbourhood of ω_0.

Using the property that

$$\mathrm{tr} \int_{\partial V} AB \, \mathrm{d}\omega = \mathrm{tr} \int_{\partial V} BA \, \mathrm{d}\omega$$

for finitely meromorphic operators A and B in a neighbourhood of ω_0 (see Proposition A.7), we have

$$\omega_2 - \omega_0 = \frac{\mathrm{tr}}{2\pi \mathrm{i}} \sum_{p=1}^{+\infty} \sum_{q=1}^{p-1} \frac{1}{p} \binom{p}{q} \int_{\partial V} \frac{L^{p-q} R^q}{(\omega - \omega_0)^{p-q}} \, \mathrm{d}\omega, \tag{4.133}$$

which in turn shows that $\omega_2 - \omega_0$ is equal to

$$\frac{1}{2\pi \mathrm{i}} \sum_{p=1}^{+\infty} \sum_{q=0}^{p-1} \frac{(-1)^{p-q}}{p} \binom{p}{q} \left(\frac{1}{2\omega_0 |D_1|} \right)^{p-q} \int_{\partial V} \frac{r(\omega)^q \langle (\mathcal{A}_0^\omega - \mathcal{A}^\omega)[\eta_0], \eta_0 \rangle^{p-q}}{(\omega - \omega_0)^{p-q}} \, \mathrm{d}\omega,$$

or equivalently is equal to

$$\sum_{p=1}^{+\infty} \sum_{q=0}^{p-1} \frac{1}{(p-q-1)! p} \binom{p}{q} \left(\frac{1}{2\omega_0 |D_1|} \right)^{p-q} \frac{\mathrm{d}^{p-q-1}}{\mathrm{d}\omega^{p-q-1}} \left(r(\omega)^q \langle \mathcal{A}^\omega [\eta_0], \eta_0 \rangle^{p-q} \right) \Big|_{\omega = \omega_0}.$$

Next, we show that, under the symmetry assumptions $\mathcal{P}_0 D_i = D_i$ and for $\alpha_0 = 0$, the factor $\langle \mathcal{A}^\omega [\eta_0], \eta_0 \rangle$ is real for all $\omega \in \mathbb{R}$. This, together with (4.133), shows that ω_2 is real.

We write $\mathrm{Im} \, \langle \mathcal{A}^\omega [\eta_0], \eta_0 \rangle = \mathcal{I}_1 + \mathcal{I}_2$, where

$$\mathcal{I}_1 = \mathrm{Im} \left\langle \left(-\frac{1}{2} I + \mathcal{K}_D^{\omega,*} \right) (\mathcal{S}_D^\omega)^{-1} [\eta_0], \eta_0 \right\rangle,$$

$$\mathcal{I}_2 = -\delta \, \mathrm{Im} \left\langle \left(\frac{1}{2} I + (\mathcal{K}_D^{-\omega\alpha_0,\omega})^* \right) (\mathcal{S}_D^{\omega\alpha_0,\omega})^{-1} [\eta_0], \eta_0 \right\rangle.$$

We begin by studying $\mathcal{I}_1$. We let v be the solution to

$$\begin{cases} \Delta v + \omega^2 v = 0 & \text{in } D, \\ v = \eta_0 & \text{on } \partial D. \end{cases}$$

Then $\partial v / \partial \nu$ is real-valued, and it follows that

$$\left\langle \left(-\frac{1}{2} I + \mathcal{K}_D^{\omega,*} \right) (\mathcal{S}_D^\omega)^{-1} [\eta_0], \eta_0 \right\rangle = \int_{\partial D_1} \frac{\partial v}{\partial \nu} \, \mathrm{d}\sigma - \int_{\partial D_2} \frac{\partial v}{\partial \nu} \, \mathrm{d}\sigma$$

is real-valued. Hence, $\mathcal{I}_1 = 0$.

We next turn to $\mathcal{I}_2$. From (4.61), we observe that

$$\overline{G^{0,\omega}}(x) = G^{0,\omega}(x) - \frac{\cos(\omega |x_0|)}{\mathrm{i}\omega L^2}.$$

We define $\overline{\mathcal{S}_D^{0,\omega}}$ and $\overline{(\mathcal{K}_D^{0,\omega})^*}$ as the operators corresponding to the kernel $\overline{G^{0,\omega}}$. Let $\mathcal{P}_{12} : \mathbb{R}^3 \to \mathbb{R}^3$ be defined by

$$\mathcal{P}_{12}(x_1, x_2, x_3) = (-x_1, -x_2, x_3).$$

Note that $\mathcal{P}_{12} = \mathcal{P}_0 \mathcal{P}$ and define for a function φ,

$$\mathcal{P}_{12}\varphi(x) := \varphi(\mathcal{P}_{12}(x)).$$

Then, for functions φ which are such that $\mathcal{P}_{12}\varphi(x) = -\varphi(x)$, we have that

$$\overline{\mathcal{S}_D^{0,\omega}}[\varphi] = \mathcal{S}_D^{0,\omega}[\varphi], \qquad \overline{(\mathcal{K}_D^{-0,\omega})^*}[\varphi] = (\mathcal{K}_D^{-0,\omega})^*[\varphi]. \tag{4.134}$$

Denote by $\psi = \left(\mathcal{S}_D^{0,\omega}\right)^{-1}[\eta_0]$. Since

$$\mathcal{S}_D^{0,\omega}[\mathcal{P}_{12}\psi](x) = \mathcal{S}_D^{0,\omega}[\psi](\mathcal{P}_{12}x) = -\mathcal{S}_D^{0,\omega}[\psi](x), \tag{4.135}$$

we conclude that $\psi(\mathcal{P}_{12}x) = -\psi(x)$. Therefore, using (4.134) and since η_0 is real-valued, we have that

$$\eta_0 = \overline{\mathcal{S}_D^{0,\omega}}[\overline{\psi}] = \mathcal{S}_D^{0,\omega}[\overline{\psi}],$$

which means that $\overline{\psi} = \psi$. Again using (4.134), we therefore have

$$\overline{\left(\frac{1}{2}I + (\mathcal{K}_D^{-0,\omega})^*\right)[\psi]} = \left(\frac{1}{2}I + (\mathcal{K}_D^{-0,\omega})^*\right)[\psi],$$

from which we can see that

$$\mathcal{I}_2 = 0.$$

To conclude, we have proved that $\langle \mathcal{A}^\omega[\eta_0], \eta_0\rangle$ is real for all real ω. This, together with (4.133), proves that ω_2 is real. □

Next, we characterise the eigenmodes corresponding to the real resonance ω_2 and show that they are bound states in the continuum. Specifically, we show that they do not radiate to the far field and, reciprocally, cannot be excited from the far field.

PROPOSITION 4.82. *Assume that $\mathcal{P}_0 D_i = D_i$ for $i = 1, 2$ and that $\alpha_0 = 0$. Let u be a solution to (4.4) with $u_{\text{in}} = 0$ and $\omega = \omega_2$. For sufficiently small δ, we have as $x_0 \to \pm\infty$ that*

$$u \sim 0.$$

PROOF. Let $\mathcal{A}^{\omega_2}[\eta] = 0$. From the $\mathcal{P}_0$-symmetry, it follows that $\mathcal{P}_{12}\eta$ is in the kernel of $\mathcal{A}^{\omega_2}$ at $-\alpha_0$. Since $\alpha_0 = 0$, and since ω_2 is a simple characteristic value of $\mathcal{A}^\omega$, we find that

$$\mathcal{P}_{12}\eta = K\eta$$

for some constant K. For small δ, we know that

$$\eta = \chi_{\partial D_1} - \chi_{\partial D_2} + \mathcal{O}(\delta).$$

We conclude that $K = -1$, or in other words, that η is odd under $\mathcal{P}_{12}$ (we emphasise that this holds exactly, *i.e.*, that also the $\mathcal{O}(\delta)$-term of η is odd). Then, as $x_0 \to \pm\infty$, we have

$$u \sim \frac{e^{\mathrm{i}\mathbf{k}_\pm \cdot x}}{2\mathrm{i}\omega w_0 L^2} \int_{\partial D} e^{-\mathrm{i}\mathbf{k}_\pm \cdot y} \left(\mathcal{S}_D^{0,\omega}\right)^{-1}[\eta](y)\, d\sigma(y).$$

From the symmetry, we have that $\left(\mathcal{S}_D^{0,\omega}\right)^{-1}[e^{-\mathrm{i}\mathbf{k}_\pm \cdot y}]$ is an odd function under $\mathcal{P}_{12}$ (as was the case in (4.135)). Since $\alpha_0 = 0$, the functions $e^{-\mathrm{i}\mathbf{k}_\pm \cdot y}$ are even, so $u \sim 0$, which proves the claim. □

The following result (which is the reciprocal result of Proposition 4.82) shows that there will be no transmission peak at $\omega = \omega_2$ (corresponding to the fact that the bound state in the continuum cannot be excited from the far field). However, we remark that there might be some small but nonzero transmission $t = \mathcal{O}(\delta^{1/2})$ originating from the first broad resonance.

PROPOSITION 4.83. *Assume that $\mathcal{P}_0 D_i = D_i$ for $i = 1, 2$, and that $\alpha_0 = 0$. Then, at $\omega = \omega_2$, the scattering matrix $S(\omega)$, defined by (4.126), satisfies*

$$S(\omega_2) = -\begin{pmatrix} 1 & 0 \\ 0 & 1 \end{pmatrix} + \mathcal{O}(\delta^{1/2}).$$

PROOF. Recall that the scattering problem (4.80) (for $\alpha = 0$) at the operating frequency ω is described by

$$\mathcal{A}^\omega[\eta] = F[u_{\text{in}}], \tag{4.136}$$

where $\mathcal{A}^\omega$ is defined by (4.131) (with $\alpha_0 = 0$) and

$$F[u_{\text{in}}] := \delta \frac{\partial u_{\text{in}}}{\partial \nu} - \left(-\frac{1}{2}I + \mathcal{K}_D^{\omega,*}\right) \left(\mathcal{S}_D^\omega\right)^{-1} [u_{\text{in}}]. \tag{4.137}$$

Even though the equation (4.136) at $\omega = \omega_2$ is not uniquely solvable, the scattering matrix is well-defined since (by Proposition 4.82) any $\eta_{\text{k}} \in \operatorname{Ker} \mathcal{A}^{\omega_2}$ corresponds to an exponentially decaying part of the solution. We can decompose the solution η as

$$\eta = \eta_0 + \eta_{\text{k}} - u_{\text{in}},$$

where $\mathcal{P}_{12}\eta_0 = \eta_0$ and η_{k} is an element in $\operatorname{Ker} \mathcal{A}^{\omega_2}$. Then, we can show that

$$\eta_0 = K\chi_{\partial D} + \mathcal{O}(\delta),$$

for some constant K. From

$$\frac{1}{2\mathrm{i}\omega w_0 L^2} \int_{\partial D} e^{-\mathrm{i}\mathbf{k}_\pm \cdot y} \left(\mathcal{S}_D^{\omega\alpha_0,\omega}\right)^{-1} [u_{\text{in}}](y) \, d\sigma(y) = 1 + \mathcal{O}(\omega), \tag{4.138}$$

we have

$$\begin{aligned} u - u_{\text{in}} &\sim \frac{e^{\mathrm{i}\mathbf{k}_\pm \cdot x}}{2\mathrm{i}\omega w_0 L^2} \int_{\partial D} e^{-\mathrm{i}\mathbf{k}_\pm \cdot y} \left(\mathcal{S}_D^{0,\omega}\right)^{-1} [\eta](y) \, d\sigma(y) \\ &\sim \left(K - 1 + \mathcal{O}(\delta^{1/2})\right) e^{\mathrm{i}\mathbf{k}_\pm \cdot x}. \end{aligned} \tag{4.139}$$

Then, we can find from $\mathcal{A}^{\omega_2}[\eta_0 - u_{\text{in}}] = F[u_{\text{in}}]$ that

$$K\left(C^0 - \lambda_2^0 I\right) \begin{pmatrix} 1 \\ 1 \end{pmatrix} = \mathcal{O}(\delta^{1/2}),$$

where λ_2^0 is the nonzero eigenvalue of the periodic capacitance matrix C^0 defined in Definition 4.42. Since $\begin{pmatrix} 1 \\ 1 \end{pmatrix}$ is not in the kernel of $(C^0 - \lambda_2^0 I)$, we find that $K = \mathcal{O}(\delta^{1/2})$. The expression for $S(\omega_2)$ then follows from (4.139). □

In order to compute the scattering matrix $S(\omega)$ of the metascreen and to demonstrate an asymmetric transmission peak around the second resonance ω_2, which is characteristic of a Fano-type transmission anomaly, we begin with a characterisation of the solution to the scattering problem. We define the functions

$$S_j^{\alpha,\omega}(x) = \begin{cases} \mathcal{S}_D^{\alpha,\omega}[\psi_j^0 + \omega\psi_j^{1,\alpha_0}](x), & x \in \mathbb{R}^3 \setminus \overline{\mathcal{D}}, \\ \mathcal{S}_{\mathcal{D}}^\omega[\psi_j](x), & x \in \mathcal{D}, \end{cases} \tag{4.140}$$

where $\psi_j = (\mathcal{S}_D^0)^{-1}[\chi_{\partial D_j}]$ and $\psi_j^0, \psi_j^{1,\alpha_0}$ are defined respectively by (4.73) and (4.78). We then have the following result.

PROPOSITION 4.84. *Assume that $c_0 \neq 0$ and that ω is real with $0 \leq \omega \leq K\sqrt{\delta}$ for some constant $K > 0$. Then, as $\delta \to 0$,*

$$u - u_{\text{in}} = q_1 S_1^{\alpha,\omega} + q_2 S_2^{\alpha,\omega} - \mathcal{S}_D^{\alpha,\omega} \left(\mathcal{S}_D^{\alpha,\omega}\right)^{-1} [u_{\text{in}}] + \mathcal{O}(\omega^2), \tag{4.141}$$

where $q := \begin{pmatrix} q_1 \\ q_2 \end{pmatrix}$ satisfies the problem

$$\left(C^0 + \omega C^{1,\alpha_0} - \frac{\omega^2 |D_1|}{\delta} I\right) \begin{pmatrix} q_1 \\ q_2 \end{pmatrix} = -\begin{pmatrix} p_1 \\ p_2 \end{pmatrix} + \mathcal{O}(\delta), \quad p_i = \int_{\partial D_i} \left(\mathcal{S}_D^{\omega\alpha_0,\omega}\right)^{-1} [u_{\text{in}}] \, \mathrm{d}\sigma,$$

for $i = 1, 2$.

PROOF. We solve the equation (4.136). When $c_0 \neq 0$ and ω is real, we know that $\mathcal{A}^\omega$ is invertible with bounded inverse. Then, we have that η satisfies

$$\eta = q_1 \chi_{\partial D_1} + q_2 \chi_{\partial D_2} - u_{\text{in}} + \mathcal{O}(\omega^2 + \delta),$$

which proves (4.141). To prove that the equation for q holds, we proceed by expanding $\mathcal{A}^\omega[\eta]$ and integrating over ∂D_i to get

$$\begin{aligned} \int_{\partial D_i} \mathcal{A}^\omega[\eta] \, \mathrm{d}\sigma = &-q_i |D_1| \omega^2 - \delta \int_{\partial D_i} \left(\mathcal{S}_D^{\omega\alpha_0,\omega}\right)^{-1} [\eta] \, \mathrm{d}\sigma + \int_{D_i} \omega^2 u_{\text{in}} \, \mathrm{d}x \\ &+ \mathcal{O}(\delta\omega^2 + \omega^4). \end{aligned}$$

Turning to the right-hand side of (4.136), we have

$$\begin{aligned} \int_{\partial D_i} F[u_{\text{in}}] \, \mathrm{d}\sigma &= -\int_{\partial D_i} \left(-\frac{1}{2} I + \mathcal{K}_D^{\omega,*}\right) \left(\mathcal{S}_D^\omega\right)^{-1} [u_{\text{in}}] \, \mathrm{d}\sigma + \mathcal{O}(\delta^2) \\ &= -\int_{\partial D_i} \frac{\partial}{\partial \nu} \left(\mathcal{S}_D^\omega\right)^{-1} [u_{\text{in}}] \bigg|_- \mathrm{d}\sigma + \mathcal{O}(\delta^2) \\ &= \omega^2 \int_{D_i} u_{\text{in}} \, \mathrm{d}x + \mathcal{O}(\delta^2). \end{aligned}$$

Therefore, we have the equation

$$-q_i |D_1| \omega^2 - \delta \int_{\partial D_i} \left(\mathcal{S}_D^{\omega\alpha_0,\omega}\right)^{-1} [\eta] \, \mathrm{d}\sigma = \mathcal{O}(\delta\omega^2 + \omega^4).$$

Defining

$$p_i = \int_{\partial D_i} \left(\mathcal{S}_D^{\omega\alpha_0,\omega}\right)^{-1} [u_{\text{in}}] \, \mathrm{d}\sigma,$$

the above equation can be rewritten in matrix form as

$$\left(\delta C^0 + \omega\delta C^{1,\alpha_0} - \frac{\omega^2 |D_1|}{\delta} I\right) \begin{pmatrix} q_1 \\ q_2 \end{pmatrix} = -\delta \begin{pmatrix} p_1 \\ p_2 \end{pmatrix} + \mathcal{O}(\delta\omega^2 + \omega^4),$$

which proves the claim. □

We are now ready to compute the scattering matrix $S(\omega)$ and demonstrate the Fano-type asymmetric transmission line. We define the coefficients

$$R_{j,\pm} := \frac{1}{2\mathrm{i}\omega w_0 L^2} \int_{\partial D} e^{-\mathrm{i}\mathbf{k}_\pm \cdot y} \left(\psi_j^0(y) + \omega \psi_j^{1,\alpha_0}(y)\right) \mathrm{d}\sigma(y), \quad j = 1, 2,$$

which describe the radiation of $S_j^{\alpha,\omega}$ given by (4.140) as $x_0 \to \pm\infty$. As $\omega \to 0$, we have the following asymptotic behaviour:

(4.142)
$$\begin{aligned} R_{j,\pm} &= \frac{1}{2\mathrm{i}\omega w_0 L^2}\left(\int_{\partial D} \psi_j^0(y)\, \mathrm{d}\sigma(y) - \int_{\partial D} \mathrm{i}\mathbf{k}_\pm \cdot y\psi_j^0(y)\, \mathrm{d}\sigma(y) + \omega \int_{\partial D} \psi_j^{1,\alpha_0}\, \mathrm{d}\sigma\right) \\ &\quad + \mathcal{O}(\omega) \\ &= -\frac{\mathbf{k}_\pm \cdot \mathbf{c}_j}{2\omega w_0 L^2} + \frac{1}{2\mathrm{i} w_0 L^2}\int_{\partial D} \psi_j^{1,\alpha_0}\, \mathrm{d}\sigma + \mathcal{O}(\omega) \\ &= \frac{1}{2} - \frac{1}{2\omega w_0 L^2}(\mathbf{k}_\pm - \alpha)\cdot \mathbf{c}_j + \mathcal{O}(\omega) \\ &= \frac{1}{2} \pm \frac{(-1)^j c_0}{2L^2} + \mathcal{O}(\omega), \end{aligned}$$

where c_0 and w_0 are defined respectively by (4.128) and (4.130).

The following theorem was obtained in [**26**].

THEOREM 4.85. *Assume that $c_0 \neq 0$ and that $0 \leq \omega \leq K\sqrt{\delta}$ for some constant K. Then, we have the following asymptotic expansion of the scattering matrix as $\delta \to 0$:*

$$S(\omega) = \frac{\omega_1^\alpha}{\omega_1^\alpha - \omega}\begin{pmatrix}1 & 1\\ 1 & 1\end{pmatrix} + \frac{2\mathrm{i}\omega\, \mathrm{Im}\,(\omega_2^\alpha)}{(\omega_2^\alpha)^2 - \omega^2}\begin{pmatrix}1 & -1\\ -1 & 1\end{pmatrix} - \begin{pmatrix}1 & 0\\ 0 & 1\end{pmatrix} + \mathcal{O}(\delta^{1/2}), \tag{4.143}$$

where the error term is uniform with respect to ω.

PROOF. We begin by computing $p := \begin{pmatrix}p_1\\ p_2\end{pmatrix}$. Recall that $u_{\mathrm{in}}(x) = e^{\mathrm{i}\mathbf{k}_+\cdot x}$. We then have

$$\begin{aligned} p_i = \int_{\partial D_i} (\mathcal{S}_D^{\alpha,\omega})^{-1}[u_{\mathrm{in}}]\, \mathrm{d}\sigma &= \int_{\partial D} u_{\mathrm{in}} \left(\mathcal{S}_D^{-\alpha,\omega}\right)^{-1}[\chi_{\partial D_i}]\, \mathrm{d}\sigma \\ &= \int_{\partial D} \mathrm{i}\mathbf{k}_+ \cdot x\psi_i^0\, \mathrm{d}\sigma + \omega\int_{\partial D} \psi_i^{1,-\alpha_0}\, \mathrm{d}\sigma + \mathcal{O}(\omega^2) \\ &= \mathrm{i}\omega w_0 L^2 + \mathrm{i}(\mathbf{k}_+ - \alpha)\cdot \mathbf{c}_i + \mathcal{O}(\delta), \end{aligned}$$

so that

$$p = \mathrm{i}\omega w_0 L^2 \begin{pmatrix}1\\ 1\end{pmatrix} + \mathrm{i}\omega w_0 c_0 \begin{pmatrix}1\\ -1\end{pmatrix} + \mathcal{O}(\delta).$$

As $x_0 \to \pm\infty$, we have, using (4.142) and (4.138), that

$$\begin{aligned} (u - u_{\mathrm{in}})(x) &\sim (q_1 R_{1,\pm} + q_2 R_{2,\pm} - 1)\, e^{\mathrm{i}\mathbf{k}_\pm\cdot x} \\ &\sim \left(\frac{q_2 + q_1}{2} \pm \frac{c_0}{L^2}\frac{q_2 - q_1}{2} - 1 + \mathcal{O}(\omega)\right) e^{\mathrm{i}\mathbf{k}_\pm\cdot x}. \end{aligned}$$

For all real ω with $0 \leq \omega \leq K\sqrt{\delta}$ and for fixed $c_0 \neq 0$, there exists a constant $A > 0$ such that $|\omega - \omega_i| > A\delta$ for $i = 1, 2$. Then, $\left(C^0 + \omega C^{1,\alpha_0} - \frac{\omega^2|D_1|}{\delta} I\right)$ is invertible and its inverse satisfies

$$\left(C^0 + \omega C^{1,\alpha_0} - \frac{\omega^2|D_1|}{\delta} I\right)^{-1} = \mathcal{O}(\delta^{-1/2}),$$

uniformly in ω. We have

$$
\begin{aligned}
d &:= \det\left(C^0 + \omega C^{1,\alpha_0} - \frac{\omega^2|D_1|}{\delta}I\right) \\
&= \left(-\mathrm{i}\omega w_0 L^2 - \frac{\omega^2|D_1|}{\delta}\right)\left(2C^0_{11} - \frac{\mathrm{i}\omega w_0 c_0^2}{L^2} - \frac{\omega^2|D_1|}{\delta}\right) + \mathcal{O}(\delta).
\end{aligned}
$$

Hence,

$$
\begin{aligned}
\left(C^0 + \omega C^{1,\alpha_0} - \frac{\omega^2|D_1|}{\delta}I\right)^{-1} &= \frac{1}{d}\Bigg(\left(C^0_{11} - \frac{\mathrm{i}\omega w_0 c_0^2}{2L^2}\right)\begin{pmatrix}1 & 1\\ 1 & 1\end{pmatrix} \\
&\quad - \frac{\mathrm{i}\omega w_0 L^2}{2}\begin{pmatrix}1 & -1\\ -1 & 1\end{pmatrix} + \mathrm{i}\alpha\cdot\mathbf{c}_1\begin{pmatrix}0 & -1\\ 1 & 0\end{pmatrix} - \frac{\omega^2|D_1|}{\delta}I\Bigg).
\end{aligned}
$$

Using Proposition 4.84, we therefore have

$$
\begin{aligned}
q = &-\frac{1}{d}\left(\mathrm{i}\omega w_0 L^2\left(2C^0_{11} - \frac{\mathrm{i}\omega w_0 c_0^2}{L^2} - \frac{\omega^2|D_1|}{\delta}\right) + \mathrm{i}\alpha\cdot\mathbf{c}_1(\mathrm{i}\omega w_0 c_0)\right)\begin{pmatrix}1\\ 1\end{pmatrix} \\
&- \frac{1}{d}\left(\mathrm{i}\omega w_0 c_0\left(-\mathrm{i}\omega w_0 L^2 - \frac{\omega^2|D_1|}{\delta}\right) - \mathrm{i}\alpha\cdot\mathbf{c}_1(\mathrm{i}\omega w_0 L^2)\right)\begin{pmatrix}1\\ -1\end{pmatrix} + \mathcal{O}(\delta^{1/2}),
\end{aligned}
$$

uniformly in ω. Simplifying the above expression, we obtain

$$
\begin{aligned}
q = &\frac{-\mathrm{i}\omega w_0 L^2}{-\mathrm{i}\omega w_0 L^2 - \frac{\omega^2|D_1|}{\delta}}(1+\varepsilon_1)\begin{pmatrix}1\\ 1\end{pmatrix} + \frac{-\mathrm{i}\omega w_0 c_0}{2C^0_{11} - \frac{\mathrm{i}\omega w_0 c_0^2}{L^2} - \frac{\omega^2|D_1|}{\delta}}(1-\varepsilon_2)\begin{pmatrix}1\\ -1\end{pmatrix} \\
&+ \mathcal{O}(\delta^{1/2}),
\end{aligned}
$$

where

$$
\varepsilon_1 = \frac{\mathrm{i}\alpha\cdot\mathbf{c}_1\frac{c_0}{L^2}}{\left(2C^0_{11} - \frac{\mathrm{i}\omega w_0 c_0^2}{L^2} - \frac{\omega^2|D_1|}{\delta}\right)}, \qquad \varepsilon_2 = \frac{\mathrm{i}\alpha\cdot\mathbf{c}_1\frac{L^2}{c_0}}{\left(-\mathrm{i}\omega w_0 L^2 - \frac{\omega^2|D_1|}{\delta}\right)}.
$$

When ω is away from ω_1, we have $\varepsilon_2 = \mathcal{O}(\delta^{1/2})$. For $\omega - \omega_1 = \mathcal{O}(\delta)$, we have

$$
\frac{\mathrm{i}\omega w_0 c_0}{\left(2C^0_{11} - \frac{\mathrm{i}\omega w_0 c_0^2}{L^2} - \frac{\omega^2|D_1|}{\delta}\right)}(1-\varepsilon_2) = \mathcal{O}(\delta^{1/2}).
$$

An analogous argument for ω close to ω_2 shows that for $0 \le \omega \le K\sqrt{\delta}$, it holds that

$$
q = \frac{-\mathrm{i}\omega w_0 L^2}{\left(-\mathrm{i}\omega w_0 L^2 - \frac{\omega^2|D_1|}{\delta}\right)}\begin{pmatrix}1\\ 1\end{pmatrix} + \frac{-\mathrm{i}\omega w_0 c_0}{\left(2C^0_{11} - \frac{\mathrm{i}\omega w_0 c_0^2}{L^2} - \frac{\omega^2|D_1|}{\delta}\right)}\begin{pmatrix}1\\ -1\end{pmatrix} + \mathcal{O}(\delta^{1/2}).
$$

Then, as $x_0 \to \pm\infty$, we can see that

$$
\begin{aligned}
u - u_{\mathrm{in}} &\sim \left(\frac{q_2+q_1}{2} \pm \frac{c_3}{L^2}\frac{q_2-q_1}{2} - 1 + \mathcal{O}(\omega)\right)e^{\mathrm{i}\mathbf{k}_\pm\cdot x} \\
&\sim \left(\frac{-\mathrm{i}\omega w_0 L^2}{-\mathrm{i}\omega w_0 L^2 - \frac{\omega^2|D_1|}{\delta}} \pm \frac{\mathrm{i}\omega w_0 c_0^2}{L^2\left(2C^0_{11} - \frac{\mathrm{i}\omega w_0 c_0^2}{L^2} - \frac{\omega^2|D_1|}{\delta}\right)} - 1 + \mathcal{O}(\delta^{1/2})\right)e^{\mathrm{i}\mathbf{k}_\pm\cdot x},
\end{aligned}
$$

which gives

$$
t = \frac{-\mathrm{i}\omega w_0 L^2}{-\mathrm{i}\omega w_0 L^2 - \frac{\omega^2|D_1|}{\delta}} + \frac{\mathrm{i}\omega w_0 c_0^2}{L^2\left(2C^0_{11} - \frac{\mathrm{i}\omega w_0 c_0^2}{L^2} - \frac{\omega^2|D_1|}{\delta}\right)} + \mathcal{O}(\delta^{1/2}),
$$

and

$$r = \frac{-\mathrm{i}\omega w_0 L^2}{-\mathrm{i}\omega w_0 L^2 - \frac{\omega^2 |D_1|}{\delta}} - \frac{\mathrm{i}\omega w_0 c_0^2}{L^2\left(2C_{11}^0 - \frac{\mathrm{i}\omega w_0 c_0^2}{L^2} - \frac{\omega^2|D_1|}{\delta}\right)} - 1 + \mathcal{O}(\delta^{1/2}).$$

In light of Theorem 4.80, we can rewrite this as

$$S(\omega) = \frac{\omega_1^\alpha}{\omega_1^\alpha - \omega}\begin{pmatrix} 1 & 1 \\ 1 & 1 \end{pmatrix} + \frac{2\mathrm{i}\omega \operatorname{Im}(\omega_2^\alpha)}{(\omega_2^\alpha)^2 - \omega^2}\begin{pmatrix} 1 & -1 \\ -1 & 1 \end{pmatrix} - \begin{pmatrix} 1 & 0 \\ 0 & 1 \end{pmatrix} + \mathcal{O}(\delta^{1/2}),$$

which proves the claim. □

At the resonances, *i.e.*, when $\omega = 0$ or $\omega = \operatorname{Re}(\omega_2)$, the scattering matrix is given by

$$S(0) = \begin{pmatrix} 0 & 1 \\ 1 & 0 \end{pmatrix} + \mathcal{O}(\delta^{1/2}) \quad \text{and} \quad S(\operatorname{Re}(\omega_2)) = \begin{pmatrix} 0 & -1 \\ -1 & 0 \end{pmatrix} + \mathcal{O}(\delta^{1/2}),$$

corresponding to transmission peaks where the transmittance is close to 1. The widths of these peaks are specified by the corresponding imaginary parts $\operatorname{Im}(\omega_1)$ and $\operatorname{Im}(\omega_2)$.

Theorem 4.85 can be used to demonstrate the Fano-type transmission anomaly. If we tune the parameters of the system so that $\operatorname{Im}(\omega_1)$ is large while $\operatorname{Im}(\omega_2)$ is small, then we can have, for small ω^*, that

$$\frac{\omega_1}{\omega_1 - (\operatorname{Re}(\omega_2) - \omega^*)} \approx \frac{\omega_1}{\omega_1 - (\operatorname{Re}(\omega_2) + \omega^*)} \approx \frac{\omega_1}{\omega_1 - \operatorname{Re}(\omega_2)} =: t_1,$$

where t_1 is not too small. In this case, the transmission coefficient is given by

$$t(\operatorname{Re}(\omega_2) + \omega^*) \approx \frac{1}{1 - \frac{\operatorname{Re}(\omega_2)}{\omega_1}} - \frac{1}{1 - \frac{\omega^*}{\mathrm{i}\operatorname{Im}(\omega_2)}}.$$

In particular, at $\omega^* = \operatorname{Re}(\omega_2)\frac{\operatorname{Im}(\omega_2)}{\operatorname{Im}(\omega_1)}$, we can see that

$$t(\operatorname{Re}(\omega_2) + \omega^*) \approx 0, \quad t(\operatorname{Re}(\omega_2) - \omega^*) \approx 2t_1.$$

We emphasise that $\omega^* > 0$ and that t is close to zero at $\omega = \operatorname{Re}(\omega_2) + \omega^*$ and not at $\omega = \operatorname{Re}(\omega_2) - \omega^*$. In other words, we have an asymmetric transmission peak at $\omega = \operatorname{Re}(\omega_2)$. For some frequency slightly larger than $\operatorname{Re}(\omega_2)$ the transmittance will be close to zero, but for all frequencies slightly lower than $\operatorname{Re}(\omega_2)$ the transmittance will be nonzero.

To conclude this section, we compute numerically the transmission and reflection spectra of a two-dimensional analogue of the structure studied above. We assume that D consists of two disks D_1 and D_2, both of radius R_D and separated by a distance d. Moreover, we denote by θ the angle at which D is rotated from being parallel to the x_1-axis, as depicted in Figure 4.27. In order to achieve a Fano-type resonance, we will choose a small but nonzero θ. We use the material parameters $L = 1$, $R_D = 0.05$, $d = 0.3$, $\alpha_0 = 0$ and $v = v_1 = v_2 = 1$. We compute the transmittance $T = |t|^2$ and reflectance $R = |r|^2$ in two different ways: using the asymptotic formulas derived in Theorem 4.80 (denoted by T_f and R_f) and by discretising the Helmholtz operator (denoted by T and R) using the multipole expansion method described in Appendix B.6.

Figures 4.28–4.30 show numerically computed transmission spectra for different values of δ. In Figure 4.28, we choose $\theta = 0.025\pi$ while in Figures 4.29 and 4.30 we choose $\theta = 0.05\pi$. All cases demonstrate Fano-type transmission anomalies, where

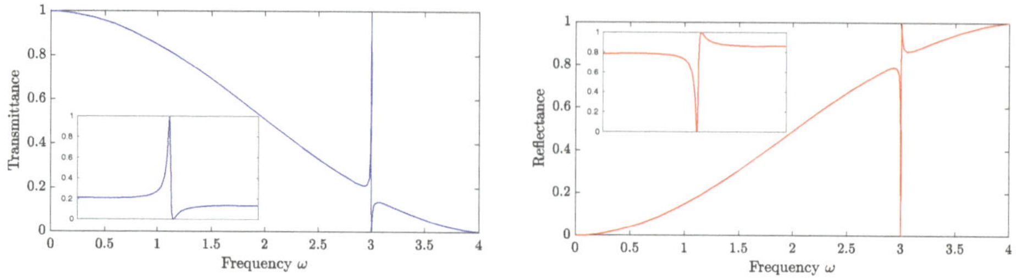

(A) Transmission spectrum. (B) Reflection spectrum.

FIGURE 4.28. Transmittance (a) and reflectance (b) in the case $\delta = 0.02$, computed using the multipole expansion method. The asymmetric Fano-type transmission curve is clearly visible. Choosing δ rather large makes the Fano-type asymmetry more pronounced. Here, we use $\theta = 0.025\pi$.

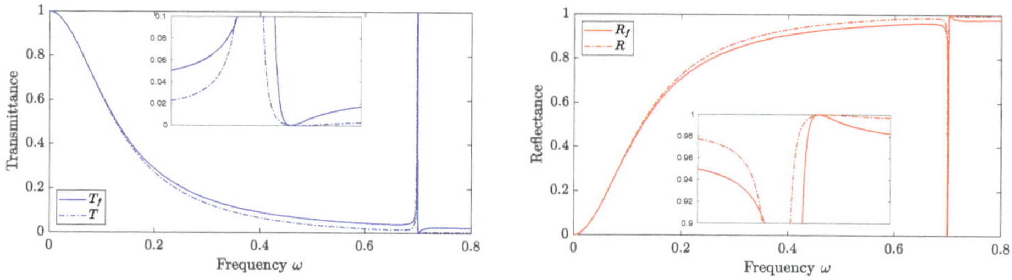

(A) Transmission spectrum. (B) Reflection spectrum.

FIGURE 4.29. Transmittance (a) and reflectance (b) in the case $\delta = 10^{-3}$, computed using the multipole expansion method (dashed line) and the asymptotic formulas (solid line). Compared to Figure 4.28, the smaller value of δ here means that the resonances occur in the subwavelength regime and that the asymptotic formulas provide a good approximation. Here, we use $\theta = 0.05\pi$.

the transmittance is close to zero for frequencies slightly above the second peak. For larger values of δ, the widths of the first broad transmission peaks are larger and the Fano-type transmission more pronounced. As expected, the asymptotic formulas are more accurate for smaller values of δ. In Figure 4.31, we show the transmission spectra for the same parameters as Figure 4.29 but with $\theta = 0$, which means that the structure is $\mathcal{P}_0$-symmetric. We observe that there is no sharp transmission peak in this case, which is expected since the eigenmodes are bound states in the continuum and are not excited by incident waves from the far field.

4.8. Anderson localisation

As seen in the previous sections, *locally resonant* materials, whose constituents exhibit resonance at small (*i.e.*, *subwavelength*) frequencies, can be used to achieve wave localisation at subwavelength scales. A periodic crystal of N repeated high-contrast resonators is shown to have N subwavelength resonant band functions. A bandgap is guaranteed to exist above these bands and other gaps can exist between them, depending on the choice of geometry. Compact perturbations can create

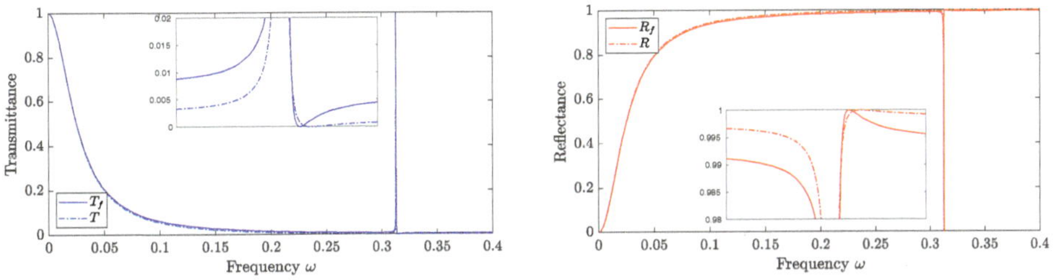

(A) Transmission spectrum (B) Reflection spectrum

FIGURE 4.30. Transmittance (a) and reflectance (b) in the case $\delta = 2 \cdot 10^{-4}$, computed using the multipole expansion method (dashed line) and the asymptotic formulas (solid line). Here, we use $\theta = 0.05\pi$.

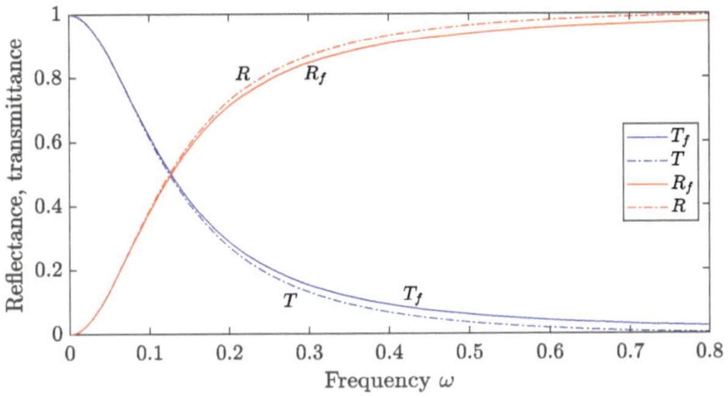

FIGURE 4.31. In a symmetric structure, there is a bound state in the continuum. Here, we repeat the analysis from Figure 4.29 (with $\delta = 1/1000$) but with $\theta = 0$. We observe that the transmission peak around $\omega \approx 0.7$ is not present in this case, consistent with Proposition 4.83 and the fact that corresponding resonant mode is a bound state in the continuum. This was computed using both the multipole expansion method (dashed line in blue for the transmittance and in red for the reflectance) and the asymptotic formulas (solid line in blue for the transmittance and in red for the reflectance).

eigenmodes with eigenfrequencies within these bandgaps. These "midgap" modes are elements of the point spectrum and cannot couple with the Bloch modes, so they decay exponentially quickly away from the defect. Some examples of localised modes, for a planar square lattice of high-contrast three-dimensional resonators with randomly chosen material parameters, are shown in Figure 4.32.

In this section, we derive concise formulas for the midgap frequencies that exist when there are finitely many random perturbations and describe the localised modes of a fully random structure by increasing the number of perturbations. The behaviour of the randomly perturbed systems can be understood by considering the simple phenomena of *hybridisation* and *level repulsion*. That is, when two eigenmodes are coupled, one of the eigenfrequencies of the resulting hybridised modes will be shifted up while the other will be shifted down. Meanwhile, level repulsion means that when disorder is added to the entries of a matrix, the eigenvalues tend to separate. These can be used to explain, firstly, the observation that the stronger

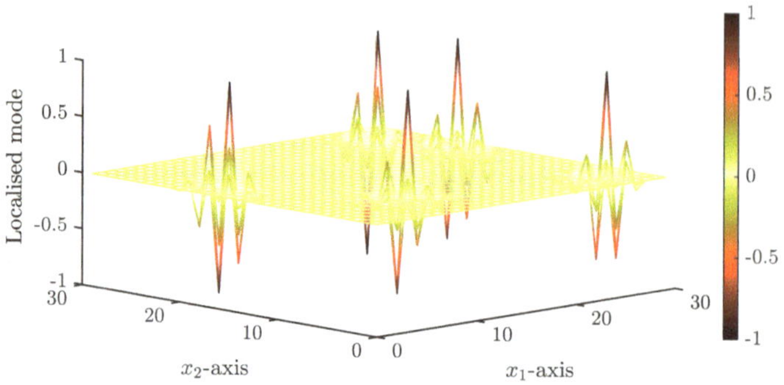

FIGURE 4.32. Examples of localised eigenmodes in a two-dimensional, finite, square lattice. Here shown in the case of $N = 900$ resonators and random perturbations drawn from a uniform distribution with standard deviation $\sigma = 0.05$.

the disorder, the more localised modes are created, as well as the tendency for the eigenmodes to become more strongly localised as the strength of the disorder increases.

The phase transition whereby disordered media changes from being conducting (exhibiting only weak localisation) to being insulating (the absence of diffusion) is sometimes referred to as the *Anderson transition* [**57, 58, 108**]. Here, we similarly observe a phase transition at a positive value of the perturbation strength. At this transition, eigenfrequencies become degenerate and the symmetries of the corresponding modes swap (*i.e.*, even modes change to odd modes, and vice versa). In general, when eigenmodes hybridise, the volume of their support increases, decreasing the extent to which they can be said to be localised. However, this trend breaks down at the phase transition. At this point, any small perturbation will lift the degeneracy and cause the modes to decouple, producing strongly localised modes. A sharp peak in the average degree of localisation is observed here, which is a key signature of the transition.

The analysis in this section is again based on the generalised capacitance matrix. It demonstrates localisation in systems with long-range interactions. The results presented here are from [**24**].

4.8.1. Formulation of the resonance problem. We assume that the resonator system consists of periodically arranged resonators, whose material parameters may be non-periodic. Inside the unit cell Y, we assume that there are N resonators, $D_i \subset Y$, for $i = 1, \dots, N$. These are disjoint connected domains with boundaries in $C^{1,s}$ for some $0 < s < 1$. For $m \in \Lambda$, we let D_i^m denote the translated resonator, and we let ∂D denote the total collection of resonators:

$$\mathcal{D} = \bigcup_{m\in\Lambda} \bigcup_{i\in\{1,\dots,N\}} D_i^m, \qquad D_i^m = D_i + m, \qquad D = \bigcup_{i\in\{1,\dots,N\}} D_i,$$

where Λ is a lattice of dimension $1 \leq d_l \leq d$. We let v_i^m denote the wave speed in D_i^m, and v denote the wave speed in the surrounding medium. Each resonator is additionally associated with a parameter $\delta_i^m = \mathcal{O}(\delta)$ for some small parameter

$\delta \to 0$. We will consider the equation

$$
(4.144) \quad \begin{cases} \Delta u + \frac{\omega^2}{v^2} u = 0 & \text{in } \mathbb{R}^d \setminus \mathcal{D}, \\ \Delta u + \frac{\omega^2}{(v_i^m)^2} u = 0 & \text{in } D_i^m, \ i = 1, \ldots, N, \\ u|_+ - u|_- = 0 & \text{on } \partial \mathcal{D}, \\ \delta_i^m \frac{\partial u}{\partial \nu}\Big|_+ - \frac{\partial u}{\partial \nu}\Big|_- = 0 & \text{on } \partial D_i^m, \ i = 1, \ldots, N, \\ u(x_l, x_0) \text{ satisfies the outgoing radiation condition as } |x_0| \to +\infty. \end{cases}
$$

We generalise the framework introduced in Section 4.5 to a general, possibly random, parameter distribution. We characterise the defect modes in terms of the generalised capacitance matrix. In the case of a compact perturbation of a periodic system, the capacitance formalism reduces to an eigenvalue problem for a Toeplitz matrix.

4.8.2. Discrete–operator formulation. The goal is to derive a capacitance formulation that is valid even in the case where the material parameters are not distributed periodically.

Assume that u is a localised eigenmode to (4.144) in the subwavelength regime, *i.e.*, u corresponds to an eigenvalue ω which scales as $\mathcal{O}(\delta^{1/2})$. Here, we refer to localisation in the L^2-sense and, in addition, we assume that u is normalised, in the sense that

$$
\int_{\mathbb{R}^{d_l}} |u(x_l, 0)|^2 \, dx_l = 1 \quad \text{and} \quad \int_{\mathbb{R}^{d_l}} |u(x_l, x_0)|^2 \, dx_l < +\infty,
$$

for any $x_0 \in \mathbb{R}^{d-d_l}$. From [**40**], we have the following result.

LEMMA 4.86. *Let u be a localised, normalised eigenmode to* (4.144) *corresponding to an eigenvalue ω which satisfies $\omega = \mathcal{O}(\delta^{1/2})$ as $\delta \to 0$. Then, uniformly in $x \in \mathcal{D}$,*

$$
u(x) = u_i^m + \mathcal{O}(\delta^{1/2}), \quad x \in D_i^m, i = 1, \ldots, N, m \in \Lambda,
$$

for some u_i^m that are constant with respect to x and δ.

PROOF. Intuitively, the idea is that as $\delta \to 0$, the limiting problem is a homogeneous Neumann problem inside D_i^m. To show this, we let Ω be a neighbourhood of $D^m := \cup_{i=1,\ldots,N} D_i^m$ and consider u satisfying

$$
(4.145) \quad \begin{cases} \Delta u + \frac{\omega^2}{v^2} u = 0 & \text{in } \Omega \setminus \overline{D^m}, \\ \Delta u + \frac{\omega^2}{(v_i^m)^2} u = 0 & \text{in } D_i^m, \\ u|_+ - u|_- = 0 & \text{on } \partial D_i^m, \\ \delta_i^m \frac{\partial u}{\partial \nu}\Big|_+ - \frac{\partial u}{\partial \nu}\Big|_- = 0 & \text{on } \partial D_i^m, \end{cases}
$$

for $i = 1, \ldots, N$. Then

$$
u = \begin{cases} H + \mathcal{S}_{D^m}^{k}[\phi] & \text{in } \Omega \setminus \overline{D^m}, \\ \mathcal{S}_{D^m}^{k_i^m}[\psi] & \text{in } D_i^m, \end{cases}
$$

for some densities ϕ and ψ and H satisfying $(\Delta + k^2)H = 0$ in Ω. Here, $k = \omega/v$ and $k_i^m = \omega/v_i^m$. Clearly, if u satisfies (4.145), then cu satisfies (4.145) for any $c \in \mathbb{C}$, so we can assume that

$$
\|\phi\|_{L^2(\partial D^m)} = \mathcal{O}(1), \quad \|\psi\|_{L^2(\partial D^m)} = \mathcal{O}(1) \quad \text{and } \|H\|_{H^1(\partial D^m)} = \mathcal{O}(1),
$$

as $\delta \to 0$. From the boundary conditions and using the jump relations (1.13) and (1.14), we have

$$\left(-\frac{1}{2}I + \widetilde{\mathcal{K}}_{D^m}^{\omega,*}\right)[\psi] = \mathcal{O}(\delta),$$

where $\widetilde{\mathcal{K}}_{D^m}^{\omega,*}$ is defined as in (1.12). Since $\omega = \mathcal{O}(\delta^{1/2})$, it then follows from (B.58) that $\psi = \psi^{(0)} + \mathcal{O}(\delta^{1/2})$ for some $\psi^{(0)} \in \operatorname{Ker}\left(-\frac{1}{2}I + \mathcal{K}_{D^m}^{-0,*}\right)$. Hence, $\mathcal{S}_{D^m}^{0}[\psi^{(0)}](x)$ is constant for $x \in D^m$; see Lemma 1.4. Moreover, from the low-frequency expansion (B.55), we have

$$\widetilde{\mathcal{S}}_{D^m}^{\omega}[\psi] = \mathcal{S}_{D^m}^{0}[\psi^{(0)}] + \mathcal{O}(\delta^{1/2}),$$

with $\widetilde{\mathcal{S}}_{D^m}^{\omega}$ being defined as in (1.10), which proves the claim. □

We define $\widehat{\mathbf{u}}^m = \begin{pmatrix} u_1^m \\ \vdots \\ u_N^m \end{pmatrix} \in \mathbb{R}^N$. Since ω is a subwavelength frequency, we can write

$$\omega = \omega_0 + \mathcal{O}(\delta), \quad \omega_0 = \beta\delta^{1/2}, \tag{4.146}$$

for some constant β. For simplicity, we assume that $\delta_i^m = \delta_i$ for all $m \in \Lambda$ and denote the material parameter v_i^m as

$$(v_i^m)^2 = v_i^2 b_i^m, \quad \forall\, i = 1, \dots, N,$$

for some (possibly random) variables b_i^m that satisfy $b_i^m = \mathcal{O}(1)$ as $\delta \to 0$. We define $\mathcal{B}_m$ as the $N \times N$ diagonal matrix whose i^{th} entry is given by b_i^m.

We introduce the discrete Floquet transform $\mathcal{U} : \left(\ell^2(\Lambda)\right)^N \to \left(L^2(Y^*)\right)^N$ and its inverse

$$\mathcal{U}^{-1} : \left(L^2(Y^*)\right)^N \to \left(\ell^2(\Lambda)\right)^N,$$

which are given by

$$\mathcal{U}[\phi](\alpha) := \sum_{m\in\Lambda} \phi(m) e^{\mathrm{i}\alpha\cdot m}, \qquad \mathcal{U}^{-1}[\psi](m) := \frac{1}{|Y^*|}\int_{Y^*} \psi(\alpha) e^{-\mathrm{i}\alpha\cdot m}\,\mathrm{d}\alpha. \tag{4.147}$$

For $\alpha \in Y^*$, we let C^α and $\mathcal{C}^\alpha$ be respectively the capacitance matrix and the generalised capacitance matrix corresponding to the system of periodically arranged resonators.

In order to reframe the capacitance formulation to the real-space variable, we define the real-space capacitance coefficients and generalised capacitance coefficients, respectively, as

$$\widehat{C}^m = \mathcal{U}^{-1}[C^\alpha](m), \quad \widehat{\mathcal{C}}^m = \mathcal{U}^{-1}[\mathcal{C}^\alpha](m), \tag{4.148}$$

indexed by the real-space variable $m \in \Lambda$. In this section, we use the superscript $\widehat{\ }$ to denote the quantities posed in terms of the space variable $m \in \Lambda$.

The discrete Floquet transform is the tool that allows us to show the following result, by transforming into the dual space and then back again. Taking the Floquet transform, we find from (4.144) that

$$
\begin{cases}
\Delta u^\alpha + k^2 u^\alpha = 0 \quad \text{in } Y \setminus \overline{D}, \\
\Delta u^\alpha + \omega^2 \sum_{m\in\Lambda} \frac{e^{\mathrm{i}\alpha\cdot m}}{(v_i^m)^2} u(x+m\Lambda) = 0 \quad \text{in } D_i, \\
u^\alpha|_+ - u^\alpha|_- = 0 \quad \text{on } \partial D_i, \\
\delta_i \frac{\partial u^\alpha}{\partial \nu}\bigg|_+ - \frac{\partial u^\alpha}{\partial \nu}\bigg|_- = 0 \quad \text{on } \partial D_i, \\
e^{-\mathrm{i}\alpha\cdot m} u^\alpha(x_l, x_0) \text{ is } \Lambda\text{-periodic in } x_l, \\
u^\alpha(x_l, x_0) \text{ satisfies the } \alpha\text{-quasiperiodic outgoing radiation condition} \\
\quad \text{as } |x_0| \to +\infty.
\end{cases}
\tag{4.149}
$$

Moreover, inside D_i, we have from Lemma 4.86 that

$$
u^\alpha(x) = u_i^\alpha + \mathcal{O}(\delta^{1/2}), \quad x \in D_i, \qquad u_i^\alpha = \sum_{m\in\Lambda} u_i^m e^{\mathrm{i}\alpha\cdot m},
$$

for some sequences $u_i^m \in \ell^2(\mathbb{C})$ for $i = 1, \ldots, N$. It then follows that

$$
u^\alpha(x) = u_1^\alpha V_1^\alpha(x) + \ldots + u_N^\alpha V_N^\alpha(x) + \mathcal{O}(\delta^{1/2}), \quad x \in Y \setminus D,
$$

where V_i^α, $i = 1, \ldots, N$, are defined by (4.60). On the one hand, using the transmission conditions in (4.149) and integration by parts, we obtain

$$
\begin{aligned}
\int_{\partial D_i} \frac{\partial u^\alpha}{\partial \nu}\bigg|_+ \mathrm{d}\sigma &= \frac{1}{\delta_i} \int_{\partial D_i} \frac{\partial u^\alpha}{\partial \nu}\bigg|_- \mathrm{d}\sigma \\
&= -\frac{\omega^2}{\delta_i v_i^2} \int_{D_i} \sum_{m\in\Lambda} \frac{e^{\mathrm{i}\alpha\cdot m}}{b_i^m} u^\alpha(x)\,\mathrm{d}x \\
&= -\frac{\omega^2 |D_i|}{\delta_i v_i^2} \sum_{m\in\Lambda} \frac{u_i^m e^{\mathrm{i}\alpha\cdot m}}{b_i^m} + \mathcal{O}(\delta).
\end{aligned}
\tag{4.150}
$$

On the other hand, we have

$$
\begin{aligned}
\int_{\partial D_i} \frac{\partial u^\alpha}{\partial \nu}\bigg|_+ \mathrm{d}\sigma &= u_1^\alpha \int_{\partial D_i} \frac{\partial V_1^\alpha}{\partial \nu}\,\mathrm{d}\sigma + \ldots + u_N^\alpha \int_{\partial D_i} \frac{\partial V_N^\alpha}{\partial \nu}\,\mathrm{d}\sigma + \mathcal{O}(\delta^{1/2}) \\
&= -u_1^\alpha C_{i1}^\alpha - \ldots - u_N^\alpha C_{iN}^\alpha + \mathcal{O}(\delta^{1/2}),
\end{aligned}
\tag{4.151}
$$

where C_{ij}^α for $i, j = 1, \ldots, N$, are the entries of C^α defined in (4.59).

The following characterisation of localised modes from [**40**] follows.

PROPOSITION 4.87. *Any localised solution u to (4.144), corresponding to a subwavelength frequency $\omega = \omega_0 + \mathcal{O}(\delta)$, satisfies*

$$
\mathcal{B}_m \sum_{n\in\Lambda} \widehat{\mathcal{C}}^{m-n} \widehat{\mathbf{u}}^n = \omega_0^2 \widehat{\mathbf{u}}^m,
\tag{4.152}
$$

for every $m \in \Lambda$, where $\widehat{\mathcal{C}}^m = \mathcal{U}^{-1}[\mathcal{C}^\alpha](m)$ with $\mathcal{C}^\alpha$ being the generalised quasiperiodic capacitance matrix introduced in Definition 4.34.

PROOF. We can combine the fact from Lemma 4.86 that the eigenmodes are approximately constant on the resonators with the fact that the eigenmodes are given by eigenvectors of the generalised quasiperiodic capacitance matrix to obtain a discrete-operator formulation of the subwavelength resonance problem. In fact, from (4.150) and (4.151) it follows that

$$\mathcal{C}^\alpha \sum_{m\in\Lambda} \widehat{\mathbf{u}}^m e^{\mathrm{i}\alpha\cdot m} = \omega_0^2 \sum_{m\in\Lambda} (\mathcal{B}_m)^{-1}\widehat{\mathbf{u}}^m e^{\mathrm{i}\alpha\cdot m}, \tag{4.153}$$

for all $\alpha \in Y^*$. We can choose to pose the problem either in the real (spatial) variable m or in the dual (momentum) variable α. In this framework, equation (4.153), posed in the dual variable α, reads

$$\mathcal{C}^\alpha \mathbf{u}^\alpha = \omega_0^2 \left(\sum_{m\in\Lambda} (\mathcal{B}_m)^{-1}\mathcal{U}^{-1}[\mathbf{u}^\alpha](m) e^{\mathrm{i}\alpha\cdot m} \right), \qquad \mathbf{u}^\alpha = \mathcal{U}[\widehat{\mathbf{u}}^m](\alpha). \tag{4.154}$$

We can reframe this equation in the real variable m by taking the inverse discrete Floquet transform defined in (4.147). Since $\widehat{\mathbf{u}}^m = \mathcal{U}^{-1}[\mathbf{u}^\alpha](m)$, we can apply $\mathcal{U}^{-1}$ to (4.154) to obtain (4.152). □

We can write (4.152) in terms of discrete lattice operators. If $\widehat{\mathfrak{u}} \in \ell^2(\Lambda^N, \mathbb{C}^N)$, we define the operator $\mathfrak{C}$ on $\ell^2(\Lambda^N, \mathbb{C}^N)$ by

$$\mathfrak{C}\mathfrak{u}(m) = \sum_{n\in\Lambda} \widehat{\mathcal{C}}^{m-n}\mathfrak{u}(n).$$

Moreover, we define the operator $\mathfrak{B}$ as the diagonal multiplication by $\mathcal{B}_m$. With this notation, equation (4.152) reads

$$\mathfrak{B}\mathfrak{C}\mathfrak{u} = \omega_0^2\mathfrak{u}, \tag{4.155}$$

where

$$\mathfrak{u} = \begin{pmatrix} \vdots \\ \mathbf{u}^{-1} \\ \mathbf{u}^0 \\ \mathbf{u}^1 \\ \mathbf{u}^2 \\ \vdots \end{pmatrix}.$$

Equation (4.155) characterises the subwavelength resonant frequencies and their corresponding resonant modes. A localised mode corresponds to an eigenvalue of the operator $\mathfrak{B}\mathfrak{C}$. In the periodic case (when $\mathfrak{B} = I$), the spectrum of the lattice operator $\mathfrak{C}$ is continuous and does not contain eigenvalues, so there are no localised modes. The spectral bands are given by Theorem 4.35 in this case. As we shall see, the operator $\mathfrak{B}\mathfrak{C}$ might have a pure-point spectrum in the non-periodic case.

For a linear lattice, $d_l = 1$ (and $d = 3$), these operators amount to doubly infinite matrices:

$$\mathfrak{C} = \begin{pmatrix} \ddots & \vdots & \vdots & \vdots & \vdots & ⋰ \\ \cdots & \widehat{\mathcal{C}}^0 & \widehat{\mathcal{C}}^1 & \widehat{\mathcal{C}}^2 & \widehat{\mathcal{C}}^3 & \cdots \\ \cdots & \widehat{\mathcal{C}}^{-1} & \widehat{\mathcal{C}}^0 & \widehat{\mathcal{C}}^1 & \widehat{\mathcal{C}}^2 & \cdots \\ \cdots & \widehat{\mathcal{C}}^{-2} & \widehat{\mathcal{C}}^{-1} & \widehat{\mathcal{C}}^0 & \widehat{\mathcal{C}}^1 & \cdots \\ \cdots & \widehat{\mathcal{C}}^{-3} & \widehat{\mathcal{C}}^{-2} & \widehat{\mathcal{C}}^{-1} & \widehat{\mathcal{C}}^0 & \cdots \\ ⋰ & \vdots & \vdots & \vdots & \vdots & \ddots \end{pmatrix} \text{ and } \mathfrak{B} = \begin{pmatrix} \ddots & \vdots & \vdots & \vdots & \vdots & ⋰ \\ \cdots & \mathcal{B}_{-1} & 0 & 0 & 0 & \cdots \\ \cdots & 0 & \mathcal{B}_0 & 0 & 0 & \cdots \\ \cdots & 0 & 0 & \mathcal{B}_1 & 0 & \cdots \\ \cdots & 0 & 0 & 0 & \mathcal{B}_2 & \cdots \\ ⋰ & \vdots & \vdots & \vdots & \vdots & \ddots \end{pmatrix}. \tag{4.156}$$

Here, $\mathfrak{C}$ is the (block) Laurent operator corresponding to the symbol $\mathcal{C}^\alpha$.

4.8.3. Toeplitz matrix formulation for compact defects. In this section, we assume that the operators $\mathcal{B}_m$ are identity for all but finitely many m. For simplicity, we phrase this analysis in the case $\Lambda = \mathbb{Z}$, where we assume that the resonators corresponding to $0 \leq m \leq M$ have differing parameters, *i.e.*,

$$b_i^m = \begin{cases} 1, & m < 0 \text{ or } m > M, \\ 1 + x_i^m, & 0 \leq m \leq M, \end{cases} \tag{4.157}$$

for some parameters x_i^m. We let X_m be the diagonal matrix with entries x_i^m. In this setting, we obtain a (block) Toeplitz matrix formulation. For $\omega \notin \sigma(\mathfrak{C})$, we define

$$\mathcal{T}(\omega) = \begin{pmatrix} T^0 & T^1 & T^2 & \cdots & T^M \\ T^{-1} & T^0 & T^1 & \cdots & T^{M-1} \\ T^{-2} & T^{-1} & T^0 & \cdots & T^{M-2} \\ \vdots & \vdots & \vdots & \ddots & \vdots \\ T^{-M} & T^{-(M-1)} & T^{-(M-2)} & \cdots & T^0 \end{pmatrix},$$

where

$$T^m = -\frac{1}{|Y^*|} \int_{Y^*} e^{\mathrm{i}\alpha m} \mathcal{C}^\alpha \left(\mathcal{C}^\alpha - \omega^2 I\right)^{-1} \mathrm{d}\alpha.$$

PROPOSITION 4.88. *Assume that $\Lambda = \mathbb{Z}$ and that b_i^m are given by* (4.157). *Then $\omega_0 \notin \sigma(\mathfrak{C})$ is an eigenvalue of* (4.155) *if and only if*

$$\det\big(I - \mathcal{X}\mathcal{T}(\omega_0)\big) = 0,$$

where $\mathcal{X}$ is the block-diagonal matrix with entries X_m.

PROOF. From (4.154) it follows that, for each m, we have

$$\mathcal{U}^{-1}\big[\mathcal{B}_m \mathcal{C}^\alpha \mathbf{u}^\alpha\big](m) = \omega_0^2 \mathcal{U}^{-1}\big[\mathbf{u}^\alpha\big](m).$$

Since X_m is only nonzero for $m = 0, 1, \ldots, M$, we have

$$\mathcal{U}^{-1}\Big[(\mathcal{C}^\alpha - \omega_0^2 I)\mathbf{u}^\alpha + X_m \mathcal{C}^\alpha \mathbf{u}^\alpha\Big](m) = 0, \qquad 0 \leq m \leq M, \tag{4.158}$$

$$\mathcal{U}^{-1}\Big[(\mathcal{C}^\alpha - \omega_0^2 I)\mathbf{u}^\alpha\Big](m) = 0, \qquad m < 0 \text{ or } m > M. \tag{4.159}$$

In other words, $(\mathcal{C}^\alpha - \omega_0^2 I)\mathbf{u}^\alpha$ is given by a finite Fourier series in α:

$$(\mathcal{C}^\alpha - \omega_0^2 I)\mathbf{u}^\alpha = \sum_{m=0}^{M} \mathbf{c}_m e^{\mathrm{i}\alpha m},$$

where $\mathbf{c}_m = \mathcal{U}^{-1}[X_m \mathcal{C}^\alpha \mathbf{u}^\alpha](m)$. Together with (4.158) we find that

$$\mathbf{c}_n = -X_n \sum_{m=0}^{M} \frac{1}{|Y^*|} \int_{Y^*} e^{\mathrm{i}\alpha(m-n)} \mathcal{C}^\alpha (\mathcal{C}^\alpha - \omega_0^2 I)^{-1} \mathbf{c}_m \, \mathrm{d}\alpha.$$

If we denote $\mathbf{c} = \begin{pmatrix} \mathbf{c}_0 \\ \vdots \\ \mathbf{c}_M \end{pmatrix}$, we then obtain the following nonlinear eigenvalue problem in ω_0:

$$\mathcal{X}\mathcal{T}(\omega_0)\mathbf{c} = \mathbf{c}, \tag{4.160}$$

where $\mathcal{X}$ is the block-diagonal matrix with entries X_m. □

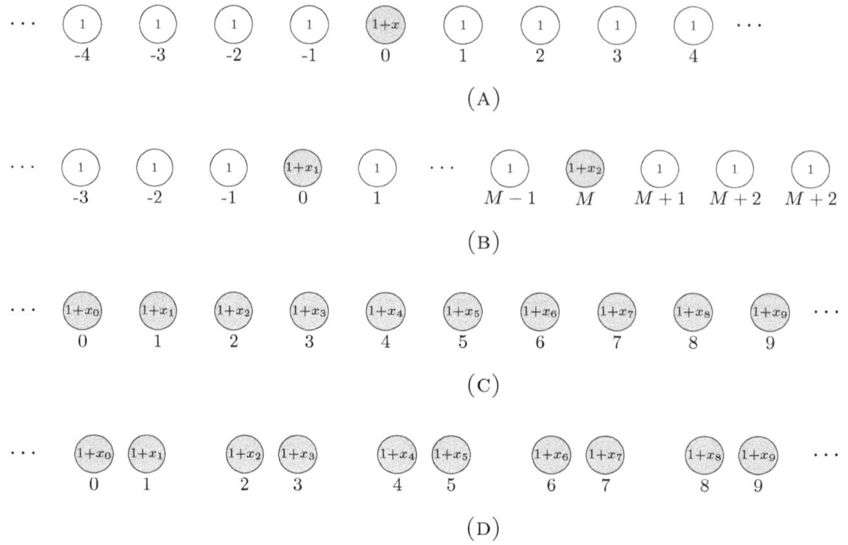

FIGURE 4.33. Arrays of resonators with (a) a single defect, on the 0th resonator, (b) two defects, on the 0th and the Mth resonators, (c) defects on every resonator, and (d) defects on every resonator in an array of dimers. In each case, the material parameters on defective resonators are perturbed $1 \mapsto 1 + x_i$ for some $x_i \in (-1, +\infty)$. We depict circular resonators here. In this section, we conduct analytic studies of structures (a) and (b) to make predictions about the behaviour of (c) and (d).

REMARK 4.89. *The localised modes can, equivalently, be described by the Toeplitz matrix with entries*

$$\widetilde{T}^m = \frac{1}{|Y^*|} \int_{Y^*} e^{\mathrm{i}\alpha m} \left(\mathcal{C}^\alpha - \omega^2 I\right)^{-1} \mathrm{d}\alpha,$$

leading to the nonlinear eigenvalue problem

$$\omega_0^2 \mathcal{Y} \widetilde{\mathcal{T}}(\omega_0)\mathbf{c} = \mathbf{c},$$

where $\mathcal{Y} = (I + \mathcal{X})^{-1} - I$.

4.8.4. Finite perturbations of a chain. In this section, we take the specific example of a chain of equidistant resonators where a finite number of resonators are perturbed, as illustrated in Figure 4.33.

Specifically, we have the lattice dimension $d_l = 1$, the space dimension $d = 3$ and we take a unit cell with just a single resonator: $N = 1$. As we shall see, a single defect may induce one localised mode. When multiple defects hybridise, level repulsion will guarantee the existence of at least one localised mode. We emphasise that the phenomena highlighted in this section are not specific to the dimensionality and will be qualitatively similar in $d_l = 2$ or $d_l = 3$; see, for example, Figure 4.32.

4.8.4.1. *Single defect.* We assume that $N = 1$ and that the resonator corresponding to $m = 0$ has a different parameter, *i.e.*,

$$b_1^m = \begin{cases} 1, & m \neq 0, \\ 1 + x, & m = 0, \end{cases} \tag{4.161}$$

for some parameter $x > -1$. In the single resonator case, $\mathcal{C}^\alpha$ is just a scalar $\mathcal{C}^\alpha = \lambda_1^\alpha$ given by

$$\lambda_1^\alpha := \sqrt{\frac{\delta_1 v_1^2}{|D_1|}(\mathcal{S}_{D_1}^{0,\alpha})^{-1}[\chi_{\partial D_1}]}.$$

It follows from Proposition 4.88 that any eigenvalue ω satisfies $xT^0(\omega) = 1$; in other words,

$$\frac{x}{|Y^*|}\int_{Y^*} \frac{\lambda_1^\alpha}{\omega^2 - \lambda_1^\alpha}\, \mathrm{d}\alpha = 1. \tag{4.162}$$

We let $\omega_* = \max_{\alpha \in Y^*} \sqrt{\lambda_1^\alpha}$. For $\omega > \omega_*$, it is clear that $T^0(\omega) > 0$. Therefore, there are no solutions in the case $x \le 0$. Moreover,

$$\lim_{\omega \to \omega_*} T^0(\omega) = +\infty, \qquad \lim_{\omega \to +\infty} T^0(\omega) = 0.$$

Therefore, if $x > 0$, then there is a solution $\omega = \omega_0$ of (4.162). In other words, the defect induces an eigenvalue ω_0^2 in the pure-point spectrum of $\mathfrak{BC}$, corresponding to an exponentially localised eigenmode.

The expected localised eigenmode is shown in Figure 4.34(a) for a finite system of 31 circular resonators.

4.8.4.2. *Double defect.* Next, we continue with the case $N = 1$, but now assume that there are two defects at the resonators corresponding to $m = 0$ and $m = M$;

$$b_1^m = 1,\ m \notin \{0, M\}, \qquad b_1^0 = 1 + x_1, \qquad b_1^M = 1 + x_2, \tag{4.163}$$

for some parameters $x_1, x_2 \neq 0$. In this case, the localised modes are characterised by the Toeplitz matrix formulation

$$\mathcal{X}\mathcal{T}(\omega)\mathbf{c} = \mathbf{c},$$

where

$$\mathcal{X} = \begin{pmatrix} x_1 & 0 & 0 & \cdots & 0 \\ 0 & 0 & 0 & \cdots & 0 \\ 0 & 0 & 0 & \cdots & 0 \\ \vdots & \vdots & \vdots & \ddots & \vdots \\ 0 & 0 & 0 & \cdots & x_2 \end{pmatrix}, \qquad \mathcal{T}(\omega) = \begin{pmatrix} T^0 & T^1 & T^2 & \cdots & T^M \\ T^{-1} & T^0 & T^1 & \cdots & T^{M-1} \\ T^{-2} & T^{-1} & T^0 & \cdots & T^{M-2} \\ \vdots & \vdots & \vdots & \ddots & \vdots \\ T^{-M} & T^{-(M-1)} & T^{-(M-2)} & \cdots & T^0 \end{pmatrix},$$

and

$$T^m = -\frac{1}{|Y^*|}\int_{Y^*} \frac{\lambda_1^\alpha e^{\mathrm{i}\alpha m}}{\lambda_1^\alpha - \omega^2}\, \mathrm{d}\alpha.$$

In other words, we are solving the equation

$$\det(I - \mathcal{X}\mathcal{T}(\omega)) = 0.$$

We observe that

$$\det(I - \mathcal{X}\mathcal{T}(\omega)) = \det\left(I - \begin{pmatrix} x_1 T^0 & x_1 T^M \\ x_2 T^{-M} & x_2 T^0 \end{pmatrix}\right),$$

so we seek the solutions ω to

$$\left(1 - x_1 T^0(\omega)\right)\left(1 - x_2 T^0(\omega)\right) - x_1 x_2 T^{-M}(\omega) T^M(\omega) = 0. \tag{4.164}$$

In Figure 4.35, we show the subwavelength resonant frequencies of defect modes in a system with two defects. These are obtained by solving (4.164) for $x_1 = x_2 = x$ and varying either x or the separation distance M. We see that for any $x > 0$ there is at least one localised eigenmode. If either M or x is small, then there is only one solution, as the hybridisation means the smaller solution is not greater than the

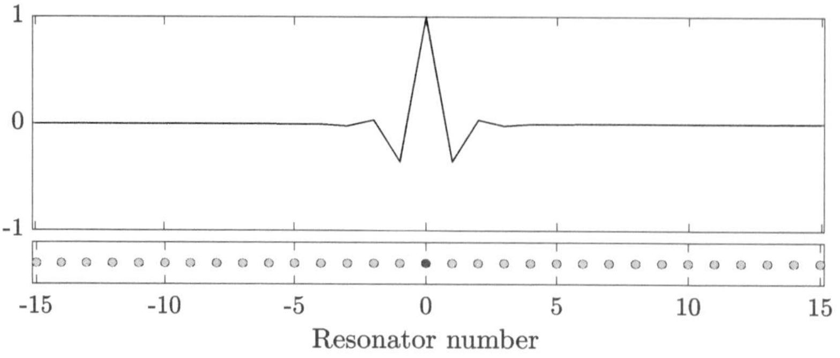

(A)

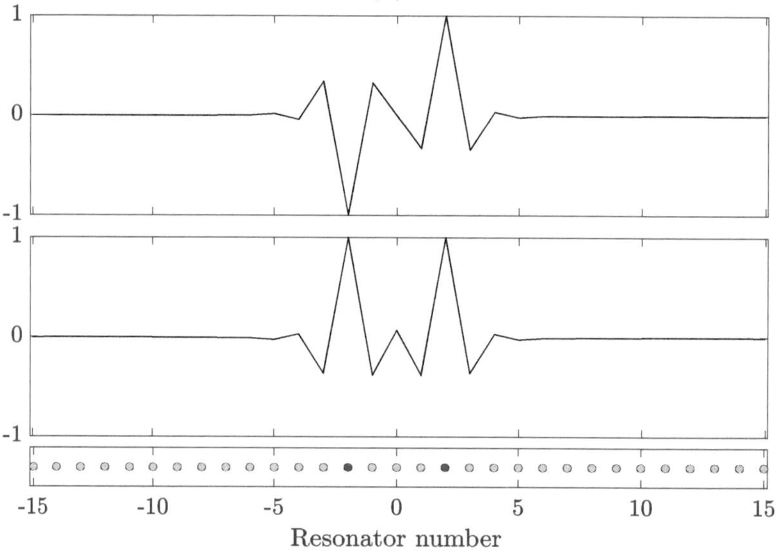

(B)

FIGURE 4.34. (a) Plot of the localised mode along the x_1-axis in the case of a single defect with $x = 0.15$. Here, we assume the resonators to be arranged collinearly along the x_1-axis. A sketch of the resonator structure is shown below: here we have a chain of $N = 31$ equally spaced circular resonators with the perturbation $1 \mapsto 1 + x$ made to the material parameters of centre resonator. We observe a single localised mode even in the case of small $x > 0$, with larger x leading to a stronger degree of localisation (a faster decaying mode). However, in the case $-1 < x < 0$ there are no localised modes. (b) The two localised modes that exist when two defects are introduced with perturbation strengths $x_1 = x_2 = 0.15$ and separation distance $M = 4$. In this case, the higher mode has a dipole (odd) symmetry whole the lower mode has a monopole (even) symmetry.

maximum of the spectral band, so there is no localised eigenmode. For any fixed x, there is a critical minimum value M_{sep} of M in order for two localised eigenmodes to exist. In Figure 4.35(c), we show the relation between the critical values for the distance M and the perturbation strength x.

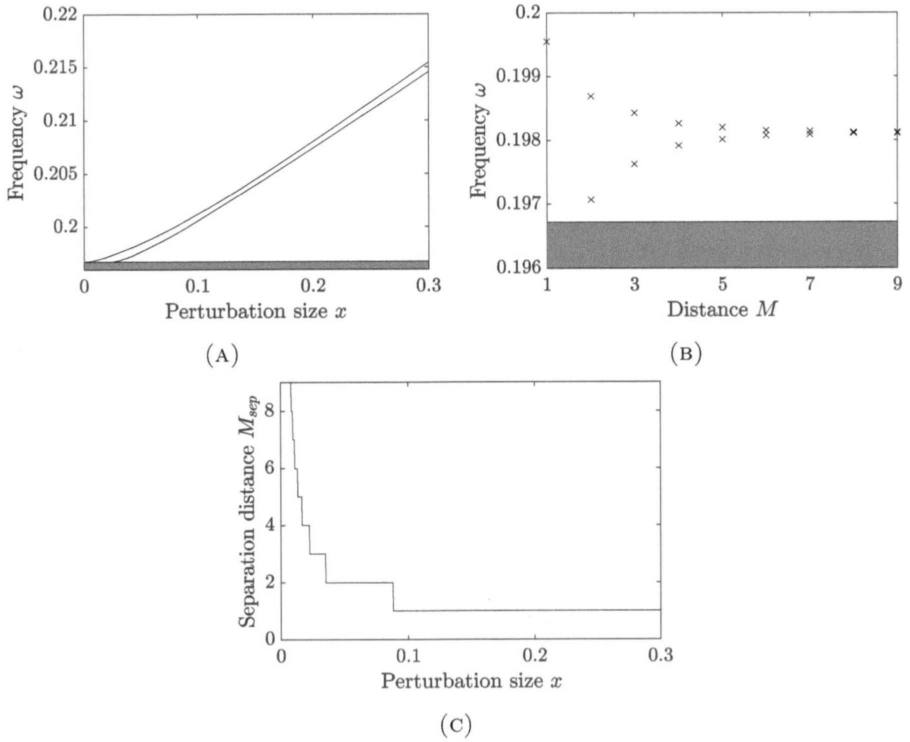

FIGURE 4.35. Localised frequencies for two defects, for the case of identical defects $x_1 = x_2 = x$. (a) Solutions ω of (4.164) as function of the perturbation size x for fixed distance $M = 3$. (b) Solutions ω of (4.164) as function of the distance M for fixed perturbation size $x = 0.05$. If either M or x is small, then there is only one solution, as the level repulsion causes the second solution to fall within the spectral band (denoted by the shaded area). (c) The minimum separation distance M_{sep} required to obtain two solutions to (4.164), plotted as a function of x in the case $x_1 = x_2 = x$.

4.8.4.3. *Level repulsion in a two-defect structure.* In Figure 4.36, we model the resonant frequencies of a system with two random defects. In particular, we assume that the values of x_1 and x_2 are drawn independently from the uniform distribution $\mathcal{U}[\mu-\sqrt{3}\sigma, \mu+\sqrt{3}\sigma]$. We see that for small σ there are still two localised modes. However, for larger values of σ, the eigenfrequencies can be lost into the spectral band. Nevertheless, looking at Figure 4.36(b) we see that the introduction of random perturbations causes the average value of each midgap frequency to move further apart (and further from the edge of the bandgap). This is the phenomenon of level repulsion. One consequence of this is that the eigenmodes, on average, become more strongly localised as the variance σ^2 increases (since the strength of the localisation increases when the resonant frequency is further from the edge of the bandgap).

Another consequence of level repulsion is that the existence of localised modes is much more robust in the case of two defects than in the case of a single defect. Figure 4.37(a) shows the probability of at least one localised eigenmode existing for

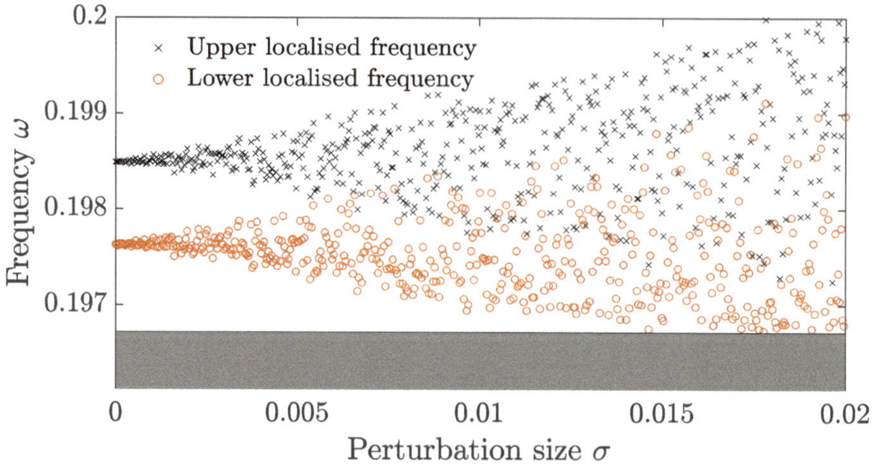

(A)

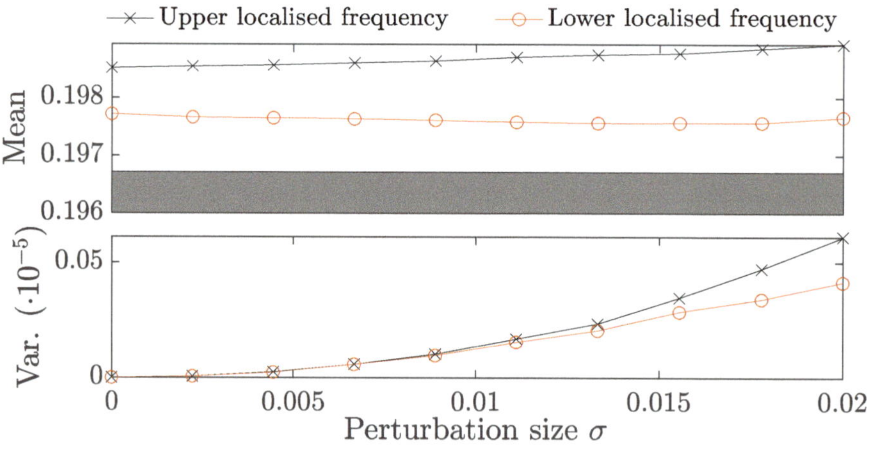

(B)

FIGURE 4.36. Localised frequencies for two random defects. (a) Solutions ω of (4.164) with x_1 and x_2 drawn from a uniform $\mathcal{U}[\mu - \sqrt{3}\sigma, \mu + \sqrt{3}\sigma]$ distribution, with varying σ, fixed mean $\mu = 0.05$ and fixed separation distance $M = 6$. (b) The mean (upper) and variance (lower) of the localised frequencies ω for different values of σ, fixed mean $\mu = 0.05$ and fixed $M = 3$. For each simulated value of σ, 1000 simulated values are used to compute the mean and variance, with values being left empty when no such frequency exists (*e.g.* when there is only one solution to (4.164), as the hybridisation causes the second solution to fall within the spectral band).

three different cases: a single defect, a double defect with small separation distance $M = 1$ and a double defect with larger separation distance $M = 6$. The defects are uniformly distributed with means chosen to normalise the distributions in the sense that the largest defect frequency when $\sigma = 0$ is the same in all three cases

($\omega = 0.198$). We see that the probability of localisation drops off quickly for a single defect. When $\sigma = 0.1$, the probability of localisation for a single defect is 65.6%, compared to 93.6% for the double defect with small separation distance and 99.3% when the separation distance is larger. The intuitive explanation for the difference between the two double defect cases is that the larger distance causes the two localised eigenfrequencies to be closer together in the deterministic case of $\sigma = 0$, meaning that they interact with one another more strongly, so the level repulsion has a greater effect. When at least one localised eigenmode exists, the distributions of the largest localised eigenvalue, shown in Figure 4.37(b), are similar in all three cases.

4.8.4.4. *Phase transition and eigenmode symmetry swapping.* Analogously to Figure 4.35(a), Figure 4.38(a) shows the two localised frequencies, but here in the case $M = 2$. When M is even, the two curves will intersect (here around $x \approx 0.21$), corresponding to a doubly degenerate frequency. Crucially, this is a transition point whereby the symmetries of the corresponding eigenmodes swap. Below the transition point, the first mode has dipole (odd) symmetry, while the second mode has monopole (even) symmetry; above the transition point, the first mode has monopole (even) symmetry while the second mode has dipole (odd) symmetry.

We now study the degree of localisation around this transition point. Throughout, we define the degree of localisation $l(u)$ of an eigenmode $u \in \ell^2(\Lambda^N, \mathbb{C}^N)$ as

$$l(u) = \frac{\|u\|_\infty}{\|u\|_2}. \tag{4.165}$$

We remark that $0 < l(u) \leq 1$: small $l(u)$ corresponds to a delocalised mode while $l(u) = 1$ corresponds to a mode that is perfectly localised to a single resonator. Other measures to distinguish Anderson localised and extended modes, such as the inverse participation ratio [**124, 132**] defined as $\|u\|_4/\|u\|_2$, yield qualitatively similar results.

In Figure 4.38(b), we plot the average degree of localisation when the parameters are drawn independently from $\mathcal{U}[x - \sqrt{3}\sigma, x + \sqrt{3}\sigma]$ for $\sigma = 10^{-4}$. The plot shows a sharp peak at the transition point; a remarkably high degree of localisation is achieved even when the variance is small. This can be understood as a decoupling of the (otherwise coupled) modes, which, at the transition point, become concentrated to the individual defects as opposed to both defects.

For higher values of σ, we expect to achieve a strong degree of localisation for a larger range of x. As seen in Figure 4.38(c), there is a region in the x-σ-plane where we achieve optimal localisation. When x is large relative to σ, the modes will couple and the degree of localisation is low. When σ is large relative to x, we are in a regime with, on average, only a single localised mode and a low degree of localisation.

4.8.4.5. *Many defects.* The motivation for the above analyses of systems with one or two local defects is to be able to infer properties of fully random systems, which will be studied in the next subsection. To further this analysis, we again assume that $N = 1$ and consider a system with N_d independent defects. That is, we assume that

$$b_1^m = 1,\ m \notin \{0, M, 2M, \dots, N_d M\} \quad \text{and} \quad b_1^{kM} = 1 + x_k,\ k \in \{0, 1, 2, \dots, N_d\},$$

for some parameters $x_1, \dots, x_{N_d} \neq 0$. In this case, we can use the Toeplitz matrix formulation from before to see that the localised modes are characterised by the

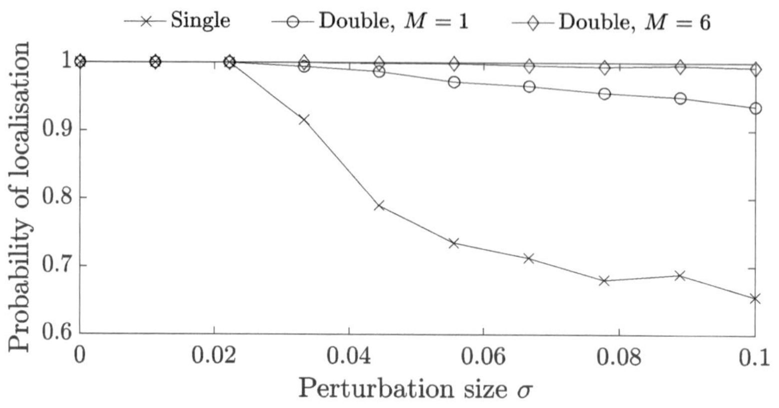

(A)

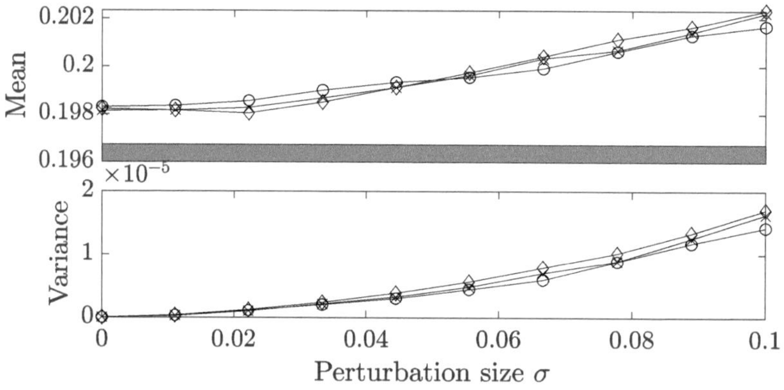

(B)

FIGURE 4.37. Comparison of the robustness of systems with one or two defects. (a) The probability of at least one localised mode existing for the case of a single defect, a double defect with separation distance $M = 1$ and a double defect with separation distance $M = 6$. In each case, the perturbations are drawn at random from a uniform distribution $\mathcal{U}[\mu - \sqrt{3}\sigma, \mu + \sqrt{3}\sigma]$ and, for very large values of σ we enforce $1 + x_i \geq 10^{-4} > 0$, so that the material parameters are all strictly positive. The means are chosen as $\mu = 0.05$, $\mu = 0.0325$ and $\mu = 0.05$, respectively, so that the largest localised eigenfrequency is the same in each case when $\sigma = 0$. (b) The mean and variance of the largest localised eigenfrequency for the three different models. Each value was calculated using 1000 independent random realisations.

equation

$$\det\left(I - \mathcal{A}_{\mathcal{X}\mathcal{T}}(\omega)\right) = 0, \tag{4.166}$$

where $\mathcal{A}_{\mathcal{X}\mathcal{T}}(\omega)$ is the $N_d \times N_d$ matrix given by

$$\mathcal{A}_{\mathcal{X}\mathcal{T}}(\omega)_{ij} = x_i T^{M(j-i)}(\omega) \quad \text{with} \quad T^m = -\frac{1}{|Y^*|}\int_{Y^*} \frac{\lambda_1^\alpha e^{\mathrm{i}\alpha m}}{\lambda_1^\alpha - \omega^2}\,\mathrm{d}\alpha.$$

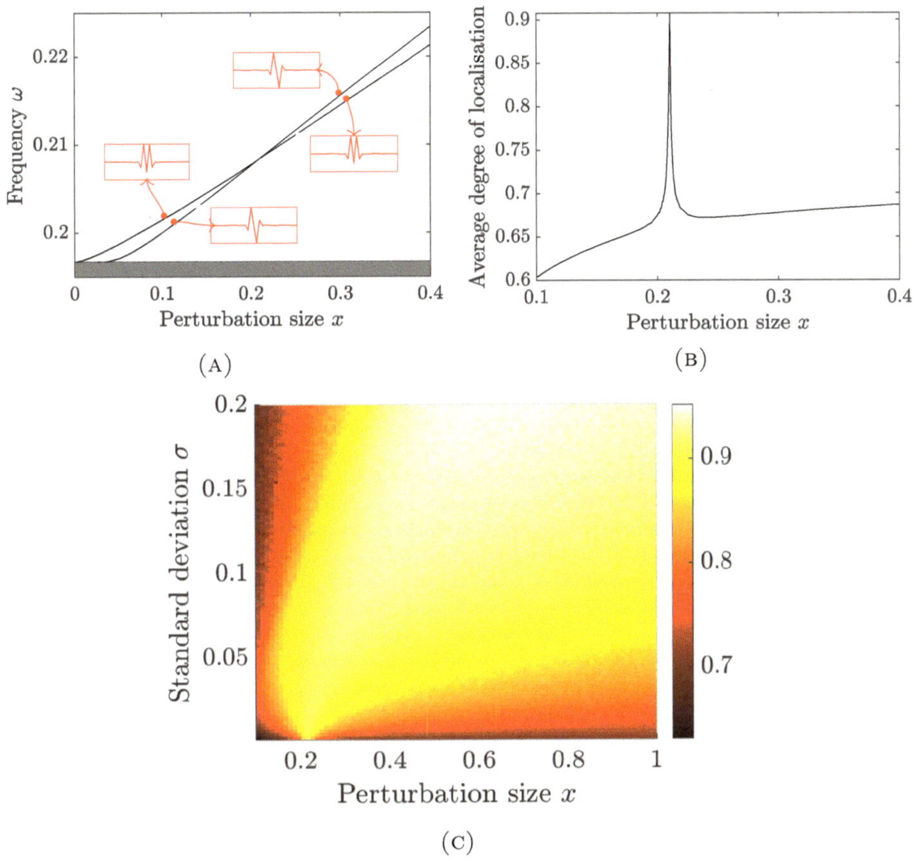

FIGURE 4.38. Localised frequencies for two defects, for the case of $M = 2$. (a) Solutions ω of (4.164), in the case of identical defects $x_1 = x_2 = x$, as function of the perturbation size x. The two curves intersect around $x \approx 0.21$. The symmetries of the eigenmodes (shown as inlays) are opposite on either side of the intersection. (b) Localisation length $l(u)$ of the two modes, with parameters drawn independently from $\mathcal{U}[x - \sqrt{3}\sigma, x + \sqrt{3}\sigma]$ for $\sigma = 10^{-4}$. At the intersection, the monopole/dipole modes decouple, resulting in a sharp peak of $l(u)$. (c) Average degree of localisation for different x and σ. The peak seen in *(b)* becomes wider for larger σ, giving rise to a region in the x-σ-plane with high degree of localisation.

We can solve the system (4.166) to find the localised modes for a system of N_d defects. In Figure 4.39, we show how the number of localised modes varies as the number of defects N_d increases. The number of localised modes is normalised by dividing by the number of defects N_d, to give a measure of the density of localised modes (the number of defects is proportional to the length of the part of the resonator chain that contains the defects, which has length $N_d M$). Different values of M and σ are compared. In all cases, we observe that as N_d becomes large, the system appears to converge to a state whereby between 30% and 50% of the modes are localised.

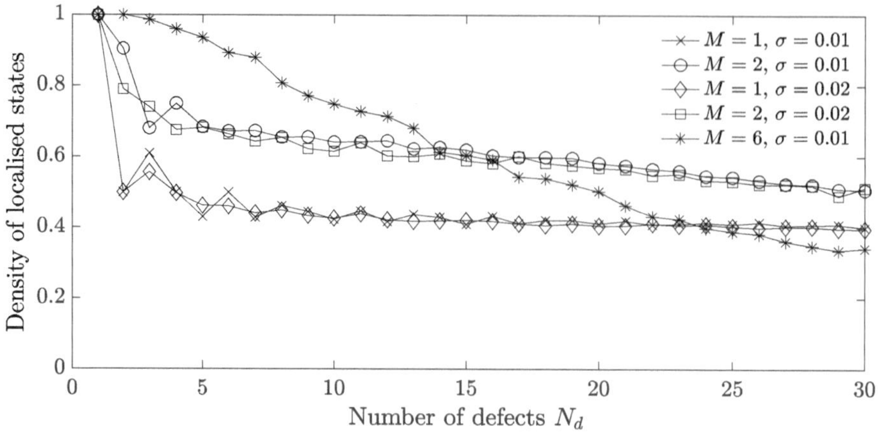

FIGURE 4.39. Localised modes for increasingly many defects. We plot the average number of localised modes divided by the number of defects for increasing numbers of defects. The defects are drawn from a uniform $\mathcal{U}[0.05 - \sqrt{3}\sigma, 0.05 + \sqrt{3}\sigma]$ distribution and have separation distance M. Each point is an average over 100 independent random realisations.

4.8.5. Fully random parameter distribution. Here we adopt the setting of Section 4.8.4 but additionally assume that all parameters b_i^m are randomly distributed. We assume that

$$b_i^m = 1 + \sigma X_i^m, \qquad \mathfrak{B} = I + \sigma \mathfrak{X},$$

where $\mathfrak{X}$ is a diagonal operator. We take X_i^m to be independent and uniformly distributed on $[-\sqrt{3}, \sqrt{3}]$. In order to guarantee that the material parameters remain positive, we require $\sigma < \frac{1}{\sqrt{3}}$.

Localised modes are characterised by the equation

$$\mathfrak{B}\mathfrak{C} = (\mathfrak{C} + \sigma \mathfrak{X}\mathfrak{C})\, \mathfrak{u} = \omega^2 \mathfrak{u}. \tag{4.167}$$

In other words, we have characterised the problem in terms of the spectral problem for a random perturbation of a discrete operator. We emphasise that the unperturbed part of the operator, namely $\mathfrak{C}$, is a dense matrix, whose off-diagonal terms describe long-range interactions with slow decay like $1/|i-j|$. Additionally, the perturbation $\sigma \mathfrak{X}\mathfrak{C}$ is also a dense matrix, containing random perturbations to both the diagonal- and off-diagonal terms. These two properties make the problem considerably more challenging than Anderson's original model (see Remark 4.90), and precise statements on the spectrum of $\mathfrak{C} + \sigma \mathfrak{X}\mathfrak{C}$ are generally beyond reach.

4.8.5.1. *Band structure and emergence of localised modes.* In Figure 4.40, we show the numerically computed density of states (see Chapter 7 for a definition), along with the average degree of localisation. Here, we take a chain of equally spaced resonators, which in the case $\sigma = 0$ has a single spectral band. Localised modes emerge as σ becomes nonzero, whose frequencies lie around the edge of the spectral band. Increasing σ leads to more localised modes and a stronger degree of localisation.

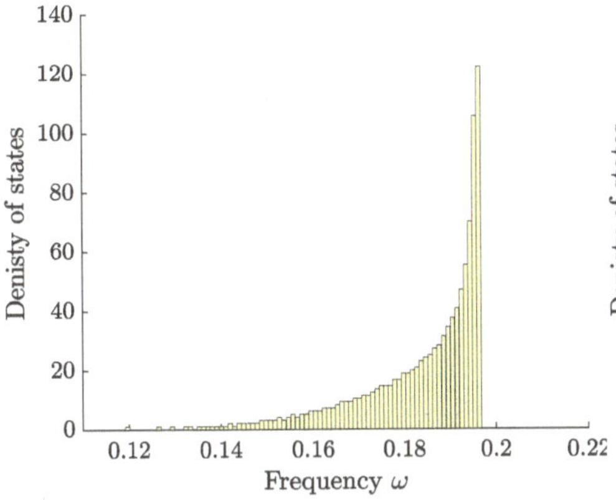

(A) $\sigma = 0$

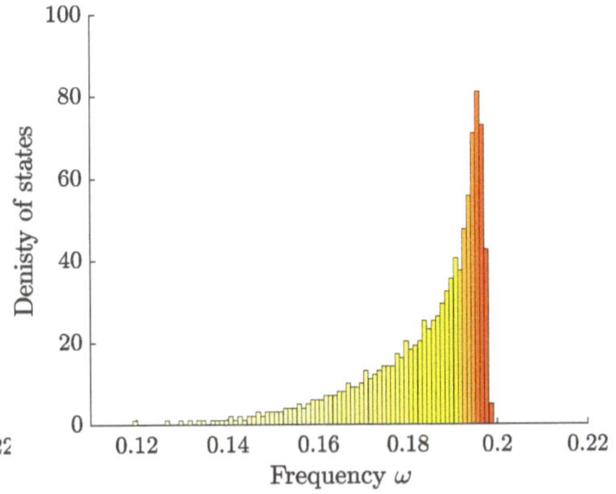

(B) $\sigma = 0.02$.

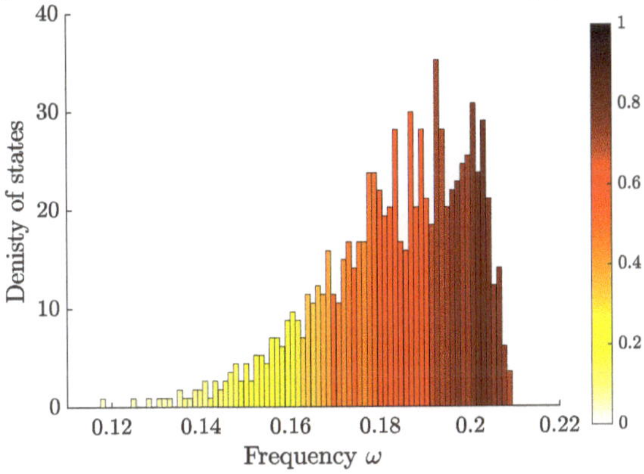

(C) $\sigma = 0.1$.

FIGURE 4.40. Example outcome of the density of states in a chain of equally spaced resonators with uniformly distributed random parameters. Additionally, the colour shows the average degree of localisation $l(u) = \|u\|_\infty / \|u\|_2$, where large $l(u)$ corresponds to a strongly localised mode u. In this case, there is a single spectral band, between 0 and ~ 0.2, when $\sigma = 0$. For small but nonzero σ, localised modes emerge at the edge of this band. The number of localised modes increase when σ increases further. Here shown for a finite chain of $N = 1000$ resonators.

Next, we turn to the case $N = 2$. Here, we take a chain of "dimers", *i.e.*, a chain of resonators with alternating, unequal separation distance, as depicted in Figure 4.33(d). In the case $\sigma = 0$, which is studied in Section 4.6, we have two spectral bands separated by a bandgap. For small but nonzero σ, the bandgap persists and localised modes emerge at either edge of the bands. Increasing σ further, more modes become localised and the band-gap structure disappears, as illustrated in Figure 4.41.

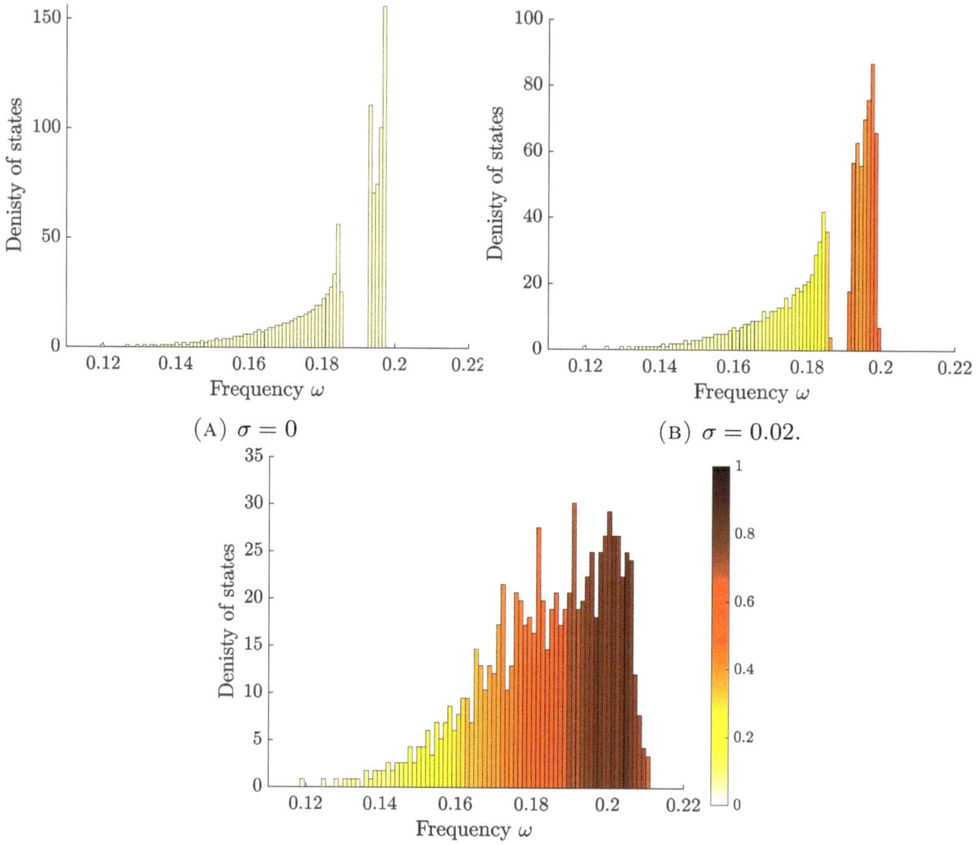

(A) $\sigma = 0$

(B) $\sigma = 0.02$.

(C) $\sigma = 0.1$.

FIGURE 4.41. Example outcome of the density of states in a chain of resonator dimers with uniformly distributed random parameters. Additionally, the colour shows the (normalised) average degree of localisation $l(u)$. The case $\sigma = 0$ exhibits two spectral bands, separated by a bandgap. For small but nonzero σ, localised modes emerge around the edges of the bands: above the first band, and below and above the second band. When σ increases, the bandgap vanishes and the density of localised modes increases across the spectrum. Here shown for a finite chain of $N = 1000$ resonators.

4.8.5.2. *Level repulsion in the fully random case.* Many of the arguments presented in this section involve the principle of level repulsion: that eigenvalues tend to separate when increasingly large random errors are added. This was demonstrated in the double defect case in Figure 4.36(b). We can also study the level repulsion in a finite chain of equally spaced resonators. In particular, Figure 4.42(a) shows that the average separation increases as the perturbation strength σ increases. We also show the distributions of the separations for two specific values of σ in Figures 4.42(b) and 4.42(c). In Figure 4.42(b), the perturbations are so small that this distribution is (visually) the same as that of the unperturbed structure. In Figure 4.42(c), the larger perturbations have affected the shape of the distribution.

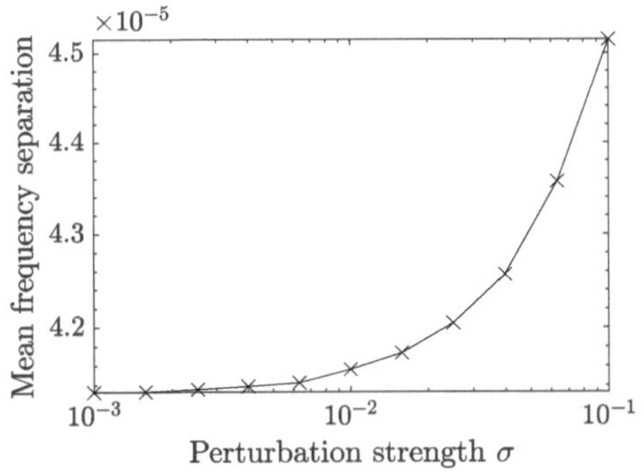

(A)

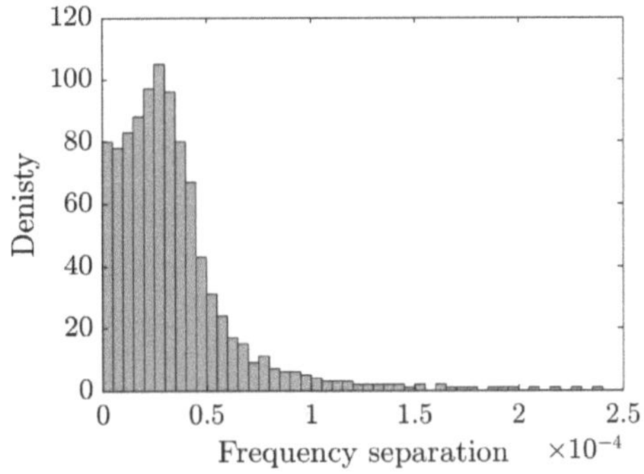

(B) $\sigma = 0.001$

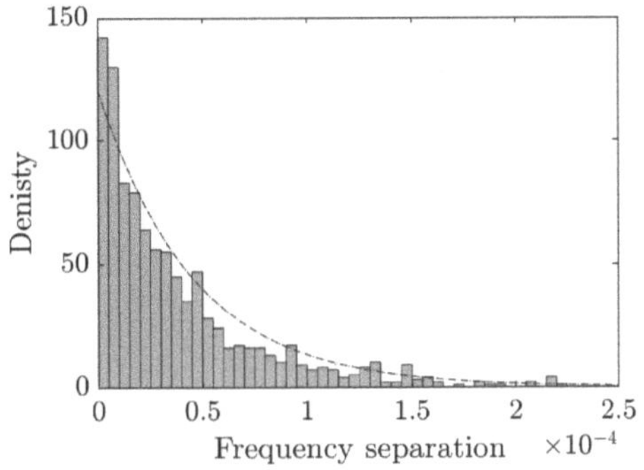

(C) $\sigma = 0.1$

FIGURE 4.42. Level repulsion in a finite chain of $N = 1000$ resonators with random perturbations. (a) The average nearest-neighbour separation between the (real parts of the) resonant frequencies for different perturbation strengths σ, which is the standard deviation of the uniform random variables. (b) and (c) are histograms showing the distributions of the normalised frequency separations (frequency separation divided by mean frequency separation) for $\sigma = 10^{-3}$ and $\sigma = 10^{-1}$, respectively. A distribution of the form e^{-x} is shown with a dashed line on (c), for comparison.

For comparison, a distribution of the form e^{-x} is shown, which is also seen in other randomly perturbed spectral problems (see *e.g.* Figures 2 and 3 of [**63**]). It is worth noting that much of the standard theory of eigenvalues of random matrices is not useful in this case. First, we are making a very specific choice of unperturbed structure, which is periodic and has a spectral structure (which is the main factor determining the distribution in Figure 4.42(b)). Additionally, the random perturbations are encoded in the diagonal matrix $\mathfrak{X}$, such that the entries of $\mathfrak{C}+\sigma\mathfrak{X}\mathfrak{C}$ are not independent from one another (entries in the same row have nonzero covariance).

REMARK 4.90. *It is interesting to compare and contrast* (4.167) *to the Anderson model* [**57**,**58**,**155**]. *While* (4.167) *intrinsically accounts for long-range interactions in the system, the Anderson model is a tight-binding model of the form* (1.79), *with a Hamiltonian* H_{tb} *whose diagonal matrix elements are the atomic site energies* e_i, *and whose off diagonal elements* V_{ij}, *the "hopping matrix elements", are equal to a negative constant* $-V$ *if the atomic sites* i *and* j *are nearest neighbours and are zero otherwise. The disorder is supplied by the site energies, which are taken to be independent, uniformly distributed random variables, while in* (4.167), *when the disorder is supplied, the entries of the operator* $\mathfrak{BC}$ *are correlated. Both models display a transition between localised and nonlocalised eigenmodes but in different physical settings.*

CHAPTER 5

Non-Hermitian systems of subwavelength resonators

Non-Hermitian systems that include sources of energy input and output (*i.e.* damping) are interesting because standard energy conservation relations no longer apply. This can result in exotic scattering and resonance behaviour. For example, in Chapter 2, we saw that finite arrays of non-Hermitian subwavelength resonators can support exceptional points. Similarly, periodic non-Hermitian systems have been observed to display a variety of behaviours that are not possible in Hermitian systems. These include *unidirectional reflectionless transmission*, *extraordinary transmission*, *eigenmode condensation at one edge*, *double-near-zero effective properties around the origin of the Brillouin zone*, *non-reciprocal wave propagation* and *opening of momentum gaps*.

In this chapter, we study three classes of non-Hermitian systems of subwavelength resonators: (i) $\mathcal{PT}$-symmetric resonators, (ii) systems where an imaginary gauge potential (in the form of a first-order directional derivative) is added inside the resonators to break Hermiticity, and (iii) time-modulated resonators where the time modulation of the material parameters only occurs inside the resonators.

We first study a $\mathcal{PT}$-symmetric metascreen of resonators and prove that it exhibits unidirectional reflectionless transmission at certain frequencies and demonstrate extraordinary transmission close to these frequencies. We also study interface modes introduced in terms of the resonator wave speeds and not as considered in Section 4.6 in terms of a defect in the geometry of the structure.

Then we consider the skin effect in a non-Hermitian resonator system with an imaginary gauge potential and provide a discrete approximation of the resonance problem in terms of a gauge capacitance matrix. When the gauge potential vanishes, this capacitance matrix reduces to the capacitance matrix previously derived for Hermitian systems of subwavelength resonators. Due to long-range interactions in the system, the gauge capacitance matrix is dense. By considering a range-k approximation to keep the long-range interactions to a certain extent, we obtain a k-banded gauge capacitance matrix and prove the exponential decay of its pseudo-eigenvectors through Toeplitz matrix theory. Hence, we deduce the condensation of the eigenmodes at one edge of the structure. This gives mathematical foundations of the skin effect in non-Hermitian systems with long-range coupling in three dimensions.

Finally, we provide a capacitance matrix characterisation of the band structure of a time-modulated system and demonstrate non-reciprocal wave propagation, momentum gaps, exceptional points and Dirac degeneracies at the origin of the Brillouin zone, the latter which enables zero-refractive index materials in the subwavelength regime.

5.1. Systems with complex material parameters

5.1.1. Band structure and exceptional points. In this section, we study the resonance problem in a first non-Hermitian setting. We consider dimers in a two-dimensional square lattice with a period L in $\mathbb{R}^3$. For simplicity, we let $v = 1$. We assume that the dimer $D = D_1 \cup D_2$ satisfies the parity-time symmetry assumption (2.3), that is,

$$\mathcal{P}D_1 = D_2, \quad \delta_1 v_1^2 = \mathcal{T}(\delta_2 v_2^2),$$

where $\mathcal{P}$ is the parity operator and $\mathcal{T}$ is the time-reversal operator defined by (2.4).

We consider the regime when the frequency $\omega \to 0$ while the quasiperiodicity α satisfies $|\alpha| > c > 0$ for some c independent of ω. We let $C^\alpha = (C_{ij}^\alpha)_{i,j=1,2}$ be the quasiperiodic capacitance matrix corresponding to the $\mathcal{PT}$-symmetric metascreen and define the generalised quasiperiodic capacitance matrix $\mathcal{C}^\alpha$ by $\mathcal{C}^\alpha = VC^\alpha$, with V being given by

$$V := \begin{pmatrix} \frac{v_1^2\delta_1}{|D_1|} & 0 \\ 0 & \frac{v_2^2\delta_2}{|D_2|} \end{pmatrix}.$$

Then, analogously to Theorem 4.35, we can show that as $\delta \to 0$, the quasiperiodic resonant frequencies satisfy the asymptotic formula

$$\omega_i^\alpha = \sqrt{\lambda_i^\alpha} + \mathcal{O}(\delta^{3/2}), \quad i = 1, 2,$$

where λ_i^α are the eigenvalues of the generalised quasiperiodic capacitance matrix $\mathcal{C}^\alpha$.

As in (2.5), we write

$$\delta_1 v_1^2 = a + \mathrm{i}b, \quad \delta_2 v_2^2 = a - \mathrm{i}b,$$

with positive, real-valued parameters a and b. Then, the eigenvalues of the generalised quasiperiodic capacitance matrix are given by

$$\lambda_i^\alpha = aC_{11}^\alpha \pm \sqrt{a^2|C_{12}^\alpha|^2 - b^2\big((C_{11}^\alpha)^2 - |C_{12}^\alpha|^2\big)}. \tag{5.1}$$

Therefore, an exceptional point (see Definition 2.1) for the $\mathcal{PT}$-symmetric metascreen occurs when $b = b^*(\alpha)$, given by

$$b_0(\alpha) = \frac{a|C_{12}^\alpha|}{\sqrt{(C_{11}^\alpha)^2 - |C_{12}^\alpha|^2}}.$$

The exceptional point now depends both on the geometry and on the quasiperiodicity α, and will therefore correspond to a point in the band structure. This is illustrated in Figure 5.1, which shows the band structure of a $\mathcal{PT}$-symmetric metascreen. Close to the origin of the Brillouin zone, the system is always below the asymptotic exceptional point. For larger α and for large enough b, there will be a point α_0 where $b = b^*(\alpha_0)$. For α above this point, the band structure of the system has a nonzero imaginary part, and the two bands are complex-conjugated to leading-order in δ.

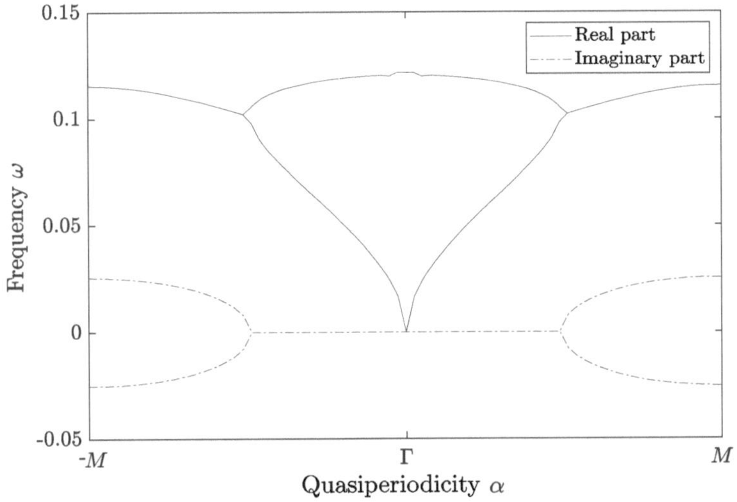

FIGURE 5.1. Plot of the real parts (blue) and imaginary parts (red) of the band structure of the $\mathcal{PT}$-symmetric metascreen. The exceptional point is a point (α, ω) at which the frequencies coincide to leading-order and their corresponding eigenvectors are parallel. Here, we simulate a two-dimensional problem using circular resonators with period $L = 1$, separation distance $0.5L$, radius $0.15L$ and material parameters $a = 2 \times 10^{-4}, b = 1 \times 10^{-4}$ and $v = 1$. These points correspond to $\delta \approx 2.2 \cdot 10^{-4}$. Moreover, Γ and M indicate, respectively, the points $\alpha = 0$ and $\alpha = \pi/L$.

5.1.2. Extraordinary transmission and unidirectional reflection. We now investigate scattering phenomena of metascreens in the non-Hermitian case. In particular, we will extend the analysis in Section 4.7 (valid in the Hermitian case where $v_i \in \mathbb{R}$) to the non-Hermitian case with a $\mathcal{PT}$-symmetric metascreen. We assume that the following extra symmetry condition holds:

$$\mathcal{P}_2 D_i = D_i, \ i = 1, 2,$$

where $\mathcal{P}_2$ is the in-plane parity symmetry defined by

$$\mathcal{P}_2(x_1, x_2, x_3) = (-x_1, -x_2, x_3).$$

Next, we define the generalised periodic capacitance matrix as

$$\mathcal{C}^0 = VC^0,$$

where C^0 is the periodic capacitance matrix, which from Proposition 4.78 is independent of α_0. Moreover, we find that the eigenvalues λ_1^0, λ_2^0 and corresponding eigenvectors $\mathbf{v}_1^0, \mathbf{v}_2^0$ of $\mathcal{C}^0$ are given by

$$\lambda_1^0 = 0, \quad \lambda_2^0 = 2aC_{11}^0, \qquad \mathbf{v}_1^0 = \begin{pmatrix} 1 \\ 1 \end{pmatrix}, \quad \mathbf{v}_2^0 = \begin{pmatrix} -(a+ib) \\ a - ib \end{pmatrix}.$$

For α_0 such that $|\alpha_0| < 1$, the "higher-order" matrix $C^{1,\alpha_0} = (C^{1,\alpha_0})_{i,j=1,2}$ defined in (4.78) (see also (4.129)) is in fact given by

$$C_{ij}^{1,\alpha_0} = -\frac{\mathrm{i}w_0 L^2}{2}\begin{pmatrix} 1 & 1 \\ 1 & 1 \end{pmatrix} - \frac{\mathrm{i}w_0 c_0^2}{2L^2}\begin{pmatrix} 1 & -1 \\ -1 & 1 \end{pmatrix}, \quad i, j = 1, 2,$$

since $\mathbf{c}_1$ defined by (4.79) satisfies in this case $\mathbf{c}_1 = (0, 0, c_0)^\top$. Here, c_0 and w_0 are respectively defined by (4.128) and (4.130).

We let $\mathcal{C}^{1,\alpha_0} := VC^{1,\alpha_0}$. Then we can apply Theorem 4.48 to conclude that the second band in the first radiation continuum , $|\alpha| < \omega < \inf_{q\in\Lambda^*\setminus\{0\}} |\alpha + q|$, is approximated by

$$\omega_2^\alpha = \sqrt{\frac{2aC_{11}^0}{|D_1|}} + \frac{\mathrm{i}aw_0}{4C_{11}^0}\left(\frac{b^2L^2}{a^2} - \frac{c_0^2}{L^2}\right) + \mathcal{O}(\delta^{3/2}).$$

We assume that the incident field is the sum of plane waves given by (4.124) with wave vectors $\mathbf{k}_\pm$ defined in (4.123) for $d = 3$ and seek the behaviour of the solution u of the scattering problem (4.80). Analogously to Proposition 4.84, we obtain the following asymptotic behaviour of the scattered wave in the subwavelength regime.

PROPOSITION 5.1. *Let ω be real with $0 \leq \omega \leq K\sqrt{\delta}$ for some constant K positive, and let $\alpha = \omega\alpha_0$. Let $(\lambda_2^0, \mathbf{v}_2^0)$ be the second eigenpair of the generalised capacitance matrix $\mathcal{C}^0$ and let $\lambda = \omega^2$. Assume that*

$$\mathrm{Im}(d^\top \mathcal{C}^{1,\alpha_0}\mathbf{v}_2^0) \neq 0, \quad \textit{where } d = (1, -1)^\top.$$

Then, for ω such that $\lambda = \lambda_2^0 + \lambda^*$, where $\lambda^* = \mathcal{O}(\omega^3)$, the solution to the scattering problem (4.80) can be written as

$$u - u_{\mathrm{in}} = -(a + \mathrm{i}b)\mu S_1^{\alpha,\omega} + (a - \mathrm{i}b)\mu S_2^{\alpha,\omega} - \mathcal{S}_D^{\alpha,\omega}\left(\mathcal{S}_D^{\alpha,\omega}\right)^{-1}[u_{\mathrm{in}}] + \mathcal{O}(\omega^2),$$

where $S_j^{\alpha,\omega}, j = 1, 2$, are defined by (4.140) and μ is given by

$$\mu = \frac{d^\top p}{d^\top\left(\omega\mathcal{C}^{1,\alpha_0} - \lambda^* I\right)\mathbf{v}_2^0} + \mathcal{O}(\omega), \qquad p = -\begin{pmatrix} \frac{v_1^2\delta_1}{|D_1|}\int_{\partial D_1}\left(\mathcal{S}_D^{\alpha,k}\right)^{-1}[u_{\mathrm{in}}]\,\mathrm{d}\sigma \\ \frac{v_2^2\delta_2}{|D_2|}\int_{\partial D_2}\left(\mathcal{S}_D^{\alpha,k}\right)^{-1}[u_{\mathrm{in}}]\,\mathrm{d}\sigma \end{pmatrix}.$$

Here, the error terms are uniform with respect to λ^* in a neighbourhood of 0.

REMARK 5.2. *If ω is instead close to the first resonant frequency $\omega_1^0 = 0$, then we can show that the incoming wave is largely unaffected by the metascreen and the reflected field is small; see* [**28**].

The following theorem, proved in [**28**], describes the scattering behaviour of this $\mathcal{PT}$-symmetrical screen of resonators. It can be considered as a generalisation of Theorem 4.85 to the non-Hermitian setting and is proved analogously.

THEOREM 5.3. *Assume that D is $\mathcal{PT}$-symmetric, so that $\mathcal{P}D_1 = D_2$ while*

$$\delta_1 v_1^2 = a + \mathrm{i}b \quad \textit{and } \delta_2 v_2^2 = a - \mathrm{i}b \quad \textit{for } a > 0, b \geq 0.$$

Moreover, assume that $\mathcal{P}_2 D_i = D_i$ for $i = 1, 2$. Let $\omega_* = \sqrt{\frac{2aC_{11}^0}{|D_1|}}$. Assume that $bL^2 \neq ac_0$ and that $\omega \in \mathbb{R}$ is in the subwavelength regime such that $\omega - \omega_* = \mathcal{O}(\delta)$. We then have the following asymptotic expansion of the scattering matrix defined by (4.126)*:*

$$S(\omega) = \frac{2\mathrm{i}\omega\,\mathrm{Im}\left(\omega_2^\alpha\right)}{(\omega_2^\alpha)^2 - \omega^2}\begin{pmatrix} 1 & -1 \\ -1 & 1 \end{pmatrix} + \frac{2\omega w_0 b}{a|D_1|((\omega_2^\alpha)^2 - \omega^2)}\begin{pmatrix} -ac_0 & \mathrm{i}bL^2 \\ \mathrm{i}bL^2 & ac_0 \end{pmatrix} - \begin{pmatrix} 1 & 0 \\ 0 & 1 \end{pmatrix}$$
$$+ \mathcal{O}(\delta^{1/2}),$$

where the error term is uniform with respect to ω in a neighbourhood of ω_*. In particular, we have

$$r_\pm(\omega) = -\frac{\omega_*^2 - \omega^2 \pm \frac{2\omega w_0 b c_0}{|D_1|}}{(\omega_2^\alpha)^2 - \omega^2} + \mathcal{O}(\delta^{1/2}),$$

and, at leading-order, the reflection coefficients $r_\pm$ vanish at some frequencies $\omega_\pm$.

REMARK 5.4. *There are two subwavelength frequency regimes not covered in Theorem 5.3: when ω is close to the first band at $\omega_1^0 = 0$ or when ω is well separated from the two bands. When ω is close to ω_1^0, Remark 5.2 tells us that $t_+ = t_- = 1$ and $r_+ = r_- = 0$. When ω is well separated from ω_1^0 and ω_2^α, it is easy to show that t_+ and t_- will be small, while r_+ and r_- have magnitude close to 1. These regimes are demonstrated in Figure 5.2.*

REMARK 5.5. *The assumption $bL^2 \neq ac_0$ comes from the condition*

$$\operatorname{Im}(d^\top C^{1,\alpha_0} \mathbf{v}_2^0) \neq 0$$

in Proposition 5.1. At the critical point $b = ac_0/L^2$, the denominator of S will vanish at $\lambda^* = 0$. Around this point, we therefore expect the transmittance and reflectance to be very large, corresponding to extraordinary transmission. This is demonstrated numerically in Figure 5.3.

Comparing Theorem 4.85 (which is valid only for $b = 0$) and Theorem 5.3, we see that there is an extra term in Theorem 5.3 corresponding to the nonzero gain and loss $b \neq 0$. This term is responsible for the approximate zeros of r_+ and r_- that lead to unidirectional reflection. This is numerically verified in Figure 5.2, where we observe that the two reflectances vanish on different sides of the critical frequency ω_*. Moreover, at least formally, we see that when $\operatorname{Im}(\omega_2^\alpha) = 0$, the singularity of the scattering matrix S will not vanish in the case $b \neq 0$. In the Hermitian case, real resonances correspond to bound states in the continuum, which decouple from the far field and therefore cannot be excited by incoming waves. In the non-Hermitian case, however, we can have real resonances with modes which are excited by incoming waves. Such resonances correspond to extraordinary transmission, where the transmitted field is greatly amplified. This amplification, which is impossible in the Hermitian case due to energy conservation, is possible due to the energy input in the non-Hermitian case.

5.1.3. Non-Hermitian band inversion and edge modes. In Sections 4.6.1 and 4.6.2, we assumed that the wave speeds satisfy $v = v_1 = v_2 = 1$, and only considered topological phenomena originating from the geometry of the resonator chain. In this section, we do the opposite; namely, we study topologically protected modes in structures where the geometry is periodic and where topological edges are introduced in terms of the resonator wave speeds. Most importantly, we allow the wave speeds to be complex, giving a non-Hermitian capacitance formulation, similarly to that in the previous section. The results presented in this section are from [**40**].

We let D be the dimer $D = D_1 \cup D_2$ as previously defined in Section 4.6.1. For $i = 1, 2$ and $m \in \mathbb{Z}$, we introduce the notation

$$D_i^m := D_i + (mL, 0, 0), \qquad k_i^m := \frac{\omega}{v_i^m}.$$

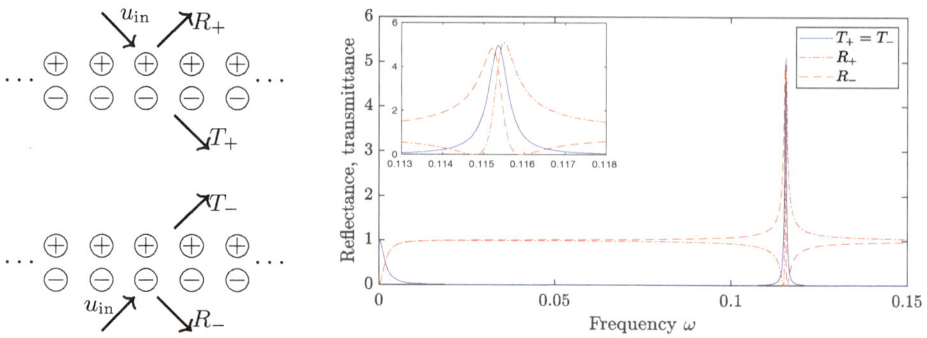

FIGURE 5.2. Plot of the transmittance $T_\pm = |t_\pm|^2$ (blue) and reflectance $R_\pm = |r_\pm|^2$ (red) as functions of the frequency. Due to reciprocity we have $T_+ = T_-$. However, we observe that $R_+ \neq R_-$. In particular, there are points when $R_\pm$ vanish while $R_\mp$ is nonzero, known as unidirectional reflection. Moreover, in this non-Hermitian case the transmission coefficients are not bounded by unity and can attain large peak values, known as extraordinary transmission.

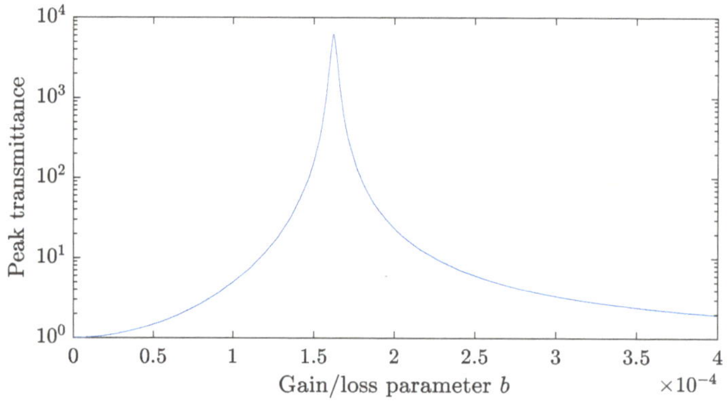

FIGURE 5.3. Plot of the maximum transmittance as a function of the gain/loss parameter. The extraordinarily high transmittance at $b = ac_0/L^2$ is clearly demonstrated. Here, we simulate a two-dimensional problem with the same parameters as those in Figures 5.1 and 5.2.

Here, v_i^m denotes the wave speed inside D_i^m. We will assume that the contrast parameter is given by $\delta \in \mathbb{R}$ for all resonators. We then consider

$$
\begin{cases}
\Delta u + k^2 u = 0 \quad \text{in } \mathbb{R}^3 \setminus \mathcal{D}, \\
\Delta u + (k_i^m)^2 u = 0 \quad \text{in } D_i^m, \\
u|_+ - u|_- = 0 \quad \text{on } \partial \mathcal{D}, \\
\delta \dfrac{\partial u}{\partial \nu}\Big|_+ - \dfrac{\partial u}{\partial \nu}\Big|_- = 0 \quad \text{on } \partial \mathcal{D}, \\
u(x_1, x_2, x_3) \text{ satisfies the outgoing radiation condition} \\
\quad \text{as } \sqrt{x_2^2 + x_3^2} \to +\infty.
\end{cases}
\tag{5.2}
$$

In general, this equation is not periodic and, as in Section 4.8, cannot be approached using the Floquet-Bloch theory. For prescribed values $v_1, v_2 \in \mathbb{C}$ we will consider two different cases, depending on v_i^m:

$$\text{(5.3)} \qquad \text{Periodic structure:} \qquad v_1^m = v_1, \ m \in \mathbb{Z}, \qquad v_2^m = v_2, \ m \in \mathbb{Z},$$

$$\text{(5.4)} \qquad \text{Defect structure:} \qquad v_1^m = \begin{cases} v_1, & m \leq 0, \\ v_2, & m > 0, \end{cases} \qquad v_2^m = \begin{cases} v_2, & m \leq 0, \\ v_1, & m > 0. \end{cases}$$

We emphasise that complex values of v_i correspond to non-Hermitian structures. In the case $v_1 = \overline{v_2}$, the periodic structure is $\mathcal{PT}$-symmetric.

We begin by studying the periodic structure. In Section 4.6.1, we have described the concept of band inversion in the Hermitian case. In the present case, since the quasiperiodic generalised capacitance matrix $\mathcal{C}^\alpha$ is non-Hermitian, the band inversion phenomenon is formulated in a slightly different manner. For $j = 1, 2$, we let $\mathbf{u}_j$ and $\mathbf{w}_j$ denote a bi-orthogonal system of eigenvectors of $\mathcal{C}^\alpha$. In other words, $\mathbf{u}_j$ and $\mathbf{w}_j$ are eigenvectors of $\mathcal{C}^\alpha$ and $(\mathcal{C}^\alpha)^*$, respectively, and satisfy $\mathbf{w}_i \cdot \mathbf{u}_j = \delta_{ij}$.

We let the right Bloch eigenmodes u_j^α be the modes of (5.2), and the left Bloch eigenmodes w_j^α be the modes of (5.2) with wave speeds given by $\overline{v_1}$ and $\overline{v_2}$, respectively. Analogously to Corollary 4.37, we then have

$$u_j^\alpha = \mathbf{u}_j \cdot \mathbf{S}_D^{\alpha,k}(x) + \mathcal{O}(\delta^{1/2}),$$
$$w_j^\alpha = \mathbf{w}_j \cdot \mathbf{S}_D^{\alpha,k}(x) + \mathcal{O}(\delta^{1/2}).$$

DEFINITION 5.6 (Non-Hermitian Zak phase). *For a nondegenerate band ω_j^α of the periodic structure* (5.3), *we let u_j^α and w_j^α be a family of normalised right, respectively, left, eigenmodes that depend continuously on α. We define the (non-Hermitian) Zak phase φ_j^{zak} by*

$$\varphi_j^{\text{zak}} := \frac{\mathrm{i}}{2} \int_{Y^*} \left(\left\langle w_j^\alpha, \frac{\partial u_j^\alpha}{\partial \alpha} \right\rangle + \left\langle u_j^\alpha, \frac{\partial w_j^\alpha}{\partial \alpha} \right\rangle \right) \mathrm{d}\alpha.$$

In the case $v_1 = v_2 \in \mathbb{R}$, this definition coincides with Definition 4.72, and therefore we choose the same notation for these two definitions. In the sequel, we will occasionally write $\varphi_j^{\text{zak}}(v_1, v_2)$ to denote the Zak phase defined with wave speed v_1 inside D_1 and v_2 inside D_2.

In the Hermitian case, a nonzero Zak phase is equivalent to an inverted band structure. The fact that the Hermitian Zak phase is quantised originates from the fact that the eigenmodes are purely monopole and dipole modes at the points Γ and M in the Brillouin zone. Unlike the Hermitian case, the non-Hermitian Zak phase is not quantised. A non-integer value of the Zak phase can be attributed to a "partial" band inversion, where the eigenmodes are expressed as (complex) linear combinations of monopole and dipole modes, which swap as α traverses Y^*.

For the next result, we assume that the Hermitian counterpart of the structure is topologically trivial. More precisely, we assume

$$\text{(5.5)} \qquad \varphi_j^{\text{zak}}\big(\mathrm{Re}(v_1), \mathrm{Re}(v_2)\big) = 0.$$

Then the following result from [**40**] holds.

PROPOSITION 5.7. *Assume that the chain is periodic and topologically trivial, in other words, that it satisfies* (5.3) *and* (5.5). *Then we have*

$$\varphi_j^{\text{zak}}(v_1, v_2) = -\varphi_j^{\text{zak}}(v_2, v_1) + \mathcal{O}(\delta) \quad \text{and} \quad \varphi_j^{\text{zak}}(\overline{v_1}, \overline{v_2}) = \varphi_j^{\text{zak}}(v_1, v_2) + \mathcal{O}(\delta).$$

In particular, if $v_2 = \overline{v_1}$*, we have* $\varphi_j^{\mathrm{zak}}(v_1, \overline{v_1}) = \mathcal{O}(\delta)$.

REMARK 5.8. *Proposition* 5.7 *provides intuition on how to create structures that support edge modes. Proposition* 5.7 *shows that distinct Zak phases can, in general, be achieved by swapping* v_1 *and* v_2 *while keeping the geometry fixed. Therefore, the defect specified in* (5.4) *results in Zak phases with opposite signs on the different sides of the edge. The reason we assume* (5.5) *is to emphasise that distinct Zak phases can originate as a pure non-Hermitian effect, which disappears in the Hermitian limit as* $\mathrm{Im}\,(v_1), \mathrm{Im}\,(v_2) \to 0$.

We now turn to the analysis of the defect problem, specified in (5.4) and illustrated in Figure 5.4.

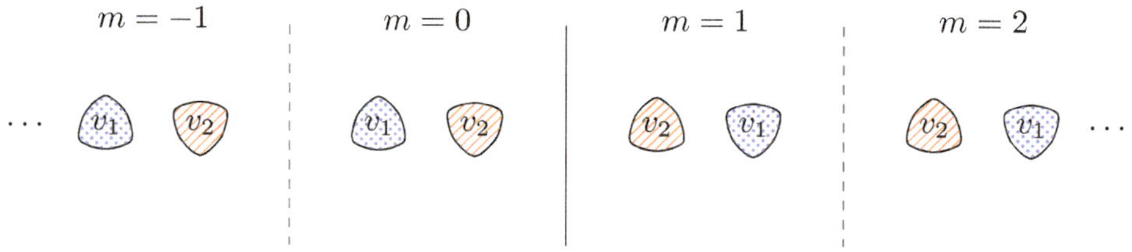

FIGURE 5.4. Illustration of the edge. The special case $v_1 = \overline{v_2}$ corresponds to local $\mathcal{PT}$ symmetry. Legend: material parameter v_1, material parameter v_2.

Since the geometry of this defect chain is periodic, we can utilise, as in Section 4.8, a discrete Floquet approach to derive a capacitance matrix characterisation of any localised mode. The discrete lattice-operator formulation derived in Proposition 4.87 can be used to first obtain the following result, which is valid for general values of v_i^m.

PROPOSITION 5.9. *Any localised solution* u *to* (5.2)*, corresponding to a subwavelength frequency* $\omega = \omega_0 + \mathcal{O}(\delta)$*, satisfies*

$$u(x) = \sum_{m\in\mathbb{Z}} u_i^m e^{\mathrm{i}\alpha m} + \mathcal{O}(\delta^{1/2}) \qquad \text{in } D_i, i = 1, 2,$$

with

$$\begin{pmatrix} C_{11}^\alpha & C_{12}^\alpha \\ C_{21}^\alpha & C_{22}^\alpha \end{pmatrix} \begin{pmatrix} \sum_{m\in\mathbb{Z}} u_1^m e^{\mathrm{i}\alpha m L} \\ \sum_{m\in\mathbb{Z}} u_2^m e^{\mathrm{i}\alpha m L} \end{pmatrix} = \omega_0^2 \frac{|D_1|}{\delta} \begin{pmatrix} \sum_{m\in\mathbb{Z}} \frac{u_1^m e^{\mathrm{i}\alpha m L}}{(v_1^m)^2} \\ \sum_{m\in\mathbb{Z}} \frac{u_2^m e^{\mathrm{i}\alpha m L}}{(v_2^m)^2} \end{pmatrix}. \tag{5.6}$$

Next, we define

$$U_1 = \sum_{m\le 0} u_1^m e^{\mathrm{i}\alpha m L}, \quad U_2 = \sum_{m>0} u_1^m e^{\mathrm{i}\alpha m L}, \quad U_3 = \sum_{m\le 0} u_2^m e^{\mathrm{i}\alpha m L},$$

and

$$U_4 = \sum_{m>0} u_2^m e^{\mathrm{i}\alpha m L}.$$

LEMMA 5.10. *We have*

$$U_1 = bU_3, \quad U_4 = bU_2,$$

for some $b \in \mathbb{C}$*, independent of* α*, satisfying* $|b| < 1$.

PROOF. Observe first that, due to the $\mathcal{P}$-symmetry of the structure, we have $U_1 = bU_3$ and $U_4 = bU_2$ for some $b = b(\alpha)$ which might depend on α. The factor b being constant in α is equivalent to u_1^n/u_2^n being independent of n for $n \leq 0$. To prove this, we apply the inverse discrete Floquet transform to (5.6). We first define

$$\widehat{C}_{ij}^m = \frac{L}{2\pi}\int_{Y^*} C_{ij}^\alpha e^{-\mathrm{i}\alpha m L}\, d\alpha, \qquad m \in \mathbb{Z}.$$

We then find that

$$\begin{aligned}\sum_{m\in\mathbb{Z}} \left(\widehat{C}_{11}^{m-n} u_1^m + \widehat{C}_{12}^{m-n} u_2^m\right) &= \frac{\mu}{(v_1^m)^2} u_1^n,\\ \sum_{m\in\mathbb{Z}} \left(\widehat{C}_{21}^{m-n} u_1^m + \widehat{C}_{22}^{m-n} u_2^m\right) &= \frac{\mu}{(v_2^m)^2} u_2^n,\end{aligned}$$

for $n \in \mathbb{Z}$, where $\mu = \omega_0^2|D_1|/\delta$. We define

$$U^m = \begin{pmatrix} u_1^m \\ u_2^m \end{pmatrix}, \quad K = \begin{pmatrix} v_1^2 & 0 \\ 0 & v_2^2 \end{pmatrix}, \quad J = \begin{pmatrix} 0 & 1 \\ 1 & 0 \end{pmatrix}, \quad \widehat{C}^m = \begin{pmatrix} \widehat{C}_{11}^m & \widehat{C}_{12}^m \\ \widehat{C}_{21}^m & \widehat{C}_{22}^m \end{pmatrix},$$

and introduce as in (4.156) the doubly infinite vectors and matrices

$$\mathfrak{u} = \begin{pmatrix} \vdots \\ U^{-1} \\ U^0 \\ U^1 \\ U^2 \\ \vdots \end{pmatrix}, \qquad \mathfrak{K} = \begin{pmatrix} \ddots & \vdots & \vdots & \vdots & \vdots & \iddots \\ \cdots & K & 0 & 0 & 0 & \cdots \\ \cdots & 0 & K & 0 & 0 & \cdots \\ \cdots & 0 & 0 & JK & 0 & \cdots \\ \cdots & 0 & 0 & 0 & JK & \cdots \\ \iddots & \vdots & \vdots & \vdots & \vdots & \ddots \end{pmatrix},$$

and

$$\mathfrak{C} = \begin{pmatrix} \ddots & \vdots & \vdots & \vdots & \vdots & \iddots \\ \cdots & \widehat{C}^0 & \widehat{C}^1 & \widehat{C}^2 & \widehat{C}^3 & \cdots \\ \cdots & \widehat{C}^{-1} & \widehat{C}^0 & \widehat{C}^1 & \widehat{C}^2 & \cdots \\ \cdots & \widehat{C}^{-2} & \widehat{C}^{-1} & \widehat{C}^0 & \widehat{C}^1 & \cdots \\ \cdots & \widehat{C}^{-3} & \widehat{C}^{-2} & \widehat{C}^{-1} & \widehat{C}^0 & \cdots \\ \iddots & \vdots & \vdots & \vdots & \vdots & \ddots \end{pmatrix}.$$

Observe that $\mathfrak{C}$ is the Laurent operator whose symbol is the quasiperiodic capacitance matrix C^α. Here, $\mathfrak{K}$ and $\mathfrak{C}$ are operators on $\ell_2^2(\mathbb{C})$, which is defined as the space of square-summable sequences of vectors in $\mathbb{C}^2$. We then find that $(\mu, \mathfrak{u})$ is an eigenpair solution to the spectral problem

$$\mathfrak{K}\mathfrak{C}\mathfrak{u} = \mu\mathfrak{u}.$$

Moreover, we see that $\widetilde{\mathfrak{u}} := \begin{pmatrix} \cdots & U^{-2} & U^{-1} & U^2 & U^3 & \cdots \end{pmatrix}^\top$ is an eigenvector corresponding to the same eigenvalue μ. Since μ is a simple eigenvalue, it follows that

$$\widetilde{\mathfrak{u}} = a\mathfrak{u}$$

for some $a \in \mathbb{C}$. In other words, we have that u_1^n/u_2^n is constant for all $n \leq 0$ and that $b = u_1^n/u_2^n$ is constant in α. Moreover, since $\mathfrak{u}$ is square-summable, it follows that $b < 1$ which proves the claim. $\square$

The constant $b \in \mathbb{C}$ can be interpreted as the decay of the localised mode between two consecutive resonators. Using Lemma 5.10, we have

$$\sum_{m\in\mathbb{Z}} u_1^m e^{\mathrm{i}\alpha mL} = U_2 + bU_3, \qquad \sum_{m\in\mathbb{Z}} u_2^m e^{\mathrm{i}\alpha mL} = bU_2 + U_3.$$

Moreover, by using (5.4) we obtain that

$$\sum_{m\in\mathbb{Z}} \frac{u_1^m e^{\mathrm{i}\alpha mL}}{(v_1^m)^2} = \frac{U_2}{v_2^2} + \frac{bU_3}{v_1^2}, \qquad \sum_{m\in\mathbb{Z}} \frac{u_2^m e^{\mathrm{i}\alpha mL}}{(v_2^m)^2} = \frac{bU_2}{v_1^2} + \frac{U_3}{v_2^2}.$$

In total, (5.6) reads

$$\begin{pmatrix} C_{11}^\alpha & C_{12}^\alpha \\ C_{21}^\alpha & C_{22}^\alpha \end{pmatrix}\begin{pmatrix} 1 & b \\ b & 1 \end{pmatrix}\begin{pmatrix} U_2 \\ U_3 \end{pmatrix} = \mu \begin{pmatrix} \frac{1}{v_2^2} & \frac{b}{v_1^2} \\ \frac{b}{v_1^2} & \frac{1}{v_2^2} \end{pmatrix}\begin{pmatrix} U_2 \\ U_3 \end{pmatrix}.$$

Then, we define the matrices

$$A = \begin{pmatrix} 1 & b \\ b & 1 \end{pmatrix}, \quad B = |D_1| \begin{pmatrix} \left(\delta v_2^2\right)^{-1} & b\left(\delta v_1^2\right)^{-1} \\ b\left(\delta v_1^2\right)^{-1} & \left(\delta v_2^2\right)^{-1} \end{pmatrix}.$$

The corresponding capacitance matrix, whose eigenvalue problem provides a discrete approximation to the localised mode and the corresponding frequency, is therefore given by

$$\mathcal{C}_{\text{edge}} = B^{-1} C^\alpha A.$$

The eigenvalues of this matrix depend on $\alpha \in Y^*$. They are given by

$$\mu_j^\alpha(b) = \frac{v_1^2 v_2^2}{(b^2 v_2^4 - v_1^4)}\left(C_{11}^\alpha\left(b^2 v_2^2 - v_1^2\right) + b\left(v_2^2 - v_1^2\right)\mathrm{Re}(C_{12}^\alpha) + (-1)^j \tau(\alpha)\right),$$

where $\tau(\alpha)$ denotes

$$\sqrt{\left(C_{11}^\alpha\left(b^2 v_2^2 - v_1^2\right) + b\left(v_2^2 - v_1^2\right)\mathrm{Re}(C_{12}^\alpha)\right)^2 - (b^2-1)(b^2 v_2^4 - v_1^4)\left((C_{11}^\alpha)^2 - |C_{12}^\alpha|^2\right)}.$$

Here, we choose a holomorphic branch of the square root around a neighbourhood of the curve $f(\alpha)$ for $\alpha \in (-\pi/L, \pi/L]$, where

$$f(\alpha) = \left(C_{11}^\alpha\left(b^2 v_2^2 - v_1^2\right) + b\left(v_2^2 - v_1^2\right)\mathrm{Re}(C_{12}^\alpha)\right)^2 - (b^2-1)(b^2 v_2^4 - v_1^4)\left((C_{11}^\alpha)^2 - |C_{12}^\alpha|^2\right).$$

We also define the curve $g(\alpha)$ as

$$g(\alpha) = \left(C_{12}^\alpha\left(b^2 v_2^2 - v_1^2\right) + C_{11}^\alpha b\left(v_2^2 - v_1^2\right)\right)^2.$$

At points α where C_{12}^α is real, we have $f(\alpha) = g(\alpha)$, and therefore $\mu_j^\alpha = \mu_\pm^\alpha$, where

$$\mu_\pm^\alpha = \frac{(v_1 v_2)^2 (b \pm 1)}{b v_2^2 \pm v_1^2}\left(C_{11}^\alpha \pm C_{12}^\alpha\right).$$

In particular, the above formula holds at $\alpha = 0$ and at $\alpha = \pi/L$. Here, the sign depends on the branch of the square root. Since the capacitance coefficients C_{ij}^α depend on α, the eigenvalue μ_j^α can only be constant in α if the points $\alpha = 0$ and $\alpha = \pi/L$ correspond to opposite signs. In other words, the concatenation γ of the curves defined by f and by g must have an odd winding number around the origin, resulting in a swap of branches.

At $\alpha = 0$, we have from the proof of Lemma 4.44 that $C_{12}^0 = -C_{11}^0$, and so

$$\mu_-^0 = 2C_{11}^0 \frac{(v_1 v_2)^2 (b-1)}{b v_2^2 - v_1^2}, \qquad \mu_+^0 = 0.$$

Therefore, we always have $\mu_+^0 \neq \mu_-^{\pi/L}$. The condition $\mu_-^0 = \mu_+^{\pi/L}$ is equivalent to

$$\lambda_2^0 \frac{b-1}{b v_2^2 - v_1^2} = \lambda_2^{\pi/L} \frac{b+1}{b v_2^2 + v_1^2}.$$

Here, $\lambda_2^0 = 2C_{11}^0$ and $\lambda_2^{\pi/L} = C_{11}^{\pi/L} + C_{12}^{\pi/L}$, which, according to Lemma 4.70, satisfy in the dilute regime $0 < \lambda_2^{\pi/L} < \lambda_2^0$. From this we find two possible values for b:
(5.7)

$$b_{\pm} = \frac{1}{2}\left(l\left(1 - \frac{v_1^2}{v_2^2}\right) \pm \sqrt{l^2\left(1 - \frac{v_1^2}{v_2^2}\right)^2 + \frac{4v_1^2}{v_2^2}} \right), \qquad \text{where } l = \frac{\lambda_2^0 + \lambda_2^{\pi/L}}{\lambda_2^0 - \lambda_2^{\pi/L}}.$$

In the case where the structure is $\mathcal{PT}$-symmetric with unbroken $\mathcal{PT}$-symmetry, *i.e.*, if $v_1 = \overline{v_2}$, with

$$|\operatorname{Im}(v_1^2)| < \frac{\operatorname{Re}(v_1^2)}{\sqrt{l^2 - 1}},$$

the two values of b have unit modulus: $|b_-| = |b_+| = 1$. In the case of broken $\mathcal{PT}$-symmetry, that is,

$$|\operatorname{Im}(v_1^2)| \geq \frac{\operatorname{Re}(v_1^2)}{\sqrt{l^2 - 1}},$$

or in the case $v_1 \neq \overline{v_2}$, there will always be a value b_0 with $|b_0| < 1$ and a value b_1 with $|b_1| > 1$.

We have now proved the following result from [**40**].

THEOREM 5.11. *Suppose that the array of resonators has a defect in the material parameters specified by* (5.4)*. Then, for small* δ*,*

- *if* $v_1 = \overline{v_2}$ *with* $|\operatorname{Im}(v_1^2)| \leq \operatorname{Re}(v_1^2)/\sqrt{l^2-1}$ *(unbroken* $\mathcal{PT}$*-symmetry), the structure does not support simple localised modes in the subwavelength regime.*
- *if* $v_1 = \overline{v_2}$ *with* $|\operatorname{Im}(v_1^2)| > \operatorname{Re}(v_1^2)/\sqrt{l^2-1}$ *(broken* $\mathcal{PT}$*-symmetry) or if* $v_1 \neq \overline{v_2}$ *(no* $\mathcal{PT}$*-symmetry), the frequency* ω *of a simple localised mode in the subwavelength regime must satisfy*

$$\omega = \sqrt{\mu_j^\alpha(b_0)} + O(\delta).$$

Here, b_0 *is the value of* b *specified by* (5.7) *satisfying* $|b_0| < 1$*.*

REMARK 5.12. *Theorem* 5.11 *provides a characterisation of the possible frequency* ω *and decay length* b *of simple localised modes in the subwavelength regime. In order to prove that such a mode indeed exists, we would have to prove that one eigenvalue* $\mu_j^\alpha(b_0)$ *is indeed constant in* α*. Analytically, this is obscured by the fact that the capacitance coefficients have a complicated dependency on* α*. Numerically, however, the eigenvalues* μ *are straightforward to compute, providing convincing evidence that such modes exist. Moreover, the localised modes can easily be computed in large finite chains (see Figure* 5.5*), demonstrating excellent agreement between the numerical and theoretically predicted values.*

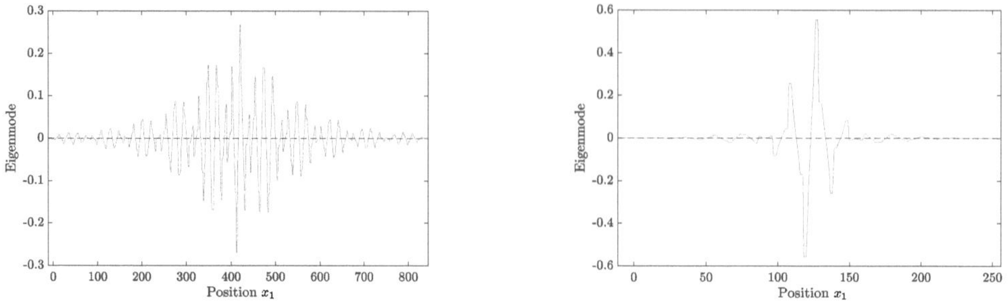

(A) Localised mode in the case of small imaginary part.

(B) Localised mode in the case of large imaginary part.

FIGURE 5.5. Plots of localised modes in finite, large, arrays of resonators. Observe the different x_1-axis scales in the two subfigures, demonstrating significantly different degrees of localisation. The localised modes have purely non-Hermitian origin and disappear in the Hermitian limit when $\mathrm{Im}\,(v_i) \to 0$.

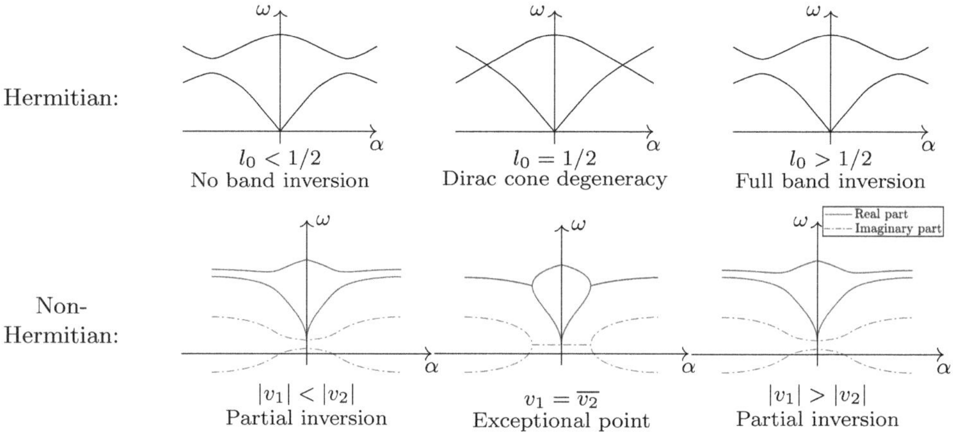

FIGURE 5.6. Comparison between the topological phase transitions studied in Sections 4.6.1 and 5.1.3. In the Hermitian case studied in Section 4.6.1, the Dirac cone found in the symmetric case $l_0 = 1/2$ can open into topologically distinct bandgaps. Here, l_0 is the ratio of the separation of the resonators to the unit cell length. In the non-Hermitian case, the degeneracy corresponds to an exceptional point. This exceptional point can open into separable bands which have distinct, albeit non-quantised, Zak phases.

In both the Hermitian and non-Hermitian cases, there are degeneracies associated to symmetric structures. Breaking the symmetry can open the degeneracy into topologically distinct bandgaps as illustrated in Figure 5.6. In the Hermitian case, the degeneracy is a *Dirac cone.* In the non-Hermitian, $\mathcal{PT}$-symmetric case, there can be exceptional point degeneracies which open when the $\mathcal{PT}$-symmetry is detuned. In both the Hermitian and the non-Hermitian cases, edge modes are created when the two topologically different phases are joined along an interface.

5.2. Non-Hermitian skin effect

The non-Hermitian skin effect is the phenomenon whereby a large proportion of the bulk eigenmodes of a non-Hermitian system is localised at one edge of an open chain of resonators. It has applications in topological photonics, phononics and other condensed matter systems [**88, 91, 122, 165**]. It is one of the most promising and exciting research topics in physics in recent years [**60, 111, 119, 135, 164, 172, 177**].

In this section, we study the non-Hermitian skin effect in a three-dimensional system of finitely many subwavelength resonators with an imaginary gauge potential. We introduce a discrete approximation of the eigenmodes and eigenfrequencies of the system in terms of the eigenvectors and eigenvalues of the so-called gauge capacitance matrix $\mathcal{C}_N^\gamma$, which is a dense matrix due to long-range interactions in the system. Based on translational invariance of this matrix and the decay of its off-diagonal entries, we prove the condensation of the eigenmodes at one edge of the structure by showing the exponential decay of its pseudo-eigenvectors. In particular, we consider a range-k approximation to keep the long-range interaction to a certain extent, thus obtaining a k-banded gauge capacitance matrix $\mathcal{C}_{N,k}^\gamma$. Using techniques for Toeplitz matrices and operators, we establish the exponential decay of the pseudo-eigenvectors of $\mathcal{C}_{N,k}^\gamma$ and demonstrate that they approximate those of the gauge capacitance matrix $\mathcal{C}_N^\gamma$ well. We numerically show that long-range interactions affect only the first eigenmodes in the system. As a result, a tridiagonal approximation of the gauge capacitance matrix, similar to the nearest-neighbour approximation in quantum mechanics, provides a good approximation for the higher modes. Moreover, we also numerically illustrate the behaviour of the eigenmodes and the stability of the non-Hermitian skin effect with respect to disorder. The results presented in this section are from [**7**].

5.2.1. Problem setting. We consider a one-dimensional array of subwavelength resonators given by $\mathcal{D} := \bigcup_{i=1}^N D_i$, where D_i is the translated domain $D_i := D + (i, 0, 0)$. The non-reciprocal model problem reads

$$
\begin{cases}
\Delta u + \omega^2 u = 0 & \text{in } \mathbb{R}^3 \setminus \overline{\mathcal{D}}, \\
\Delta u + \omega^2 u + \gamma \partial_1 u = 0 & \text{in } \mathcal{D}, \\
u|_+ - u|_- = 0 & \text{on } \partial\mathcal{D}, \\
\delta \frac{\partial u}{\partial \nu}\Big|_+ - \frac{\partial u}{\partial \nu}\Big|_- = 0 & \text{on } \partial\mathcal{D}, \\
u \text{ satisfies an outgoing radiation condition,}
\end{cases}
\tag{5.8}
$$

where the first-order directional derivative with real coefficient $\gamma \in \mathbb{R}, \gamma \neq 0$ corresponds to an imaginary gauge potential (or in other terms to the introduction of a directional damping term) within the resonators; see Figure 5.7. Here, the directional first-order damping term is motivated by imaginary gauge potentials from condensed matter theory. It breaks the time-reversal symmetry of the problem and is the crucial mechanism responsible for the condensation effects that will be analysed here. It is worth emphasising that this term is added only inside the subwavelength resonators, and its effect would be negligible without exciting the structure's subwavelength resonant frequencies. The high-contrast regime ($\delta \to 0$) is therefore essential to observe the non-Hermitian skin effect in the subwavelength regime.

As for the case $\gamma = 0$, in order to approximate the subwavelength resonant frequencies and eigenmodes in (5.8), we perform an asymptotic analysis in a high-contrast limit, given by $\delta \to 0$. This limit recovers subwavelength resonant frequencies and eigenmodes, while keeping the characteristic size of the resonators fixed. It leads to a discrete approximation of (5.8) in terms of a so-called gauge capacitance matrix. Investigating the Toeplitz structure of the gauge capacitance matrix, we prove the condensation of subwavelength eigenmodes at one edge of the structure $\mathcal{D}$.

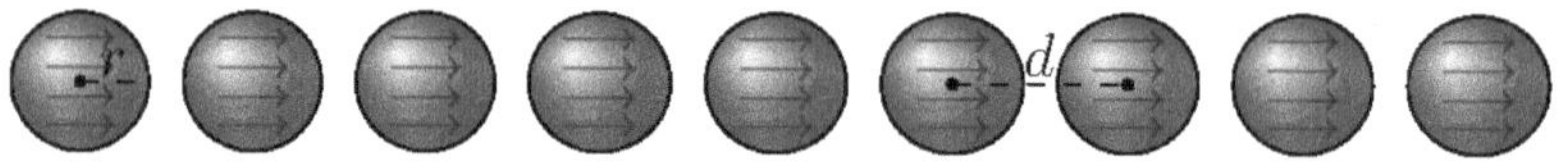

FIGURE 5.7. A chain of identical spherical resonators with radius r, equally spaced with a distance d between each centre. The arrows indicate the directionality introduced by the first-order damping term (with strength γ) in the model (5.8).

5.2.2. Gauge capacitance matrix approximation. We are looking for nontrivial solutions to the non-reciprocal Helmholtz problem (5.8) in the subwavelength regime. With the notation and results of Appendix B.5 in hand, problem (5.8) can be reformulated using layer potential techniques. In fact, there exist density functions ψ and ϕ such that the solution u of (5.8) can be represented as

$$u(x) = \begin{cases} \widetilde{\mathcal{S}}_{\mathcal{D},\gamma}^{\omega}[\psi], & x \in \mathcal{D}, \\ \mathcal{S}_{\mathcal{D}}^{\omega}[\phi], & x \in \mathbb{R}^3 \backslash \mathcal{D}. \end{cases} \tag{5.9}$$

Hence, for ω small enough, (5.8) is equivalent to

$$\mathcal{A}(\omega, \delta) \begin{pmatrix} \psi \\ \phi \end{pmatrix} = 0, \tag{5.10}$$

with $\mathcal{A}(\omega, \delta) : L^2(\partial\mathcal{D}) \times L^2(\partial\mathcal{D}) \to H^1(\partial\mathcal{D}) \times L^2(\partial\mathcal{D})$ being given by

$$\mathcal{A}(\omega, \delta) = \begin{pmatrix} \widetilde{\mathcal{S}}_{\mathcal{D},\gamma}^{\omega} & -\mathcal{S}_{\mathcal{D}}^{\omega} \\ -\frac{1}{2}I + \widetilde{\mathcal{K}}_{\mathcal{D},\gamma}^{\omega,*} & -\delta(\frac{1}{2}I + \mathcal{K}_{\mathcal{D}}^{\omega,*}) \end{pmatrix}. \tag{5.11}$$

Note that for $\omega = \delta = 0$, we have

$$\mathcal{A}(0,0) = \begin{pmatrix} \widetilde{\mathcal{S}}_{\mathcal{D},\gamma}^{0} & -\mathcal{S}_{\mathcal{D}}^{0} \\ -\frac{1}{2}I + \widetilde{\mathcal{K}}_{\mathcal{D},\gamma}^{0,*} & 0 \end{pmatrix}. \tag{5.12}$$

Analogously to Chapter 1, the subwavelength resonant modes arise as perturbations of the kernel functions of $-\frac{1}{2}I + \widetilde{\mathcal{K}}_{\mathcal{D},\gamma}^{0,*}$ for small (but nonzero) δ. In order to reduce the problem of finding the subwavelength eigenfrequencies and eigenmodes to a finite-dimensional eigenvalue problem, we introduce the gauge capacitance matrix.

DEFINITION 5.13 (Gauge capacitance matrix). *The gauge capacitance matrix* $\mathcal{C}_N^{\gamma}$ *reads*

$$(\mathcal{C}_N^{\gamma})_{ij} = -\frac{\delta}{\int_{D_i} e^{\gamma x_1} \, \mathrm{d}x} \int_{\partial D_i} e^{\gamma x_1} (\mathcal{S}_{\mathcal{D}}^0)^{-1}[\chi_{\partial D_j}](x) \, \mathrm{d}\sigma(x). \tag{5.13}$$

The following results can be proved analogously to those obtained for $\gamma = 0$ in Chapter 1. They show that the gauge capacitance matrices can be used to provide asymptotic characterisations of the resonant eigenfrequencies and associated eigenmodes of systems of subwavelength resonators with an imaginary gauge potential.

THEOREM 5.14 (Discrete approximations). *The N subwavelength eigenfrequencies ω_i of (5.8) satisfy, as $\delta \to 0$,*

$$\omega_n = \sqrt{\lambda_n} + \mathcal{O}(\delta), \tag{5.14}$$

where $(\lambda_n)_{1\le n\le N}$ are the eigenvalues of $\mathcal{C}_N^\gamma$. Moreover, if we let $\mathbf{v}_n$ to be the eigenvector of $\mathcal{C}_N^\gamma$ associated to λ_n, then the normalised resonant mode u_n associated to the resonant frequency ω_n is given by

$$u_n(x) = \begin{cases} \mathbf{v}_n \cdot \mathbf{S}^{\omega_n}(x) + \mathcal{O}(\delta) & \text{in } \mathbb{R}^3 \setminus \bar{\mathcal{D}}, \\ \mathbf{v}_n \cdot \widetilde{\mathbf{S}}^{\omega_n}(x) + \mathcal{O}(\delta) & \text{in } D_i, \end{cases} \tag{5.15}$$

where

$$\widetilde{\mathbf{S}}^{\omega_n}(x) = \begin{pmatrix} \widetilde{\mathcal{S}}_{\mathcal{D},\gamma}^{\omega_n}[\psi_1](x) \\ \vdots \\ \widetilde{\mathcal{S}}_{\mathcal{D},\gamma}^{\omega_n}[\psi_N](x) \end{pmatrix}, \qquad \mathbf{S}^{\omega_n}(x) = \begin{pmatrix} \mathcal{S}_{\mathcal{D}}^{\omega_n}[\psi_1](x) \\ \vdots \\ \mathcal{S}_{\mathcal{D}}^{\omega_n}[\psi_N](x) \end{pmatrix}, \tag{5.16}$$

with $\psi_i = (\widetilde{\mathcal{S}}_{\mathcal{D},\gamma}^{0})^{-1}[\chi_{\partial D_i}]$ for $i = 1, \dots, N$.

5.2.3. Toeplitz structure of the gauge capacitance matrix. The capacitance matrix formulation in Theorem 5.14 is very effective not only to efficiently solve the problem, but also to provide valuable insight into the solution. In fact, to demonstrate the non-Hermitian skin effect in three-dimensional systems of subwavelength resonators, we derive some properties of the gauge capacitance matrix $\mathcal{C}_N^\gamma$. In particular, we show that it is, up to a small perturbation, a Toeplitz matrix. In order to do that, we need to establish some convergence results, characterising the matrix in the limit when the number of resonators goes to infinity. In particular, we look at a periodic chain of resonators with only one single resonator in the unit cell. We let $l_1 \in \mathbb{R}^3$ denote the lattice vector that generates the linear lattice $\Lambda := \{ml_1 | m \in \mathbb{Z}\}$. Without loss of generality, we assume that l_1 is aligned with the x_1-axis. We denote by $Y \subset \mathbb{R}^3$ the fundamental domain of the lattice. The dual lattice of Λ, denoted by Λ^*, is generated by α_1, satisfying $\alpha_1 \cdot l_1 = 2\pi$. The Brillouin zone Y^* is defined as $Y^* := (\mathbb{R} \times \{\mathbf{0}\})/\Lambda^*$. As before, we consider the infinite structure of resonators $\bigcup_{m\in\Lambda}(D+m)$. Then, we extend the set of equations (5.8) to the infinite structure and apply the Floquet transform with $u^\alpha(x) := \mathcal{U}[u](x,\alpha)$ to obtain

$$\begin{cases} \Delta u^\alpha + \omega^2 u^\alpha = 0 & \text{in } Y\backslash D, \\ \Delta u^\alpha + \omega^2 u^\alpha + \gamma \partial_1 u^\alpha = 0 & \text{in } D, \\ u^\alpha|_+ - u^\alpha|_- = 0 & \text{on } \partial D, \\ \delta \left.\frac{\partial u^\alpha}{\partial \nu}\right|_+ - \left.\frac{\partial u^\alpha}{\partial \nu}\right|_- = 0 & \text{on } \partial D, \\ u^\alpha \text{ is } \alpha\text{-quasiperiodic}, \\ u^\alpha \text{ satisfies the } \alpha\text{-quasiperiodic radiation condition}. \end{cases} \tag{5.17}$$

We can proceed as in the usual case and define the quasiperiodic gauge capacity as follows.

DEFINITION 5.15. *For $\alpha \neq 0$, the gauge quasiperiodic capacity is given by*

$$\widehat{\mathcal{C}}^{\alpha,\gamma} = -\frac{\delta}{\int_D e^{\gamma x_1} \,\mathrm{d}x} \int_{\partial D} e^{\gamma x_1} (\mathcal{S}_D^{\alpha,0})^{-1}[\chi_{\partial D}](x) \,\mathrm{d}\sigma(x). \tag{5.18}$$

The above introduced gauge capacity is a "dual-space" capacitance representation. The infinitely periodic system with only one resonator in a unit cell is equivalent to an infinite chain of equidistant resonators which has a "real-space" capacitance representation through the inverse Floquet-Bloch transformation given by

$$\mathcal{C}_{ij}^{\gamma} = \frac{1}{|Y^*|} \int_{Y^*} \widehat{\mathcal{C}}^{\alpha,\gamma} e^{-\mathrm{i}\alpha(i-j)} \mathrm{d}\alpha. \tag{5.19}$$

It is clear from the definition above that $\mathcal{C}^\gamma$ is a Toeplitz matrix. We now return to the finite system. Let $\mathcal{D}_N$ denote a finite chain of N equidistant resonators and $\mathcal{C}_N^\gamma$ be the corresponding gauge capacitance matrix. We can extend $\mathcal{D}_N$ to a larger chain $\mathcal{D}_R$ with R resonators and denote by $\mathcal{C}_R^\gamma$ the corresponding gauge capacitance matrix. We then let $\widetilde{\mathcal{C}}_N^\gamma$ be the embedded $N \times N$-size centre block of $\mathcal{C}_R^\gamma$. We call $\widetilde{\mathcal{C}}_N^\gamma$ the truncated gauge capacitance matrix of size $N \times N$. Then, we have the following theorem.

THEOREM 5.16. *Given fixed N and $i, j = 1, \ldots, N$, as $R \to +\infty$, we have*

$$\lim_{R \to +\infty} (\widetilde{\mathcal{C}}_N^\gamma)_{ij} = \mathcal{C}_{ij}^\gamma. \tag{5.20}$$

PROOF. For given $\phi \in L^2(\partial D)$, we have

$$\mathcal{S}_D^{\alpha,0}[\phi] = \mathcal{S}_{\mathcal{D}_R}^0[\phi] + \mathcal{R}^\alpha[\phi], \tag{5.21}$$

where $\mathcal{R}^\alpha = o(1)$ as $R \to +\infty$. From the Neumann series, we have

$$(\mathcal{S}_D^{\alpha,0})^{-1}[\chi_{\partial D}] = (\mathcal{S}_{\mathcal{D}_R}^0)^{-1}[\chi^\alpha] + o(1)\,, \tag{5.22}$$

with

$$\chi^\alpha = \sum_{|m|<N|l_1|} \chi_{\partial D_m} e^{\mathrm{i}\alpha\cdot m}. \tag{5.23}$$

Here, $D_m = D + m l_1$ for $m \in \mathbb{Z}$ and $|l_1|$ is the norm of the lattice generating vector l_1. Finally, as $R \to +\infty$, we have

$$\begin{aligned}
\mathcal{C}_{ij}^\gamma &= -\frac{1}{|Y^*|} \int_{Y^*} \frac{\delta}{\int_D e^{\gamma x_1} \,\mathrm{d}x} \int_{\partial D} e^{\gamma x_1} (\mathcal{S}_D^{\alpha,0})^{-1}[\chi^\alpha] e^{-\mathrm{i}\alpha(i-j)} \,\mathrm{d}\sigma \,\mathrm{d}\alpha + o(1) \\
&= \frac{\delta}{\int_{D_i} e^{\gamma x_1} \,\mathrm{d}x} \int_{\partial D_i} e^{\gamma x_1} (\mathcal{S}_{\mathcal{D}_R}^0)^{-1}[\chi_{\partial D_j}] \,\mathrm{d}\sigma + o(1) \qquad (5.24) \\
&= (\widetilde{\mathcal{C}}_N^\gamma)_{ij} + o(1)\,.
\end{aligned}$$

□

The next results show that the matrix $\mathcal{C}_N^\gamma$ has a Toeplitz structure up to some small perturbation.

LEMMA 5.17. *We have the following estimate of the matrix entries:*

$$\left|(\mathcal{C}_N^\gamma)_{ij} - (\widetilde{\mathcal{C}}_N^\gamma)_{ij}\right| \leq \frac{\delta K}{(1 + \min(i, N-i))(1 + \min(j, N-j))}, \tag{5.25}$$

for some constant K independent of N.

PROOF. Defining x_1^i to be the x_1-coordinate of the centre of D_i, we can compute

$$\begin{aligned} &\left|(\mathcal{C}_N^\gamma)_{ij} - (\widetilde{\mathcal{C}}_N^\gamma)_{ij}\right| \\ &= \left|\frac{\delta}{\int_{D_i} e^{\gamma x_1} \mathrm{d}x} \int_{\partial D_i} e^{\gamma x_1} \left((\mathcal{S}_{\mathcal{D}_N}^0)^{-1} - (\mathcal{S}_{\mathcal{D}_R}^0)^{-1}\right)[\chi_{\partial D_j}]\, \mathrm{d}\sigma\right| \\ &= \left|\frac{\delta}{\int_{D_i} e^{\gamma (x_1 - x_1^i)} \mathrm{d}x} \int_{\partial D_i} e^{\gamma (x_1 - x_1^i)} \left((\mathcal{S}_{\mathcal{D}_N}^0)^{-1} - (\mathcal{S}_{\mathcal{D}_R}^0)^{-1}\right)[\chi_{\partial D_j}]\, \mathrm{d}\sigma\right| \\ &\leq \frac{\delta K}{(1+\min(i, N-i))(1+\min(j, N-j))}, \end{aligned} \tag{5.26}$$

for some constant K independent of N. □

LEMMA 5.18. *We have the following estimate:*

$$\left|(\mathcal{C}_N^\gamma)_{ij} - \mathcal{C}_{ij}^\gamma\right| \leq \frac{\delta K}{(1+\min(i, N-i))(1+\min(j, N-j))}, \quad i, j = 1, \ldots, N, \tag{5.27}$$

for some constant K independent of N.

PROOF. From Theorem 5.16, it follows that for every $\varepsilon > 0$, there exists a large enough R such that $\left|(\widetilde{\mathcal{C}}_N^\gamma)_{ij} - \mathcal{C}_{ij}^\gamma\right| < \varepsilon$. The desired estimate follows from Lemma 5.17 and the triangular inequality. □

The following classical result allows us to describe the decay property of the entries of the gauge capacitance matrix. We refer, for instance, to [**105**] for its proof.

LEMMA 5.19. *Let $f \in BV(\mathbb{T})$. Then the nth Fourier coefficient $\widehat{f}(n)$ of f satisfies*

$$|\widehat{f}(n)| \leq \frac{\mathrm{Var}_{\mathbb{T}}(f)}{2\pi |n|}, \quad n \neq 0. \tag{5.28}$$

Here, $\mathbb{T}$ is the unit circle, $BV(\mathbb{T})$ is the set of functions of bounded variation on $\mathbb{T}$, and $\mathrm{Var}_{\mathbb{T}}(f)$ is the total variation of f, i.e., the supremum *of $\sum_k |f(x_k) - f(x_{k-1})|$, where the* supremum *is taken over all finite sequences (x_k) such that $0 = x_0 < x_1 < \ldots < x_{n-1} < x_n = 2\pi$.*

From Lemma 5.19, we obtain the following decay estimate for the entries of the gauge capacitance matrix.

PROPOSITION 5.20. *For $i \neq j$, there exists a constant K, independent of N, so that*

$$|\mathcal{C}_{ij}^\gamma| \leq \frac{\delta K}{|i-j|}. \tag{5.29}$$

PROOF. We first note that by exactly the same arguments as those in Section 4.4, we can show that $\alpha \mapsto \widehat{\mathcal{C}}^{\alpha,\gamma}$ is piecewise differentiable whose derivative is absolutely integrable over Y^*. It follows that it has bounded variation. By definition (5.19), $\mathcal{C}_{ij}^\gamma$ is the $(i-j)^{\text{th}}$ Fourier coefficient of $\widehat{\mathcal{C}}^{\alpha,\gamma}$, and so Lemma 5.19 shows that

$$\left|\mathcal{C}_{ij}^\gamma\right| \leq \frac{\delta K}{|i-j|}. \tag{5.30}$$

□

We now generalise this decay estimate to the finite gauge capacitance matrix. From Lemma 1.21, we have the following decay estimate of the gauge capacitance coefficients.

PROPOSITION 5.21. *Let d_{ij} be the distance between D_i and D_j. For $i \neq j$, there exists a constant K, depending only on the shape of D_i and D_j, such that*

$$|(\mathcal{C}_N)^\gamma_{ij}| \leq \delta K d_{ij}^{-1}.$$

In particular, we have a constant K', independent of N, such that

$$|(\mathcal{C}_N)^\gamma_{ij}| \leq \frac{\delta K'}{|i-j|}. \tag{5.31}$$

PROOF. We begin by observing that

$$\begin{aligned} \left|(\mathcal{C}_N)^\gamma_{ij}\right| &= \left| \frac{\delta_i v_i^2}{\int_{D_i} e^{\gamma x_1} \mathrm{d}x} \int_{\partial D_i} e^{\gamma x_1} \psi_j \mathrm{d}\sigma \right| \\ &\leq \delta K_1 \int_{\partial D_i} |\psi_j| \mathrm{d}\sigma, \end{aligned} \tag{5.32}$$

for some constant K_1. We, moreover, have

$$\psi_j = \left(\frac{1}{2} + \mathcal{K}_{\mathcal{D}_N}^{0,*}\right)[\psi_j] = \frac{\partial}{\partial \nu} V_j(x),$$

where V_j is defined as in (1.30). In the same way as in the proof of Lemma 1.21, the maximum principle implies that $\partial V_j / \partial \nu \geq 0$ on $\partial \mathcal{D}_N \setminus \partial D_j$. From (5.32), it therefore follows that

$$\left|(\mathcal{C}_N)^\gamma_{ij}\right| \leq \delta K_1 \left|(C_N)_{ij}\right|, \tag{5.33}$$

where $(C_N)_{ij}$ are the regular capacitance coefficients, defined as in (1.26). Now, let $\mathcal{D}_2 = D_i \cup D_j$ be the resonator structure consisting solely of the two components D_i and D_j, and let C_2 be the 2×2 capacitance matrix associated to $\mathcal{D}_2$. Following Lemma 1.19, we then have

$$|(C_2)_{12}| < \frac{K_2}{d_{ij}},$$

for some constant K_2. Moreover, since $(C_N)_{ij}$ and $(C_2)_{12}$ are both negative, we have from Lemma 1.21 that

$$|(C_N)_{ij}| \leq |(C_2)_{12}| < \frac{K_2}{d_{ij}}.$$

This, combined with (5.33), proves the claim. □

Figure 5.8 shows the decay rate of the entries of the gauge capacitance matrix, which is in accordance with Proposition 5.21. We note that the decay rate is close to $|i-j|^{-r}$ for $r \approx 2$. The precise decay rate depends on the geometry of the resonators. In the dilute limit (when the resonators are asymptotically small), the decay rate will approach $r = 1$; see Lemma 1.19.

5.2.4. Non-Hermitian skin effect. As shown in Theorem 5.14, at leading-order in δ, each subwavelength resonant mode u_n of (5.8) is determined by an eigenvector of the gauge capacitance matrix $\mathcal{C}_N^\gamma$. Thus, the non-Hermitian skin effect is related to the exponential decay of the eigenvectors of the gauge capacitance matrix $\mathcal{C}_N^\gamma$.

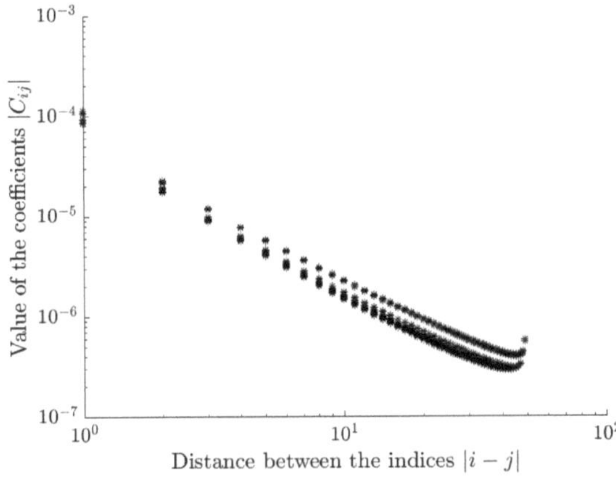

FIGURE 5.8. Decay of the entries of the gauge capacitance matrix for $N = 100$ and $\gamma = 1$ in log-log scale. The horizontal axis represents the index $|i-j|$ and the y-axis displays the absolute value of $(\mathcal{C}_N^\gamma)_{ij}$.

5.2.4.1. *Exponential decay of pseudo-eigenvectors of k-banded gauge capacitance matrices.* We first define a k-banded gauge capacitance matrix $\mathcal{C}_{N,k}^\gamma$ from $\mathcal{C}_N^\gamma$ as

$$\left(\mathcal{C}_{N,k}^\gamma\right)_{ij} = \begin{cases} (\mathcal{C}_N^\gamma)_{ij}, & \text{for } |i-j| < k, \\ 0, & \text{for } |i-j| \geq k, \end{cases} \qquad 1 \leq i,j \leq N, \tag{5.34}$$

and demonstrate the exponential decay of its pseudo-eigenvectors. We let, for $|i-j| < k$,

$$a_{j-i} = \mathcal{C}_{ij}^\gamma \tag{5.35}$$

with $\mathcal{C}_{ij}^\gamma$ being defined by (5.19). The matrix whose $(i,j)^{\text{th}}$ entry is a_{i-j} is a Toeplitz matrix, whose symbol is given by

$$f(z) = \sum_{j=-(k-1)}^{k-1} a_j z^j. \tag{5.36}$$

Define $\mathbb{T} := \{z \in \mathbb{C} : |z| = 1\}$ and $\mathbb{T}_r := \{z \in \mathbb{C} : |z| = r\}$. Let $I(f(\mathbb{T}), \lambda)$ be the winding number of $f(\mathbb{T})$ around λ in the usual positive (counterclockwise) sense.

Define

$$\xi(i,j) = \sum_{q=i}^{j} \frac{1}{q}. \tag{5.37}$$

Then, according to Abel's criteria,

$$\sum_{j=1}^{+\infty} \frac{\xi(j+k_1, j+k_2)^2}{j^2}$$

is convergent for any $k_1 < k_2$ and $j + k_1 \geq 1$. The following result on the pseudo-eigenvectors of $\mathcal{C}_{N,k}^\gamma$ holds. We refer to Appendix C.11 for its proof.

THEOREM 5.22. *For any $\varepsilon > 0$, we can choose an integer $k > 0$ such that*

$$\max\left(\frac{1}{k}, \sum_{j=k+1}^{+\infty} \frac{\xi(j+2-k, j+k)^2}{(1+j)^2}\right) \leq \varepsilon$$

with $\xi(\cdot)$ being defined by (5.37). Let λ be any complex number with $I(f(\mathbb{T}), \lambda) \neq 0$ for f defined by (5.36). For some $\rho < 1$ and any sufficiently large $N > 4k$, there exist nonzero pseudo-eigenvectors $\mathbf{v}^{(N)}$ of $\mathcal{C}_{N,k}^{\gamma}$ with $\left\|\mathbf{v}^{(N)}\right\|_2 = 1$ satisfying

$$\left\|\left(\mathcal{C}_{N,k}^{\gamma} - \lambda\right)\mathbf{v}^{(N)}\right\|_2 < \max\left(C_1, C_2 N^{k-1}\right)\rho^{N-2k+2} + \delta K\sqrt{10\varepsilon} \tag{5.38}$$

such that

$$\frac{\left|\left(\mathbf{v}^{(N)}\right)_j\right|}{\max_j \left|\left(\mathbf{v}^{(N)}\right)_j\right|} \leq \begin{cases} C_3\rho^{j-1}, & \text{if } I(f(\mathbb{T}), \lambda) > 0, \\ C_3\rho^{N-j}, & \text{if } I(f(\mathbb{T}), \lambda) < 0, \end{cases} \quad 1 \leq j \leq N, \tag{5.39}$$

where C_1, C_2, C_3 are independent of N. In particular, the constant ρ can be taken as any number such that $1 > \rho \geq r$ with $I\left(f\left(\mathbb{T}_r\right), \lambda\right) > 0$ or $1/\rho \geq 1/r > 1$ with $I\left(f\left(\mathbb{T}_{\frac{1}{r}}\right), \lambda\right) < 0$. Here, $\mathbb{T}_s = s\mathbb{T}$ for $s > 0$.

Figures 5.9 and 5.10 show respectively the symbol functions of the k-banded Toeplitz matrices $\mathcal{C}_{N,k}^{\gamma}$ for $\gamma = \pm 1$ evaluated on the unit circle $\mathbb{T}$. As predicted by Theorem 5.22, all points inside the circle are pseudo-eigenvalues of $\mathcal{C}_{N,k}^{\gamma}$ for large enough N, including the black dots, which are the eigenvalues of the gauge capacitance matrix $\mathcal{C}_N^{\gamma}$. Figure 5.9b shows the eigenvectors of $\mathcal{C}_N^{\gamma}$ corresponding to the black and red dots. In particular, the localised eigenmodes in gray are the pseudo-eigenmodes of $\mathcal{C}_{N,k}^{\gamma}$. As seen in Figures 5.9 and 5.10, the winding numbers and the positions of the pseudo-eigenvalues correctly predict the exponential decay of the corresponding pseudo-eigenmodes. As seen in red in Figures 5.9b and 5.10b, the non-localised mode corresponds to the eigenvalue of $\mathcal{C}_N^{\gamma}$ outside the enclosed region with nonzero winding.

5.2.4.2. *Exponential decay of pseudo-eigenvectors of gauge capacitance matrices.* In this subsection, we demonstrate that, for sufficiently large k, the pseudo-eigenvectors of $\mathcal{C}_{N,k}^{\gamma}$ having the exponential decay property are also pseudo-eigenvectors of $\mathcal{C}_N^{\gamma}$. In particular, we have the following theorem. We refer to Appendix C.12 for its proof.

THEOREM 5.23. *For any $\varepsilon_1 > 0$, we can choose an integer $k > 0$ such that*

$$\max\left(\frac{1}{k}, \sum_{j=k}^{+\infty} \frac{1}{j^2}\right) < \varepsilon_1.$$

Then, for any unit pseudo-eigenvector of $\mathcal{C}_{N,k}^{\gamma}$ satisfying

$$\left\|\left(\mathcal{C}_{N,k}^{\gamma} - \lambda I\right)\mathbf{v}^{(N)}\right\|_2 \leq \varepsilon_2, \tag{5.40}$$

for some $\varepsilon_2 > 0$ small enough and for $j = 1, \ldots, N$,

$$\frac{\left|\left(\mathbf{v}^{(N)}\right)_j\right|}{\max_j \left|\left(\mathbf{v}^{(N)}\right)_j\right|} \leq C\rho^{j-1}, j = 1, \ldots, N, \text{ or } \frac{\left|\left(\mathbf{v}^{(N)}\right)_j\right|}{\max_j \left|\left(\mathbf{v}^{(N)}\right)_j\right|} \leq C\rho^{N-j}, \tag{5.41}$$

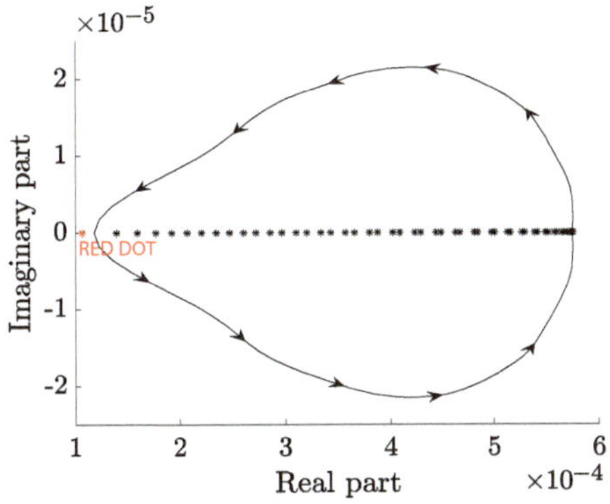

(A) Symbol function $f(\mathbb{T})$ and eigenvalues (asterisks) plotted in the complex plane.

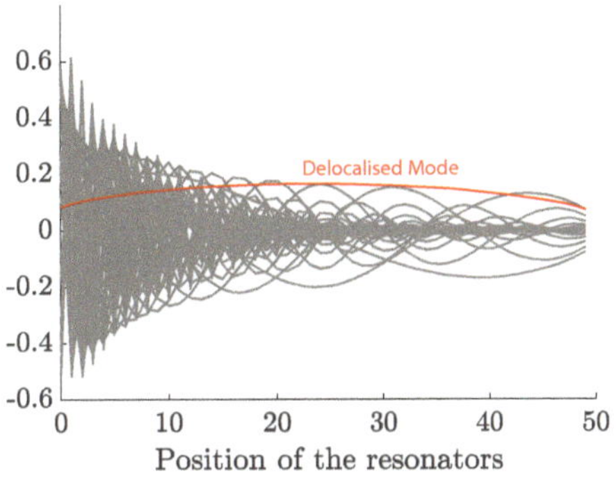

(B) Eigenmodes of $\mathcal{C}_N^\gamma$.

FIGURE 5.9. Chain of resonators with $N = 50$ and $\gamma = 1$. Here, we choose $k = 10$. In (a), the symbol function of $\mathcal{C}_{N,k}^\gamma$ corresponds to the interior of the solid curve and shows a winding number equal to -1. The black dots are the eigenvalues of $\mathcal{C}_N^\gamma$ corresponding to localised modes in (b) and the red dot is the eigenvalue of $\mathcal{C}_N^\gamma$ that corresponds to the non-localised mode shown in red in (b).

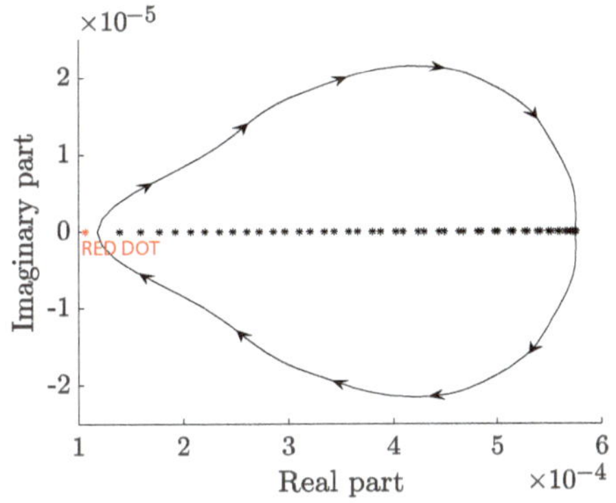

(A) Symbol function $f(\mathbb{T})$ and eigenvalues (asterisks) plotted in the complex plane.

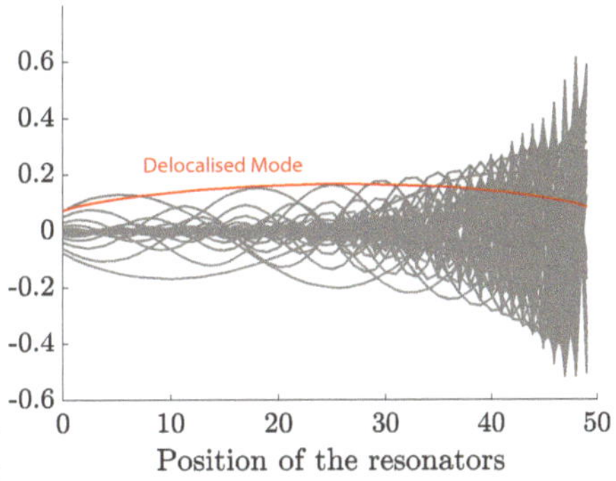

(B) Eigenmodes of $\mathcal{C}_N^\gamma$.

FIGURE 5.10. Chain of resonators with $N = 50$ and $\gamma = -1$. Here, we again choose $k = 10$. Compared to (a) in fig. 5.9, the symbol function of $\mathcal{C}_{N,k}^\gamma$ shows a winding number equal to 1. The modes corresponding to the black dots in (a) are now localised at the right edge of the system.

for some constants C and $0 < \rho < 1 - \varepsilon_1$, we have

$$\left\| \left(\mathcal{C}_N^\gamma - \lambda I\right) \mathbf{v}^{(N)} \right\|_2 \leq \varepsilon_2 + \delta K C \left(\frac{1-\varepsilon_1}{1-\varepsilon_1-\rho}\sqrt{\varepsilon_1} + \varepsilon_1 \rho^k \frac{1}{1-\rho}\sqrt{\frac{1}{1-\rho^2}} \right). \tag{5.42}$$

Theorem 5.23 elucidates that, for sufficiently large $k > 0$, an exponentially decaying pseudo-eigenvector of the k-banded matrix $\mathcal{C}_{N,k}^\gamma$ is also a pseudo-eigenvector

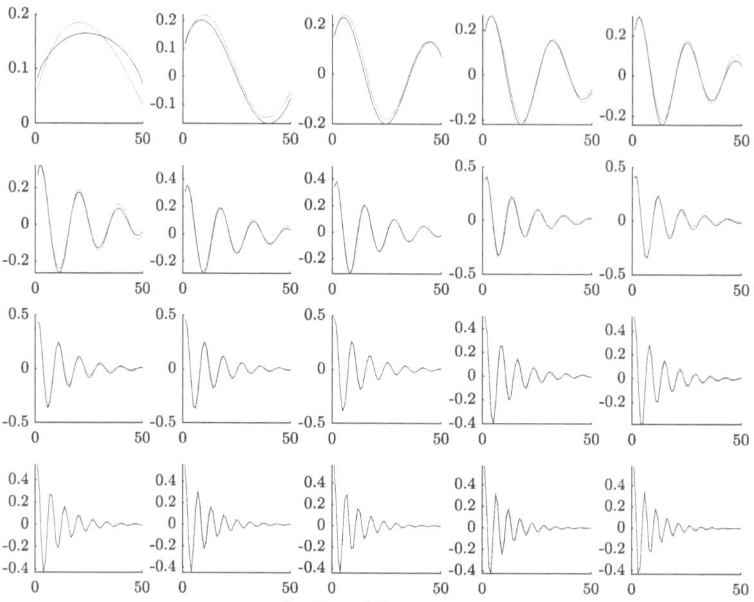

FIGURE 5.11. The first 20 eigenmodes of the gauge capacitance matrix $\mathcal{C}_N^\gamma$ are in black. The blue modes are the pseudo-eigenmodes of the 10-banded matrix $\mathcal{C}_{N,k}^\gamma$. The eigenmodes are normalised.

of the gauge capacitance matrix $\mathcal{C}_N^\gamma$. Theorems 5.22 and 5.23 indicate that the pseudo-eigenvectors/eigenvectors of the gauge capacitance matrix are exponentially decaying when $I(f(\mathbb{T}), \lambda)$ is not zero with f being the symbol associated with a sufficiently large k-banded submatrix. In particular, together with the numerical results in Figures 5.9 and 5.10, it is shown that most of the eigenmodes of $\mathcal{C}_N^\gamma$ have the exponential decay property, as most of the eigenvalues of $\mathcal{C}_N^\gamma$ lie inside $f(\mathbb{T})$. This verifies the non-Hermitian skin effect of the resonating system (5.8).

Figure 5.11 shows the first 20 eigenvectors $\mathbf{v}_j$'s of the gauge capacitance matrix $\mathcal{C}_N^\gamma$ (black line) and the corresponding 20 pseudo-eigenvectors $\mathbf{v}_j(\varepsilon)$'s of the k-banded matrix $\mathcal{C}_{N,k}^\gamma$ for $k = 10$ (blue line). We observe that, as predicted by Theorem 5.23, pseudo-eigenvectors (and hence eigenvectors) with stronger exponential decay can be better approximated by the corresponding pseudo-eigenvectors of the k-banded matrix.

In Figure 5.12, we compare the eigenvectors of the tridiagonal capacitance matrix $\mathcal{C}_{N,2}^\gamma$ and those of $\mathcal{C}_N^\gamma$. It is first shown that although the nearest-neighbour approximation is insufficient to approximate all exponentially decaying eigenvectors of $\mathcal{C}_N^\gamma$, it does approximate those that decay sufficiently fast. Secondly, we observe that, for the first several modes, due to long-range interactions, the non-Hermitian skin effect in three-dimensional resonator systems is less pronounced than that predicted by the nearest-neighbour approximation. Thus, long-range interactions mainly affect the first several eigenmodes in systems of subwavelength resonators.

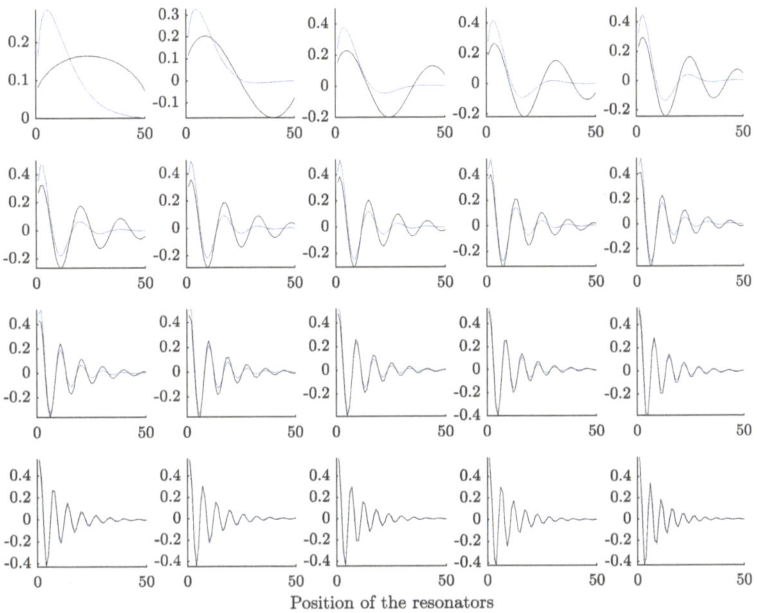

FIGURE 5.12. The first 20 modes of the tridiagonal part of the matrix (in blue) and the full matrix (in black).

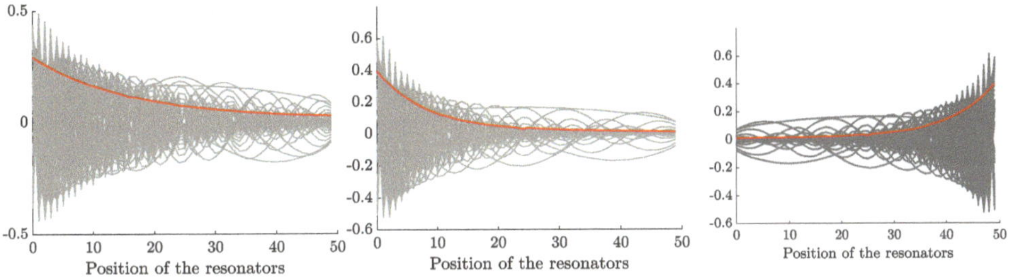

(A) Eigenmodes for $\gamma = 0.5$. (B) Eigenmodes for $\gamma = 1$. (C) Eigenmodes for $\gamma = -1$.

FIGURE 5.13. We plot all the eigenmodes for different values of γ. The red line symbolises the average of the absolute value of the amplitudes.

5.2.4.3. *Numerical illustrations of the non-Hermitian skin effect.* In this subsection, we provide a variety of additional numerical illustrations of the skin effect. In particular, we study the stability of the skin effect in the presence of disorder.

In Figure 5.13, we consider a chain of 100 identical equidistant spherical resonators in one line aligned with the x_1-axis. We plot all the eigenmodes in one graph and demonstrate the condensation of the eigenvectors at one edge of the chain. The condensation becomes more pronounced with increasing γ. Furthermore, changing the sign of γ changes the direction of condensation.

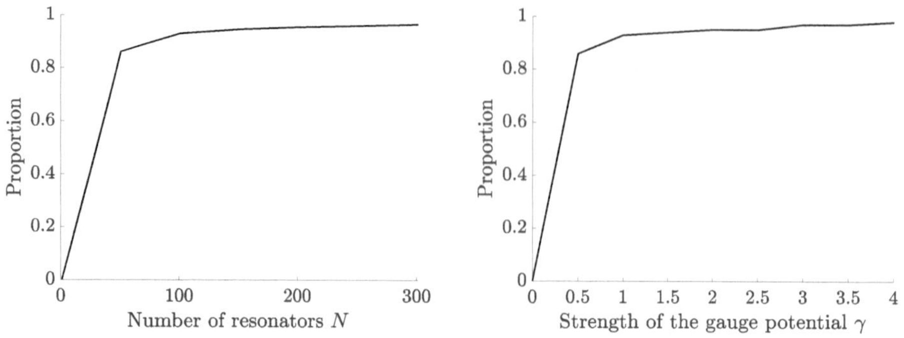

(A) Proportion of condensated eigenmodes with $\gamma = 1$ and increasing N.

(B) Proportion of condensated eigenmodes with $N = 100$ and increasing γ.

FIGURE 5.14. Proportions of condensated eigenmodes.

We say that an eigenmode is condensated if the ratio between the norm of its restriction to the first 20% of the resonators and its entire norm is greater than 80%. We then count the number of condensated eigenmodes and compute the ratio to N (the total number of eigenmodes). In Figure 5.14, we plot the portion of condensanted eigenmodes when increasing γ or N.

To further quantify the non-Hermitian skin effect, we introduce the concept of degree of condensation. We consider a chain of N spherical resonators along the x_1-direction with centres at distinct x_1-coordinates $\{{x_1}^i\}_{i=1,\ldots,n}$. For every eigenmode $\mathbf{v}$, we define the degree of condensation as the vector $d_{\mathbf{v}} \in \mathbb{R}^N$, where

$$(d_{\mathbf{v}})_i = \frac{\|\mathbf{v}|_{\{x \leq x_1^i\}}\|_2}{\|\mathbf{v}\|_2} \quad \text{for } i = 1, \ldots, N. \tag{5.43}$$

Here, the vector $\mathbf{v}|_{\{x \leq x_1^i\}}$ has the same first i entries as $\mathbf{v}$ but all the others are set to 0.

We now consider in Figure 5.15 the stability of the non-Hermitian skin effect in terms of the positions of the resonators. Define the uniformly distributed random variables $\varepsilon_i \sim \mathcal{U}[-\varepsilon, \varepsilon]$ for $i = 1, \ldots, N$. We perturb the centre of the resonators x_1^i to $x_1^i(1 + \varepsilon_i)$ and repeat the experiment 100 times, while fixing $\gamma = 1$. For each perturbation, we compute the average degrees of condensation. The stability result is similar to the one obtained in the one-dimensional case [**5**, **10**].

Next, in Figure 5.16, we consider the stability of the non-Hermitian skin effect in terms of γ. Define the uniformly distributed random variables $\varepsilon_i \sim \mathcal{U}[-\varepsilon, \varepsilon]$ for $i = 1, \ldots, N$. We perturb γ to $\gamma(1 + \varepsilon_i)$ in the ith resonator and compute the average degrees of condensation over 100 runs while fixing the equidistant resonator structure.

REMARK 5.24. *It is worth emphasising that systems with non-Hermiticity arising from complex material parameters are reducible to Hermitian systems away from their exceptional points* [**5**]. *In order to see this, we consider the following ordinary differential equation:*

$$\frac{d^2\psi}{dt^2} + VC\psi = 0, \tag{5.44}$$

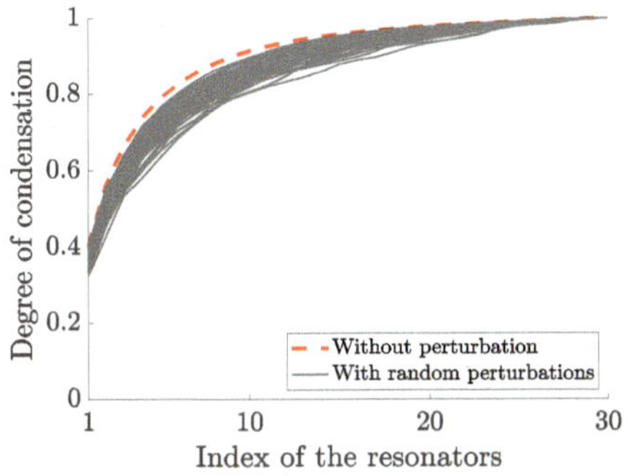

(A) Average degrees of condensation computed over 100 realisations with $\varepsilon = 0.1$.

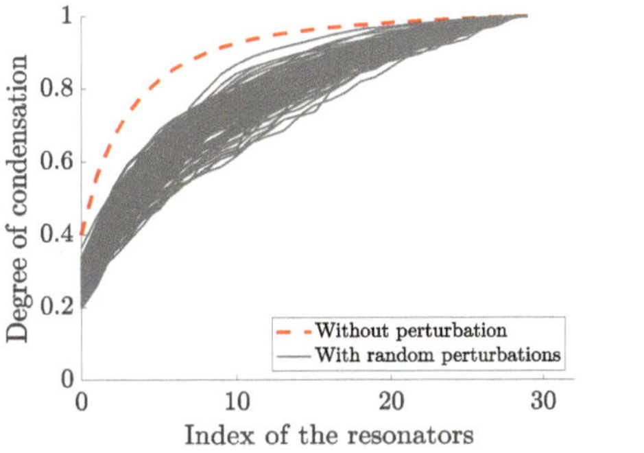

(B) Average degrees of condensation computed over 100 realisations with $\varepsilon = 0.2$.

FIGURE 5.15. Average degrees of condensation for different strengths of geometric disorder. Grey lines are the randomly perturbed modes while the red line represents the unperturbed case.

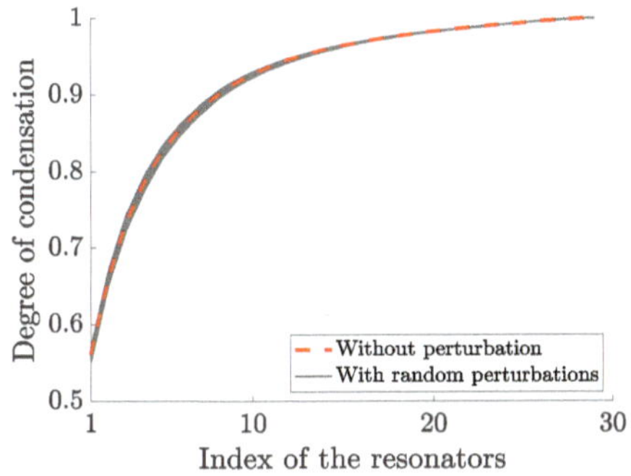

(A) Average degrees of condensation computed over 100 random realisations with $\varepsilon = 0.1$.

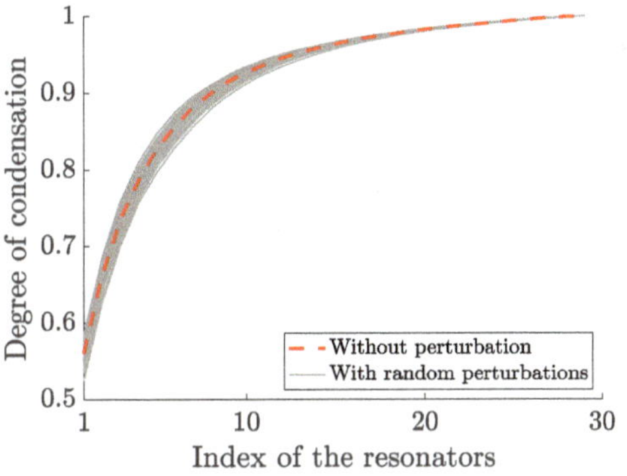

(B) Average degrees of condensation computed over 100 realisations with $\varepsilon = 0.2$.

FIGURE 5.16. Average degrees of condensation for different strengths of disorder in γ. Grey lines are the randomly perturbed modes while the red line represents the unperturbed case.

where V is a diagonal matrix encoding the (complex) material parameters and C the (Hermitian) capacitance matrix. We assume that VC is diagonalisable and invertible, i.e., we are away from exceptional points. As will be seen in the next section, (5.44) *describes the set of subwavelength eigenfrequencies of the resonator system as quasifrequencies of the differential equation. After a change of basis, we can thus assume that $VC = D = \operatorname{diag}(\lambda_1, \lambda_2)$. Introduce the transformation*

$$\begin{aligned} G : \mathbb{R} &\to \mathbb{C}^2 \\ t &\mapsto (\lambda_1^{-\frac{1}{2}} t, \lambda_2^{-\frac{1}{2}} t) \end{aligned}$$

and then consider

$$\phi(t) = \psi \circ G(t),$$

which satisfies

$$\frac{d^2\phi}{dt^2}(t) = D^{-1} \cdot (\frac{d^2\psi}{dt^2} \circ G(t)).$$

Therefore, $\phi(t)$ satisfies the Hermitian ordinary differential equation

$$\frac{d^2\phi}{dt^2} + \phi = 0,$$

as

$$D(\frac{d^2\phi}{dt^2}(t) + \phi(t)) = DD^{-1}(\frac{d^2\psi}{dt^2} \circ G(t)) + D(\psi \circ G(t)) = (\frac{d^2\psi}{dt^2} + D\psi) \circ G(t) = 0.$$

Furthermore, as will be shown in Chapter 7, the spectra of the finite structures converge to the band structure of the infinite ones. Conversely, systems with non-Hermiticity arising from imaginary gauge potentials are fundamentally distinct from the Hermitian systems, and a generalised (or complex) Brillouin zone must be considered to understand their limiting spectra. We refer the reader to [**5**].

REMARK 5.25. *It is also worth noting that the skin effect model studied in this section can be seen as the classical analogue of the non-Hermitian Anderson model* [**101**]*. The non-Hermitian Anderson model is a tight-binding model of the form* (1.79)*, with a (non-Hermitian) Hamiltonian H_{tb} whose diagonal matrix elements are identical and whose off diagonal elements $V_{i(i+1)} = e^{\gamma}, V_{(i-1)i} = e^{-\gamma}$ for some $\gamma \in \mathbb{R}, \gamma \neq 0$, and are zero otherwise. While in the one-dimensional case the two models coincide* [**5**]*, here in three dimensions, the gauge capacitance matrix is a dense matrix. This significant difference makes the analysis of the condensation of the eigenmodes at one of the edges of the system in three dimensions more elaborated than in Anderson's original non-Hermitian model. It should also be noted that, in both the one- and three-dimensional cases, random perturbations of the system's physical parameters affect not only the diagonal entries of the gauge capacitance matrix, but also the off-diagonal entries, which considerably complicates the analysis of the stability of the condensation effect* [**10**].

5.3. Systems with time-modulated material parameters

5.3.1. Time-modulated metamaterials. We have now seen a variety of phenomena that can occur when the material parameters periodically vary in the spatial variable x. Mathematically, we can treat the time variable t in a similar way. If the material parameters are periodically dependent on t, there can be conceptually similar phenomena that nevertheless have fundamentally different physical implications. Structures based on these principles have emerged in the field of *space–time metamaterials*.

The Helmholtz equation we have studied so far originates from the scalar wave equation when posed in the frequency domain and is valid only when the material parameters are constant in time. To study the time-dependent case, we return to the wave equation

$$\left(\frac{\partial}{\partial t}\frac{1}{\kappa(x,t)}\frac{\partial}{\partial t} - \nabla \cdot \frac{1}{\rho(x,t)}\nabla\right) u(x,t) = 0, \quad x \in \mathbb{R}^d, t \in \mathbb{R}. \tag{5.45}$$

Here, $\kappa(x,t)$ and $\rho(x,t)$ are the material parameters. Let $D = D_1 \cup \cdots \cup D_N$ be a finite collection of resonators in $d = 3$ spatial dimensions. We assume that the modulation is only performed inside the resonators, so that

$$\kappa(x,t) = \begin{cases} \kappa_0, & x \in \mathbb{R}^3 \setminus \overline{D}, \\ \kappa_i(t)\kappa_r, & x \in D_i, \end{cases}, \qquad \rho(x,t) = \begin{cases} \rho_0, & x \in \mathbb{R}^3 \setminus \overline{D}, \\ \rho_i(t)\rho_r, & x \in D_i. \end{cases} \tag{5.46}$$

We assume that κ_i, ρ_i, for $i = 1, \ldots, N$, are periodic in t with period T and frequency $\Omega = 2\pi/T$ and consider only the case where the modulation of $1/\rho$ and $1/\kappa$ consists of a finite Fourier series with a large number of Fourier coefficients

$$\frac{1}{\rho_i(t)} = \sum_{n=-M}^{M} r_{i,n} e^{\mathrm{i}n\Omega t}, \qquad \frac{1}{\kappa_i(t)} = \sum_{n=-M}^{M} k_{i,n} e^{\mathrm{i}n\Omega t},$$

for some $M \in \mathbb{N}$.

As before, we define the (time-dependent) contrast parameter and wave speeds as

$$\delta_i(t) = \frac{\rho_i(t)\rho_r}{\rho_0} = \rho_i(t)\delta, \quad v_i(t) = \sqrt{\frac{\kappa_i(t)}{\rho_i(t)}} v_r, \quad v_r = \sqrt{\frac{\kappa_r}{\rho_r}}, \quad v = \sqrt{\frac{\kappa_0}{\rho_0}},$$

and assume that for all $t \in \mathbb{R}$ and for all $i = 1, \ldots, N$,

$$\delta_i(t) = \mathcal{O}(\delta), \quad v_i(t) = \mathcal{O}(1) \quad v, v_r = \mathcal{O}(1),$$

where $\delta \ll 1$.

The notion of frequency is slightly altered in this time-modulated setting. Since the wave equation (5.45) is periodic in t, we can apply the Floquet transform in t and obtain the differential problem

$$\begin{cases} \left(\frac{\partial}{\partial t} \frac{1}{\kappa(x,t)} \frac{\partial}{\partial t} - \nabla \cdot \frac{1}{\rho(x,t)} \nabla \right) u(x,t) = 0, \\ u(x,t)e^{-\mathrm{i}\omega t} \text{ is } T\text{-periodic in } t, \end{cases} \tag{5.47}$$

for the finite system of resonators D. If D is repeated periodically with respect to a lattice Λ, then we arrive at

$$\begin{cases} \left(\frac{\partial}{\partial t} \frac{1}{\kappa(x,t)} \frac{\partial}{\partial t} - \nabla \cdot \frac{1}{\rho(x,t)} \nabla \right) u(x,t) = 0, \\ u(x,t)e^{-\mathrm{i}\alpha\cdot x} \text{ is } \Lambda\text{-periodic in } x, \\ u(x,t)e^{-\mathrm{i}\omega t} \text{ is } T\text{-periodic in } t, \end{cases} \tag{5.48}$$

for each $\alpha \in Y^*$. In (5.47) and (5.48), the transmission conditions are given by

$$u\big|_+ = u\big|_-, \quad \delta \frac{\partial u}{\partial \nu}\Big|_+ = \frac{1}{\rho_i} \frac{\partial u}{\partial \nu}\Big|_- \quad \text{on } \partial D_i \times [0,T], i = 1, \ldots, N. \tag{5.49}$$

It is apparent that ω plays the role of quasiperiodicity in Floquet-Bloch theory, and we will refer to ω as a quasifrequency. Since ω is defined modulo Ω, we define the time Brillouin zone $Y_t^* := \mathbb{C}/(\Omega\mathbb{Z})$. Observe that we allow complex quasifrequencies; real ω correspond to solutions which are oscillating in t, whereas non-real ω correspond to exponentially increasing or decaying solutions.

5.3.2. Capacitance matrix formulation of subwavelength quasifrequencies. Due to the periodic nature of Y_t^*, the usual definition of subwavelength frequencies does not apply to quasifrequencies. For example, in the particular case where $\Omega = \mathcal{O}(\delta^{1/2})$ (which will be of interest later), the whole Brillouin zone scales as $\mathcal{O}(\delta^{1/2})$, which means that *all* quasifrequencies tend to zero as $\delta \to 0$. In order to distinguish between these quasifrequencies, we introduce the following definition.

DEFINITION 5.26 (Subwavelength quasifrequency). *A quasifrequency $\omega = \omega(\delta) \in Y_t^*$ of (5.47) or (5.48) is said to be a subwavelength quasifrequency if there is a corresponding solution $u(x,t)$, continuously depending on δ, which can be written as*

$$u(x,t) = e^{\mathrm{i}\omega t} \sum_{n=-\infty}^{+\infty} v_n(x) e^{\mathrm{i}n\Omega t},$$

where

$$\omega \to 0 \text{ and } M\Omega \to 0 \text{ as } \delta \to 0,$$

for some integer-valued function $M = M(\delta)$ satisfying $M = \mathcal{O}\left(\delta^{-\gamma/2}\right)$ for some $0 < \gamma < 1$ such that

$$\sum_{n=-\infty}^{+\infty} \|v_n\|_{L^2(D)} = \sum_{n=-M}^{M} \|v_n\|_{L^2(D)} + o(1),$$

as $\delta \to 0$.

Definition 5.26 says that a subwavelength quasifrequency corresponds to an eigenmode which is essentially supported in the low-frequency regime.

The following theorem, proved in [**41**], gives the capacitance matrix approximation to the subwavelength quasifrequencies of a finite system of time-modulated resonators as $\delta \to 0$.

THEOREM 5.27. *Assume that the material parameters are given by (5.46). Then, as $\delta \to 0$, the quasifrequencies $\omega \in Y_t^*$ to the wave equation (5.45) in the subwavelength regime are, to leading-order in δ, given by the quasifrequencies of the system of ordinary differential equations for $y_i(t)$,*

$$\sum_{j=1}^{N} C_{ij} y_j(t) = -|D_i| \frac{\mathrm{d}}{\mathrm{d}t}\left(\frac{1}{\delta_i v_i^2}\frac{\mathrm{d}y_i}{\mathrm{d}t}\right), \tag{5.50}$$

for $i = 1, \ldots, N$.

The matrix appearing in the left-hand side of (5.50) is the capacitance matrix. We can rewrite (5.50) into the following system of Hill equations:

$$\frac{d^2\Psi}{dt^2}(t) + M(t)\Psi(t) = 0, \tag{5.51}$$

where the vector Ψ and the matrix M are defined as

$$\Psi(t) = \left(\frac{y_i(t)}{\sqrt{\delta_i(t) v_i^2(t)}}\right)_{i=1}^{N}, \quad M(t) = W_1(t) C W_2(t) + W_3(t),$$

with W_1, W_2 and W_3 being the diagonal matrices with diagonal entries

$$(W_1)_{ii} = \frac{\sqrt{\delta_i v_i^2}}{|D_i|}, \qquad (W_2)_{ii} = \sqrt{\delta_i v_i^2}, \qquad (W_3)_{ii} = \sqrt{\delta_i v_i^2}\frac{\mathrm{d}}{\mathrm{d}t}\frac{1}{\delta_i v_i^2}\frac{\mathrm{d}\sqrt{\delta_i v_i^2}}{\mathrm{d}t},$$

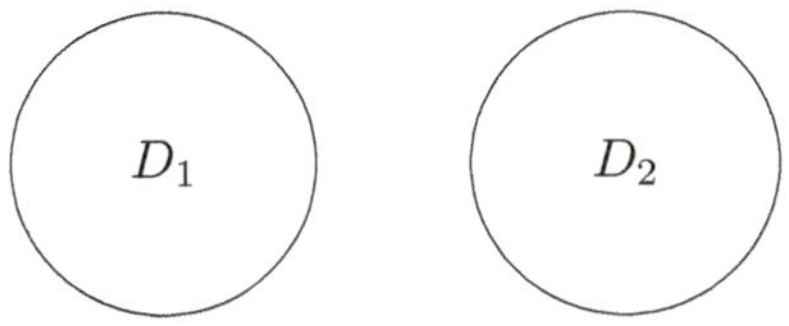

$\kappa_1(t) = 1 + \varepsilon \sin(\Omega t)$ $\kappa_2(t) = 1 - \varepsilon \sin(\Omega t)$

(A) A $\mathcal{PT}$-symmetric pair of spherical resonators.

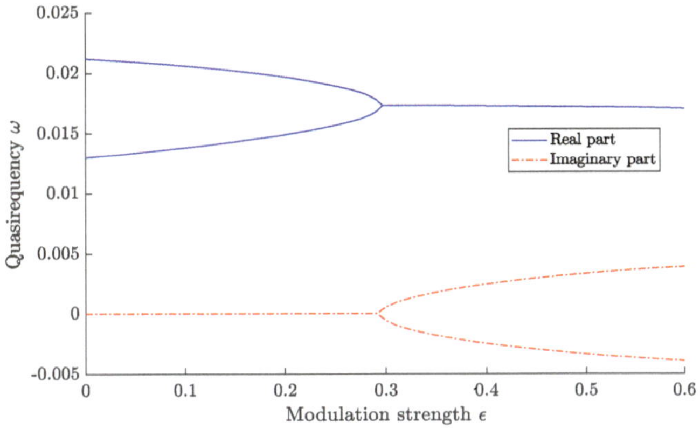

(B) The two subwavelength resonant frequencies.

FIGURE 5.17. Similarly as in the non-Hermitian case demonstrated in Figure 2.1, there can be exceptional points in the time-modulated case. The two subwavelength quasifrequencies of a pair of $\mathcal{PT}$-symmetric, time-modulated resonators can be approximated through the capacitance formulation, and an asymptotic exceptional point occurs at $\varepsilon \approx 0.3$.

for $i = 1, \ldots, N$.

The time-modulated case shares many similarities with the non-Hermitian case (with complex parameters) studied before. Both of these cases have energy input and output to the system, and, not surprisingly, we can find exceptional points in the time-modulated systems. Figure 5.17 shows the emergence of an exceptional point in a dimer structure, similar to Figure 2.1 but instead due to time modulation [**42**].

Analogously to Theorem 5.27, the quasifrequencies of the wave equation (5.48) in the subwavelength regime are, to leading-order in δ, given by the quasifrequencies of the system of ordinary differential equations for $y_i(t)$,

$$\sum_{j=1}^{N} C_{ij}^{\alpha} y_j(t) = -|D_i| \frac{\mathrm{d}}{\mathrm{d}t} \left(\frac{1}{\delta_i v_i^2} \frac{\mathrm{d}y_i}{\mathrm{d}t} \right), \tag{5.52}$$

for $i = 1, \ldots, N$. Here, C^{α} is the quasiperiodic capacitance matrix corresponding to the periodic system $\cup_{m \in \Lambda}(D + m)$ of resonators.

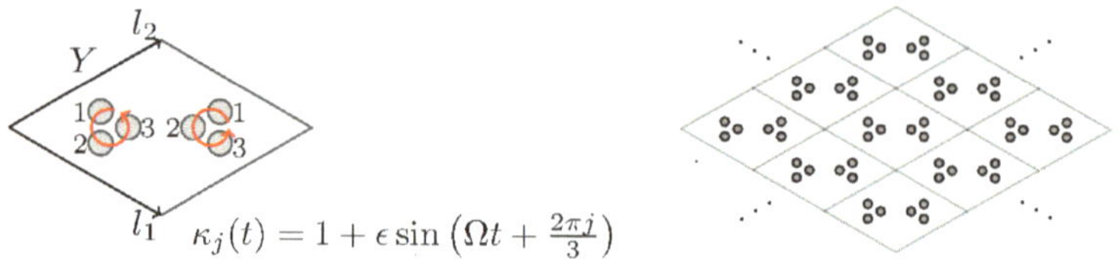

FIGURE 5.18. Illustration of the trimer honeycomb lattice, with phase-shifted time modulations inside the trimers.

REMARK 5.28. *Note that* $\delta_i v_i^2 = \delta v_r^2 \kappa_i(t)$. *Hence, in view of* (5.50) *and* (5.52), *time modulations of the material parameter* ρ *inside the resonators do not affect the subwavelength quasifrequencies of the system at leading-order in* δ.

5.3.3. Double-near-zero materials in time-modulated systems. In Section 4.3.5 we saw the emergence of exotic parameter values, namely negative effective parameters, due to the small-scale structure of the metamaterial. In the present subsection, we observe effective material parameters which are close to zero, in which case wave propagation in time-modulated structures occurs without phase change (corresponding to "infinite" phase velocity).

The Dirac cone observed in Section 4.3.5 is located at the corner of the Brillouin zone, which means that the corresponding modes do not lie in the radiation continuum. For physical structures of finite dimensions, these waves will be confined to the structure and will not interact with incoming waves. Physically, this can be seen as an impedance mismatch, leading to poor transmission because this structure corresponds to a *single*-near-zero material, where only one of the material parameters vanishes.

In order to overcome this low transmission, it is desirable to create structures with Dirac cones at the origin of the Brillouin zone. The corresponding modes lie in the radiation continuum and therefore will interact with incoming waves. To achieve this, we study a time-modulated honeycomb structure as illustrated in Figure 5.18. Using the theory from Section 5.3.1, we can compute the (quasi-)band structure associated with this material. In (5.50), we replace the capacitance matrix C with the quasiperiodic capacitance matrix C^α studied in Section 4.3. In doing so, we can compute the quasiperiodic subwavelength quasifrequencies. In Figure 5.19, we see the static (folded) band structure, with six bands in the subwavelength regime. For a particular modulation strength, we see that a Dirac cone degeneracy may appear at the origin of the Brillouin zone, enabling a double-near-zero material around this point.

5.3.4. Non-reciprocal wave propagation. In this subsection, we discuss the most fundamental mechanisms of non-reciprocity in metamaterials based on time modulation. By using lattices of spatiotemporally modulated subwavelength resonators where time-reversal symmetry is broken, unidirectional excitation of waves guided at subwavelength scales can be proved [**16**].

We say that waves propagate reciprocally if for every $\alpha \in Y^*$, the set of quasifrequencies of (5.48) at α coincides with the set of quasifrequencies at $-\alpha$. The reciprocal equation associated with (5.48) defined with $\alpha \in Y^*$ is that with $-\alpha$.

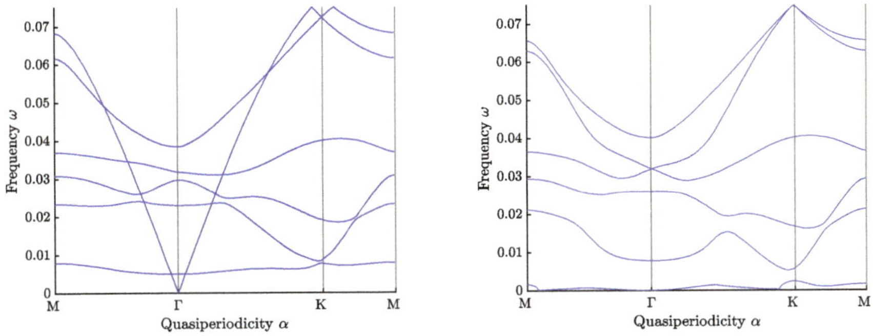

(A) Static band structure. (B) Modulated band structure.

FIGURE 5.19. In time-modulated structures, Dirac cones may appear at the origin of the Brillouin zone. Such points are associated to double near-zero materials, enabling wave transmission without phase changes and with strong interactions with incoming waves.

In the presence of only spatial modulation, the time-reversal symmetry is not broken, and consequently the band functions are always symmetric for opposite directions.

Note that in view of Remark 5.28, ρ does not affect the band functions at the leading order in δ. Then, in order to achieve non-reciprocal wave propagation, we assume that the resonators have time-modulated boundary-layer sources $s_i(t)$ instead of time-modulating ρ. This alters the transmission conditions posed on the boundaries of the resonators and yields the slightly modified model

$$
\begin{cases}
\left(\frac{1}{\kappa_0}\frac{\partial^2}{\partial t^2} - \frac{1}{\rho_0}\Delta\right) u(x,t) = 0 \quad \text{in } Y \setminus \overline{D} \times [0,T], \\
\left(\frac{1}{\kappa_r}\frac{\partial}{\partial t}\frac{1}{\kappa_i(t)}\frac{\partial}{\partial t} - \frac{1}{\rho_r}\Delta\right) u(x,t) = 0 \quad \text{in } D_i \times [0,T], \\
u(x,t)\big|_+ = \frac{1}{s_i(t)} u(x,t)\big|_- \quad \text{on } \partial D_i \times [0,T], \\
\delta \frac{\partial u}{\partial \nu}(x,t)\big|_+ = \frac{1}{s_i(t)}\frac{\partial u}{\partial \nu}(x,t)\big|_- \quad \text{on } \partial D_i \times [0,T], \\
u(x,t)e^{-\mathrm{i}\alpha\cdot x} \text{ is } \Lambda\text{-periodic in } x, \\
u(x,t)e^{-\mathrm{i}\omega t} \text{ is } T\text{-periodic in } t,
\end{cases}
\tag{5.53}
$$

for $i = 1, \ldots, N$ and $\alpha \in Y^*$.

Analogously to Theorem 5.27, we can prove that the quasifrequencies $\omega \in Y_t^*$ to the wave equation (5.53) in the subwavelength regime are, to leading-order in δ, given by the quasifrequencies of the system of ordinary differential equations for $y_i(t)$,

$$
\sum_{j=1}^{N} C_{ij} y_j(t) = -\frac{|D_i|\rho_r}{\delta\kappa_r s_i(t)}\frac{\mathrm{d}}{\mathrm{d}t}\left(\frac{1}{\kappa_i(t)}\frac{\mathrm{d}(s_i(t)y_i(t))}{\mathrm{d}t}\right),
\tag{5.54}
$$

for $i = 1, \ldots, N$.

In contrast with (5.50), (5.54) shows that for the modified model (5.53), time modulations of s affect the subwavelength quasifrequencies of the resonator system at the leading-order in δ. (5.54) can be reduced to a first-order ordinary differential

FIGURE 5.20. A chain of trimers.

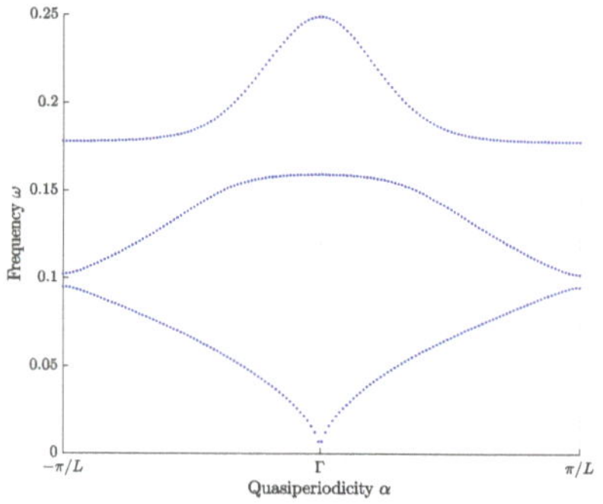

(A) Unfolded band structure.

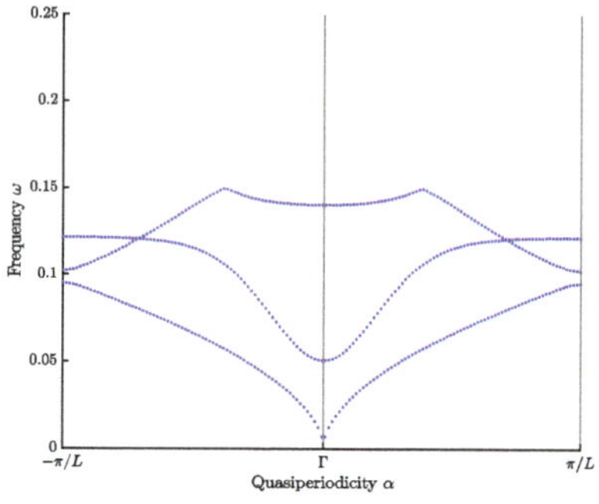

(B) Folded band structure with $\Omega = 0.3$.

FIGURE 5.21. Subwavelength band functions of the one-dimensional lattice of unit cells containing three resonators in the static (*i.e.*, unmodulated) case. In (b) there is a degenerate point around $\alpha = 2.23$.

equation. Assume that κ_i is constant and s_i is given by

$$s_i(t) = \frac{1}{1 + \varepsilon \cos(\Omega t + \phi_i)}, \quad i = 1, \ldots, N, \tag{5.55}$$

where Ω is the modulation frequency and ϕ_i is a phase shift between the resonators. Then, using the asymptotic results (with respect to the modulation amplitude ε) in Appendix C.13, we can prove that time modulations may break the time-reversal symmetry and open degenerate points (see Definition C.36) of the folded band structure into non-symmetric bandgaps; opposite propagation directions are subject to distinct bandgaps [**16**]. If the excitation frequency falls inside the bandgap for only one propagation direction, wave transmission is then prohibited in this direction but not in the opposite one, leading to non-reciprocal transmission properties. The fundamental fact here is again that phase-shifted ("rotation like") time modulations of subwavelength resonators can provide a kind of "artificial spin". Such an artificial spin cannot be achieved in systems of one or two resonators. In fact, non-reciprocity requires at least three resonators inside one unit cell of the material; see [**16**].

Consider a one-dimensional lattice with unit cell containing three resonators with phase shifts $\phi_1 = 0$, $\phi_2 = \pi/2$ and $\phi_3 = \pi$; see Figure 5.20.

Suppose that the modulation frequency Ω is chosen so that the static system has a degenerate point. The band structure of this structure is depicted in Figure 5.21, where we have a doubly degenerate point at $\alpha = \pm 2.23$ with different folding numbers: $m_1 = 0$ and $m_2 = 1$.

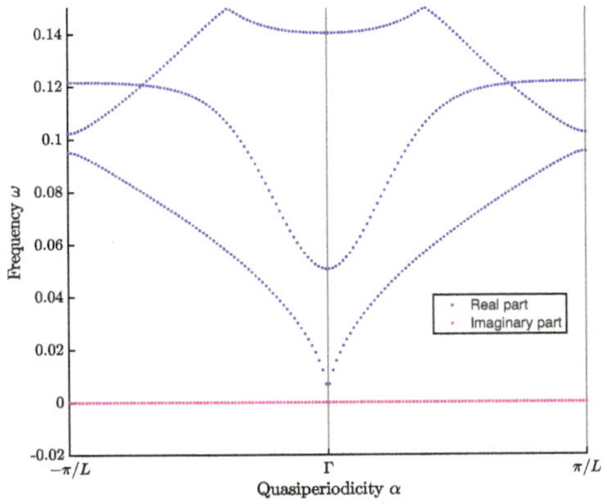

(A) Static ($\varepsilon = 0$) band structure of the one-dimensional lattice of resonators, folded with $\Omega = 0.3$.

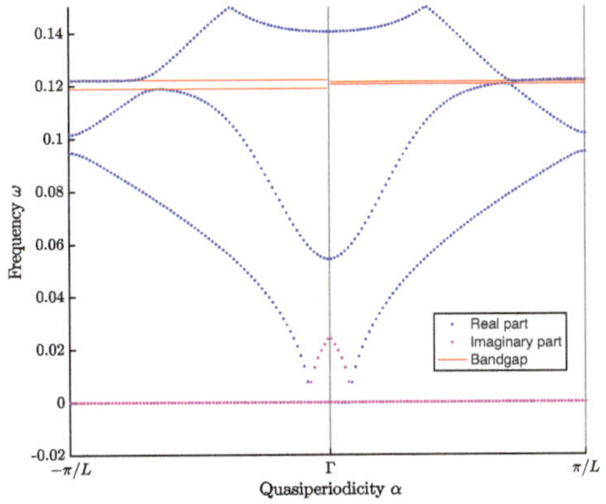

(B) Time-modulated ($\varepsilon = 0.1$) band structure of the one-dimensional lattice of resonators, folded with $\Omega = 0.3$.

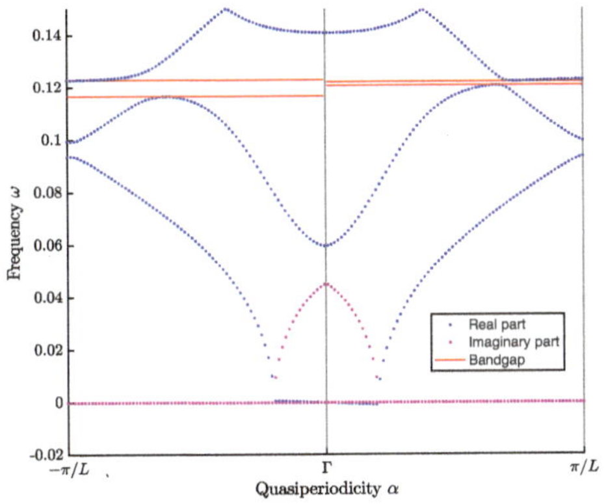

(C) Time-modulated ($\varepsilon = 0.2$) band structure of the one-dimensional lattice of resonators, folded with $\Omega = 0.3$.

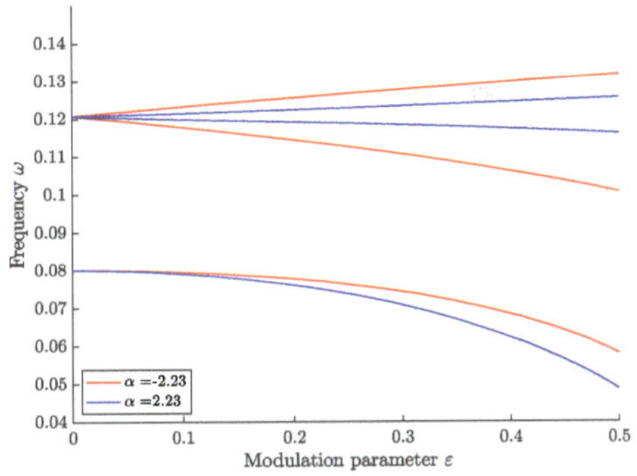

(D) $\Omega = 0.3$, $\alpha = \pm 2.23$ showing perturbations of degenerate points (depending linearly on ε), and nondegenerate points (depending quadratically on ε).

FIGURE 5.22. Band structure of one-dimensional trimers ($N = 3$) of subwavelength resonators with time modulation in s. As ε increases from 0, the degenerate points open into asymmetric bandgaps, indicated by solid red lines in (b) and (c). Waves with frequencies inside the disjunctive union of these bandgaps can only propagate in one direction.

In Figure 5.22, we demonstrate the subwavelength band functions of Figure 5.21 as ε increases from 0. Note that in the absence of time modulation, the band functions are symmetric for opposite directions, while the time modulation of s opens non-symmetric bandgaps at degenerate points, as a consequence of breaking time-reversal symmetry.

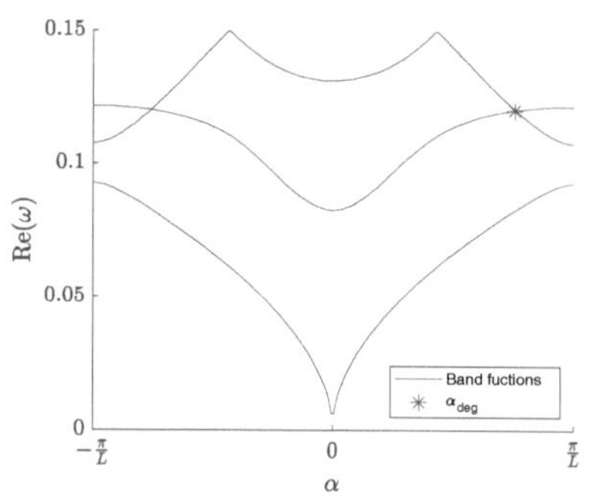

(A) Band functions of the static trimer chain with $\Omega = 0.3$, exhibiting degenerate points.

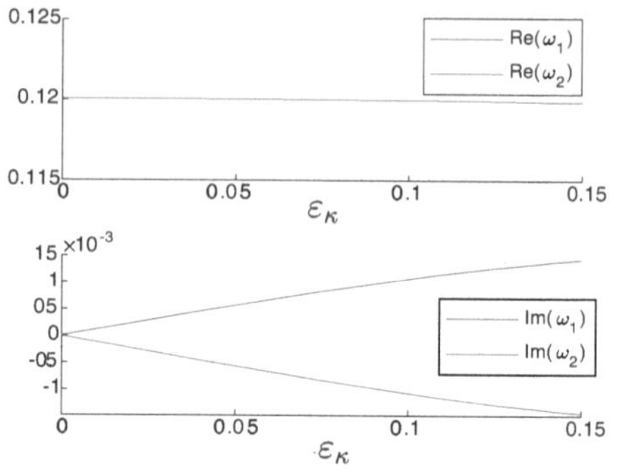

(B) Quasifrequencies at α_{deg} when modulating κ. Here, ε denotes the modulation amplitude.

FIGURE 5.23. Quasifrequencies of the trimer chain under time modulation of κ at the degenerate point. The modulation frequency is $\Omega = 0.3$.

5.3.5. Opening of momentum gaps. Consider the time-modulated model in (5.47)–(5.49). In [**18**], it was shown that around doubly degenerate points, by time modulating κ, one of the two Bloch modes is exponentially decaying while the other is exponentially increasing over time. The momentum gaps where waves exhibit this exponential behaviour are known as *k-gaps*.

We again consider an infinite chain of trimers, as shown in Figure 5.20 and assume that ρ_i is constant and κ_i is given by

$$\kappa_i(t) = \frac{1}{1 + \varepsilon \cos(\Omega t + \phi_i)}, \quad i = 1, 2, 3, \tag{5.56}$$

where Ω is the modulation frequency and ϕ_i is a phase shift between the resonators.

The band structure of the trimer chain depends on the modulation frequency Ω. It exhibits degenerate points in the static case as a result of the folding in the time Brillouin zone. This is depicted in Figure 5.23a. We then fix a positive degenerate point $\alpha = \alpha_{\text{deg}}$ and turn on the time modulations of κ to observe the emergence of the quasifrequencies ω_1 and ω_2 from the degenerate point. Their real and imaginary parts are plotted against the modulation amplitude ε.

Note that there is no bandgap when modulating only κ, since the first-order term is purely imaginary at the degenerate point. Nevertheless, as shown in Figure 5.24, there is a k-gap opening in this case and the structure can support exponentially growing waves with these quasiperiodicities.

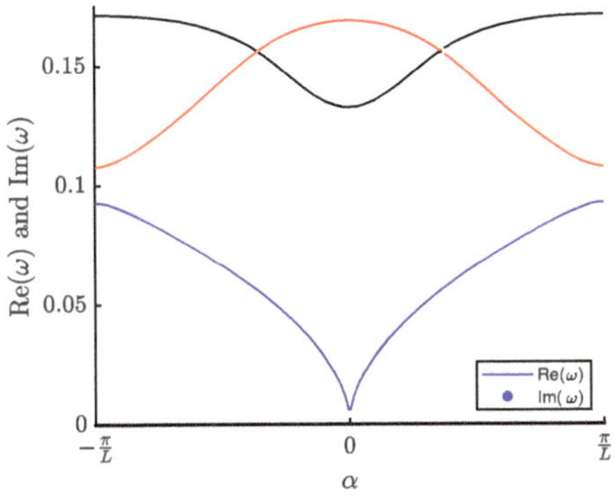

(A) Degenerate points for the static case.

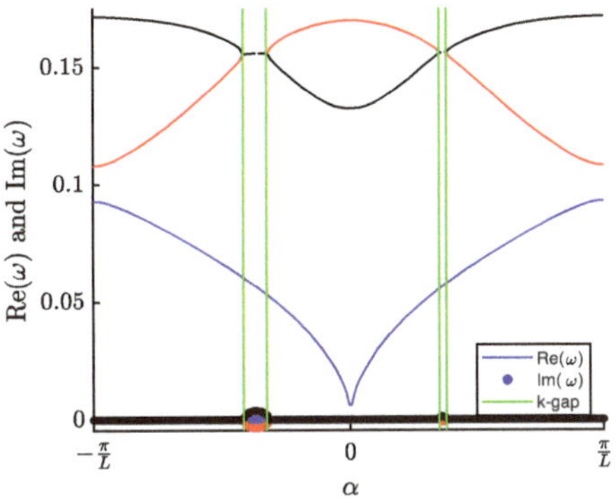

(B) Appearance of k-gaps by modulating κ.

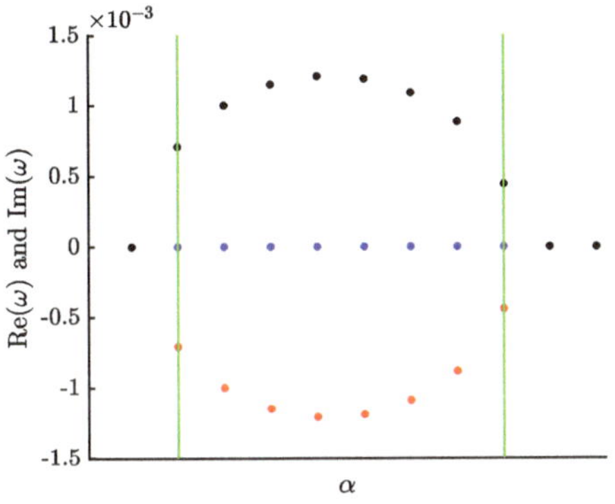

(C) Zoom in of the second k-gap in Figure 5.24b.

FIGURE 5.24. Transmission properties of a time-modulated one-dimensional lattice of trimers. We set $\Omega = 0.35$ and the amplitude of the modulation to be $\varepsilon = 0.1$. In (b), the green line marks the k-gaps where the second (in red) and third (in black) Bloch band functions have nonzero imaginary parts. In (c), The dots in red (respectively in black) correspond to negative (respectively positive) imaginary part of the second (respectively third) Bloch band function inside the k gap. The first Bloch band function (in blue) is real.

Part 3

Convergence results for large finite systems of subwavelength resonators

CHAPTER 6

Spectral convergence of defect modes

In this chapter, we will focus on localised modes which decay rapidly outside of some compact region. As a result, they are less severely affected by edge effects in finite-sized arrays. Existing results have shown that in certain one- and two-dimensional systems, any *defect mode* of the infinite structure will have a corresponding mode in the truncated structure converging to the defect mode as the size tends to infinity. In [**114–116, 123**], it was shown that the convergence of the defect mode eigenfrequency to the pure point spectrum of the infinite periodic operator is exponential with respect to the size of the truncated array. These results concerned either one-dimensional systems or two-dimensional systems with two-dimensional lattices.

Our goal here is to study these convergence questions in the setting of the generalised capacitance matrix. As we have seen in previous chapters, this is a dense matrix model that includes long-range interactions. In this setting, we will prove that any defect mode eigenfrequency of the infinite structure has a sequence of eigenvalues of the truncated structures converging to it. This expands on the one- and two-dimensional models explored previously by showing that this convergence is algebraic when the dimension of the periodic lattice is less than the dimension of the differential operator (which is three in this chapter). This is due to the long-range interactions that arise, since waves are able to radiate in the "spare" dimensions and couple with the far field. When the lattice is three-dimensional, with no spare dimensions, we see the same exponential convergence as for one-dimensional lattices in one-dimensional differential problems and two-dimensional lattices in two-dimensional differential problems. The results presented in this chapter are from [**23**]. We emphasise that they hold in the Hermitian case of real-valued, positive material parameters.

6.1. The generalised capacitance matrix model

In this section, we recall the generalised capacitance matrix model that will be the object of this chapter. Then, we will prove a convergence result for the capacitance coefficients, which will be the basis of the theorems in subsequent sections.

6.1.1. Notation. We study a system of periodically repeated resonators in a lattice Λ in $\mathbb{R}^3$ with periodicity dimension d_l. We take $Y \subset \mathbb{R}^3$ to be a single unit cell and let $D \Subset Y$ be a collection of N resonators contained in Y,

$$D = \bigcup_{i=1}^{N} D_i,$$

where D_i are disjoint domains in Y with boundary $\partial D_i \in C^{1,s}$ for some $s > 0$. In the periodic lattice, we let $D_i^m = D_i + m$, for $m \in \Lambda$, and then denote the full lattice as

$$\mathcal{D} = \bigcup_{m\in\Lambda} \bigcup_{i=1}^{N} D_i^m.$$

We will define a finite system of resonators resulting from truncation of the periodic lattice. Let $I_r \subset \Lambda$ be all lattice points within distance r from the origin

$$I_r = \{m \in \Lambda \mid |m| < r\}.$$

We define the finite collection of resonators $\mathcal{D}_\mathrm{f} = \mathcal{D}_\mathrm{f}(r)$ as

$$\mathcal{D}_\mathrm{f}(r) = \bigcup_{m\in I_r} D + m = \bigcup_{m\in I_r} \bigcup_{i=1}^{N} D_i^m. \tag{6.1}$$

The goal is to clarify in which sense the spectral properties of a finite, but large, lattice can be approximated by the corresponding infinite one.

We denote by

$$(C_\mathrm{f}^{mn})_{ij}(r) = \int_{\partial D_i^m} (\mathcal{S}_{\mathcal{D}_\mathrm{f}}^0)^{-1}[\chi_{\partial D_j^n}] \,\mathrm{d}\sigma, \tag{6.2}$$

for $1 \le i, j \le N$ and $m, n \in I_r$, the capacitance matrix for a finite lattice I_r. Here, we have explicitly indicated the dependence on the size r of the truncated lattice. For $m, n \in I_r$, we observe that $C_\mathrm{f}^{mn}(r)$ is a matrix of size $N \times N$, while the block matrix $C_\mathrm{f} = (C_\mathrm{f}^{mn})$ is a matrix of size $N|I_r| \times N|I_r|$.

We next recall from Section 4.8 that, for $\alpha \in Y^*$ and for $1 \le i, j \le N$, the quasiperiodic capacitance matrix ("dual-space" representation) for the infinite lattice is the $N \times N$-matrix defined as

$$C_{ij}^\alpha = \int_{\partial D_i} (\mathcal{S}_D^{0,\alpha})^{-1}[\chi_{\partial D_j}] \,\mathrm{d}\sigma, \tag{6.3}$$

while the "real-space" capacitance coefficients at the lattice point m are given by

$$\widehat{C}_{ij}^m = \frac{1}{|Y^*|} \int_{Y^*} C_{ij}^\alpha e^{-\mathrm{i}\alpha\cdot m} \,\mathrm{d}\alpha, \tag{6.4}$$

for $1 \le i, j \le N$ and $m \in \Lambda$. Here, $\widehat{C}_{ij}^0$ corresponds to the diagonal block that contains the capacitance coefficients of the resonators within a single unit cell. We will use the notation $\mathfrak{C}$ to denote the infinite matrix that contains all the $\widehat{C}_{ij}^m$ coefficients for all $1 \le i, j \le N$ and all $m \in \Lambda$.

A new and important quantity for the analysis in this chapter is the truncated capacitance matrix C_t. This is obtained by keeping only $N|I_r| \times N|I_r|$ coefficients from $\mathfrak{C}$, to give a matrix that is the same size as C_f. We shall emphasise that C_t is "nonphysical" in the sense that it does not correspond to a capacitance matrix associated with any physical structure, but rather to the finite matrix obtained by simply truncating the infinite matrix $\mathfrak{C}$.

A schematic of the various pieces of notation used in this chapter and how they relate to each other is given in Figure 6.1.

The proof strategy deployed in this chapter is to compare the spectra of C_f with that of C_t, and then let $r \to +\infty$ in order to approximate the spectrum of $\mathfrak{C}$. In particular, the modes we will compare are *defect modes*, which are spatially

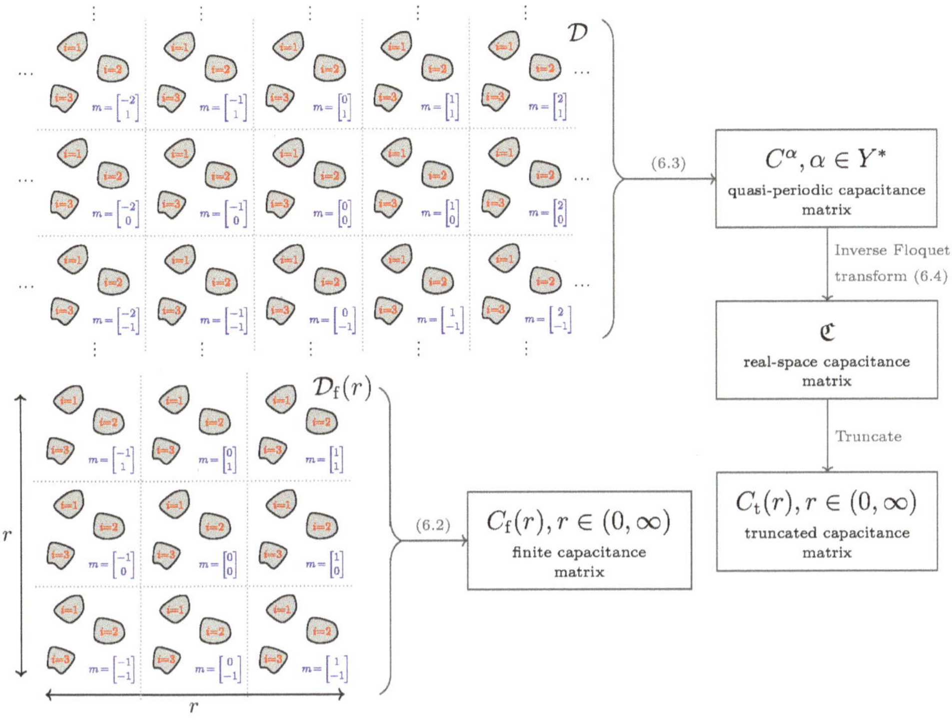

FIGURE 6.1. This chapter studies the convergence of the eigenfrequencies of defect modes in a truncated periodic material to the spectrum of the corresponding infinite material. We use capacitance matrices as a canonical model for many-body scattering of time-harmonic waves. The aim of this chapter is to show how eigenvalues of the finite capacitance matrix $C_f(r)$ converge to those of the real-space capacitance matrix $\mathfrak{C}$. The calligraphic font for $\mathfrak{C}$ denotes the fact that this is an infinite matrix. Our strategy is to compare the spectrum of $C_f(r)$ with the truncated capacitance matrix $C_t(r)$, which is obtained by truncating all but a finite $\mathcal{O}(r)$ number of rows in $\mathfrak{C}$, before letting $r \to +\infty$. Throughout this chapter, we use the block matrix notation $(C^{mn})_{ij}$ to refer to the $i, j \in \{1, \dots, N\}$ entry of the $n, m \in \Lambda$ block in a matrix C.

localised modes that exist due to the presence of defects in the otherwise periodic material, an example of which is shown in Figure 6.2.

As in Section 4.8, we will model the defect modes through premultiplication by a defect matrix $\mathfrak{B}$. For each $m \in \Lambda$, we let $\mathcal{B}_m$ be an $N \times N$ diagonal matrix

$$\mathcal{B}_m = \begin{pmatrix} b_1^m & 0 & \cdots & 0 \\ 0 & b_2^m & \cdots & 0 \\ \vdots & \vdots & \ddots & \vdots \\ 0 & 0 & \cdots & b_N^m \end{pmatrix}, \tag{6.5}$$

where the diagonal entries b_i^m are real-valued parameters. In this chapter, we only consider *compact* defects, where $b_i^m = 1$ for all but finitely many i and m. For the infinite structure, we let $\mathfrak{B}$ be the infinite block-diagonal matrix that contains

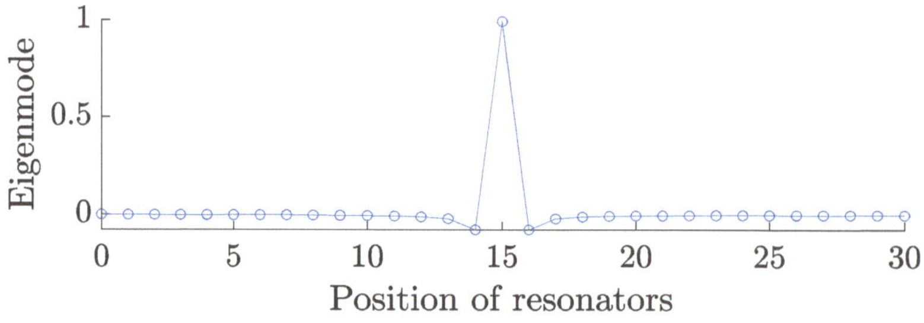

FIGURE 6.2. An example of a localised defect mode for a system of 31 resonators. The eigenvalues of the finite matrix $B_\mathrm{t}C_\mathrm{f}$ are computed, where C_f is the capacitance matrix for a system of evenly spaces resonators and B_t is the identity matrix but with the central entry $(B_\mathrm{t})_{11}^0 = 2$.

$\mathcal{B}_m$ for all $m \in \Lambda$. Under the assumption on the b_i^m, $\mathfrak{B}$ is said to be a compact perturbation of the identity. As shown in Proposition 4.87, the spectrum of the infinite structure is given by the solutions to the spectral problem

$$\mathfrak{B}\mathfrak{C}\mathfrak{u} = \lambda \mathfrak{u}, \tag{6.6}$$

where, for ease of notation, λ denotes ω^2. For the finite structure of size r, we let B_t be the block-diagonal matrix $(\mathcal{B}_m), m \in I_r$ and consider the spectral problem

$$B_\mathrm{t}C_\mathrm{f}u = \lambda u.$$

An example of such a defect mode is shown in Figure 6.2. A system of 31 resonators is modelled, with the finite defect matrix B_t chosen to be the identity, perturbed so that its central element is $(B_\mathrm{t})_{11}^0 = 2$.

6.1.2. Convergence of capacitance coefficients. Based on the layer potential characterisations of the capacitance matrices for finite and infinite systems in Chapters 1 and 4, we prove in this section that the capacitance coefficients of a large but finite structure converge, as the size grows, to the corresponding coefficients of the infinite structure.

The following result shows that a fixed block of the infinite capacitance matrix is approximately equal to a corresponding block of the capacitance matrix of the finite structure. In other words, the capacitance coefficients of the finite structure can be approximated through the infinite structure as long as we are sufficiently far away from the edges of the finite structure. The proof is similar to that of Theorem 5.16.

THEOREM 6.1. *For fixed $m, n \in \Lambda$, we have, as $r \to +\infty$,*

$$\lim_{r\to+\infty} C_\mathrm{f}^{mn}(r) = \widehat{C}^{m-n}.$$

PROOF. First, observe that

$$\mathcal{S}_{\mathcal{D}_\mathrm{f}}^0[\psi] = \sum_{m\in I_r} \mathcal{S}_{D+m}^0[\psi_m],$$

where $\psi_m = \psi|_{\partial(D+m)}$. Recall that the quasiperiodic single layer potential is defined as

$$\mathcal{S}_D^{0,\alpha}[\phi] = \int_{\partial D} \sum_{m\in\Lambda} G^0(x-y-m)e^{\mathrm{i}\alpha\cdot m}\phi(y)\ \mathrm{d}\sigma.$$

Given $\phi \in L^2(\partial D)$, we define $\phi_m^\alpha \in L^2(\partial(D+m))$ as

$$\phi_m^\alpha(y) = \phi(y-m)e^{\mathrm{i}\alpha\cdot m}.$$

Then it is clear that

$$\mathcal{S}_D^{0,\alpha}[\phi] = \sum_{m\in\Lambda} \mathcal{S}_{D+m}^0[\phi_m^\alpha].$$

We can then decompose

$$\begin{aligned}\mathcal{S}_D^{0,\alpha}[\phi] &= \sum_{m\in I_r} \mathcal{S}_{D+m}^0[\phi_m^\alpha] + \int_{\partial D} \sum_{m\in\Lambda\setminus I_r} G^0(x-y-m)e^{\mathrm{i}\alpha\cdot m}\phi(y)\ \mathrm{d}\sigma \\ &= \mathcal{S}_{\mathcal{D}_\mathrm{f}}^0[\phi^\alpha] + \mathcal{R}^\alpha[\phi],\end{aligned}$$

where, in the operator norm, $\mathcal{R}^\alpha = o(1)$ as $r \to +\infty$. From the Neumann series, we now have

$$(\mathcal{S}_D^{0,\alpha})^{-1}[\chi_{\partial D_i}] = (\mathcal{S}_{\mathcal{D}_\mathrm{f}}^0)^{-1}[\chi_i^\alpha] + o(1)\,, \tag{6.7}$$

where χ_i^α is defined as

$$\chi_i^\alpha = \sum_{m\in I_r} \chi_{\partial D_i^m} e^{\mathrm{i}\alpha\cdot m}.$$

Moreover, from Lemma B.28 we know that the error term in (6.7) holds uniformly in α. If $m, n \in I_r$ are fixed and $i, j = 1, \ldots, N$, we then have from (6.7) that

$$\begin{aligned}\widehat{C}_{ij}^{m-n} &= \frac{1}{|Y^*|}\int_{Y^*}\int_{\partial D_i} e^{-\mathrm{i}\alpha\cdot(m-n)}(\mathcal{S}_{\mathcal{D}_\mathrm{f}}^0)^{-1}[\chi_j^\alpha]\ \mathrm{d}\sigma\ \mathrm{d}\alpha + o(1) \\ &= \int_{\partial D_i^m} (\mathcal{S}_{\mathcal{D}_\mathrm{f}}^0)^{-1}[\chi_{\partial D_j^n}]\ \mathrm{d}\sigma + o(1) \\ &= (C_\mathrm{f}^{mn})_{ij}(r) + o(1)\,.\end{aligned}$$

This proves the claim. □

The numerical results presented in Figure 6.3 demonstrate the convergence of the capacitance coefficients, as established by Theorem 6.1. We plot $|(C_\mathrm{f})_{11}^0 - \widehat{C}_{11}^0|$ for a one-dimensional, two-dimensional, and three-dimensional lattice. Here, $(C_\mathrm{f})_{11}^0$ is the central element of C_f. When $d = 1$ or $d = 2$, the convergence is algebraic. The crucial property here is that the dimension of the lattice is less than the three dimensions of the underlying differential problem. As a result, waves can propagate away from the structure in the "spare" dimensions. This introduces long-range interactions to the system, as nonadjacent resonators are able to interact by coupling with the far field. Conversely, when $d = 3$ the dimension of the lattice is maximal (in the sense that we have a three-dimensional lattice in the three-dimensional space). In this case, we see exponential convergence, as there are no "spare" dimensions that allow waves to propagate away from the structure and couple with the far field.

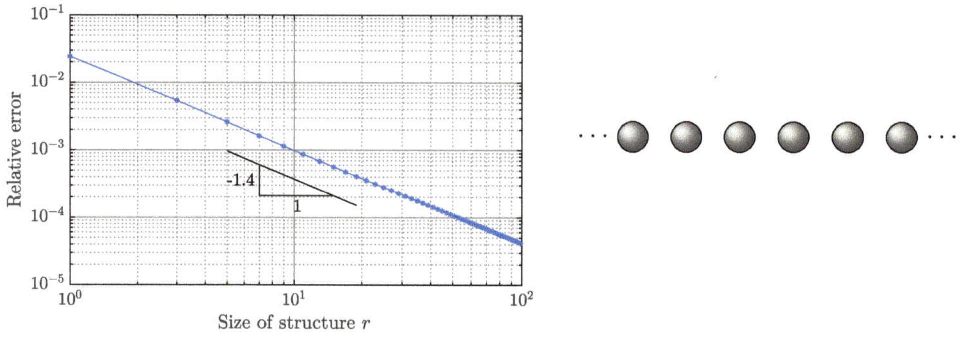

(A) One-dimensional lattice.

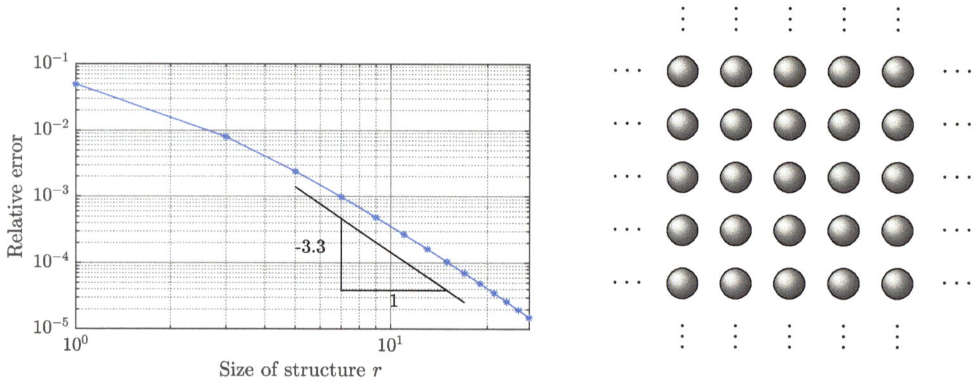

(B) Two-dimensional square lattice.

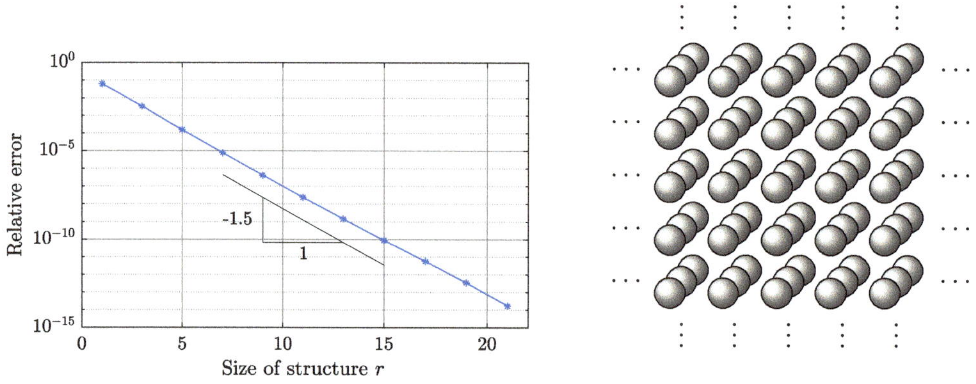

(C) Three-dimensional cubic lattice.

FIGURE 6.3. Convergence of the capacitance coefficient of large finite lattices. For each lattice, there is a single resonator in the unit cell ($N = 1$). In each case, we plot $|(C_{\mathrm{f}})^0_{11} - \widehat{C}^0_{11}|$ for increasing size r of the finite structure. Observe the log-log scales in (a) and (b) and the semi-log scale in (c), which correspond to algebraic and exponential convergence, respectively.

6.2. Convergence to pure point spectrum

In this section, we study a problem in which the infinite structure has a pure point spectrum corresponding to a localised mode. We introduce a defect to the model to create such a mode. For a finite, truncated structure, there will be an eigenvalue arbitrarily close to the pure point spectrum.

6.2.1. Example of a defect structure. Before developing any convergence theory, we present an example of a defect structure that exhibits a pure point spectrum corresponding to a localised mode. We take a lattice with a single resonator $N = 1$ inside each unit cell. We take a single resonator with perturbed ("defect") material parameter. In other words,

$$b_1^m = \begin{cases} 1, & m \neq 0, \\ 1 + x, & m = 0, \end{cases} \tag{6.8}$$

for some parameter $x > -1$. Observe that $x = 0$ corresponds to the unperturbed case and $|x|$ describes the magnitude of the perturbation. The eigenvalues of the (infinite-dimensional) generalised capacitance matrix $\mathfrak{B}\mathfrak{C}$ in this setting were studied in Section 4.8. It was found that $\lambda(=\omega^2)$ is an eigenvalue of $\mathfrak{B}\mathfrak{C}$ if and only if it is a root of the equation (4.162), where λ_1^α is the single eigenvalue of the quasiperiodic capacitance matrix C^α of the unperturbed periodic structure. This equation has a solution $\lambda = \lambda_0(=\omega_0^2)$ precisely in the case $x > 0$. In other words, the defect induces an eigenvalue λ_0 in the pure point spectrum of $\mathfrak{B}\mathfrak{C}$, corresponding to an exponentially localised eigenmode. An example of such a localised eigenmode is shown in Figure 6.2.

6.2.2. Convergence of defect modes. In this section, we prove that if the infinite structure has a localised mode, there will be an eigenvalue of the truncated structure arbitrarily close to the localised frequency.

We let $\mathfrak{C}$ denote the infinite capacitance matrix. As before, we let C_f denote the capacitance matrix of a finite structure of size $N|I_r| \times N|I_r|$. Furthermore, we let C_t denote the truncated matrix of $\mathfrak{C}$ of size $N|I_r| \times N|I_r|$, and similarly let B_t be the truncation of $\mathfrak{B}$.

We assume that $\mathfrak{B}\mathfrak{C}$ has a localised eigenmode $\mathfrak{u}$, and let u_t be the truncation of $\mathfrak{u}$ of size $N|I_r|$. The first result follows only from the decay of the localised mode.

LEMMA 6.2. *Assume that $\mathfrak{B}$ is a compact perturbation of the identity such that $\mathfrak{B}\mathfrak{C}$ has a localised eigenmode $\mathfrak{u}$ with corresponding eigenvalue λ. Then there is an eigenvalue $\widetilde{\lambda} = \widetilde{\lambda}(r)$ of $B_\mathrm{t}C_\mathrm{t}$ satisfying*

$$\lim_{r\to+\infty} \widetilde{\lambda}(r) = \lambda.$$

PROOF. We let $\mathfrak{u}_\mathrm{t}$ be the infinite vector obtained by padding u_t with 0. Since $\mathfrak{u}$ is in $\ell^2(\Lambda)$, for any $\varepsilon > 0$ we can choose large enough r so that

$$\|\mathfrak{u} - \mathfrak{u}_\mathrm{t}\|_{\ell^2} < \varepsilon.$$

Since $\mathfrak{B}\mathfrak{C}$ is a bounded operator, we then have

$$\|\lambda\mathfrak{u} - \mathfrak{B}\mathfrak{C}\mathfrak{u}_\mathrm{t}\|_{\ell^2} < K\varepsilon,$$

for some $K > 0$. Restricting to the finite block of size r, we have

$$\|\lambda u_\mathrm{t} - B_\mathrm{t}C_\mathrm{t}u_\mathrm{t}\|_2 < K\varepsilon.$$

In other words, λ is in the $K\varepsilon$-pseudospectrum of B_tC_t, and since B_tC_t is normal, we have an eigenvalue $\widetilde{\lambda}$ of B_tC_t satisfying

$$|\widetilde{\lambda}(r) - \lambda| = K\varepsilon.$$

This proves the claim. □

Next, we show the following result.

LEMMA 6.3. *As $r \to +\infty$, we have $\|C_f\|_2 < K$ for some K independent of r.*

PROOF. We know from Lemma 1.15 that the capacitance matrix C_f is diagonally dominant:

$$(C_f^{mm})_{ii} > \sum_{n \in I_r, j \neq i} \left|(C_f^{mn})_{ij}\right|,$$

for any i, m. For fixed i and m, we know from Lemma 1.21 that $(C_f^{mm})_{ii}(r)$ is increasing in r, and for all r we have

$$(C_f^{mm})_{ii}(r) < \widehat{C}_{ii}^0,$$

where, as before, $\widehat{C}_{ii}^0$ is the corresponding entry of the infinite capacitance matrix $\mathfrak{C}$. In particular, the eigenvalues of $C_f(r)$ are bounded as $r \to +\infty$, which shows the claim. □

As discussed above, the matrix C_t appearing in Lemma 6.2 is nonphysical, as it is a truncation of the matrix for the infinite system. Instead, we need to phrase the result for the matrix C_f, which describes the finite system. The following theorem is the main result of this chapter.

THEOREM 6.4. *Assume that $\mathfrak{B}$ is a compact perturbation of the identity such that $\mathfrak{B}\mathfrak{C}$ has a localised eigenmode $\mathfrak{u}$ with corresponding eigenvalue λ. Then there is an eigenvalue $\widehat{\lambda} = \widehat{\lambda}(r)$ of B_tC_f satisfying*

$$\lim_{r \to +\infty} \widehat{\lambda}(r) = \lambda.$$

PROOF. We let

$$K_1 = \sup_{r>0} \|C_f(r) - C_t\|_2$$

and observe from Lemma 6.3 that $K < +\infty$. We also let

$$K_2 = \|B_f\|_2.$$

Given $\varepsilon > 0$, we pick $r_0 > 0$ such that the following four terms are small:

$$\|C_{0,f} - C_{0,t}\|_2 < \frac{\varepsilon}{4K_2}, \quad \|u_t - u_{0,t}\|_2 < \frac{\varepsilon}{4K_1K_2}, \quad \|B_f(C_t - C_{0,t})u_{0,t}\|_2 < \frac{\varepsilon}{4},$$

and

$$\|B_f(C_f - C_{0,f})u_{0,t}\|_2 < \frac{\varepsilon}{4},$$

for all r large enough; the first inequality follows from Theorem 6.1 while the subsequent inequalities follow from the $\ell^2(\Lambda)$-decay of u. Here, $C_{0,t}, u_{0,t}$ and $C_{0,f}$ are the truncations of C_t, u_t and C_f to the smaller lattice of radius r_0 (padded with zero where needed for matrix operations). We know from Lemma 6.2 that we can take r large enough so that B_tC_t has an eigenvalue $\widetilde{\lambda}$ of distance ε from λ. We then have

$$\begin{aligned} B_fC_fu_t = {}& B_fC_tu_t + B_f(C_f - C_t)(u_t - u_{0,t}) + B_f(C_{0,f} - C_{0,t})u_{0,t} \\ &+ B_f(C_f - C_{0,f})u_{0,t} - B_f(C_t - C_{0,t})u_{0,t}. \end{aligned}$$

Then

$$\|(B_t C_f - B_t C_t)u_t\|_2 < \varepsilon,$$

which means that there is an eigenvalue $\widehat{\lambda}$ of distance ε from $\widetilde{\lambda}$, and hence

$$|\widehat{\lambda} - \lambda| < 2\varepsilon.$$

□

REMARK 6.5. *As an example,* $\mathfrak{B}$ *and* $\mathfrak{C}$ *as given in Section* 6.2.1 *satisfy the assumptions of Theorem* 6.4.

6.2.3. Numerical illustration. Figure 6.4 shows the convergence of the difference between the defect frequency computed for a finite structure and for the corresponding infinite structure, computed analytically using (4.162).

In Figure 6.4a, we consider a one-dimensional lattice with a single resonator in the unit cell ($N = 1$). The difference between the defect frequency computed for a finite structure and for the corresponding infinite structure scales as $\mathcal{O}(r^{-1.4})$, where r is the length of the truncated structure. The lower right plot shows the spectrum of successively larger lattices. In the geometry sketch on the right, the corresponding entry b_1^m from the matrix $\mathfrak{B}$ is shown above each resonator. In Figure 6.4b, we consider a two-dimensional square lattice with a single resonator in the unit cell ($N = 1$). Here, the error scales as $\mathcal{O}(r^{-3.3})$, where r is the width of the (square) truncated structure. In Figure 6.4c, we consider a three-dimensional cubic lattice with a single resonator in the unit cell ($N = 1$). Here, the error scales as $\mathcal{O}(e^{-2.4r})$, where r is the width of the (cubic) truncated structure.

Comparing Figure 6.4 with Figure 6.3, it appears that the error of the frequency of the defect mode is inheriting the convergence rate of the capacitance coefficients. In other words, when $d_l = 1$ or $d_l = 2$, there are long-range interactions through coupling with the far field, leading to algebraic convergence. In $d_l = 3$, there are no "spare" dimensions, and the convergence is exponential. This is consistent with the results for one-dimensional models [**114, 115, 123**] and for two-dimensional lattices in two-dimensional problems [**116**].

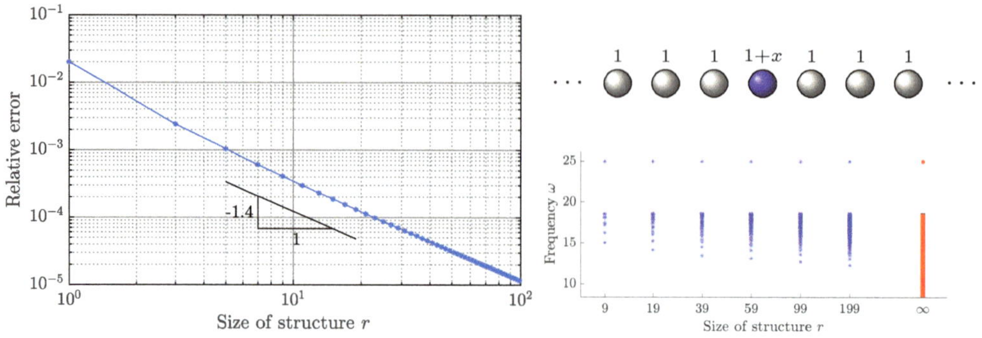

(A) One-dimensional lattice.

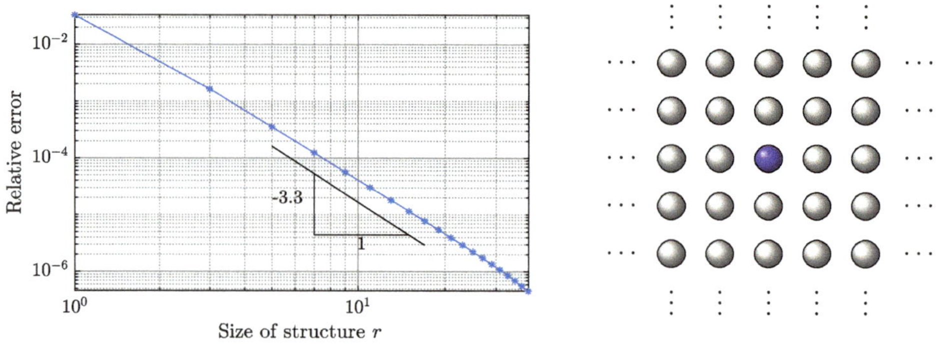

(B) Two-dimensional square lattice.

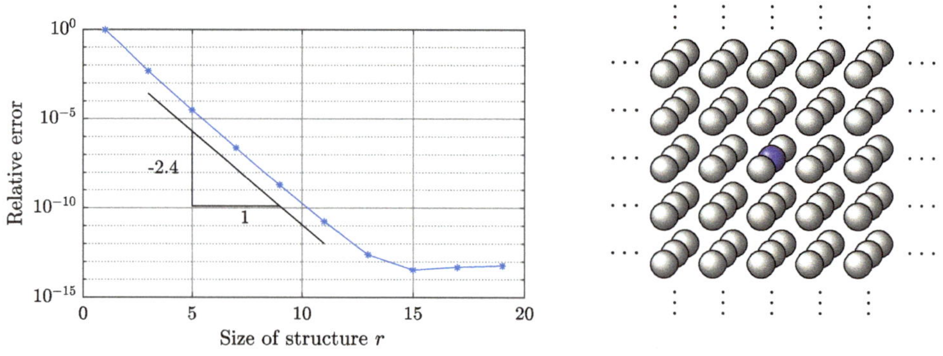

(C) Three-dimensional cubic lattice.

FIGURE 6.4. Convergence of the frequency of the defect modes, for a defect on the central resonator (with $x = 1$) created by perturbing a single entry of $\mathfrak{B}$.

CHAPTER 7

Convergence to the essential spectrum and band structure

In this chapter, we show that subwavelength resonant frequencies of a system of coupled resonators in a truncated periodic lattice converge to the essential spectrum of the corresponding infinite lattice. As in Chapter 6, we use the capacitance matrix as a model for fully coupled resonators with long-range interactions in three spatial dimensions. As the system of finitely many resonators becomes large, the corresponding capacitance matrix also grows in size. Thus, the problem at hand is to understand the asymptotic distribution of the eigenvalues of the capacitance matrix, in the limit that its size becomes arbitrarily large. For one-, two- or three-dimensional lattices embedded in three-dimensional space, we show that the (discrete) density of states for the finite system converges in distribution to the (continuous) density of states of the infinite system. We achieve this by proving a weak convergence of the finite capacitance matrix to the corresponding (translationally invariant) Toeplitz matrix of the infinite structure. With this characterisation at hand, we use the truncated Floquet-Bloch transform to introduce a notion of spectral band structure for finite materials. We refer the reader to [**15**] for its mathematical foundations. This principle is also applicable to structures that are not translationally invariant and have interfaces. We demonstrate this by considering examples of perturbed systems with defect modes, such as the classical analogue of the interface Su-Schrieffer-Heeger model studied in Section 4.6. Again, the crux of this chapter is to use the properties of the capacitance matrix, as described in Chapter 1, to develop an asymptotic eigenvalue distribution theory analogous to the ones for banded or Toeplitz matrices [**62, 90, 98, 156, 158**]. In particular, we use the fact that on the one hand the capacitance matrix is "almost" banded in the sense that the entries decay in successive off-diagonals as proved in Lemma 5.17 and on the other hand, for a periodic resonator array, it converges to a doubly infinite matrix that has constant (block) diagonals as shown in Section 4.8. The results presented in this chapter are from [**25**].

7.1. Eigenvalue distribution and essential spectral convergence

The main objective of this section is to prove the distributional convergence of the density of states of finite and infinite materials. The main technical result is Lemma 7.1, which establishes the convergence of the finite and truncated capacitance matrices in a certain weak "averaged" norm. We emphasise that the finite and truncated capacitance matrices are not expected to converge strongly in the matrix operator norm. This is because, regardless of its size, the finite structure will always exhibit edge effects.

7.1.1. Density of states. We let $Y = [-1/2, 1/2]^{d_l} \times \{0\}$. For an N-level system with band functions $\widehat{\omega}_k(\alpha)$, $k = 1, \dots, N$, we define the "finite material" density of states $D_\mathrm{f}(\omega)$ and "infinite material" density of states $D(\omega)$ as the distributions (see, for example, [**75, 178**])

(7.1)
$$D_\mathrm{f}(\omega) = \frac{1}{N|I_r|}\sum_{i=1}^{N|I_r|}\delta(\omega-\omega_i) \quad \text{and} \quad D(\omega) = \frac{1}{(2\pi)^{d_l}}\int_{Y^*}\frac{1}{N}\sum_{k=1}^{N}\delta\big(\omega-\widehat{\omega}_k(\alpha)\big)\,\mathrm{d}\alpha.$$

Let $L(\omega) := \big\{\alpha \in Y^* \mid \widehat{\omega}_k(\alpha) = \omega \text{ for some } k\big\}$ be the level set of $\widehat{\omega}_k(\alpha)$ at ω. Carrying out the integral in (7.1), we can rewrite D as

$$D(\omega) = \frac{1}{(2\pi)^{d_l}}\int_{L(\omega)}\frac{1}{N}\sum_{k=1}^{N}\frac{1}{|\nabla_\alpha\widehat{\omega}_k(\alpha)|}\,\mathrm{d}\sigma(\alpha), \tag{7.2}$$

where $\mathrm{d}\sigma(\alpha)$ is the surface measure on $L(\omega)$. Points α where $|\nabla_\alpha\widehat{\omega}_k(\alpha)| = 0$ are known as *van Hove singularities* and occur around any band edge [**159**].

7.1.2. Finite material eigenvalue distribution. The truncated matrix C_t of size r is a multilevel block Toeplitz matrix, with known asymptotic eigenvalue distribution in terms of the eigenvalues $\widehat{\omega}_k(\alpha)$ of the quasiperiodic capacitance matrix C^α (see, *e.g.*, [**89, 98**] for the one-dimensional case and [**157, 158**] for the two- and three-dimensional cases). Based on these results, we will show that the finite capacitance matrix C_f has identical eigenvalue distribution to C_t as the size tends to infinity.

For an $r\times r$ matrix $M = (m_{ij})$, let $|M|$ denote the normalised Frobenius norm

$$|M|^2 = \frac{1}{r}\sum_{i,j=1}^{r}|m_{ij}|^2. \tag{7.3}$$

We will use $\|M\|_2$ for the standard Euclidean matrix norm. The following lemma will be needed.

LEMMA 7.1. *As $r \to +\infty$, the matrices C_t and C_f are asymptotically equivalent, in other words, it holds that*

- $\lim_{r\to+\infty}|C_\mathrm{f} - C_\mathrm{t}| = 0$;
- $\|C_\mathrm{f}\|_2$ *and* $\|C_\mathrm{t}\|_2$ *are uniformly bounded as* $r \to +\infty$.

PROOF. $\|C_\mathrm{f}\|_2$ is uniformly bounded by Lemma 6.3, while $\|C_\mathrm{t}\|_2$ is uniformly bounded since C_t is a Toeplitz matrix of an essentially bounded symbol.

Let $\mathcal{D}_\mathrm{f}$ denote the finite lattice of width r. Theorem 6.1 tells us how to extend this finite lattice $\mathcal{D}_\mathrm{f}$ to a larger lattice $\widetilde{\mathcal{D}}_\mathrm{f}$, which has width $\widetilde{r} > r$. In particular, Theorem 6.1 shows that we can make this extension in such a way that the corresponding "r-sized" block $\widetilde{C}_{\mathrm{f},0}$ of the finite capacitance matrix corresponding to $\widetilde{\mathcal{D}}_\mathrm{f}$ is arbitrarily close to the "r-sized" truncated matrix C_t. That is, given an $\varepsilon > 0$, we can make a sufficiently large extension such that

$$\|C_\mathrm{t} - \widetilde{C}_{\mathrm{f},0}\|_2 < \varepsilon. \tag{7.4}$$

Observe that

$$\left|C_\mathrm{f} - \widetilde{C}_{\mathrm{f},0}\right|^2 = \frac{1}{N|I_r|}\sum_{\substack{m,n\in I_r\\ 1\le i,j\le N}}\left(\int_{\partial D_i^m}\left((\mathcal{S}^0_{\mathcal{D}_f})^{-1} - (\mathcal{S}^0_{\widetilde{\mathcal{D}}_f})^{-1}\right)[\chi_{\partial D_j^n}]\,\mathrm{d}\sigma\right)^2. \tag{7.5}$$

We define the "tail" $\mathcal{D}_0$ of the extended lattice as

$$\mathcal{D}_0 = \widetilde{\mathcal{D}_\mathrm{f}} \setminus \mathcal{D}_\mathrm{f}.$$

Observe that we have a block-structure of the single layer potential on the extended lattice:

$$\mathcal{S}^0_{\widetilde{\mathcal{D}_\mathrm{f}}} = \begin{pmatrix} \mathcal{S}^0_{\mathcal{D}_f} & \mathcal{S}^0_{\mathcal{D}_0}\big|_{\partial\mathcal{D}_\mathrm{f}} \\ \mathcal{S}^0_{\mathcal{D}_\mathrm{f}}\big|_{\partial\mathcal{D}_0} & \mathcal{S}^0_{\mathcal{D}_0} \end{pmatrix}.$$

We then have a block inverse of $\mathcal{S}^0_{\widetilde{\mathcal{D}_\mathrm{f}}}$ as follows:

$$(\mathcal{S}^0_{\widetilde{\mathcal{D}_\mathrm{f}}})^{-1} = \begin{pmatrix} \left(\mathcal{S}^0_{\mathcal{D}_\mathrm{f}} - \mathcal{S}^0_{\mathcal{D}_0}\big|_{\partial\mathcal{D}_\mathrm{f}}(\mathcal{S}^0_{\mathcal{D}_0})^{-1}\mathcal{S}^0_{\mathcal{D}_\mathrm{f}}\big|_{\partial\mathcal{D}_0}\right)^{-1} & A_1 \\ A_2 & A_3 \end{pmatrix}, \tag{7.6}$$

where A_i are bounded operators that are immaterial for our analysis. We are now ready to start estimating the difference between C_f and C_t. We want to estimate the term

$$\mathcal{S}^0_{\mathcal{D}_0}\big|_{\partial\mathcal{D}_\mathrm{f}}(\mathcal{S}^0_{\mathcal{D}_0})^{-1}\mathcal{S}^0_{\mathcal{D}_\mathrm{f}}\big|_{\partial\mathcal{D}_0}(\mathcal{S}^0_{\mathcal{D}_\mathrm{f}})^{-1}[\chi_{\partial D^n_j}].$$

Define

$$U(x) = \mathcal{S}^0_{\mathcal{D}_\mathrm{f}}(\mathcal{S}^0_{\mathcal{D}_\mathrm{f}})^{-1}[\chi_{\partial D^n_j}] \quad \text{and} \quad V = \mathcal{S}^0_{\mathcal{D}_0}(\mathcal{S}^0_{\mathcal{D}_0})^{-1}[U|_{\partial\mathcal{D}_0}];$$

these functions satisfy the systems of equations

$$\begin{cases} \Delta U(x) = 0, & x \in \mathbb{R}^3 \setminus \mathcal{D}_\mathrm{f}, \\ U\big|_{\partial\mathcal{D}_\mathrm{f}} = \chi_{\partial D^m_i}, \\ U(x) = \mathcal{O}(\frac{1}{|x|}) \text{ as } |x| \to +\infty, \end{cases} \quad \text{and} \quad \begin{cases} \Delta V(x) = 0, & x \in \mathbb{R}^3 \setminus \mathcal{D}_0, \\ V(x) = U(x), & x \in \partial\mathcal{D}_0, \\ V(x) = \mathcal{O}(\frac{1}{|x|}) \text{ as } |x| \to +\infty. \end{cases}$$

Observe that the boundary conditions satisfied by U are imposed on $\partial\mathcal{D}_\mathrm{f}$ while the boundary conditions satisfied by V are imposed on $\partial\mathcal{D}_0$. In particular, $U(x)$ scales like $|m|^{-1}$ for $x \in \partial\mathcal{D}_0$ while $V(x)$ scales like $(|m||n|)^{-1}$ for $x \in \partial D^n_i$. As $r \to +\infty$, we therefore have

$$\int_{\partial D^m_i} \left((\mathcal{S}^0_{\mathcal{D}_f})^{-1} - (\mathcal{S}^0_{\widetilde{\mathcal{D}_f}})^{-1}\right)[\chi_{\partial D^n_j}]\,\mathrm{d}\sigma \le \frac{K_1}{(1+|m|)(1+|n|)},$$

for some constant K_1. From (7.5) and (7.6), we use the Neumann series to find that

$$\left|C_\mathrm{f} - \widetilde{C}_{\mathrm{f},0}\right|^2 \le \frac{K_2}{r^{d_l}} \sum_{m,n\in I_r} \frac{1}{(1+|m|)^2(1+|n|)^2}, \tag{7.7}$$

for some constant K_2. Depending on the dimension d_l of the lattice, we have the following bounds on the above sum as $r \to \infty$:

$$\sum_{m,n\in I_r} \frac{1}{(1+|m|)^2(1+|n|)^2} \le \begin{cases} K_3, & d_l = 1, \\ K_3(\ln r)^2, & d_l = 2, \\ K_3 r^2, & d_l = 3, \end{cases} \tag{7.8}$$

for some constant K_3. In all cases, we have from (7.7) that $\left|C_\mathrm{f} - \widetilde{C}_{\mathrm{f},0}\right| \to 0$, which together with (7.4) concludes the proof.

In all cases, $\left|C_\mathrm{f} - \widetilde{C}_{\mathrm{f},0}\right| \to 0$, which together with (7.4) concludes the proof. □

From [**98**], we know that asymptotically equivalent matrices have identical eigenvalue distributions, as their sizes tend to infinity. This gives the following result on the distributional convergence of the discrete density of states $D_{\mathrm{f}}(\omega)$ to the continuous density of states $D(\omega)$.

THEOREM 7.2. *As $r \to +\infty$, $D_{\mathrm{f}}(\omega)$ converges to $D(\omega)$ in the sense of distributions. In other words, for any smooth function F with compact support, we have*

$$\lim_{r\to+\infty} \int_{-\infty}^{+\infty} D_{\mathrm{f}}(\omega)F(\omega)\,\mathrm{d}\omega = \int_{-\infty}^{+\infty} D(\omega)F(\omega)\,\mathrm{d}\omega.$$

PROOF. Since C_{t} and C_{f} are asymptotically equivalent, we have from [**98**] that

$$\lim_{r\to+\infty} \frac{1}{N|I_r|} \sum_{i=1}^{N|I_r|} F(\omega_i) = \frac{1}{(2\pi)^{d_l}} \int_{Y^*} \frac{1}{N} \sum_{k=1}^{N} F(\widehat{\omega}_k(\alpha))\,\mathrm{d}\alpha,$$

from which the theorem follows. □

From Theorem 7.2 we have that the frequencies $\omega_i, i = 1,\dots,N|I_r|$, of the finite capacitance matrix are distributed according to

$$\omega_i \sim \widehat{\omega}_k(\alpha),$$

where α is uniformly distributed in the Brillouin zone Y^* and $\widehat{\omega}_k(\alpha), k = 1,\dots,N$, are the eigenvalues of the quasiperiodic capacitance matrix. The proportion of modes with eigenfrequencies in the infinitesimal interval between ω and $\omega + \mathrm{d}\omega$ is then approximated by

$$D_{\mathrm{f}}(\omega)\,\mathrm{d}\omega \approx \frac{\mathrm{d}\omega}{(2\pi)^{d_l}} \int_{L(\omega)} \frac{1}{N} \sum_{k=1}^{N} \frac{1}{|\nabla_\alpha \omega_k(\alpha)|}\,\mathrm{d}\sigma(\alpha).$$

For a one-dimensional chain of single resonators ($d_l = 1, N = 1$), we obtain the following:

$$D_{\mathrm{f}}(\omega)\,\mathrm{d}\omega \approx \frac{\mathrm{d}\omega}{2\pi} \frac{1}{|\widehat{\omega}'(\alpha)|}\Bigg|_{\alpha\in\alpha(\omega)}, \tag{7.9}$$

where $\alpha(\omega) = \big\{\alpha \in Y^* \mid \widehat{\omega}_1(\alpha) = \omega\big\}$, which is shown by the solid lines in Figure 7.1a.

7.1.3. Numerical results. The convergence of the distribution of the resonant frequencies can be studied numerically. In Figure 7.1, we plot histograms of the discrete finite set of subwavelength resonant frequencies for truncated structures and compare this to the density of states (DOS) for the infinite array, given by (7.9). In each case, the histograms and the DOS are normalised so that the area under the curves is equal to 1. We can see that the distribution of the truncated eigenvalues closely resembles the DOS as the structure becomes sufficiently large (for 1000 resonators, the curve is difficult to distinguish from the histogram in our plot). We can quantify the error by computing the area between the two curves (the histograms being viewed as curves for these purposes). This is shown in the lower plot of Figure 7.1 and we observe linear convergence as the size of the array increases. In other words, the discrete DOS of the truncated resonant frequencies is converging in distribution to the DOS at a linear rate.

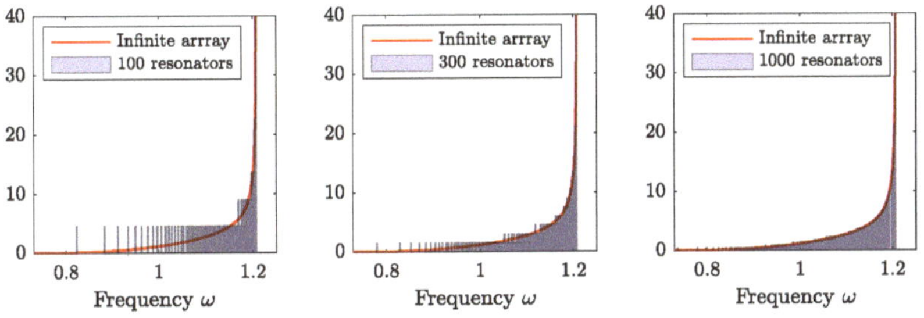

(A) Histograms of the discrete, finite set of subwavelength resonant frequencies for truncated structures and the density of states for the infinite array.

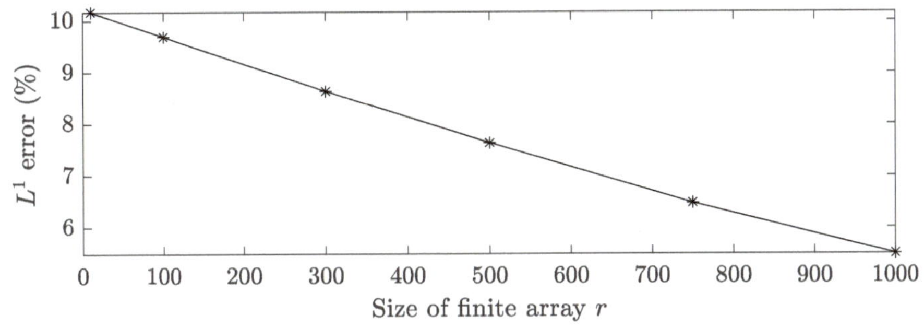

(B) L_1 error between the histograms of the discrete, finite set of subwavelength resonant frequencies and the density of state curves.

FIGURE 7.1. Convergence in distribution of the resonant frequencies of the truncated linear array to the density of states (DOS) for the infinite array. (a) The resonant frequencies of the truncated arrays are shown in histograms and the DOS for the infinite array (7.2) as a solid red line. Both the histograms and the DOS have been normalised to have unit area under the curves. (b) The L^1 error between the histogram plots and the DOS curve, which converges to zero as the size of the finite array increases.

Similar histograms can be produced for multi-dimensional lattices. For example, in Figure 7.2 we plot the same histograms for successively larger square (two-dimensional) lattices. Once again, we can see that the distribution of the eigenfrequencies converges to a fixed distribution as the size of the finite lattice increases. One notable difference from the one-dimensional lattice shown in Figure 7.1 is that the distribution is not singular at the edge of the first band.

7.2. Pointwise convergence and discrete band structure

Since edge effects persist in the limit $r \to +\infty$, we should not expect that all eigenvalues of C_{f} converge to the spectrum of $\mathfrak{C}$. Nevertheless, for any point in the continuous spectrum, there will always be arbitrarily close eigenvalues. We can repeat the arguments used in the proof of Lemma 7.1 to obtain the following theorem on pointwise convergence of the essential spectrum and Bloch modes.

THEOREM 7.3. *Let $\omega = \widehat{\omega}_k(\alpha)$ be an eigenvalue of the quasiperiodic capacitance matrix C^α for some $\alpha \in Y^*$, corresponding to the normalised eigenvector $u \in \mathbb{R}^N$.*

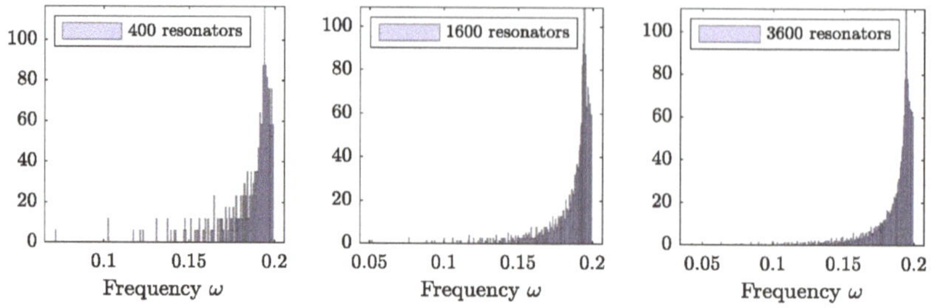

FIGURE 7.2. Distribution of the resonant frequencies of a truncated square (two-dimensional) array, plotted as histograms.

Then, for any $\varepsilon > 0$, we can choose $r > 0$ such that the finite capacitance matrix $C_{\mathrm{f}}(r)$ has a family of eigenvalues $\omega_i^2, i \in I$ and associated normalised eigenvectors $u_i \in \mathbb{R}^{N|I_r|}$ satisfying for some $c_i \in \mathbb{C}$,

$$|\omega^2 - \omega_i^2| < \varepsilon, \qquad \left\| \widetilde{u} - \sum_{i\in I} c_i u_i \right\|_2 < \varepsilon,$$

where $\widetilde{u} \in \mathbb{R}^{N|I_r|}$ is the normalised quasiperiodic extension of u.

REMARK 7.4. *Although any eigenvalue and eigenvector of the quasiperiodic capacitance matrix can be approximated by eigenvalues and eigenvectors of the finite capacitance matrix, the converse need not hold. Indeed, due to edge effects, C_{f} might have eigenvalues that do not approach those of C_{t} in the limit $r \to +\infty$. Although C_{f} converges to C_{t} in the (weak) norm $|\cdot|$, it does not converge in the (strong) Euclidean operator norm.*

The discrete band function calculation introduced in Chapter 6 provides a notion of how well an eigenmode of C_{f}, for large r, is approximated by Bloch modes of the infinite structure. Given an eigenmode u_j, we can take the truncated Floquet-Bloch transform of u_j as

$$(\widehat{u}_j)_\alpha = \sum_{m\in I_r} (u_j)_m e^{\mathrm{i}\alpha\cdot m}, \qquad \alpha \in Y^*. \tag{7.10}$$

Here, we denote by $(u_j)_m$ the vector of length N associated to the cell $m \in \Lambda$. Observe that u_j is a vector of length $N|I_r|$ while $(\widehat{u}_j)_\alpha$ is a vector of length N. Looking at the Euclidean 2-norm $\|(\widehat{u}_j)_\alpha\|_2$ as a function of α, this function has distinct peaks, which are the quasiperiodicities α associated with u_j. An example is shown in Figure 7.3. We can then define a discrete band structure whereby the eigenvalues ω_j of C_{f} are associated to a quasiperiodicity α_j given by

$$\alpha_j = \underset{\alpha\in Y^*}{\operatorname{argmax}} \|(\widehat{u}_j)_\alpha\|_2. \tag{7.11}$$

Note that the symmetry of the problem means that if α is an approximate quasiperiodicity, then so will $-\alpha$ be. In cases of additional symmetries of the lattice, we expect additional symmetries of the quasiperiodicities.

As a demonstrative example of this process, we consider the case of a single resonator periodically repeated in one direction. This has a single continuous band of eigenfrequencies, as shown in Figure 7.4a. For a truncated version of the structure,

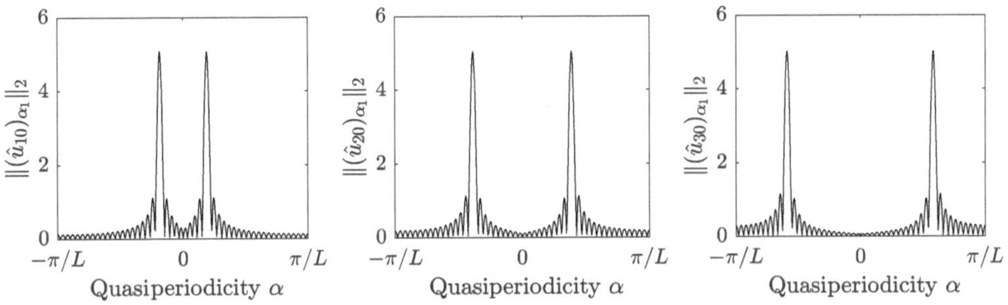

FIGURE 7.3. A given eigenmode u_j of a truncated periodic structure can be associated with a quasiperiodicity α. Here, we plot the norm of the truncated Floquet-Bloch transform of the 10th, 20th and 30th eigenmodes of an array of 50 resonators. In each case, there are clear peaks that we can assign as the quasiperiodicities of the eigenmodes. These values can be used to reconstruct approximate spectral band functions.

the approximate band functions can be reconstructed as above. In Figure 7.3, we show the norm of the truncated Floquet-Bloch transform of the 10th, 20th and 30th eigenmodes in an array of 50 resonators. In each case, the function is even about zero and has a clear peak, which allows us to identify an appropriate quasiperiodicity α_j via (7.11). These values can be used to plot the 50 resonant frequencies alongside the continuous bands of the limiting infinite structure, which is shown in Figure 7.4a. We see that even for a set of only 50 resonators, the approximate band function closely resembles that of the infinite structure.

In Figure 7.4b, we compare the continuous and truncated spectra of an array of resonators arranged in pairs (dimers). The truncated structure has 100 resonators arranged in 50 pairs. This geometry is an example of the Su-Schrieffer-Heeger chain studied in Section 4.6. This system has two subwavelength spectral bands and the truncated modes are split evenly between approximating the two bands.

Additionally, we can consider this method for lattices of higher dimensions. Figure 7.5a shows the case of a square lattice of resonator dimers. Similarly to Figure 7.4b, there is a bandgap between the first and second bands, and we see a close agreement between the discrete and continuous band structure. Figure 7.5b shows a similar property in the case of a honeycomb lattice, where the finite lattice is truncated along zig-zag edges of the lattice. As shown in Section 4.3, there are Dirac cones on each corner of the Brillouin zone. In the truncated structure, in addition to the "bulk modes" whose frequencies closely agree with the continuous spectrum, there are "edge modes" which are localised around the edges and whose points in the band structure lie away from the continuous bands.

7.3. Non-periodic band structure

The method described in Section 7.2 can also be applied to aperiodic structures that have been perturbed to introduce defects. Two examples of this are shown in Figure 7.6. In each case, the defects have induced localised eigenmodes which do not have well-defined associated quasiperiodicities. These are shown with dashed lines. The rest of the truncated spectrum still agrees well with the continuous spectrum of the limiting infinite operator.

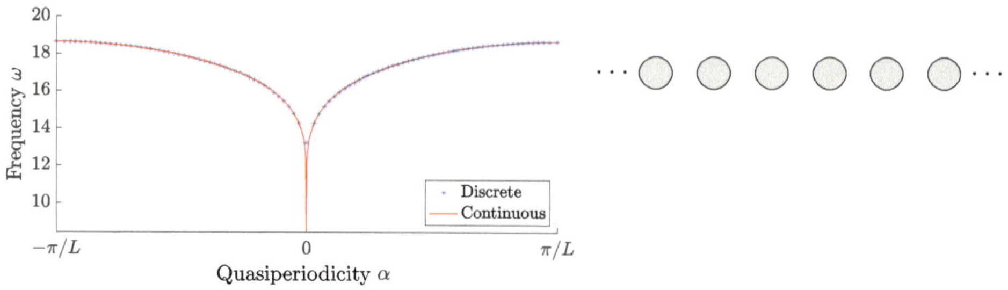

(A) Single periodic resonators ($N = 1$).

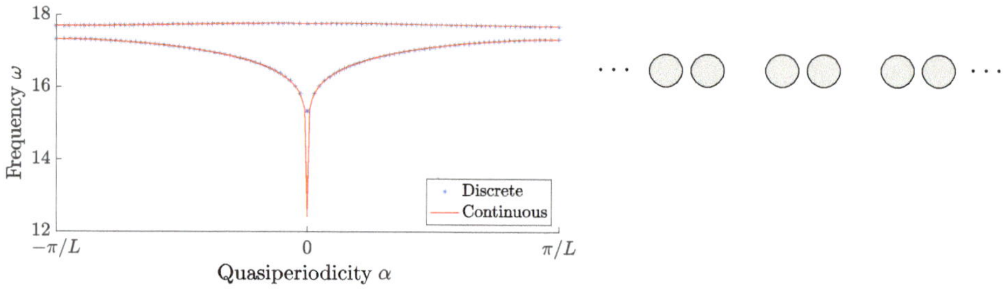

(B) Periodic pairs of resonators ($N = 2$).

FIGURE 7.4. The continuous spectrum of the infinite structure and the discrete spectrum of the truncated structure for one-dimensional lattices. (a) Single periodic resonators ($N = 1$) with a truncated structure consisting of 50 resonators. (b) Periodic pairs of resonators ($N = 2$) with a truncated structure containing 100 resonators. In both cases, the truncated Floquet-Bloch transform (7.10) is used to approximate the quasiperiodicity of the truncated modes.

The example shown in Figure 7.6a is that of a local defect, where the material parameters are changed on the central resonator. As was studied in Section 4.8, this corresponds to multiplying the capacitance matrix by a diagonal matrix that is equal to the identity matrix other than a value greater than 1 in the central entry. A formula for the eigenfrequency of this defect mode in the infinite structure was derived in Section 4.8 and proved in Chapter 6 that the eigenfrequencies of the localised modes in the truncated arrays converge to that value as the size of the structure goes to infinity. Figure 7.6a resembles Figure 7.4a with the addition of a flat band, corresponding to a localised mode.

In Figure 7.6b, we study the example of an interface mode in the Su-Schrieffer-Heeger chain. This localised eigenmode exhibits enhanced robustness with respect to imperfections, a property it inherits from the underlying topological properties of the periodic structure via the notion of *topological protection*; see Section 4.6. Although the periodicity of the structure is broken due to the interface, we can still visualise the spectrum as a discrete band structure through (7.10) and (7.11).

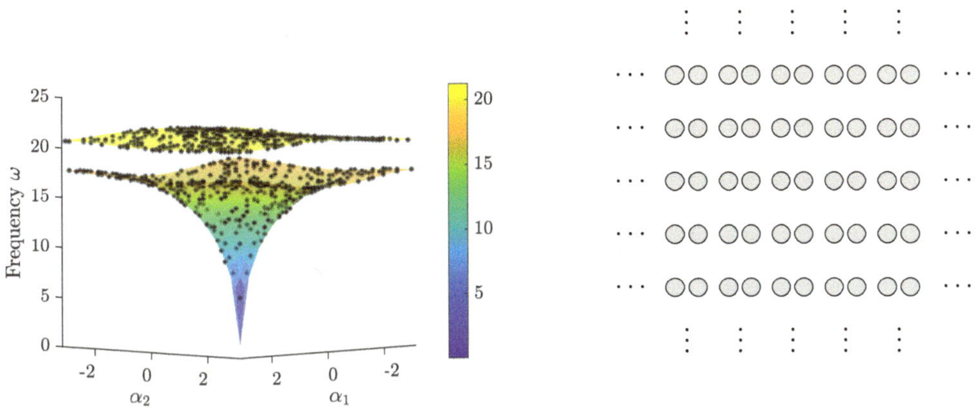

(A) Square lattice.

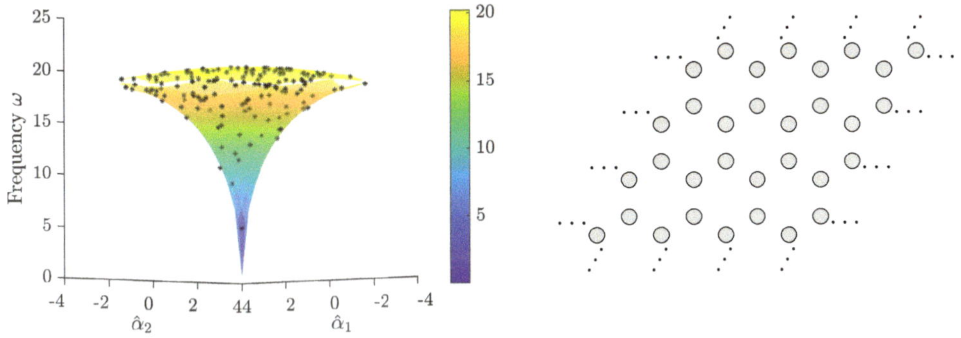

(B) Honeycomb lattice.

FIGURE 7.5. Examples of continuous and discrete spectra of the infinite and truncated structures, respectively. (a) A square lattice with two resonators per unit cell, resulting in two bands separated by a gap. (b) A honeycomb lattice with Dirac cones at the vertices of the Brillouin zone. In both cases, the truncated structures have 800 resonators and the truncated Floquet-Bloch transform is used to approximate the quasiperiodicity of the truncated modes.

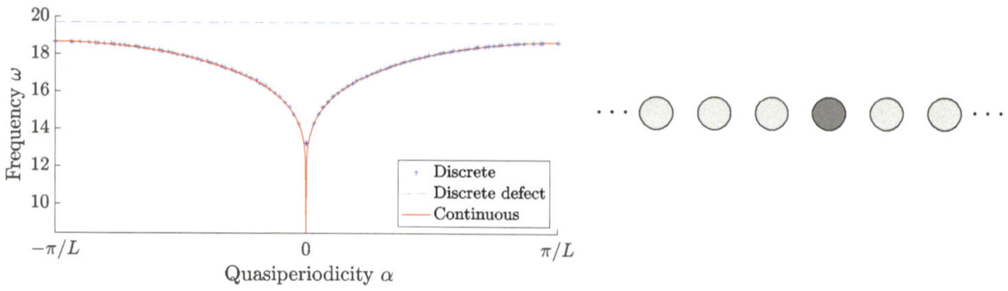

(A) Point defect.

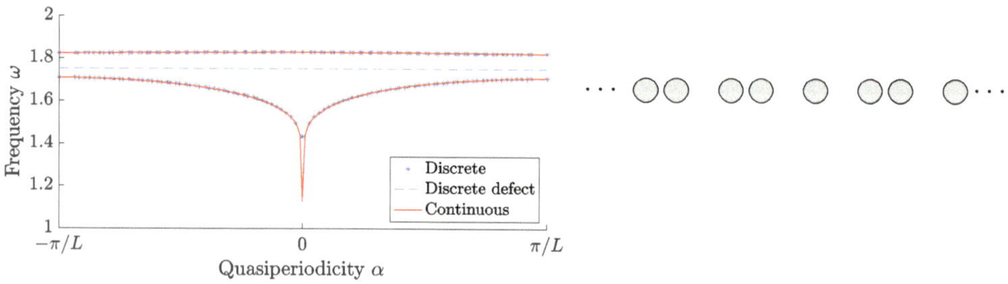

(B) Topological "Su-Schrieffer-Heeger" defect.

FIGURE 7.6. The discrete spectrum of a truncated array with defects can be related to the continuous spectrum of the unperturbed periodic structure. (a) A local defect in a periodic array of resonators. The truncated array has 51 resonators with the defect on the 26th. The continuous spectrum of the infinite array is the same as that shown in Figure 7.4a. (b) A topological (non-compact) defect in an array of resonator pairs. The truncated array has 101 resonators with the defect on the 51st. The continuous spectrum of the infinite array is the same as that shown in Figure 7.4b. In both cases, the defect introduces a localised eigenmode, which does not have a well-defined associated quasiperiodicity and is shown with a dashed line.

Concluding remarks and open problems

In these lectures, we have considered wave scattering problems posed in the subwavelength regime and repeatedly encountered the concept of capacitance. We have studied the mathematical properties of the generalised capacitance matrix, corresponding to a variety of Hermitian, non-Hermitian and time-modulated settings with both finite and infinitely periodic structures. Crucially, we established convergence results for the finite generalised capacitance matrix as its size goes to infinity, making the correspondence between finite and infinite settings precise. Our approach revealed the mathematical fundamentals of the exotic physical phenomena exhibited by metamaterials, which are responsible for their popularity for applications in nanophotonics and nanophononics. This demonstrates the power of reducing scattering and resonance PDE problems to discrete matrix approximations in this way.

Our formulation emerged from a functional analytic approach, where the generalised capacitance matrix describes the perturbation of the kernel of a nonlinear integral operator for asymptotically small parameter values. Its strength is that the capacitance matrix accounts for the strong interactions between resonant elements, thereby providing a unified rigorous model to study challenging problems in both Hermitian and non-Hermitian subwavelength physics and to explain exotic behaviours of metamaterials.

To conclude, we formulate some open questions. It would be very valuable to develop a qualitative characterisation of interface eigenmodes in finite, dimer systems of subwavelength resonators with a geometric defect (analogous to infinite systems studied in Section 4.6) and to prove the exponential decay of their associated eigenmodes. The main difficulty arises from the long-range interactions between the resonators, which lead to a slow decay of the off-diagonal terms of the corresponding capacitance matrix. Nevertheless, in view of the results on k-banded approximations of three-dimensional capacitance matrices obtained in Section 5.2 together with the convergence spectral results as the size of the structure goes to infinity proved in Chapters 6 and 7, it may be possible to show the existence and uniqueness of localised eigenmodes in finite chains of subwavelength resonators, as recently done in the one-dimensional case in [**9**]. In this direction, it would be very interesting to relate the robustness of the localisation effect at the interface to the statistics of the eigenvalues of the capacitance matrix under random perturbations in the parameters of the finite system. This would extend the well known Thouless localisation criterion [**76**] to interface eigenmodes. A closely related problem is the study of the localisation and delocalisation properties of the eigenvectors of the randomly perturbed generalised capacitance matrix. As the generalised capacitance matrix is the product of the material parameter diagonal matrix and the capacitance matrix, perturbing randomly the material parameters of the resonators or their positions or geometries makes the entries of the generalised capacitance matrix correlated. Little is known about the statistical properties of eigenvalues and eigenvectors of random matrices whose entries are correlated.

Based on Section 4.7, various attempts have also been made to use the topological properties of periodic structures to produce Fano-type responses that are robust as a result of being *topologically protected* [**175**]. With the analysis of Section 4.6 in mind, we expect that it will be possible to create robust Fano-type resonances in the subwavelength regime by coupling a "continuum" to a "discrete" state which is topologically protected. Moreover, it would be interesting to study a large system of finitely many resonators and examine the extent to which its Fano-type transmission and reflection behaviours can be approximated by the infinite system that was studied in Section 4.7.

In Chapter 6, we have applied the truncated Floquet-Bloch transform to compute the spectral properties of structures that are not translationally invariant and have interfaces. This original numerical method may find important applications in the study of properties of locally periodic structures. The results in [**15**] rigorously justify this method and provide its mathematical foundations. Based on k-Toeplitz theory, it is proved in [**15**] that the truncated Floquet-Bloch transform can recover the limiting quasiperiodicity from the bulk eigenmodes of the finite system. Moreover, the truncated Floquet-Bloch transform has been applied to three aperiodic structures: Su-Schrieffer-Heeger finite chains of resonators, compactly perturbed finite periodic systems, and dislocated chains of resonators. For each of these three aperiodic structures, it has been shown that the truncated Floquet-Bloch transform recovers the band structure and the localised eigenmodes in the bandgap. Specifying its applicability range would be a highly promising line of research with far-reaching implications. In particular, it would be very interesting to generalise this method to more general disordered structures than those previously studied.

For models with non-Hermiticity arising from imaginary gauge potentials, spectral convergence theorems similar to those in Chapters 6 and 7 are not expected to hold. The spectra are known to exhibit very high sensitivity to the choice of boundary conditions, and, in particular, the spectra of the finite and infinite models might be vastly different. Similarly to the discrete band structure calculations in Section 7.2, we expect the finite modes to be associated with a *complex* momentum and to adequately describe these, we need to extend the Brillouin zone to the complex plane. A precise study of this framework would be a valuable future work.

In [**12**], the notion of a generalised Brillouin zone is first introduced by allowing the quasiperiodicities to be complex and proving that this shift of the Brillouin zone into the complex plane accounts for the unidirectional spatial decay of the eigenmodes and leads to correct spectral convergence properties for one-dimensional systems of subwavelength resonators. Proving rigorously how the spectral properties of a finite structure in three dimensions are associated with those of the corresponding semi-infinitely or infinitely periodic lattices and giving explicit characterisations of how to extend the Hermitian theory to non-reciprocal three-dimensional settings would be an important direction of research; see [**13**]. Another very interesting problem is to apply the truncated Floquet-Bloch transform to non-Hermitian systems, in particular those exhibiting a skin effect.

A further challenging problem is to generalise the results on the non-Hermitian skin effect in Section 5.2 to dimer structures. To do so, new results on the spectra and pseudospectra of banded block-Toeplitz operators would be needed. Another challenging problem is to prove that when the strength of the disorder increases, there is a competition between the non-Hermitian skin effect and Anderson localisation in the bulk as recently shown numerically in the one-dimensional case in [**10**]. In connection with this, it would be important to prove as in the one-dimensional case (see [**6**]) that all the eigenvalues of the gauge capacitance matrix in three dimensions are real (as illustrated in Figures 5.9 and 5.10) and random perturbations of the positions of the resonators or/and the parameter γ inside the resonators leave them on the real axis, but make some of them jump outside the region with nontrivial winding numbers.

It would also be very interesting to establish an effective medium theory for time-modulated systems of subwavelength resonators similar to the one described in Chapter 3 for static dilute systems. In [**17**], a general framework for deriving effective material properties of time-modulated one-dimensional systems of subwavelength resonators is provided. It applies to one-dimensional subwavelength resonator systems with a general form of time-dependent parameters. The resonators can be accurately described by a point-scattering formulation when the width of the resonators is small. In contrast to the static setting, where this point interaction approximation yields a Lippmann-Schwinger equation for the effective material properties, the mode coupling in the time-modulated case instead yields an infinite linear system of (coupled) Lippmann-Schwinger-type equations. Generalising this effective system of coupled differential equations to three-dimensional structures is a challenging and important problem.

Another challenging problem is to consider defects in space and time in time-modulated systems of subwavelength resonators. The existence of bandgaps in both space and momentum variables may lead to spatio-temporally localised modes. This refers to a transition from a spatial delocalistion to spatial localisation, followed by a transition back to delocalisation. Showing the existence of spatio-temporally localised modes and characterising them would open the door to space-time wave manipulations at subwavelength scales.

In [**44**], defects are combined in space and time to achieve space-time localised waves in one-dimensional resonator systems. These results can be generalised not only to three-dimensional systems of subwavelength resonators but also to systems where sources of energy gain and loss represented by time-modulated complex material parameters are introduced inside the resonators [**166**]. They can also be used to provide the mathematical foundations of the non-Hermitian skin effect in subwavelength resonator systems achieved by breaking reciprocity with time modulation [**125**]. These two very interesting directions of research shall be the subjects of forthcoming works.

Another very promising direction of research to advance the field of metamaterials is to propose formulas for the (optimal) design of systems of subwavelength resonators with prescribed properties such as anomalous transmission and reflection spectra, frequency range for unidirectional transmission, broadband absorption, subwavelength defect mode eigenfrequencies or exceptional points; see, *e.g.*, [**32, 43**].

In these lectures, all of the results obtained for finite periodic structures in the subwavelength regime heavily rely on the translation invariance properties of the system of resonators (or analogously the almost Toeplitz or block Toeplitz structure of the associated capacitance matrices) together with the use of Floquet-Bloch theory for the analysis of the spectral properties of the corresponding infinite structure. A natural question is whether similar results and wave localisation mechanisms can be obtained in disordered systems. By disordered, we mean systems that are not translationally invariant.

A first attempt to construct a consistent picture of how wave localisation occurs at subwavelength scales in disordered systems of subwavelength resonators was the subject of [**14**, **53**].

As discussed in Section 4.8, one way to break local translation invariance is to randomly vary the material parameters of the resonators. Another way is to consider large aperiodic arrays of subwavelength resonators, constructed by sampling randomly or quasiperiodically from a fixed set of distinct *building blocks* consisting of (possibly multiple) subwavelength resonators.

As shown in the one-dimensional case in [**8**, **14**], where only nearest-neighbour interactions occur, such block disordered systems have quite unusual spectral behaviour: existence of hybridised bound eigenmodes; two distinct localisation mechanisms: bandgap localisation, which stems from the bandgaps of the constituent blocks, and disorder localisation, which occurs in very large systems and is significantly weaker than the bandgap one; localisation of all the eigenmodes as the size of the structure goes to infinity; fractal behaviour of eigenmodes; and universal estimates for their density of states. These recent results can be extended to one-dimensional non-Hermitian systems of subwavelength resonators with gauge potentials; see [**13**].

It would be very interesting and challenging to generalise them to the three-dimensional case and to have a complete and rigorous theory for wave localisation in general three-dimensional disorder systems in the subwavelength regime, where long-range interactions occur.

Finally, we note that a class of nonlinear artificial materials consisting of nonlinear subwavelength high-contrast resonator building blocks has been introduced. This class of materials has revived the exciting prospect of realising the long-standing goal of robust transport of waves at subwavelength scales and their amplification in only one direction, opening avenues to channel energy and information transport in quantum computing; see, *e.g.*, [**77**, **104**, **131**, **146**, **151**, **160**].

Although very significant advances in experimental and numerical modeling have been achieved in the field of nonlinear subwavelength physics during recent decades, very little is known from a mathematical point of view. To our knowledge, properties of nonlinear subwavelength resonances have not yet been analysed in the mathematical literature. In contrast to the linear response regime studied in depth in these lectures, there is a clear lack of deep understanding of the theory of nonlinear subwavelength resonances. In [**48**], a discrete approximation to the nonlinear resonance problem is provided for finite systems of subwavelength resonators where the nonlinearity is of Kerr type. As shown in [**52**], this discrete approximation allows not only deriving the effect of nonlinearities on the linear resonances but, more importantly, characterises the soliton-like solutions in crystals of subwavelength resonators. It would be very interesting to elucidate further the interplay between nonlinearities and subwavelength resonances, in particular the nonlinearity-induced topological and localisation transitions in such systems.

APPENDIX A

Some basic results

A.1. Asymptotic perturbation theory of Gohberg and Sigal

In this section, we review the main results of Gohberg and Sigal theory [**92**] concerning the generalisation to operator-valued functions of two classical results in complex analysis, the *argument principle* and *Rouché's theorem*. We also apply these results to compute asymptotic formulas for the characteristic values of perturbed Fredholm holomorphic operators.

A.1.1. Definitions and preliminaries. In this section, we introduce a few definitions and present some basic results which are useful for the statement of the generalised Rouché theorem.

A.1.1.1. *Compact operators.* If $\mathcal{B}$ and $\mathcal{B}'$ are two Banach spaces, we denote by $\mathcal{L}(\mathcal{B}, \mathcal{B}')$ the space of bounded linear operators from $\mathcal{B}$ into $\mathcal{B}'$. An operator $K \in \mathcal{L}(\mathcal{B}, \mathcal{B}')$ is said to be compact provided K takes any bounded subset of $\mathcal{B}$ to a relatively compact subset of $\mathcal{B}'$, that is, a set with compact closure.

The operator K is said to be of finite rank if $\operatorname{Range} K$, the range of K, is finite-dimensional. Clearly, every operator of finite rank is compact.

We now provide some basic results on compact operators.

LEMMA A.1. *The set of compact operators is a closed two-sided ideal in the algebra of bounded operators $\mathcal{L}(\mathcal{B}, \mathcal{B}')$ with norm topology.*

LEMMA A.2. *If $K \in \mathcal{L}(\mathcal{B}, \mathcal{B}')$ and there is a sequence $\{K_N\}_{N \in \mathbb{N}}$ of operators of finite rank such that $\|K_N - K\| \to 0$, then K is compact.*

LEMMA A.3. *The operator $K \in \mathcal{L}(\mathcal{B}, \mathcal{B}')$ is compact if and only if the dual operator $K^* \in \mathcal{L}(\mathcal{B}', \mathcal{B})$ is compact.*

The next result is called the Fredholm alternative. See, for example, [**109**].

PROPOSITION A.4 (Fredholm alternative). *Let K be a compact operator on the Banach space $\mathcal{B}$. For $\lambda \in \mathbb{C}, \lambda \neq 0$, $\lambda I - K$ is surjective if and only if it is injective.*

A.1.1.2. *Fredholm operators.* An operator $A \in \mathcal{L}(\mathcal{B}, \mathcal{B}')$ is said to be *Fredholm* provided the subspace $\operatorname{Ker} A$ is finite-dimensional and the subspace $\operatorname{Range} A$ is closed in $\mathcal{B}'$ and of finite codimension. Let $\operatorname{Fred}(\mathcal{B}, \mathcal{B}')$ denote the collection of all Fredholm operators from $\mathcal{B}$ into $\mathcal{B}'$.

Next, we define the index of $A \in \operatorname{Fred}(\mathcal{B}, \mathcal{B}')$ to be

$$\operatorname{ind} A = \dim \operatorname{Ker} A - \operatorname{codim} \operatorname{Range} A.$$

The following proposition shows that the index is stable under compact perturbations [**109**].

PROPOSITION A.5. *If $A : \mathcal{B} \to \mathcal{B}'$ is Fredholm and $K : \mathcal{B} \to \mathcal{B}'$ is compact, then their sum $A + K$ is Fredholm, and*

$$\mathrm{ind}(A + K) = \mathrm{ind}A.$$

Proposition A.5 is a consequence of the following fundamental result about the index of Fredholm operators.

PROPOSITION A.6. *The mapping $A \mapsto \mathrm{ind}A$ is continuous in* $\mathrm{Fred}(\mathcal{B}, \mathcal{B}')$*; i.e.,* ind *is constant on each connected component of* $\mathrm{Fred}(\mathcal{B}, \mathcal{B}')$.

A.1.1.3. *Characteristic value and multiplicity.* We now introduce the notions of characteristic values and root functions of holomorphic operator-valued functions.

Let $\mathfrak{U}(z_0)$ be the set of all operator-valued functions with values in $\mathcal{L}(\mathcal{B}, \mathcal{B}')$ that are holomorphic in some neighbourhood of z_0, except possibly at z_0.

The point z_0 is called a *characteristic value* of $A \in \mathfrak{U}(z_0)$ if there exists a vector-valued function $\phi(z)$ with values in $\mathcal{B}$ such that

(i) $\phi(z)$ is holomorphic at z_0 and $\phi(z_0) \neq 0$;
(ii) $A(z)\phi(z)$ is holomorphic at z_0 and vanishes at this point.

Here, $\phi(z)$ is called a *root* function of $A(z)$ associated with the characteristic value z_0. The vector $\phi_0 = \phi(z_0)$ is called an eigenvector. The closure of the linear set of eigenvectors corresponding to z_0 is denoted by $\mathrm{Ker}\, A(z_0)$.

Suppose that z_0 is a characteristic value of $A \in \mathfrak{U}(z_0)$ and ϕ is an associated root function. Then there exist a number $m(\phi) \geq 1$ and a vector-valued function ψ with values in $\mathcal{B}'$, holomorphic at z_0, such that

$$A(z)\phi(z) = (z - z_0)^{m(\phi)}\psi(z), \quad \psi(z_0) \neq 0.$$

The number $m(\phi)$ is called the *multiplicity* of the root function $\phi(z)$.

For $\phi_0 \in \mathrm{Ker}\, A(z_0)$, we define the rank of ϕ_0, denoted by $\mathrm{rank}(\phi_0)$, to be the maximum multiplicities of all root functions $\phi(z)$ with $\phi(z_0) = \phi_0$.

Suppose that $n = \dim \mathrm{Ker}\, A(z_0) < +\infty$ and that the ranks of all vectors in $\mathrm{Ker}\, A(z_0)$ are finite. A system of eigenvectors ϕ_0^j, $j = 1, \ldots, n$, is called a *canonical system of eigenvectors* of A associated with z_0 if their ranks possess the following property: for $j = 1, \ldots, n$, $rank(\phi_0^j)$ is the maximum of the ranks of all eigenvectors in the direct complement in $\mathrm{Ker}\, A(z_0)$ of the linear span of the vectors $\phi_0^1, \ldots, \phi_0^{j-1}$. We call

$$N(A(z_0)) := \sum_{j=1}^{n} rank(\phi_0^j)$$

the *null multiplicity* of the characteristic value z_0 of A. If z_0 is not a characteristic value of A, we put $N(A(z_0)) = 0$.

Suppose that $A(z)^{-1}$ exists and is holomorphic in some neighbourhood of z_0, except possibly at z_0. Then the number

$$M(A(z_0)) = N(A(z_0)) - N(A(z_0)^{-1}) \tag{A.1}$$

is called the *multiplicity* of z_0. If z_0 is a characteristic value and not a pole of $A(z)$, then $M(A(z_0)) = N(A(z_0))$ while $M(A(z_0)) = -N(A(z_0)^{-1})$ if z_0 is a pole and not a characteristic value of $A(z)$.

A.1.1.4. *Normal points.* Suppose that z_0 is a pole of $A(z)$ and the Laurent series expansion of $A(z)$ at z_0 is given by

$$A(z) = \sum_{j \geq -s} (z - z_0)^j A_j. \tag{A.2}$$

If in (A.2) the operators A_{-j}, $j = 1, \ldots, s$, have finite-dimensional ranges, then $A(z)$ is called *finitely meromorphic* at z_0.

The operator-valued function $A(z)$ is said to be of *Fredholm type* (of index zero) at the point z_0 if the operator A_0 in (A.2) is Fredholm (of index zero).

If $A(z)$ is holomorphic and invertible at z_0, then z_0 is called a *regular point* of $A(z)$. The point z_0 is called a *normal point* of $A(z)$ if $A(z)$ is finitely meromorphic, of Fredholm type at z_0, and regular in a neighbourhood of z_0 except at z_0 itself.

A.1.1.5. *Trace.* Let A be a finite-rank operator acting from $\mathcal{B}$ into itself. Then there exists a finite-dimensional invariant subspace $\mathcal{C}$ of A such that A annihilates some direct complement of $\mathcal{C}$ in $\mathcal{B}$. We define the trace of A to be that of $A|_{\mathcal{C}}$, which is given in the usual way. It is desirable to recall some results about the trace operator.

PROPOSITION A.7. *The following results hold:*

(i) $\mathrm{tr}A$ *is independent of the choice of* $\mathcal{C}$, *so that it is well-defined;*
(ii) tr *is linear;*
(iii) *If* B *is a finite-rank operator from* $\mathcal{B}$ *to itself, then*
$$\mathrm{tr}AB = \mathrm{tr}BA;$$
(iv) *If* M *is a finite-rank operator from* $\mathcal{B} \times \mathcal{B}'$ *to itself, given by*
$$M = \begin{pmatrix} A & B \\ C & D \end{pmatrix},$$
then $\mathrm{tr}M = \mathrm{tr}A + \mathrm{tr}D$.

Recall that if an operator-valued function $C(z)$ is finitely meromorphic in the neighbourhood $\overline{V}$ of z_0, which contains no poles of $C(z)$ except possibly z_0, then $\int_{\partial V} C(z)\, dz$ is a finite-rank operator. The following identity is used in the proof of Theorem A.13.

PROPOSITION A.8. *Let* $A(z)$ *and* $B(z)$ *be two operator-valued functions finitely meromorphic in the neighbourhood* $\overline{V}$ *of* z_0, *which contains no poles of* $A(z)$ *and* $B(z)$ *other than* z_0. *Then we have*

$$\mathrm{tr} \int_{\partial V} A(z)B(z)\, dz = \mathrm{tr} \int_{\partial V} B(z)A(z)\, dz. \tag{A.3}$$

A.1.2. Factorisation of operators. We say that $A \in \mathfrak{U}(z_0)$ admits a factorisation at z_0 if $A(z)$ can be written as

$$A(z) = E(z)D(z)F(z), \tag{A.4}$$

where $E(z), F(z)$ are regular at z_0 and

$$D(z) = P_0 + \sum_{j=1}^{n} (z - z_0)^{k_j} P_j. \tag{A.5}$$

Here, k_j are integers, P_j's are mutually disjoint projections, $P_1, \ldots, P_n$ are rank-one operators, and $I - \sum_{j=0}^{n} P_j$ is a finite-rank operator.

THEOREM A.9. *$A \in \mathfrak{U}(z_0)$ admits a factorisation at z_0 if and only if A is finitely meromorphic and of Fredholm type of index zero at z_0.*

COROLLARY A.10. *$A(z)$ is normal at z_0 if and only if $A(z)$ admits a factorisation at z_0 such that $I = \sum_{j=0}^{n} P_j$ in* (A.5). *Moreover, we have*

$$M(A(z_0)) = k_1 + \cdots + k_n$$

for $k_1, \ldots, k_n$, given by (A.5).

Suppose that $A(z)$ admits a factorisation at z_0. Then we have (A.4) and (A.5). Since

$$A(z)^{-1} = F(z)^{-1} D(z)^{-1} E(z)^{-1}$$

with

$$D(z)^{-1} = P_0 + \sum_{j=1}^{n} (z - z_0)^{-k_j} P_j,$$

the following result holds.

COROLLARY A.11. *Every normal point of $A(z)$ is a normal point of $A(z)^{-1}$.*

A.1.3. Generalised argument principle and Rouché's theorem. We now state the generalised argument principle and Rouché's theorem for operator-valued functions.

A.1.3.1. *Argument principle.* Let V be a simply connected bounded domain with rectifiable boundary ∂V. An operator-valued function $A(z)$ which is finitely meromorphic and of Fredholm type in V and continuous on ∂V is called *normal* with respect to ∂V if the operator $A(z)$ is invertible for $z \in \overline{V}$, except for a finite number of points of V that are normal points of $A(z)$.

LEMMA A.12. *An operator-valued function $A(z)$ is normal with respect to ∂V if it is finitely meromorphic and of Fredholm type in V, continuous on ∂V, and invertible for all $z \in \partial V$.*

Now, if $A(z)$ is normal with respect to the contour ∂V and z_i, $i = 1, \ldots, \sigma$, are all its characteristic values and poles lying in V, the full multiplicity $\mathcal{M}(A; \partial V)$ of $A(z)$ for $z \in V$ is the number of characteristic values of $A(z)$ for $z \in V$, counted with their multiplicities, minus the number of poles of $A(z)$ in V, counted with their multiplicities, namely,

$$\mathcal{M}(A; \partial V) = \sum_{i=1}^{\sigma} M(A(z_i)), \tag{A.6}$$

where $M(A(z_i))$ is defined in (A.1).

THEOREM A.13 (Generalised argument principle). *Suppose that the operator-valued function $A(z)$ is normal with respect to ∂V. Then we have*

$$\frac{1}{2\pi \mathrm{i}} \operatorname{tr} \int_{\partial V} A(z)^{-1} \frac{d}{dz} A(z) \, \mathrm{d}z = \mathcal{M}(A; \partial V). \tag{A.7}$$

The following is an immediate consequence of Lemma A.12 and identities (A.3) and (A.7).

COROLLARY A.14. *If the operator-valued functions $A(z)$ and $B(z)$ are normal with respect to ∂V, then $C(z) := A(z)B(z)$ is also normal with respect to ∂V, and*

$$\mathcal{M}(C; \partial V) = \mathcal{M}(A; \partial V) + \mathcal{M}(B; \partial V).$$

The following general form of the argument principle is useful. It can be proven by the same argument as the one in the proof of Theorem A.13.

THEOREM A.15. *Suppose that $A(z)$ is an operator-valued function that is normal with respect to ∂V. Let $f(z)$ be a scalar function that is holomorphic in V and continuous in $\overline{V}$. Then*

$$\frac{1}{2\pi \mathrm{i}} \mathrm{tr} \int_{\partial V} f(z) A(z)^{-1} \frac{d}{dz} A(z) \, \mathrm{d}z = \sum_{j=1}^{\sigma} M(A(z_j)) f(z_j),$$

where z_j, $j = 1, \ldots, \sigma$, are all the points in V which are either poles or characteristic values of $A(z)$.

A.1.3.2. *Generalisation of Rouché's theorem.* A generalisation of Rouché's theorem to operator-valued functions is stated below.

THEOREM A.16 (Generalised Rouché's theorem). *Let $A(z)$ be an operator-valued function that is normal with respect to ∂V. If an operator-valued function $S(z)$ which is finitely meromorphic in V and continuous on ∂V satisfies the condition*

$$\|A(z)^{-1} S(z)\|_{\mathcal{L}(\mathcal{B},\mathcal{B})} < 1, \quad z \in \partial V,$$

then $A + S$ is also normal with respect to ∂V and

$$\mathcal{M}(A; \partial V) = \mathcal{M}(A + S; \partial V).$$

A.1.3.3. *Generalisation of Steinberg's theorem.* Steinberg's theorem asserts that if $K(z)$ is a compact operator on a Banach space, which is holomorphic in z for z in a region V in the complex plane, then $(I + K(z))^{-1}$ is meromorphic in V; see [**148**]. A generalisation of this theorem to finitely meromorphic operators was first given by Gohberg and Sigal in [**92**]. The following important result holds.

THEOREM A.17 (Generalised Steinberg's theorem). *Suppose that $A(z)$ is an operator-valued function which is finitely meromorphic and of Fredholm type in the domain V. If the operator $A(z)$ is invertible at one point of V, then $A(z)$ has a bounded inverse for all $z \in V$, except possibly for certain isolated points.*

A.2. Floquet transform and spectra of periodic elliptic operators

In this section, the Floquet transform, which in the periodic case plays the role of the Fourier transform, is introduced, and the structure of spectra of periodic elliptic operators is briefly discussed. We refer to [**106**] for details.

A.2.1. Floquet transform.

Let the lattice Λ be given by

$$\Lambda := \{ m_1 l_1 + \cdots + m_d l_d \mid m_i \in \mathbb{Z} \}$$

with $l_1, \ldots, l_d \in \mathbb{R}^d$ being the generating vectors. Denote by $Y \subset \mathbb{R}^d$ a fundamental domain of Λ:

$$Y := \{ c_1 l_1 + \cdots + c_d l_d \mid 0 \leq c_1, \ldots, c_d \leq 1 \}.$$

The dual lattice of Λ, denoted by Λ^*, is generated by $\alpha_1, \ldots, \alpha_d$ satisfying $\alpha_i \cdot l_j = 2\pi\delta_{ij}$, for $i, j = 1, \ldots, d$. The *Brillouin zone* Y^* is defined as $Y^* := \mathbb{R}^d/\Lambda^*$ and is a fundamental domain of the dual lattice Λ^*.

Let $f(x)$ be a function that decays sufficiently fast. We define the Floquet-Bloch transform of f by (4.3). This transform is an analogue of the Fourier transform for the periodic case. The parameter α is the quasiperiodicity. It is an analogue of the dual variable in the Fourier transform. If we shift x by a period $m \in \Lambda$, then we get the Floquet condition

$$\mathcal{U}[f](x+m,\alpha) = e^{\mathrm{i}\alpha\cdot m}\mathcal{U}[f](x,\alpha), \tag{A.8}$$

which shows that it suffices to know the function $\mathcal{U}[f](x,\alpha)$ on the fundamental domain (or unit cell) Y in order to recover it completely as a function of the x-variable. Moreover, $\mathcal{U}[f](x,\alpha)$ is periodic with respect to the quasiperiodicity α:

$$\mathcal{U}[f](x,\alpha+q) = \mathcal{U}[f](x,\alpha), \quad q \in \Lambda^*. \tag{A.9}$$

Therefore, α can be considered as an element of the torus $\mathbb{R}^d/\Lambda^*$. Another way of saying this is that all information about $\mathcal{U}[f](x,\alpha)$ is contained in its values for α in the fundamental domain Y^* of the dual lattice Λ^*. This domain is referred to as the (first) Brillouin zone.

The following result is an analogue of the Plancherel theorem when using the Fourier transform. Suppose that $\mathrm{d}\alpha$ and Y^* are normalised. The following theorem holds.

THEOREM A.18 (Plancherel-type theorem). *The transform*

$$\mathcal{U} : L^2(\mathbb{R}^d) \to L^2(L^2(Y), Y^*)$$

is isometric. Its inverse is given by

$$\mathcal{U}^{-1}[g](x) = \frac{1}{|Y^*|}\int_{Y^*} g(x,\alpha)\,\mathrm{d}\alpha, \quad x \in \mathbb{R}^d,$$

where the function $g(x,\alpha) \in L^2(L^2(Y), Y^*)$ is extended from Y to all $x \in \mathbb{R}^d$ according to the Floquet condition (A.8).

A.2.2. Structure of spectra of periodic elliptic operators. Now, we briefly discuss the spectral properties of periodic elliptic operators.

Consider a linear partial differential operator $L(x,\partial_x)$, whose coefficients are periodic with respect to the lattice Λ. Due to periodicity, the operator commutes with the Floquet transform

$$\mathcal{U}[Lf](x,\alpha) = L(x,\partial_x)\mathcal{U}[f](x,\alpha).$$

For each α, the operator $L(x,\partial_x)$ now acts on functions satisfying the corresponding Floquet condition (A.8). In other words, although the differential expression of the operator remains the same, its domain changes with α. Denoting this operator by $L(\alpha)$, we see that the Floquet transform $\mathcal{U}$ expands the periodic partial differential operator L in $L^2(\mathbb{R}^d)$ into the direct integral of operators

$$\int_{Y^*}^{\oplus} L(\alpha)\,\mathrm{d}\alpha. \tag{A.10}$$

The key point in (A.10) is that the operators $L(\alpha)$ act on functions defined on a torus, while the original operator acts in $\mathbb{R}^d$.

If L is a self-adjoint operator, one can prove the main spectral statement:

$$\sigma(L) = \bigcup_{\alpha \in Y^*} \sigma(L(\alpha)), \tag{A.11}$$

where σ denotes the spectrum.

If L is elliptic, the operators $L(\alpha)$ have compact resolvents and hence discrete spectra. If L is bounded from below, the spectrum of $L(\alpha)$ accumulates only at $+\infty$. Denote by $\mu_n(\alpha)$ the nth eigenvalue of $L(\alpha)$ (counted in increasing order with their multiplicities). The function $\alpha \mapsto \mu_n(\alpha)$ is continuous in Y^*. It is one branch of the dispersion relations and is called a band function. We conclude that the spectrum $\sigma(L)$ consists of the closed intervals (called the spectral bands)

$$\left[\min_{\alpha} \mu_n(\alpha), \max_{\alpha} \mu_n(\alpha)\right],$$

where $\min_\alpha \mu_n(\alpha) \to +\infty$ when $n \to +\infty$.

APPENDIX B

Layer potential techniques

B.1. Layer potentials for the Laplace equation

In this section, we review a number of basic facts and results regarding the layer potentials associated with the Laplacian. The most important results in this section are, on the one hand, the mapping properties of the Neumann-Poincaré operator and, on the other hand, the concept of capacity.

B.1.1. Green's function. Even though the following result is elementary, we recall it for the reader's convenience.

LEMMA B.1. *A Green function G^0 of the Laplace equation in $\mathbb{R}^d$, $d = 2, 3$, is given by* (1.8).

PROOF. The Laplacian is radially symmetric in $\mathbb{R}^d$, so it is natural to seek G^0 in the form $G^0(x) = w(r)$ where $r = |x|$. Since

$$\Delta w = \frac{d^2 w}{d^2 r} + \frac{(d-1)}{r}\frac{dw}{dr} = \frac{1}{r^{d-1}}\frac{d}{dr}(r^{d-1}\frac{dw}{dr}),$$

$\Delta G^0 = 0$ in $\mathbb{R}^d \setminus \{0\}$ forces that w must satisfy

$$\frac{1}{r^{d-1}}\frac{d}{dr}(r^{d-1}\frac{dw}{dr}) = 0 \quad \text{for } r > 0,$$

and hence

$$w(r) = \begin{cases} -\frac{a_3}{r} + b_3 & \text{when } d = 3, \\ a_2 \ln r + b_2 & \text{when } d = 2, \end{cases}$$

for some constants a_d and b_d. The choice of b_d is arbitrary, but a_d is fixed by the requirement that $\Delta G^0 = \delta_0$ in $\mathbb{R}^d$, where δ_0 is the Dirac function at 0, or in other words

$$\int_{\mathbb{R}^d} G^0 \Delta\phi \; dx = \phi(0) \quad \text{for } \phi \in C_0^\infty(\mathbb{R}^d). \tag{B.1}$$

Any test function $\phi \in C_0^\infty(\mathbb{R}^d)$ has compact support, so we can apply Green's formula over the unbounded domain $\{x : |x| > \varepsilon\}$ to arrive at

$$\int_{|x|>\varepsilon} G^0(x)\Delta\phi(x)\, dx = \int_{|x|=\varepsilon} \phi(x)\frac{\partial G^0}{\partial \nu}(x) \; d\sigma(x) - \int_{|x|=\varepsilon} G^0(x)\frac{\partial \phi}{\partial \nu}(x)\, d\sigma(x), \tag{B.2}$$

where $\nu = x/|x|$ on $\{|x| = \varepsilon\}$. Since

$$\nabla G^0(x) = \frac{dw}{dr}\frac{x}{|x|} = \frac{a_d x}{|x|^d} \quad \text{for } d = 2, 3,$$

we have

$$\frac{\partial G^0}{\partial \nu}(x) = a_d \varepsilon^{1-d} \quad \text{for } |x| = \varepsilon.$$

Thus, by the continuity of ϕ,

$$\int_{|x|=\varepsilon} \phi(x)\frac{\partial G^0}{\partial \nu}(x)\, d\sigma(x) = \frac{a_d}{\varepsilon^{d-1}} \int_{|x|=\varepsilon} \phi(x)\ d\sigma(x) \to a_d\omega_d\phi(0)$$

as $\varepsilon \to 0$ with ω_d being the area of the unit sphere in $\mathbb{R}^d$, whereas

$$\int_{|x|=\varepsilon} G^0(x)\frac{\partial \phi}{\partial \nu}(x)\, d\sigma(x) = \begin{cases} \mathcal{O}(\varepsilon) & \text{if } d=3, \\ \mathcal{O}(\varepsilon|\ln\varepsilon|) & \text{if } d=2. \end{cases}$$

Thus, if $a_d = 1/\omega_d$, then (B.1) follows from (B.2) after sending $\varepsilon \to 0$. □

B.1.2. Jump relations. Let D be a bounded, connected domain in $\mathbb{R}^d, d = 2, 3$, with boundary in $C^{1,s}$ for some $0 < s < 1$. Together with the single layer potential defined by (1.6) for $\omega = 0$, we introduce the double layer potential $\mathcal{D}_D^0 : L^2(\partial D) \to H^1_{\text{loc}}(\mathbb{R}^d \setminus \partial D)$ as

$$\text{(B.3)} \qquad \mathcal{D}_D^0[\phi](x) = \int_{\partial D} \frac{\partial G^0}{\partial \nu(y)}(x-y)\phi(y)\, d\sigma(y), \quad x \in \mathbb{R}^d,\ \phi \in L^2(\partial D).$$

We define $\mathcal{K}_D^0 : L^2(\partial D) \to L^2(\partial D)$ by

$$\text{(B.4)} \qquad \mathcal{K}_D^0[\phi](x) = \int_{\partial D} \frac{\partial}{\partial \nu(y)} G^0(x-y)\phi(y)\, d\sigma(y), \quad x \in \partial D,\ \phi \in L^2(\partial D).$$

The following result holds; see, for instance, [**47**, Lemma 2.3].

LEMMA B.2. *Let $\mathcal{K}_D^{0,*}$ be the L^2-adjoint of $\mathcal{K}_D^0$, i.e.,*

$$\text{(B.5)} \qquad \mathcal{K}_D^{0,*}[\phi](x) = \int_{\partial D} \frac{\partial}{\partial \nu(x)} G^0(x-y)\phi(y)\, d\sigma(y), \quad x \in \partial D,\ \phi \in L^2(\partial D).$$

Since D is smooth, both operators are compact on $L^2(\partial D)$.

PROOF. Assuming that D is a bounded domain with boundary of class $C^{1,s}$ for some $0 < s < 1$, we have the bound

$$\text{(B.6)} \qquad \left|\frac{(x-y)\cdot\nu(x)}{|x-y|^d}\right| \le C\frac{1}{|x-y|^{d-1-s}} \quad \text{for } x, y \in \partial D, x \ne y,$$

which shows that there exists a positive constant C depending only on s and D such that

$$\text{(B.7)} \qquad \int_{\partial D} \left(\frac{|(x-y)\cdot\nu(x)|}{|x-y|^d} + \frac{|(x-y)\cdot\nu(y)|}{|x-y|^d} \right) d\sigma(y) \le C,$$

and

$$\text{(B.8)} \qquad \begin{aligned} \int_{|y-x|<\varepsilon} \left(\frac{|(x-y)\cdot\nu(x)|}{|x-y|^d} + \frac{|(x-y)\cdot\nu(y)|}{|x-y|^d} \right) d\sigma(y) &\le C \int_0^\varepsilon \frac{1}{r^{d-1-s}} r^{d-2}\, dr \\ &\le C\varepsilon^s, \end{aligned}$$

for any $x \in \partial D$, by integration in polar coordinates.

Next, note that according to Lemma A.3 it suffices to prove that $\mathcal{K}_D^0$ is compact in $L^2(\partial D)$ to assert that $\mathcal{K}_D^{0,*}$ is also compact. Now, given $\varepsilon > 0$, set $G_\varepsilon^0(x) = G^0(x)$ if $|x| > \varepsilon$, $G_\varepsilon^0(x) = 0$ otherwise, and define

$$\mathcal{K}_D^{0,\varepsilon}[\phi](x) = \int_{\partial D} \frac{\partial G_\varepsilon^0}{\partial \nu(y)}(x-y)\phi(y)\ d\sigma(y).$$

Then

$$\int_{\partial D}\int_{\partial D}\left|\frac{\partial G^0_\varepsilon}{\partial \nu(y)}(x-y)\right|^2 \, d\sigma(x)\, d\sigma(y) < +\infty,$$

hence the operator norm of $\mathcal{K}^{0,\varepsilon}_D$ on $L^2(\partial D)$ satisfies

$$\|\mathcal{K}^{0,\varepsilon}_D\| \leq \left\|\frac{\partial G^0_\varepsilon}{\partial \nu}\right\|_{L^2(\partial D\times\partial D)}.$$

Let $\{\phi_p\}_{p=1}^{+\infty}$ be an orthonormal basis for $L^2(\partial D)$. It is an easy consequence of Fubini's theorem that if $\psi_{pq}(x,y)=\phi_p(x)\phi_q(y)$, then $\{\psi_{pq}\}_{p,q=1}^{+\infty}$ is an orthonormal basis for $L^2(\partial D\times\partial D)$. Hence we can write

$$\frac{\partial G^0_\varepsilon}{\partial \nu(y)}(x-y) = \sum_{p,q=1}^{+\infty}\langle\frac{\partial G^0_\varepsilon}{\partial \nu},\psi_{pq}\rangle\psi_{pq}(x,y).$$

Here $\langle,\rangle$ denotes the L^2 scalar product. For $N\in\mathbb{N}, N\geq 2$, let

$$K^{\varepsilon,N}_D[\phi](x) = \sum_{p+q\leq N}\int_{\partial D}\langle\frac{\partial G^0_\varepsilon}{\partial \nu},\psi_{pq}\rangle\psi_{pq}(x,y)\phi(y)\, d\sigma(y).$$

It is clear that the range of $K^{\varepsilon,N}_D$ lies in the span of $\phi_1,\ldots,\phi_N$, so $K^{\varepsilon,N}_D$ is of finite rank. Moreover,

$$\left\|\mathcal{K}^{0,\varepsilon}_D - K^{\varepsilon,N}_D\right\| \leq \left\|\frac{\partial G^0_\varepsilon}{\partial \nu} - \sum_{p+q\leq N}\langle\frac{\partial G^0_\varepsilon}{\partial \nu},\psi_{pq}\rangle\psi_{pq}\right\|_{L^2(\partial D\times\partial D)} \longrightarrow 0 \quad \text{as } N\to+\infty,$$

and then $\mathcal{K}^{0,\varepsilon}_D$ is compact by Lemma A.2. On the other hand,

$$\begin{aligned}\mathcal{K}^0_D[\phi](x) &= \frac{1}{\omega_d}\int_{|y-x|>\varepsilon}\frac{(y-x)\cdot\nu(y)}{|x-y|^d}\phi(y)\, d\sigma(y)\\ &\quad+\frac{1}{\omega_d}\int_{|y-x|<\varepsilon}\frac{(y-x)\cdot\nu(y)}{|x-y|^d}\phi(y)\, d\sigma(y),\\ &= \mathcal{K}^{0,\varepsilon}_D[\phi](x) + \frac{1}{\omega_d}\int_{|y-x|<\varepsilon}\frac{(y-x)\cdot\nu(y)}{|x-y|^d}\phi(y)\, d\sigma(y),\end{aligned}$$

and then, by the estimate (B.8), the operator norm of $\mathcal{K}^0_D - \mathcal{K}^{0,\varepsilon}_D$ tends to zero as $\varepsilon\to 0$, so $\mathcal{K}^0_D$ is compact by Lemma A.1. □

Note that, for $\phi\in L^2(\partial D)$,

$$(\mathcal{S}^0_D[\phi])\big|_+(x) = (\mathcal{S}^0_D[\phi])\big|_-(x) \quad \text{a.e. } x\in\partial D.$$

The following lemma relates the traces of the double layer potential and the normal derivative of the single layer potential to the operators $\mathcal{K}^0_D$ and $\mathcal{K}^{0,*}_D$; see, for instance, [**47**, Lemmas 2.15 and 2.16].

PROPOSITION B.3 (Jump relations). *For $\phi\in L^2(\partial D)$,*

$$(\mathcal{D}^0_D[\phi])\big|_\pm(x) = \left(\mp\frac{1}{2}I + \mathcal{K}^0_D\right)[\phi](x) \quad a.e.\ x\in\partial D, \tag{B.9}$$

$$\frac{\partial}{\partial\nu}\mathcal{S}^0_D[\phi]\bigg|_\pm(x) = \left(\pm\frac{1}{2}I + \mathcal{K}^{0,*}_D\right)[\phi](x) \quad a.e.\ x\in\partial D. \tag{B.10}$$

Note that (B.10) yields the following jump relation:

$$\frac{\partial}{\partial \nu}\mathcal{S}_D^0[\phi]\Big|_+ - \frac{\partial}{\partial \nu}\mathcal{S}_D^0[\phi]\Big|_- = \phi \quad \text{on } \partial D. \tag{B.11}$$

In order to prove (B.9), we first establish that the double layer potential with constant density has a jump.

LEMMA B.4. *If D is a bounded domain with boundary of class $C^{1,s}$ for some $0 < s < 1$, then $\mathcal{D}_D^0[1](x) = 0$ for $x \in \mathbb{R}^d \setminus \overline{D}$, $\mathcal{D}_D^0[1](x) = 1$ for $x \in D$, and $\mathcal{K}_D^0[1](x) = 1/2$ for $x \in \partial D$.*

PROOF. The first equation follows immediately from Green's formula, since $G^0(x-y)$ is in $C^\infty(\overline{D})$ and harmonic in D as a function of y when $x \in \mathbb{R}^d \setminus \overline{D}$. As for the second equation, given $x \in D$, let $\varepsilon > 0$ be small enough so that $\overline{B_\varepsilon} \subset D$, where B_ε is the ball of centre x and radius ε. We can apply Green's formula to $G^0(x-y)$ on the domain $D \setminus \overline{B_\varepsilon}$ to obtain

$$\begin{aligned} 0 &= \mathcal{D}_D^0[1](x) - \frac{\varepsilon^{1-d}}{\omega_d}\int_{\partial B_\varepsilon} d\sigma(y) \\ &= \mathcal{D}_D^0[1](x) - 1. \end{aligned}$$

Now we prove the third equation. Given $x \in \partial D$, again let B_ε be the ball of centre x and radius ε. Set $\partial D_\varepsilon = \partial D \setminus (\partial D \cap B_\varepsilon)$, $\partial B'_\varepsilon = \partial B_\varepsilon \cap D$, and $\partial B''_\varepsilon = \{y \in \partial B_\varepsilon : \nu(x)\cdot y < 0\}$. (Thus $\partial B''_\varepsilon$ is the hemisphere of ∂B_ε lying on the same side of the tangent plane to ∂D at x.) A further application of Green's formula shows that

$$0 = \frac{1}{\omega_d}\int_{\partial D_\varepsilon} \frac{(y-x)\cdot\nu(y)}{|x-y|^d}\, d\sigma(y) + \int_{\partial B'_\varepsilon} \frac{\partial G^0}{\partial \nu(y)}(x-y)\, d\sigma(y).$$

Thus

$$\frac{1}{\omega_d}\int_{\partial D_\varepsilon} \frac{(y-x)\cdot\nu(y)}{|x-y|^d}\, d\sigma(y) = -\int_{\partial B'_\varepsilon} \frac{\partial G^0}{\partial \nu(y)}(x-y)\, d\sigma(y) = \frac{\varepsilon^{1-d}}{\omega_d}\int_{\partial B'_\varepsilon} d\sigma(y).$$

But on the one hand, clearly

$$\int_{\partial D} \frac{(y-x)\cdot\nu(y)}{|x-y|^d}\, d\sigma(y) = \lim_{\varepsilon\to 0}\int_{\partial D_\varepsilon} \frac{(y-x)\cdot\nu(y)}{|x-y|^d}\, d\sigma(y).$$

On the other hand, since ∂D is of class $C^{1,s}$ for some $0 < s < 1$, the distance between the tangent plane to ∂D at x and the points on ∂D at a distance ε from x is $\mathcal{O}(\varepsilon^{1+s})$, so

$$\int_{\partial B'_\varepsilon} d\sigma(y) = \int_{\partial B''_\varepsilon} d\sigma(y) + \mathcal{O}(\varepsilon^{1+s})\cdot\mathcal{O}(\varepsilon^{d-1}) = \frac{\omega_d \varepsilon^{d-1}}{2} + \mathcal{O}(\varepsilon^{d+s}),$$

and the desired result follows. □

We are now ready to prove Proposition B.3.

PROOF OF PROPOSITION B.3. In order to prove (B.9) we first consider a density $f \in C^0(\partial D)$. If $x \in \partial D$ and $t < 0$ is sufficiently small, then $x + t\nu(x) \in D$, so by Lemma B.4,

$$\mathcal{D}_D^0[f](x+t\nu(x)) = f(x) + \int_{\partial D} \frac{\partial G^0}{\partial \nu(y)}(x+t\nu(x)-y)(f(y)-f(x))\, d\sigma(y)\, . \tag{B.12}$$

To prove that the second integral is continuous as $t \to 0^-$, given $\varepsilon > 0$ let $\delta > 0$ be such that

$$|f(y) - f(x)| < \varepsilon$$

whenever $|y - x| < \delta$. Then

$$\begin{aligned}
&\int_{\partial D} \frac{\partial G^0}{\partial \nu(y)}(x + t\nu(x) - y)(f(y) - f(x))\, d\sigma(y) \\
&\quad - \int_{\partial D} \frac{\partial G^0}{\partial \nu(y)}(x - y)(f(y) - f(x))\, d\sigma(y) \\
&= \int_{\partial D \cap B_\delta} \frac{\partial G^0}{\partial \nu(y)}(x + t\nu(x) - y)(f(y) - f(x))\, d\sigma(y) \\
&\quad - \int_{\partial D \cap B_\delta} \frac{\partial G^0}{\partial \nu(y)}(x - y)(f(y) - f(x))\, d\sigma(y) \\
&\quad + \int_{\partial D \setminus B_\delta} \left(\frac{\partial G^0}{\partial \nu(y)}(x + t\nu(x) - y) - \frac{\partial G^0}{\partial \nu(y)}(x - y) \right)(f(y) - f(x))\, d\sigma(y) \\
&= I_1 + I_2 + I_3.
\end{aligned}$$

Here B_δ is the ball of centre x and radius δ. It easily follows from (B.7) that $|I_2| \le C\varepsilon$. Since

$$\left| \frac{\partial G^0}{\partial \nu(y)}(x + t\nu(x) - y) - \frac{\partial G^0}{\partial \nu(y)}(x - y) \right| \le C \frac{|t|}{|x - y|^d}, \qquad \forall\, y \in \partial D,$$

we get $|I_3| \le CM|t|$, where M is the maximum of f on ∂D. To estimate I_1, we assume that $x = 0$ and near the origin, D is given by $y = (y', y_d)$ with $y_d > \varphi(y')$, where φ is a $C^{1,s}$-function such that $\varphi(0) = 0$ and $\nabla\varphi(0) = 0$. With these local coordinates, we can show that

$$\left| \frac{\partial G^0}{\partial \nu(y)}(x + t\nu(x) - y) \right| \le C \frac{|\varphi(y')| + |t|}{(|y'|^2 + |t|^2)^{d/2}},$$

and hence $|I_1| \le C\varepsilon$. A combination of the above estimates yields

$$\begin{aligned}
\limsup_{t \to 0^-} \Bigg| &\int_{\partial D} \frac{\partial G^0}{\partial \nu(y)}(x + t\nu(x) - y)(f(y) - f(x))\, d\sigma(y) \\
&- \int_{\partial D} \frac{\partial G^0}{\partial \nu(y)}(x - y)(f(y) - f(x)\, d\sigma(y) \Bigg| \le C\varepsilon.
\end{aligned}$$

Since ε is arbitrary, we obtain that

$$\begin{aligned}
\mathcal{D}_D^0[f]|_-(x) &= f(x) + \int_{\partial D} \frac{\partial G^0}{\partial \nu(y)}(x - y)(f(y) - f(x))\, d\sigma(y) \\
&= \left(\frac{1}{2} I + \mathcal{K}_D^0 \right)[f](x) \quad \text{for } x \in \partial D.
\end{aligned}$$

If $t > 0$, the argument is the same except that

$$\int_{\partial D} \frac{\partial G^0}{\partial \nu(y)}(x + t\nu(x) - y)\, d\sigma(y) = 0,$$

and hence we write

$$\mathcal{D}_D^0[f](x + t\nu(x)) = \int_{\partial D} \frac{\partial G^0}{\partial \nu(y)}(x + t\nu(x) - y)(f(y) - f(x))\, d\sigma(y), \quad x \in \partial D,$$

instead of (B.12). We leave the rest of the proof to the reader.

Next, consider $\phi \in L^2(\partial D)$. We first note that by (B.7),

$$\lim_{t\to 0^+} \mathcal{D}_D^0[\phi](x \pm t\nu(x))$$

exists and

$$\left\| \limsup_{t\to 0^+} \mathcal{D}_D^0[\phi](x \pm t\nu(x)) \right\|_{L^2(\partial D)} \leq C\|\phi\|_{L^2(\partial D)},$$

for some positive constant C independent of ϕ.

To handle the general case, let ε be given and choose a function $f \in C^0(\partial D)$ satisfying

$$\|\phi - f\|_{L^2(\partial D)} < \varepsilon.$$

Then

$$\begin{aligned}
&\left| \mathcal{D}_D^0[\phi](x \pm t\nu(x)) - \Big(\mp \frac{1}{2} I + \mathcal{K}_D^0 \Big)[\phi](x) \right| \\
&\quad \leq \left| \mathcal{D}_D^0[f](x \pm t\nu(x)) - \Big(\mp \frac{1}{2} I + \mathcal{K}_D^0 \Big)[f](x) \right| + \left| \mathcal{D}_D^0[\phi - f](x \pm t\nu(x)) \right| \\
&\quad\quad + \left| \Big(\mp \frac{1}{2} I + \mathcal{K}_D^0 \Big)[\phi - f](x) \right|.
\end{aligned}$$

For $\lambda > 0$, let

$$A_\lambda = \Big\{ x \in \partial D : \limsup_{t\to 0^+} \left| \mathcal{D}_D^0[\phi](x \pm t\nu(x)) - (\mp \frac{1}{2} I + \mathcal{K}_D^0)[\phi](x) \right| > \lambda \Big\}.$$

For a set E let $|E|$ denote its Lebesgue measure. Then

$$\begin{aligned}
|A_\lambda| &\leq \left| \Big\{ |\mathcal{D}_D^0[\phi - f]| > \frac{\lambda}{3} \Big\} \right| + \left| \Big\{ |\phi - f| > \frac{2\lambda}{3} \Big\} \right| + \left| \Big\{ |\mathcal{K}_D^0[\phi - f]| > \frac{\lambda}{3} \Big\} \right| \\
&\leq (\frac{3}{\lambda})^2 \Big(\|\phi - f\|_{L^2(\partial D)}^2 + \frac{1}{4}\|\phi - f\|_{L^2(\partial D)}^2 + \|\mathcal{K}_D^0[\phi - f]\|_{L^2(\partial D)}^2 \Big) \\
&\leq C(\frac{3}{\lambda})^2 \varepsilon^2.
\end{aligned}$$

Here we have used the L^2-boundedness of $\mathcal{K}_D^0$ which is an obvious consequence of Lemma B.2. Since ε is arbitrary, $|A_\lambda| = 0$ for all $\lambda > 0$. This implies that

$$\lim_{t\to 0^+} \mathcal{D}_D^0[\phi](x \pm t\nu(x)) = (\mp \frac{1}{2} I + \mathcal{K}_D^0)[\phi](x) \quad \text{a.e. } x \in \partial D,$$

and completes the proof.

In a similar way, the behaviour of the gradient of the single layer potential at the boundary can be described and the connection between the traces $\partial \mathcal{S}_D^0/\partial\nu|_\pm$ and the operator $\mathcal{K}_D^{0,*}$ defined by (B.5) revealed. □

B.1.3. Mapping properties.

Let

$$L_0^2(\partial D) := \Big\{ \phi \in L^2(\partial D) : \int_{\partial D} \phi \, d\sigma = 0 \Big\}.$$

The following results hold; see, for instance, [**47**, Lemma 2.18] for their proofs.

LEMMA B.5. *The spectrum of $\mathcal{K}_D^{0,*} : L^2(\partial D) \to L^2(\partial D)$ lies in the interval $(-1/2, 1/2]$ and therefore, the operator $(1/2)\, I + \mathcal{K}_D^0$ is invertible on $L^2(\partial D)$. Moreover, the operator $-(1/2)\, I + \mathcal{K}_D^0$ is invertible on $L_0^2(\partial D)$.*

PROOF. The argument is by contradiction. Let $\lambda \in (-\infty, -1/2] \cup (1/2, +\infty)$, and assume that $\phi \in L^2(\partial D)$ satisfies $(\lambda I - \mathcal{K}_D^{0,*})[\phi] = 0$ and ϕ is not identically zero. Since $\mathcal{K}_D^0[1] = 1/2$ by Green's formula, we have

$$0 = \int_{\partial D} (\lambda I - \mathcal{K}_D^{0,*})[\phi] \, d\sigma = \int_{\partial D} \phi(\lambda - \frac{1}{2}) \, d\sigma,$$

and thus $\int_{\partial D} \phi \, d\sigma = 0$. Hence $\mathcal{S}_D^0[\phi](x) = \mathcal{O}(|x|^{1-d})$ and $\nabla \mathcal{S}_D^0[\phi](x) = \mathcal{O}(|x|^{-d})$ as $|x| \to +\infty$ for $d \geq 2$. Since ϕ is not identically zero, both of the following numbers cannot be zero:

$$A = \int_D |\nabla \mathcal{S}_D^0[\phi]|^2 \, dx \quad \text{and} \quad B = \int_{\mathbb{R}^d \setminus \overline{D}} |\nabla \mathcal{S}_D^0[\phi]|^2 \, dx.$$

In fact, if both of them are zero, then $\mathcal{S}_D^0[\phi] =$ constant in D and in $\mathbb{R}^d \setminus \overline{D}$. Hence $\phi = 0$ by (B.11) which is a contradiction.

On the other hand, using the divergence theorem and (B.10), we have

$$A = \int_{\partial D} (-\frac{1}{2} I + \mathcal{K}_D^{0,*})[\phi] \, \mathcal{S}_D^0[\phi] \, d\sigma \quad \text{and} \quad B = -\int_{\partial D} (\frac{1}{2} I + \mathcal{K}_D^{0,*})[\phi] \, \mathcal{S}_D^0[\phi] \, d\sigma.$$

Since $(\lambda I - \mathcal{K}_D^{0,*})[\phi] = 0$, it follows that

$$\lambda = \frac{1}{2} \frac{B - A}{B + A}.$$

Thus, $|\lambda| < 1/2$, which is a contradiction and so, for $\lambda \in (-\infty, -\frac{1}{2}] \cup (\frac{1}{2}, +\infty)$, $\lambda I - \mathcal{K}_D^{0,*}$ is one to one on $L^2(\partial D)$.

If $\lambda = 1/2$, then $A = 0$ and hence $\mathcal{S}_D^0[\phi] =$ constant in D. Thus $\mathcal{S}_D^0[\phi]$ is harmonic in $\mathbb{R}^d \setminus \partial D$, behaves like $\mathcal{O}(|x|^{1-d})$ as $|x| \to +\infty$ (since $\phi \in L_0^2(\partial D)$), and is constant on ∂D. By (B.10), we have $\mathcal{K}_D^{0,*}[\phi] = (1/2)\, \phi$, and hence

$$B = -\int_{\partial D} \phi \, \mathcal{S}_D^0[\phi] \, d\sigma = C \int_{\partial D} \phi \, d\sigma = 0,$$

which forces us to conclude that $\phi = 0$. This proves that $(1/2)\, I - \mathcal{K}_D^{0,*}$ is one to one on $L_0^2(\partial D)$. □

Next, we look into the kernel of $\mathcal{S}_D^0$. The following results hold. We refer to [**47**, Theorem 2.26] for their proofs.

LEMMA B.6.

(i) *If $d = 3$, then* $\operatorname{Ker} \mathcal{S}_D^0 = \{0\}$. *Moreover,* $\mathcal{S}_D^0 : L^2(\partial D) \to H^1(\partial D)$ *has a bounded inverse;*

(ii) *If $d = 2$, then* $\operatorname{Ker} \mathcal{S}_D^0$ *is either trivial or of dimension one and equal to* $\operatorname{Ker}(-\frac{1}{2} I + \mathcal{K}_D^{0,*})$. *Moreover, the operator* $A : L^2(\partial D) \times \mathbb{R} \to H^1(\partial D) \times \mathbb{R}$ *defined by*

$$A(\phi, a) = \left(\mathcal{S}_D^0[\phi] + a, \int_{\partial D} \phi \, d\sigma \right)$$

has a bounded inverse.

PROOF. Let $d = 3$. If $\mathcal{S}_D^0[\phi] = 0$ on ∂D, then $u = \mathcal{S}_D^0[\phi]$ satisfies $\Delta u = 0$ in $\mathbb{R}^3 \setminus \partial D$, $u = 0$ on ∂D and as $|x| \to +\infty$, we have $u(x) = \mathcal{O}(1/|x|)$. Therefore, for large R,

$$\int_{B_R(0) \setminus \overline{D}} |\nabla u|^2 \, dx = \int_{\partial B_R(0)} \frac{\partial u}{\partial \nu} u \, d\sigma = \mathcal{O}(\frac{1}{R}).$$

Sending $R \to +\infty$, we deduce that $\nabla u = 0$ in $\mathbb{R}^3 \setminus \overline{D}$, and thus u is constant in $\mathbb{R}^3 \setminus \overline{D}$. Since $u = 0$ on ∂D, it follows that $u = 0$ in $\mathbb{R}^3 \setminus \overline{D}$. But $\mathcal{S}_D^0[\phi] = 0$ in D, and hence $\phi = \partial \mathcal{S}_D^0[\phi]/\partial\nu|_+ - \partial \mathcal{S}_D^0[\phi]/\partial\nu|_- = 0$ on ∂D.

Now, since $H^1(\partial D) \hookrightarrow L^2(\partial D)$ is compact, it follows that the operator $\mathcal{S}_D^0 : L^2(\partial D) \to H^1(\partial D)$ is Fredholm with zero index. But, we have $\operatorname{Ker} \mathcal{S}_D^0 = \{0\}$, and therefore $\mathcal{S}_D^0$ has a bounded inverse.

Next, suppose that $d = 2$. If $\phi_0 \in \operatorname{Ker} \mathcal{S}_D^0$, then the function u defined by

$$u(x) := \mathcal{S}_D^0[\phi_0](x), \quad x \in \mathbb{R}^2$$

satisfies $u = 0$ on ∂D. Therefore, $u(x) = 0$ for all $x \in D$. It then follows from (B.10) that

$$\mathcal{K}_D^{0,*}[\phi_0] = \frac{1}{2}\phi_0 \quad \text{on } \partial D\,. \tag{B.13}$$

If $\phi_0 \in L_0^2(\partial D)$, then $u(x) \to 0$ as $|x| \to +\infty$, and hence $u(x) = 0$ for $x \in \mathbb{R}^2 \setminus D$ as well. Thus $\phi_0 = 0$. The eigenmodes of (B.13) make a one-dimensional subspace of $L^2(\partial D)$, which means that $\operatorname{Ker} \mathcal{S}_D^0$ is of at most one dimension.

Finally, we establish that A has a bounded inverse. Since $\mathcal{S}_D^0 : L^2(\partial D) \to H^1(\partial D)$ is Fredholm with zero index, we need only to prove injectivity. In fact, if $\mathcal{S}_D^0[\phi] + a = 0$ and $\int_{\partial D} \phi \, d\sigma = 0$, then $\int_{\partial D} \mathcal{S}_D^0[\phi]\, \phi \, d\sigma = 0$. But

$$\begin{aligned}\int_{\partial D} (\mathcal{S}_D^0[\phi])\phi \, d\sigma &= \int_{\partial D} \mathcal{S}_D^0[\phi]\Big(\frac{\partial}{\partial\nu}\mathcal{S}_D^0[\phi]\Big|_+ - \frac{\partial}{\partial\nu}\mathcal{S}_D^0[\phi]\Big|_-\Big)\, d\sigma \\ &= -\int_{\mathbb{R}^2} |\nabla \mathcal{S}_D^0[\phi]|^2 \, dx,\end{aligned} \tag{B.14}$$

and consequently, $\mathcal{S}_D^0[\phi] = 0$ since $\mathcal{S}_D^0[\phi] \to 0$ as $|x| \to +\infty$, which implies $\phi = 0$ and, in turn $a = 0$. □

Furthermore, we list here two useful observations in three dimensions.

LEMMA B.7. *Let $d = 3$. The following results hold:*

(i) *We have $(-\frac{1}{2}I + \mathcal{K}_D^{0,*})(\mathcal{S}_D^0)^{-1}[\chi_{\partial D}] = 0$;*

(ii) *The eigenspace corresponding to the eigenvalue $1/2$ of $\mathcal{K}_D^{0,*}$ has dimension one and is spanned by the function $(\mathcal{S}_D^0)^{-1}[\chi_{\partial D}]$.*

PROOF. In order to prove item (i) we introduce

$$u = \mathcal{S}_D^0(\mathcal{S}_D^0)^{-1}[\chi_{\partial D}] \quad \text{in } D.$$

Since $\Delta u = 0$ in D and $u = 1$ on ∂D, $u \equiv 1$ in D. Therefore, $\partial u/\partial\nu = 0$ on ∂D, which yields (i). Item (ii) follows from (i) together with the invertibility of $-(1/2)\,I + \mathcal{K}_D^{0,*} : L_0^2(\partial D) \to L_0^2(\partial D)$, proved in Lemma B.5. □

B.1.4. Concept of capacity. We now recall the concept of capacity.

If $d = 3$, then the capacity of ∂D is defined to be

$$\operatorname{Cap}_D := -\int_{\partial D} (\mathcal{S}_D^0)^{-1}[\chi_{\partial D}] \, d\sigma. \tag{B.15}$$

If u is the solution of the exterior Dirichlet problem

$$\begin{cases}\Delta u = 0 & \text{in } \mathbb{R}^3 \setminus \overline{D},\\ u = 1 & \text{on } \partial D,\\ u(x) = \mathcal{O}(\frac{1}{|x|}) & \text{as } |x| \to +\infty,\end{cases} \tag{B.16}$$

then $u := \mathcal{S}_D[(\mathcal{S}_D^0)^{-1}[\chi_{\partial D}]]$ in $\mathbb{R}^3 \setminus D$. Therefore, by the jump relation (B.10), it follows that

$$\int_{\partial D} \frac{\partial u}{\partial \nu}\bigg|_+ \, d\sigma = \int_{\partial D} (\frac{1}{2}I + \mathcal{K}_D^{0,*})[(\mathcal{S}_D^0)^{-1}[\chi_{\partial D}]] \, d\sigma = \int_{\partial D} (\mathcal{S}_D^0)^{-1}[\chi_{\partial D}] \, d\sigma,$$

since, by Lemma B.4, $\mathcal{K}_D^0[1] = 1/2$. Thus, the capacity is given by

$$\text{Cap}_D = -\int_{\partial D} \frac{\partial u}{\partial \nu}\bigg|_+ \, d\sigma = -\int_{\partial D} \frac{\partial u}{\partial \nu}\bigg|_+ u \, d\sigma = \int_{\mathbb{R}^3 \setminus \overline{D}} |\nabla u|^2 \, dx, \tag{B.17}$$

by multiplying (B.16) by u and integrating by parts over $\mathbb{R}^3 \setminus \overline{D}$.

Suppose $d = 2$. From Lemma B.6, it follows that there exists a unique pair $(\phi, a) \in L^2(\partial D) \times \mathbb{R}$ solution of the system

$$\begin{cases} \mathcal{S}_D^0[\phi] + a = 0 & \text{on } \partial D, \\ \int_{\partial D} \phi \, d\sigma = 1. \end{cases} \tag{B.18}$$

The logarithmic capacity of ∂D is defined by

$$\text{Cap}_D := e^{2\pi a}, \tag{B.19}$$

where a is given by (B.18). Moreover, $\mathcal{S}_D^0 : L^2(\partial D) \to H^1(\partial D)$ has a bounded inverse if and only if $a \neq 0$.

Note that if $a = 0$, then $\mathcal{S}_D^0$ cannot be invertible because $\mathcal{S}_D^0[\phi] = 0$. Thus, suppose that $a \neq 0$ and there exists $\varphi \in L^2(\partial D)$ such that $\mathcal{S}_D^0[\varphi] = 0$. Define

$$\phi_0 = \varphi - (\int_{\partial D} \varphi \, d\sigma)\, \phi,$$

where ϕ is defined by (B.18) and observe that

$$\mathcal{S}_D^0[\phi_0] = -\Big(\int_{\partial D} \varphi \, d\sigma\Big)\mathcal{S}_D^0[\phi] = a\int_{\partial D} \varphi \, d\sigma \text{ and } \int_{\partial D} \phi_0 \, d\sigma = 0.$$

Hence $\int_{\partial D} \mathcal{S}_D^0[\phi_0]\, \phi_0 \, d\sigma = 0$ and therefore $\phi_0 = 0$. In turn, $\int_{\partial D} \varphi \, d\sigma = 0$ because $a \neq 0$, giving $\varphi = 0$. Thus, the homogeneous equation $\mathcal{S}_D^0[\varphi] = 0$ has only the trivial solution, and $\mathcal{S}_D^0$ is invertible.

B.2. Layer potentials for the Helmholtz equation

In this section, we review a number of basic facts and results regarding the layer potentials associated with the Helmholtz equation. The most important results are on the low-frequency asymptotic expansions of layer potentials. The integral equations that apply to the Helmholtz resonance problem are obtained from a study of these layer potentials.

B.2.1. Bessel functions. We summarise some useful identities for Bessel functions. For further details, we refer the reader to [**152**, pp. 225–233].

Bessel functions of the first kind of real order ν, denoted by $J_\nu(t)$, are given in terms of the series representation

$$J_\nu(t) = (\frac{t}{2})^\nu \sum_{l=0}^{+\infty} \frac{(-t^2/4)^l}{l!\Gamma(\nu + l + 1)}, \tag{B.20}$$

where the gamma function Γ is defined by

$$\Gamma(z) = \int_0^{+\infty} e^{-y} y^{z-1}\, dy \quad \text{for } \mathrm{Re}\,(z) > 0.$$

Another definition, valid for $\mathrm{Re}\,\nu > -\frac{1}{2}$, is

$$J_\nu(t) = [\Gamma(\frac{1}{2})\Gamma(\nu + \frac{1}{2})]^{-1}(\frac{t}{2})^\nu \int_{-1}^{1} (1-y^2)^{\nu-\frac{1}{2}} e^{ity}\, \mathrm{d}y. \tag{B.21}$$

We have the recurrence relation

$$(\frac{d}{dt} + \frac{\nu}{t}) J_\nu(t) = J_{\nu-1}(t). \tag{B.22}$$

For $n \in \mathbb{Z}$, we have the integral representation

$$J_n(t) = \frac{1}{2\pi} \int_{-\pi}^{\pi} e^{it\sin\phi - in\phi}\, \mathrm{d}\phi, \tag{B.23}$$

i.e., the functions $J_n(t)$ are the Fourier coefficients of $e^{it\sin\phi}$. Therefore, the Jacobi-Anger expansion holds:

$$e^{it\sin\phi} = \sum_{n\in\mathbb{Z}} J_n(t) e^{in\phi}. \tag{B.24}$$

Formula (B.24) can be used in two dimensions to expand a plane wave as a sum of cylindrical waves. We have

$$e^{i\xi\cdot x} = \sum_{n\in\mathbb{Z}} e^{in(\frac{\pi}{2} - \theta_\xi)} J_n(|\xi||x|) e^{in\theta_x}), \tag{B.25}$$

where $x = (|x|, \theta_x)$ and $\xi = (|\xi|, \theta_\xi)$ in the polar coordinates. The function $x \mapsto J_n(|\xi||x|)e^{in\theta_x}$ is called a cylindrical wave.

For arguments $t < \nu$, the Bessel functions look qualitatively like simple powers law, with the asymptotic form for $0 < t \ll \nu$

$$J_\nu(t) \approx \frac{1}{\Gamma(\nu+1)} \left(\frac{t}{2}\right)^\nu \approx \frac{1}{\sqrt{2\pi\nu}} \left(\frac{et}{2\nu}\right)^\nu. \tag{B.26}$$

For $t > \nu$, the Bessel functions look qualitatively like cosine waves whose amplitude decay as $t^{-1/2}$. The asymptotic form for $t \gg \nu$ is

$$J_\nu(t) \approx \sqrt{\frac{2}{\pi t}} \cos\left(t - \frac{\nu\pi}{2} - \frac{\pi}{4}\right). \tag{B.27}$$

In the transition region where $t \approx \nu$, the typical amplitude of the Bessel functions is

$$J_\nu(\nu) \approx \frac{2^{1/3}}{3^{2/3}\Gamma\left(\frac{2}{3}\right)} \frac{1}{\nu^{1/3}} \approx \frac{0.4473}{\nu^{1/3}},$$

which holds asymptotically for large ν.

The Bessel function J_ν solves the ordinary differential equation, known as Bessel's equation,

$$\left(\frac{d^2}{dt^2} + \frac{1}{t}\frac{d}{dt} + (1 - \frac{\nu^2}{t^2})\right) J_\nu(t) = 0, \tag{B.28}$$

or equivalently,

$$\left(\frac{d}{dt} - \frac{\nu-1}{t}\right)\left(\frac{d}{dt} + \frac{\nu}{t}\right) J_\nu(t) = -J_\nu(t). \tag{B.29}$$

Note that adding and subtracting (B.22) and (B.29) produce the identities

$$2J'_\nu(t) = J_{\nu-1}(t) - J_{\nu+1}(t),$$
$$\frac{2\nu}{t}J_\nu(t) = J_{\nu-1}(t) + J_{\nu+1}(t).$$

Equation (B.28), for each ν, has a two-dimensional solution space. Note that $J_{-\nu}$ is also a solution. From the expression (B.20) it is clear that J_ν and $J_{-\nu}$ are linearly independent provided ν is not an integer. On the other hand, comparison of power series shows

$$J_{-n}(t) = (-1)^n J_n(t), \quad n \in \mathbb{N}.$$

A calculation of the Wronskian shows that

$$W(J_\nu, J_{-\nu})(t) = -2\frac{\sin \pi\nu}{\pi t}.$$

Therefore, J_ν and $J_{-\nu}$ are linearly independent, and consequently they form a basis of solutions to (B.28), if and only if ν is not an integer. To construct a basis of solutions uniformly good for all ν, it is natural to set

$$Y_\nu(t) = \frac{J_\nu(t)\cos \pi\nu - J_{-\nu}(t)}{\sin \pi\nu}, \tag{B.30}$$

when ν is not an integer, and define for integer n

$$Y_n(t) = \lim_{\nu \to n} Y_\nu(t).$$

We have

$$W(J_\nu, Y_\nu)(t) = \frac{2}{\pi t},$$

for all ν.

Another important pair of solutions to Bessel's equation is that of Hankel functions

$$H_\nu^{(1)}(t) = J_\nu(t) + \mathrm{i}Y_\nu(t), \quad H_\nu^{(2)}(t) = J_\nu(t) - \mathrm{i}Y_\nu(t). \tag{B.31}$$

We will need the following behaviour of $H_\nu^{(1)}$ for fixed ν and $t \to 0$. We have

$$H_\nu^{(1)}(t) \approx -\frac{\mathrm{i}2^\nu\Gamma(\nu)}{\pi}t^{-\nu}. \tag{B.32}$$

For n an integer, as $t \to 0$,

$$J_n(t) = \frac{t^n}{2^n}\left(\frac{1}{\Gamma(n+1)} - \frac{\frac{1}{4}t^2}{\Gamma(n+2)} + \frac{(\frac{1}{4}t^2)^2}{2!\Gamma(n+3)} - \frac{(\frac{1}{4}t^2)^3}{3!\Gamma(n+4)} + \cdots\right), \tag{B.33}$$

$$\begin{aligned} Y_n(t) = &-\frac{(\frac{1}{2}t)^{-n}}{\pi}\sum_{l=0}^{n-1}\frac{(n-l-1)!}{l!}(\frac{1}{4}t^2)^l + \frac{2}{\pi}\ln(\frac{1}{2}t)J_n(t) \\ &-\frac{(\frac{1}{2}t)^n}{\pi}\sum_{l=0}^{+\infty}(\psi(l+1)+\psi(n+l+1))\frac{(-\frac{1}{4}t^2)^l}{l!(n+l)!}, \end{aligned} \tag{B.34}$$

where $\psi(1) = -\gamma$ and

$$\psi(n) = -\gamma + \sum_{l=1}^{n-1}\frac{1}{l} \quad \text{for } n \geq 2$$

with γ being the Euler constant. In particular, if $n = 0$, we have

$$
\begin{aligned}
&J_0(t) = 1 - \frac{1}{4}t^2 + \frac{1}{64}t^4 + \mathcal{O}(t^6), \\
&Y_0(t) = \frac{2}{\pi}\ln t + \frac{2}{\pi}(\gamma - \ln 2) - \frac{1}{2\pi}t^2 \ln t + \Big(\frac{1}{2\pi} - \frac{1}{2\pi}(\gamma - \ln 2)\Big)t^2 \\
&\qquad + \mathcal{O}(t^4 \ln t).
\end{aligned} \tag{B.35}
$$

On the other hand, for large values of t, we have

$$
\begin{aligned}
H_0^{(1)}(t) &= \sqrt{\frac{2}{\pi t}} e^{\mathrm{i}(t - \frac{\pi}{4})}\left[1 + \mathcal{O}\Big(\frac{1}{t}\Big)\right], \\
(H_0^{(1)})'(t) &= \sqrt{\frac{2}{\pi t}} e^{\mathrm{i}(t + \frac{\pi}{4})}\left[1 + \mathcal{O}\Big(\frac{1}{t}\Big)\right]
\end{aligned} \quad \text{as } t \to +\infty. \tag{B.36}
$$

It is worth pointing out that the Bessel functions $J_{n+1/2}(t)$, for n an integer, are elementary functions. For $\nu = n + 1/2$, the integrand in (B.21) involves $(1 - y^2)^n$, so the integral can be evaluated explicitly. We have, in particular,

$$
J_{1/2}(t) = (\frac{2}{\pi t})^{1/2} \sin t.
$$

Then (B.22) gives

$$
J_{-1/2}(t) = (\frac{2}{\pi x})^{1/2} \cos t,
$$

which by (B.30) is equal to $-Y_{1/2}(t)$. Applying (B.29) and (B.22) repeatedly gives

$$
J_{n+1/2}(t) = (-1)^n \prod_{l=1}^{n} (\frac{d}{dt} - \frac{l - \frac{1}{2}}{t}) \frac{\sin t}{\sqrt{2\pi t}}
$$

and the same sort of formula for $J_{-n-1/2}(t)$, with the $(-1)^n$ removed, and $\sin t$ replaced by $\cos t$.

The functions

$$
j_n(t) := \sqrt{\frac{\pi}{2}} \frac{J_{n+\frac{1}{2}}(t)}{\sqrt{t}}, \tag{B.37}
$$

and

$$
y_n(t) := \sqrt{\frac{\pi}{2}} \frac{Y_{n+\frac{1}{2}}(t)}{\sqrt{t}} \tag{B.38}
$$

are known as the spherical Bessel functions and form a basis for the solution space of the spherical Bessel equation

$$
\Big(\frac{d^2}{dt^2} + \frac{2}{t}\frac{d}{dt} + (1 - \frac{n(n+1)}{t^2})\Big) f(t) = 0.
$$

Analogously to (B.31), we define $h_n^{(1)}$ and $h_n^{(2)}$ by

$$
h_n^{(1)}(t) = j_n(t) + \mathrm{i} y_n(t), \quad h_n^{(2)}(t) = j_n(t) - \mathrm{i} y_n(t). \tag{B.39}
$$

The following asymptotic expansions hold:

$$
\begin{cases}
h_n^{(1)}(t) = \frac{1}{t} e^{\mathrm{i}(t - \frac{n\pi}{2} - \frac{\pi}{2})}(1 + \mathcal{O}(\frac{1}{t})), \\
h_n^{(2)}(t) = \frac{1}{t} e^{-\mathrm{i}(t - \frac{n\pi}{2} - \frac{\pi}{2})}(1 + \mathcal{O}(\frac{1}{t})),
\end{cases} \tag{B.40}
$$

as $t \to +\infty$.

In three dimensions, there is an analogue to (B.25). The following plane wave expansion, also known as the Rayleigh equation, holds:

$$
e^{\mathrm{i}\xi\cdot x} = 4\pi \sum_{m=0}^{+\infty} \sum_{l=-m}^{m} \mathrm{i}^m j_m(|\xi||x|) Y_m^l(\theta_x, \phi_x) \overline{Y_m^l}(\theta_\xi, \phi_\xi), \tag{B.41}
$$

where Y_m^l are the spherical harmonics and $\xi = (|\xi|, \theta_\xi, \phi_\xi), x = (|x|, \theta_x, \phi_x)$ in the spherical coordinates.

B.2.2. Green's function.

LEMMA B.8. *For $\omega > 0$, a Green's function $G^\omega(x)$ to the Helmholtz operator $\Delta + \omega^2$ in $\mathbb{R}^d, d = 2, 3$, satisfying*

$$
\frac{x}{|x|} \cdot \nabla G^\omega(x) - \mathrm{i}\omega G^\omega(x) = \begin{cases} \mathcal{O}(|x|^{-3/2}), & d = 2, \\ \mathcal{O}(|x|^{-2}), & d = 3, \end{cases} \tag{B.42}
$$

as $|x| \to +\infty$ is given by (1.7).

REMARK B.9. *Condition* (B.42) *is exactly the Sommerfeld radiation condition* (1.2) *one should impose in order to select the physical solution to scattering problems. The Sommerfeld radiation condition is also called the outgoing radiation condition, and G^ω the outgoing Green function to the Helmholtz equation.*

PROOF OF LEMMA B.8. It is natural to look for radial solutions of the form $G^\omega(x) = w_\omega(r)$ that are subject to the Sommerfeld radiation condition (1.2) for $k = \omega$ and $u_{\text{in}} \equiv 0$. If $d = 3$, (1.7) becomes

$$
\frac{1}{r^2} \frac{d}{dr} r^2 \frac{dw_\omega}{dr} + \omega^2 w_\omega = 0, \quad r > 0,
$$

whose solution is

$$
w_\omega(r) = c_1 \frac{e^{\mathrm{i}\omega r}}{r} + c_2 \frac{e^{-\mathrm{i}\omega r}}{r}.
$$

It is easy to check that the Sommerfeld radiation condition (1.2) leads to $c_2 = 0$ and then (1.7) leads to $c_1 = -1/(4\pi)$.

If $d = 2$, equation (1.7) becomes the Bessel equation

$$
\frac{1}{r} \frac{d}{dr} r \frac{dw_\omega}{dr} + \omega^2 w_\omega = 0, \quad r > 0.
$$

The Hankel functions of the first and second kind of order 0, $H_0^{(1)}(\omega r)$ and $H_0^{(2)}(\omega r)$, form a basis for the solution space. At infinity ($r \to +\infty$), only $H_0^{(1)}(\omega r)$ satisfies the Sommerfeld radiation condition (1.2); see (B.36). At the origin ($r \to 0$), from (B.35) it follows that $H_0^{(1)}(\omega r)$ behaves like $(2\mathrm{i}/\pi)\ln(r)$. □

In two dimensions, the following full asymptotic expansion of G^ω near 0 will be useful later (see, for instance, [**110**]):

$$
G^\omega(x) = -\frac{\mathrm{i}}{4} H_0^{(1)}(\omega|x|) = \frac{1}{2\pi} \ln|x| + \eta_\omega + \sum_{j=1}^{+\infty} (b_j \ln(\omega|x|) + c_j)(\omega|x|)^{2j}, \tag{B.43}
$$

where

$$
b_j = \frac{(-1)^j}{2\pi} \frac{1}{2^{2j}(j!)^2}, \quad c_j = b_j \Big(\gamma - \ln 2 - \frac{\pi \mathrm{i}}{2} - \sum_{l=1}^{j} \frac{1}{l} \Big),
$$

and the constant

$$\eta_\omega = (1/2\pi)(\ln\omega + \gamma - \ln 2) - \mathrm{i}/4 \tag{B.44}$$

with γ being the Euler constant.

B.2.3. Explicit representation formulas. Let $R > 0$ and $B_R = \{|x| < R\}$. Let u satisfy the Helmholtz equation $\Delta u + \omega^2 u = 0$ for $|x| > R$. Moreover, assume that $u|_{\partial B_R} \in H^{1/2}(\partial B_R)$ and that u satisfies the Sommerfeld radiation condition (1.2) for $k = \omega$ and $u_{\text{in}} \equiv 0$. Then, u can be represented in $\mathbb{R}^d \setminus B_R$ as

$$u(x) = \begin{cases} \sum_{m\in\mathbb{Z}} u_m \frac{H_m^{(1)}(k|x|)}{H_m^{(1)}(kR)} e^{im\theta} & \text{if } d = 2, \\ \sum_{m=0}^{+\infty} \sum_{l=-m}^{m} u_m^l \frac{h_m^{(1)}(k|x|)}{h_m^{(1)}(kR)} Y_m^l & \text{if } d = 3, \end{cases} \tag{B.45}$$

where θ is the angular variable, $\{Y_m^l\}$ are the spherical harmonics, and u_m and u_m^l are respectively defined by

$$u|_{\partial B_R} = \begin{cases} \sum_{m\in\mathbb{Z}} u_m e^{im\theta} & \text{if } d = 2, \\ \sum_{m=0}^{+\infty} \sum_{l=-m}^{m} u_m^l Y_m^l & \text{if } d = 3. \end{cases}$$

B.2.4. Single and double layer potentials. For a bounded domain D of class $C^{1,s}$ for some $0 < s < 1$ in $\mathbb{R}^d$ and $\omega > 0$, let $\mathcal{S}_D^\omega$ and $\mathcal{D}_D^\omega$ be the single and double layer potentials defined by G^ω; that is,

$$\mathcal{S}_D^\omega[\phi](x) = \int_{\partial D} G^\omega(x-y)\phi(y)\, d\sigma(y), \quad x \in \mathbb{R}^d, \tag{B.46}$$

$$\mathcal{D}_D^\omega[\phi](x) = \int_{\partial D} \frac{\partial G^\omega(x-y)}{\partial \nu(y)} \phi(y)\, d\sigma(y)\,, \quad x \in \mathbb{R}^d \setminus \partial D, \tag{B.47}$$

for $\phi \in L^2(\partial D)$. Then $\mathcal{S}_D^\omega[\phi]$ and $\mathcal{D}_D^\omega[\phi]$ satisfy the Helmholtz equation

$$(\Delta + \omega^2)u = 0 \quad \text{in } D \text{ and } \mathbb{R}^d \setminus \overline{D}.$$

Moreover, in view of (B.42), both of them satisfy the Sommerfeld radiation condition (1.2).

Let us make note of a Green's formula to be used later. If $(\Delta + \omega^2)u = 0$ in D and $\partial u/\partial\nu \in L^2(\partial D)$, then

$$-\mathcal{S}_D^\omega\left[\frac{\partial u}{\partial \nu}\bigg|_-\right](x) + \mathcal{D}_D^\omega[u|_-](x) = \begin{cases} u(x), & x \in D, \\ 0, & x \in \mathbb{R}^d \setminus \overline{D}. \end{cases} \tag{B.48}$$

A formula similar to (B.48) holds for the solution to the Helmholtz equation in $\mathbb{R}^d \setminus \overline{D}$ subject to the Sommerfeld radiation condition (1.2):

$$\mathcal{S}_D^\omega\left[\frac{\partial u}{\partial \nu}\bigg|_+\right](x) - \mathcal{D}_D^\omega[u|_+](x) = \begin{cases} 0, & x \in D, \\ u(x), & x \in \mathbb{R}^d \setminus \overline{D}. \end{cases} \tag{B.49}$$

Analogously to (B.9) and (B.10), the following formulas give the jump relations obeyed by the double layer potential and by the normal derivative of the single layer potential on ∂D:

$$\frac{\partial(\mathcal{S}_D^\omega[\phi])}{\partial \nu}\bigg|_\pm (x) = \Big(\pm\frac{1}{2}I + \mathcal{K}_D^{\omega,*}\Big)[\phi](x) \quad \text{a.e. } x \in \partial D, \tag{B.50}$$

$$(\mathcal{D}_D^\omega[\phi])\Big|_\pm (x) = \Big(\mp\frac{1}{2}I + \mathcal{K}_D^{\omega}\Big)[\phi](x) \quad \text{a.e. } x \in \partial D, \tag{B.51}$$

for $\phi \in L^2(\partial D)$, where $\mathcal{K}_D^\omega$ is the integral operator defined by

$$\mathcal{K}_D^\omega[\phi](x) = \int_{\partial D} \frac{\partial G^\omega(x-y)}{\partial \nu(y)} \phi(y) \, \mathrm{d}\sigma(y)$$

and $\mathcal{K}_D^{\omega,*}$ (introduced in (1.11)) is the L^2-adjoint of $\mathcal{K}_D^{-\omega}$.

The jump formulas can be proved in the same way as Proposition B.3. This follows from the fact that G^ω has the same singularity as G^0. Moreover, since D is smooth, by Lemma B.2 the integral operators $\mathcal{K}_D^\omega$ and $\mathcal{K}_D^{\omega,*}$ are compact on $L^2(\partial D)$.

The following Calderón identity (also known as Plemelj's symmetrisation principle) holds.

LEMMA B.10. *We have*

$$\mathcal{S}_D^\omega \mathcal{K}_D^{\omega,*} = \mathcal{K}_D^\omega \mathcal{S}_D^\omega. \tag{B.52}$$

PROOF. Consider u of the form

$$u = \mathcal{S}_D^\omega[\phi] - \mathcal{D}_D^\omega[\psi], \tag{B.53}$$

for two densities ϕ and ψ. Then, from (B.50) and (B.51), it follows that

$$\begin{cases} u\big|_+ = \mathcal{S}_D^\omega[\phi] - \left(-\frac{1}{2}I + \mathcal{K}_D^\omega\right)[\psi], \\ \frac{\partial u}{\partial \nu}\Big|_+ = \left(\frac{1}{2}I + \mathcal{K}_D^{\omega,*}\right)[\phi] - \frac{\partial}{\partial \nu}\mathcal{D}_D^\omega[\psi]. \end{cases} \tag{B.54}$$

Let

$$w := \mathcal{S}_D^\omega\Big[\frac{\partial u}{\partial \nu}\Big|_+\Big] - \mathcal{D}_D^\omega[u|_+].$$

The representation formula (B.49) shows that $w = 0$ in D and $w = u$ in $\mathbb{R}^d \setminus \overline{D}$. Hence, by inserting (B.54) into (B.53), we arrive at

$$\left(\mathcal{S}_D^\omega \mathcal{K}_D^{\omega,*} - \mathcal{K}_D^\omega \mathcal{S}_D^\omega\right)[\phi] = 0,$$

which proves the lemma. □

B.2.5. Low-frequency asymptotic expansions of layer potentials. We recall from [**37**] some basic asymptotic expansions for the layer potentials in three and two dimensions.

B.2.5.1. *Expansions in three dimensions.* We first consider the single layer potential. We have the following asymptotic expansion as $\omega \to 0$:

$$\mathcal{S}_D^\omega = \mathcal{S}_D^0 + \sum_{j=1}^{+\infty} \omega^j \mathcal{S}_{D,j}, \tag{B.55}$$

where

$$\mathcal{S}_{D,j}[\psi](x) = -\frac{\mathrm{i}}{4\pi} \int_{\partial D} \frac{(\mathrm{i}|x-y|)^{j-1}}{j!} \psi(y) \, \mathrm{d}\sigma(y).$$

In particular, we have

$$\mathcal{S}_{D,1}[\psi](x) = -\frac{\mathrm{i}}{4\pi} \int_{\partial D} \psi(y) \, \mathrm{d}\sigma(y), \tag{B.56}$$

$$\mathcal{S}_{D,2}[\psi](x) = -\frac{1}{8\pi} \int_{\partial D} |x-y| \psi(y) \, \mathrm{d}\sigma(y). \tag{B.57}$$

LEMMA B.11. *The norm* $\|\mathcal{S}_{D,j}\|_{\mathcal{L}(L^2(\partial D),H^1(\partial D))}$ *is uniformly bounded with respect to* j. *Moreover, the series in* (B.55) *is convergent in* $\mathcal{L}(L^2(\partial D), H^1(\partial D))$.

We now consider the boundary integral operator $\mathcal{K}_D^{\omega,*}$. We have

$$\mathcal{K}_D^{\omega,*} = \mathcal{K}_D^{0,*} + \omega\mathcal{K}_{D,1} + \omega^2\mathcal{K}_{D,2} + \dots, \tag{B.58}$$

where

$$\begin{aligned}\mathcal{K}_{D,j}[\psi](x) &= -\frac{\mathrm{i}}{4\pi}\int_{\partial D} \frac{\partial(\mathrm{i}|x-y|)^{j-1}}{j!\partial\nu(x)}\psi(y)\ \mathrm{d}\sigma(y)\\ &= -\frac{\mathrm{i}^j(j-1)}{4\pi j!}\int_{\partial D}|x-y|^{j-3}(x-y)\cdot\nu(x)\psi(y)\ \mathrm{d}\sigma(y).\end{aligned}$$

In particular, we have

$$\mathcal{K}_{D,1} \equiv 0, \tag{B.59}$$

$$\mathcal{K}_{D,2}[\psi](x) = \frac{1}{8\pi}\int_{\partial D}\frac{(x-y)\cdot\nu(x)}{|x-y|}\psi(y)\ \mathrm{d}\sigma(y), \tag{B.60}$$

$$\mathcal{K}_{D,3}[\psi](x) = \frac{\mathrm{i}}{12\pi}\int_{\partial D}(x-y)\cdot\nu(x)\psi(y)\ \mathrm{d}\sigma(y). \tag{B.61}$$

LEMMA B.12. *The norm* $\|\mathcal{K}_{D,j}\|_{\mathcal{L}(L^2(\partial D))}$ *is uniformly bounded for* $j \geq 1$. *Moreover, the series in* (B.58) *is convergent in* $\mathcal{L}(L^2(\partial D), L^2(\partial D))$.

We also have the following lemma from [**39**].

LEMMA B.13. *For any* $\phi \in L^2(\partial D)$ *we have,*

$$\int_{\partial D}\mathcal{K}_{D,2}[\phi]\ \mathrm{d}\sigma = -\int_D \mathcal{S}_D^0[\phi]\ \mathrm{d}x, \qquad \int_{\partial D}\mathcal{K}_{D,3}[\phi]\ \mathrm{d}\sigma = \frac{\mathrm{i}|D|}{4\pi}\int_{\partial D}\phi\ \mathrm{d}\sigma. \tag{B.62}$$

PROOF. Note that

$$\begin{aligned}\int_{\partial D}\int_{\partial D}\frac{(x-y)\cdot\nu(x)}{|x-y|}\phi(y)\ \mathrm{d}\sigma(y)\ \mathrm{d}\sigma(x) &= \int_{\partial D}\phi(y)\int_D \nabla_x\cdot\frac{x-y}{|x-y|}\ \mathrm{d}x\ \mathrm{d}\sigma(y)\\ &= 2\int_D\int_{\partial D}\frac{\phi(y)}{|y-x|}\ \mathrm{d}\sigma(y)\ \mathrm{d}x.\end{aligned}$$

Moreover,

$$\begin{aligned}\int_{\partial D}\int_{\partial D}(x-y)\cdot\nu(x)\phi(y)\ \mathrm{d}\sigma(y)\ \mathrm{d}\sigma(x) &= \int_{\partial D}\phi(y)\int_D\nabla_x\cdot(x-y)\ \mathrm{d}x\ \mathrm{d}\sigma(y)\\ &= 3|D|\int_{\partial D}\phi(y)\ \mathrm{d}\sigma(y).\end{aligned}$$

Using the above two identities we obtain the desired results. □

Analogously to Lemma B.13, we have the following explicit expressions of $\mathcal{K}_{D,j}^*$ for $j = 2, 3$.

LEMMA B.14. *Let* $\mathcal{K}_{D,j}^*$ *be the adjoint of* $\mathcal{K}_{D,j}$ *in* $L^2(D)$. *The following identities hold:*

(i)

$$\mathcal{K}_{D,2}^*[1](x) = \frac{1}{8\pi}\int_{\partial D}\frac{(y-x)\cdot\nu(y)}{|y-x|}\ \mathrm{d}\sigma(y) = \frac{1}{8\pi}\int_D\nabla\cdot\frac{y-x}{|y-x|}dy = \frac{1}{4\pi}\int_D\frac{1}{|y-x|}dy.$$

(ii)

$$\mathcal{K}^*_{D,3}[1](x) = \frac{\mathrm{i}}{12\pi} \int_{\partial D} (y-x)\cdot\nu(y)\; d\sigma(y) = \frac{\mathrm{i}}{12\pi}\int_D \nabla\cdot(y-x)dy = \frac{\mathrm{i}}{4\pi}|D|.$$

B.2.5.2. *Expansions in two dimensions.* From (B.43), it follows that the single layer potential for the Helmholtz equation in two dimensions has the following expansion as $\omega \to 0$:

$$\mathcal{S}^\omega_D = \widehat{\mathcal{S}}^\omega_D + \sum_{j=1}^{+\infty} \left(\omega^{2j}\ln\omega\right)\mathcal{S}^{(1)}_{D,j} + \sum_{j=1}^{+\infty} \omega^{2j}\mathcal{S}^{(2)}_{D,j}, \tag{B.63}$$

where

$$\widehat{\mathcal{S}}^\omega_D[\psi](x) = \mathcal{S}^0_D[\psi](x) + \eta_\omega \int_{\partial D} \psi \; d\sigma, \tag{B.64}$$

and

$$\mathcal{S}^{(1)}_{D,j}[\psi](x) = \int_{\partial D} b_j|x-y|^{2j}\psi(y)\; d\sigma(y), \tag{B.65}$$

$$\mathcal{S}^{(2)}_{D,j}[\psi](x) = \int_{\partial D} |x-y|^{2j}(b_j \ln|x-y| + c_j)\psi(y)\; d\sigma(y). \tag{B.66}$$

Here, η_ω is given by (B.44).

Next, we consider the boundary integral operator $(\mathcal{K}^\omega_D)^*$. We have

$$\mathcal{K}^{\omega,*}_D = \mathcal{K}^{0,*}_D + \sum_{j=1}^{+\infty} \left(\omega^{2j}\ln\omega\right)\mathcal{K}^{(1)}_{D,j} + \sum_{j=1}^{+\infty} \omega^{2j}\mathcal{K}^{(2)}_{D,j}, \tag{B.67}$$

where

$$\mathcal{K}^{(1)}_{D,j}[\psi](x) = \int_{\partial D} b_j \frac{\partial |x-y|^{2j}}{\partial\nu(x)}\psi(y)\; d\sigma(y),$$
$$\mathcal{K}^{(2)}_{D,j}[\psi](x) = \int_{\partial D} \frac{\partial\left(|x-y|^{2j}(b_j\ln|x-y|+c_j)\right)}{\nu(x)}\psi(y)\; d\sigma(y).$$

LEMMA B.15. *The following expansions:*

$$\mathcal{S}^\omega_D = \widehat{\mathcal{S}}^\omega_D + \omega^2\ln\omega\; \mathcal{S}^{(1)}_{D,1} + \omega^2\mathcal{S}^{(2)}_{D,1} + \mathcal{O}(\omega^4\ln\omega);$$
$$\mathcal{K}^{\omega,*}_D = \mathcal{K}^{0,*}_D + \omega^2\ln\omega\; \mathcal{K}^{(1)}_{D,1} + \omega^2\mathcal{K}^{(2)}_{D,1} + \mathcal{O}(\omega^4\ln\omega),$$

hold in $\mathcal{L}(L^2(\partial D), H^1(\partial D))$ *and* $\mathcal{L}(L^2(\partial D), L^2(\partial D))$, *respectively.*

LEMMA B.16. *The following identities hold:*

(i)

$$(\mathcal{K}^{(1)}_{D,1})^*[1](x) = 4\bar{b}_1|D|;$$

(ii)

$$(\mathcal{K}^{(2)}_{D,1})^*[1](x) = (2\bar{b}_1 + 4\bar{c}_1)|D| + 4\bar{b}_1\int_D \ln|x-y|dy,$$

where $\bar{b}_1$ *and* $\bar{c}_1$ *are the complex conjugates of* b_1 *and* c_1. *Here,* b_1 *and* c_1 *are numbers appearing in* (B.65) *and* (B.66).

PROOF. First, we have

$$\begin{aligned}(\mathcal{K}_{D,1}^{(1)})^*[1](x) &= \bar{b}_1 \int_{\partial D} 2(y-x)\cdot\nu(y)\, d\sigma(y)\\ &= \bar{b}_1 \int_{\partial D} \frac{\partial |y-x|^2}{\partial \nu(y)}\, d\sigma(y)\\ &= \bar{b}_1 \int_D \Delta_y |y-x|^2\, dy\\ &= 4\bar{b}_1 |D|.\end{aligned}$$

We now prove the second identity. We have

$$\begin{aligned}(\mathcal{K}_{D,1}^{(2)})^*[1](x) &= \int_{\partial D} \frac{\partial\left[|y-x|^2(\bar{b}_1 \ln|x-y| + \bar{c}_1)\right]}{\partial \nu(y)}\, d\sigma(y)\\ &= \int_D \Delta_y [|y-x|^2(\bar{b}_1 \ln|x-y| + \bar{c}_1)]\, dy\\ &= 4\bar{c}_1|D| + \bar{b}_1 \int_D \Delta_y[|y-x|^2 \ln|x-y|]\, dy\\ &= 4\bar{c}_1|D| + \bar{b}_1 \int_D 4\ln|x-y|]\, dy\\ &+ \bar{b}_1 \int_D 2dy + \bar{b}_1 \int_D |y-x|^2 \Delta \ln|y-x|\, dy\\ &= (2\bar{b}_1 + 4\bar{c}_1)|D| + 4\bar{b}_1 \int_D \ln|x-y|\, dy,\end{aligned}$$

where we have used the fact that

$$\int_D |y-x|^2 \Delta_y \ln|y-x|\, dy = 0 \quad \text{for } x \in \partial D.$$

This completes the proof of the Lemma. □

B.2.6. Uniqueness results. In this subsection, we consider important uniqueness results for the Helmholtz equation.

We will need the following key result from the theory of the Helmholtz equation. It will help us prove uniqueness for exterior Helmholtz problems. For its proof, we refer to [**126**, Lemma 9.8].

LEMMA B.17 (Rellich's lemma). *Let $\omega > 0$. Let $R_0 > 0$ and $B_R = \{|x| < R\}$. Let u satisfy the Helmholtz equation $\Delta u + \omega^2 u = 0$ for $|x| > R_0$. Furthermore, assume that*

$$\lim_{R\to+\infty} \int_{\partial B_R} |u(x)|^2\, d\sigma(x) = 0. \tag{B.68}$$

Then, $u \equiv 0$ for $|x| > R_0$.

Note that the assertion of this lemma does not hold if ω is imaginary or $\omega = 0$.

PROOF. We give a proof only for $d = 3$. The proof in the two-dimensional case is similar. Representing u for $|x| > R_0$ in terms of spherical harmonics Y_m^l, we have

$$u(x) = \sum_{m=0}^{+\infty} \sum_{l=-m}^{m} \underbrace{\left(u_{m,l}^{(1)} h_m^{(1)}(\omega|x|) + u_{m,l}^{(2)} h_m^{(2)}(\omega|x|)\right)}_{:=a_{m,l}(|x|)} Y_m^l.$$

Assumption (B.68) implies that

$$\lim_{|x|\to+\infty} |x|^2 |a_{m,l}(|x|)|^2 = 0.$$

From (B.40), it follows that $u_{m,l}^{(1)} e^{it} + u_{m,l}^{(2)} = 0$ for all real t. Therefore,

$$u_{m,l}^{(1)} = u_{m,l}^{(2)} = 0,$$

which means that $u \equiv 0$ for $|x| > R_0$. □

B.2.6.1. *Exterior Helmholtz problems.* Now, using Lemma B.17, we can establish the following uniqueness result for the exterior Helmholtz problem; see, for instance, [**37**, Lemma 2.50].

LEMMA B.18. *Suppose $d = 2$ or 3. Let $\omega \in \mathbb{R}, \omega \neq 0$. Let D be a bounded smooth domain in $\mathbb{R}^d$. Let $u \in H^1_{\text{loc}}(\mathbb{R}^d \setminus \overline{D})$ satisfy*

$$\begin{cases} \Delta u + \omega^2 u = 0 \quad \text{in } \mathbb{R}^d \setminus \overline{D}, \\ \left| \frac{\partial u}{\partial r} - i\omega u \right| = \mathcal{O}\Big(r^{-(d+1)/2}\Big) \quad \text{as } r = |x| \to +\infty \quad \text{uniformly in } \frac{x}{|x|}, \\ u = 0 \text{ or } \frac{\partial u}{\partial \nu} = 0 \quad \text{on } \partial D. \end{cases}$$

Then, $u \equiv 0$ in $\mathbb{R}^d \setminus \overline{D}$.

PROOF. Let $B_R = \{|x| < R\}$. For R large enough, $D \Subset B_R$. Notice first that by multiplying $\Delta u + \omega^2 u = 0$ by $\overline{u}$ and integrating by parts over $B_R \setminus \overline{D}$, we arrive at

$$\operatorname{Im} \int_{\partial B_R} \overline{u} \frac{\partial u}{\partial \nu} \, d\sigma = 0.$$

Thus, we have

$$\operatorname{Im} \int_{\partial B_R} \overline{u} \Big(\frac{\partial u}{\partial \nu} - i\omega u \Big) \, d\sigma = -\omega \int_{\partial B_R} |u|^2 \, dx.$$

Applying the Cauchy–Schwarz inequality,

$$\begin{aligned} &\left| \operatorname{Im} \int_{\partial B_R} \overline{u} \Big(\frac{\partial u}{\partial \nu} - i\omega u \Big) \, d\sigma \right| \\ &\quad \leq \Big(\int_{\partial B_R} |u|^2 \Big)^{1/2} \Big(\int_{\partial B_R} \left| \frac{\partial u}{\partial \nu} - i\omega u \right|^2 d\sigma \Big)^{1/2}, \end{aligned}$$

and using the radiation condition (1.2), we get

$$\left| \operatorname{Im} \int_{\partial B_R} \overline{u} \Big(\frac{\partial u}{\partial \nu} - i\omega u \Big) \, d\sigma \right| \leq \frac{C}{R} \Big(\int_{\partial B_R} |u|^2 \Big)^{1/2},$$

for some positive constant C independent of R. Consequently, we obtain that

$$\Big(\int_{\partial B_R} |u|^2 \Big)^{1/2} \leq \frac{C}{R},$$

which indicates by Rellich's lemma that $u \equiv 0$ in $\mathbb{R}^d \setminus \overline{B_R}$. Hence, by the unique continuation property for $\Delta + \omega^2$, we can conclude that $u \equiv 0$ up to the boundary ∂D. This finishes the proof. □

B.2.6.2. *Transmission problem for the Helmholtz equation.* Let D be a bounded smooth domain in $\mathbb{R}^d$. Let $\widetilde{\kappa}$ and $\widetilde{\rho}$ be two piecewise constant functions such that $\widetilde{\kappa}(x) = \kappa$ and $\widetilde{\rho}(x) = \rho$ for $x \in \mathbb{R}^d \setminus \overline{D}$ and $\widetilde{\kappa}(x) = \kappa_r$ and $\widetilde{\rho}(x) = \rho_r$ for $x \in D$. Suppose that κ, ρ, κ_r, and ρ_r are positive and let $k = \omega\sqrt{\rho/\kappa}$ and $k_r = \omega\sqrt{\rho_r/\kappa_r}$. We consider the following transmission problem for the Helmholtz equation:

$$
\text{(B.69)} \qquad \begin{cases} \nabla \cdot \frac{1}{\widetilde{\rho}} \nabla u + \frac{\omega^2}{\widetilde{\kappa}} u = 0 & \text{in } \mathbb{R}^d, \\ u^s := u - u_{\text{in}} & \text{satisfies the Sommerfeld radiation condition,} \end{cases}
$$

where u_{in} is an incident wave. Notice that (B.69) can be rewritten as

$$
\text{(B.70)} \qquad \begin{cases} (\Delta + k^2) u = 0 & \text{in } \mathbb{R}^d \setminus \overline{D}, \\ (\Delta + k_r^2) u = 0 & \text{in } D, \\ u|_+ = u|_- & \text{on } \partial D, \\ \frac{1}{\rho} \frac{\partial u}{\partial \nu}\Big|_+ = \frac{1}{\rho_r} \frac{\partial u}{\partial \nu}\Big|_- & \text{on } \partial D, \\ u^s := u - u_{\text{in}} & \text{satisfies the Sommerfeld radiation condition.} \end{cases}
$$

Using Rellich's lemma, we can prove that the following uniqueness result holds [**37**, Lemma 2.50].

LEMMA B.19. *If u satisfies* (B.69) *with $u_{\text{in}} \equiv 0$, then $u \equiv 0$ in $\mathbb{R}^d$.*

PROOF. Using the fact that

$$
\int_{\partial D} \frac{\partial u}{\partial \nu}\Big|_+ \bar{u} \, d\sigma = \frac{\rho}{\rho_r} \int_{\partial D} \frac{\partial u}{\partial \nu}\Big|_- \bar{u} \, d\sigma = \frac{\rho}{\rho_r} \int_D (|\nabla u|^2 - k_r^2 |u|^2) \, dx,
$$

we find that

$$
\operatorname{Im} \int_{\partial D} \frac{\partial u}{\partial \nu}\Big|_+ \bar{u} \, d\sigma = 0 \,,
$$

which gives, by applying Lemma B.18, that $u \equiv 0$ in $\mathbb{R}^d \setminus D$. Now, u satisfies $(\Delta + k_r^2) u = 0$ in D and $u = \partial u / \partial \nu = 0$ on ∂D. By the unique continuation property of $\Delta + k_r^2$, we readily get $u \equiv 0$ in D, and hence in $\mathbb{R}^d$. □

The following result is of importance to us for establishing a representation formula for the solution u to (B.69); see [**37**, Proposition 2.51] for its proof.

PROPOSITION B.20. *Suppose that k^2 is not a Dirichlet eigenvalue for $-\Delta$ on D. For each $(F, G) \in H^1(\partial D) \times L^2(\partial D)$, there exists a unique solution $(f, g) \in L^2(\partial D) \times L^2(\partial D)$ to the system of integral equations*

$$
\text{(B.71)} \qquad \begin{cases} \mathcal{S}_D^{k_r}[f] - \mathcal{S}_D^k[g] = F \\ \frac{1}{\rho_r} \frac{\partial (\mathcal{S}_D^{k_r}[f])}{\partial \nu}\Big|_- - \frac{1}{\rho} \frac{\partial (\mathcal{S}_D^k[g])}{\partial \nu}\Big|_+ = G \end{cases} \quad \text{on } \partial D.
$$

Furthermore, there exists a constant C independent of F and G such that

$$
\text{(B.72)} \qquad \|f\|_{L^2(\partial D)} + \|g\|_{L^2(\partial D)} \leq C\Big(\|F\|_{H^1(\partial D)} + \|G\|_{L^2(\partial D)} \Big),
$$

where in the three-dimensional case the constant C can be chosen independently of k and k_r if both k and k_r are sufficiently small.

PROOF. We only give the proof for $d = 3$ and $\rho_r \neq \rho$ leaving the general case to the reader. Let $X := L^2(\partial D) \times L^2(\partial D)$ and $Y := H^1(\partial D) \times L^2(\partial D)$, and define the operator $T : X \to Y$ by

$$T(f, g) := \left(\mathcal{S}_D^{k_r}[f] - \mathcal{S}_D^k[g], \frac{1}{\rho_r} \frac{\partial (\mathcal{S}_D^{k_r}[f])}{\partial \nu}\Big|_- - \frac{1}{\rho} \frac{\partial (\mathcal{S}_D^k[g])}{\partial \nu}\Big|_+ \right).$$

We also define T_0 by

$$T_0(f, g) := \left(\mathcal{S}_D^0[f] - \mathcal{S}_D^0[g], \frac{1}{\rho_r} \frac{\partial (\mathcal{S}_D^0[f])}{\partial \nu}\Big|_- - \frac{1}{\rho} \frac{\partial (\mathcal{S}_D^0[g])}{\partial \nu}\Big|_+ \right).$$

We can easily see that $\mathcal{S}_D^k - \mathcal{S}_D^0 : L^2(\partial D) \to H^1(\partial D)$ is a compact operator, and so are $\frac{\partial}{\partial \nu}(\mathcal{S}_D^{k_r} - \mathcal{S}_D^0)|_\pm$ and $\frac{\partial}{\partial \nu}(\mathcal{S}_D^k - \mathcal{S}_D^0)|_\pm : L^2(\partial D) \to L^2(\partial D)$. Therefore, $T - T_0$ is a compact operator from X into Y. It can be proved that $T_0 : X \to Y$ is invertible. In fact, a solution (f, g) of the equation $T_0(f, g) = (F, G)$ is given by

$$f = g + (\mathcal{S}_D^0)^{-1}(F),$$
$$g = \frac{\rho \rho_r}{\rho - \rho_r}(\lambda I + \mathcal{K}_D^{0,*})^{-1} \left(G + \frac{1}{\rho_r}(\frac{1}{2} I - \mathcal{K}_D^{0,*})((\mathcal{S}_D^0)^{-1}[F]) \right),$$

where $\lambda = (\rho_r + \rho)/(2(\rho_r - \rho))$. From the invertibility of $\mathcal{S}_D^0$ and $\lambda I + \mathcal{K}_D^{0,*}$ in Lemmas B.5 and B.6, we can see, by the Fredholm alternative (Proposition A.4), that it is enough to prove that T is injective.

Suppose that $T(f, g) = 0$. Then the function u defined by

$$u(x) := \begin{cases} \mathcal{S}_D^k[g](x) & \text{if } x \in \mathbb{R}^d \setminus \overline{D}, \\ \mathcal{S}_D^{k_r}[f](x) & \text{if } x \in D, \end{cases}$$

satisfies the transmission problem (B.69) with $u_{\text{in}} \equiv 0$ and hence, by Lemma B.19, $u \equiv 0$ in $\mathbb{R}^d$. In particular, $\mathcal{S}_D^k[g] = 0$ on ∂D. Since $(\Delta + k^2)\mathcal{S}_D^k[g] = 0$ in D and k^2 is not a Dirichlet eigenvalue for $-\Delta$ on D, we have $\mathcal{S}_D^k[g] = 0$ in D, and hence in $\mathbb{R}^d$. It then follows from the jump relation (B.50) that

$$g = \frac{\partial (\mathcal{S}_D^k[g])}{\partial \nu}\Big|_+ - \frac{\partial (\mathcal{S}_D^k[g])}{\partial \nu}\Big|_- = 0 \quad \text{on } \partial D.$$

On the other hand, $\mathcal{S}_D^{k_r}[f]$ satisfies $(\Delta + k_r^2)\mathcal{S}_D^{k_r}[f] = 0$ in $\mathbb{R}^d \setminus \overline{D}$ and $\mathcal{S}_D^{k_r}[f] = 0$ on ∂D. Then it follows from Lemma B.18 that $\mathcal{S}_D^{k_r}[f] = 0$. Then, in the same way as above, we can conclude that $f = 0$. This finishes the proof of solvability of (B.71). The estimate (B.72) is an easy consequence of solvability and the closed graph theorem. Finally, it can be easily proved in the three-dimensional case that if both k and k_r are sufficiently small, then the constant C in (B.72) can be chosen independently of k and k_r. We leave the details to the reader. □

Using Proposition B.20, the following representation formula holds [**37**, Theorem 2.52].

THEOREM B.21. *Suppose that k^2 is not a Dirichlet eigenvalue for $-\Delta$ on D. Let u be the solution of* (B.69). *Then u can be represented using the single layer potentials $\mathcal{S}_D^k$ and $\mathcal{S}_D^{k_r}$ as follows:*

$$u(x) = \begin{cases} u_{\text{in}}(x) + \mathcal{S}_D^k[\phi](x), & x \in \mathbb{R}^3 \setminus \overline{D}, \\ \mathcal{S}_D^{k_r}[\psi](x), & x \in D, \end{cases} \tag{B.73}$$

where the pair $(\phi, \psi) \in L^2(\partial D) \times L^2(\partial D)$ *is the unique solution to*

$$\text{(B.74)} \qquad \begin{cases} \mathcal{S}_D^{k_r}[\psi] - \mathcal{S}_D^k[\phi] = u_{\text{in}} \\ \dfrac{1}{\rho_r} \dfrac{\partial (\mathcal{S}_D^{k_r}[\psi])}{\partial \nu}\bigg|_- - \dfrac{1}{\rho} \dfrac{\partial (\mathcal{S}_D^k[\phi])}{\partial \nu}\bigg|_+ = \dfrac{1}{\rho} \dfrac{\partial u_{\text{in}}}{\partial \nu} \end{cases} \quad \textit{on } \partial D.$$

Moreover, there exists $C > 0$ *independent of* u_{in} *such that*

$$\text{(B.75)} \qquad \|\phi\|_{L^2(\partial D)} + \|\psi\|_{L^2(\partial D)} \leq C\Big(\|u_{\text{in}}\|_{L^2(\partial D)} + \|\nabla u_{\text{in}}\|_{L^2(\partial D)} \Big).$$

B.3. Laplace eigenvalues

In this section, we transform eigenvalue problems of $-\Delta$ on an open bounded connected domain D either Neumann boundary conditions into the problem of determining the characteristic values of certain integral operator-valued functions in the complex plane. This results in a considerable advantage as it allows us to reduce the dimension of the eigenvalue problem.

B.3.1. Eigenvalue characterisation. We first restrict our attention to the three-dimensional case. We note that due to the holomorphic dependence of G^ω on ω as given in (1.7), $\mathcal{K}_D^\omega$ is an operator-valued holomorphic function in $\mathbb{C}$. Indeed, the following result holds. See, for example, [**152**].

PROPOSITION B.22 (Neumann eigenvalue characterisation). *Let* $d = 3$. *Suppose that* D *is bounded with a boundary in* $\mathcal{C}^{1,s}$ *for some* $0 < s < 1$. *Let* $\omega > 0$. *Then* ω^2 *is an eigenvalue of* $-\Delta$ *on* D *with Neumann boundary condition if and only if* ω *is a positive real characteristic value of the operator* $-(1/2)\, I + \mathcal{K}_D^\omega$ *on* $L^2(\partial D)$.

PROOF. Suppose that ω^2 is an eigenvalue of

$$\text{(B.76)} \qquad \begin{cases} \Delta u + \omega^2 u = 0 & \text{in } D, \\ \frac{\partial u}{\partial \nu} = 0 & \text{on } \partial D. \end{cases}$$

By Green's formula (B.48), we have

$$u(x) = \mathcal{D}_D^\omega[u|_{\partial D}](x), \quad x \in D.$$

It then follows from (B.51) that $(-I/2 + \mathcal{K}_D^\omega)[u|_{\partial D}] = 0$ and $u|_{\partial D} \neq 0$ since otherwise the unique continuation property for $\Delta + \omega^2$ would imply that $u \equiv 0$ in D. Thus, ω is a characteristic value of $-(1/2)\, I + \mathcal{K}_D^\omega$ on $L^2(\partial D)$.

Suppose now that ω is a characteristic value of $-(1/2)\, I + \mathcal{K}_D^\omega$; *i.e.*, there is a nonzero $\psi \in L^2(\partial D)$ such that

$$\left(-\frac{1}{2} I + \mathcal{K}_D^\omega \right)[\psi] = 0.$$

Then $u = \mathcal{D}_D^\omega[\psi]$ on $\mathbb{R}^d \setminus \overline{D}$ is a solution to the Helmholtz equation with the boundary condition $u|_+ = 0$ on ∂D and satisfies the radiation condition (1.2). The uniqueness result in Lemma B.18 implies that $\mathcal{D}_D^\omega[\psi] = 0$ in $\mathbb{R}^d \setminus \overline{D}$. Since $\partial \mathcal{D}_D^\omega[\psi]/\partial \nu$ exists and has no jump across ∂D, we get

$$\frac{\partial \mathcal{D}_D^\omega[\psi]}{\partial \nu}\bigg|_- = \frac{\partial \mathcal{D}_D^\omega[\psi]}{\partial \nu}\bigg|_+ = 0 \quad \text{on } \partial D.$$

Hence, we deduce that $\mathcal{D}_D^\omega[\psi]$ is a solution of (B.76). Note that $\mathcal{D}_D^\omega[\psi] \neq 0$ in D, since otherwise

$$\psi = \mathcal{D}_D^\omega[\psi]\big|_- - \mathcal{D}_D^\omega[\psi]\big|_+ = 0.$$

Thus, ω^2 is an eigenvalue of $-\Delta$ on D with Neumann condition, and so the proposition is proved. □

Proposition B.22 asserts that $-(1/2)\,I + \mathcal{K}_D^\omega$ is invertible on $L^2(\partial D)$ for all positive ω except for a discrete set. The following result, whose proof can be found in [**152**, Proposition 7.3], shows that $(-(1/2)\,I + \mathcal{K}_D^\omega)^{-1}$ has a continuation as an operator-valued meromorphic function in $\mathbb{C}$. This follows from the fact that the Neumann eigenvalues are of finite multiplicities.

PROPOSITION B.23. *$-(1/2)\,I + \mathcal{K}_D^\omega$ is invertible on $L^2(\partial D)$ for all $\omega \in \mathbb{C}$ except for a discrete set, and $(-(1/2)\,I+\mathcal{K}_D^\omega)^{-1}$ is an operator-valued meromorphic function on $\mathbb{C}$.*

PROOF. The operator $\mathcal{K}_D^\omega : L^2(\partial D) \to L^2(\partial D)$ is compact and holomorphic in ω. Moreover, if ω^2 (for ω real) is not an eigenvalue of $-\Delta$ on D with Neumann boundary condition, then by Proposition B.22, $-(1/2)\,I + \mathcal{K}_D^\omega$ is invertible on $L^2(\partial D)$. Therefore, by the generalised Steinberg's theorem (Theorem A.17), it follows that $-(1/2)\,I + \mathcal{K}_D^\omega$ is invertible on $L^2(\partial D)$ for all $\omega \in \mathbb{C}$ except for a discrete set, and $(-(1/2)\,I + \mathcal{K}_D^\omega)^{-1}$ is an operator-valued meromorphic function on $\mathbb{C}$. □

In the two-dimensional case, Proposition B.22 also holds true. Moreover, due to the logarithmic behaviour of the Hankel function as shown by (B.43),

$$(-(1/2)\,I + \mathcal{K}_D^\omega)^{-1}$$

has a continuation to an operator-valued meromorphic function on only $\mathbb{C} \setminus \mathrm{i}\mathbb{R}^-$.

B.3.2. Neumann function. Let $0 = \mu_1 < \mu_2 \leq \mu_3 \leq \ldots$ be the eigenvalues of $-\Delta$ on D with Neumann conditions on ∂D. Let $\{u_j\}_{j\geq 1}$ denote an orthonormal basis of $L^2(D)$ of eigenfunctions associated with the eigenvalues $\{\mu_j\}_{j\geq 1}$. Note that $\mu_1 = 0$ and $u_1(x) = 1/\sqrt{|D|}$ in D. Let $\omega \notin \{\sqrt{\mu_j}\}_{j\geq 1}$. Introduce $N_D^\omega(x,z)$ as the Neumann function for $\Delta + \omega^2$ in D corresponding to a Dirac mass at z. That is, N_D^ω is the unique solution to

$$\text{(B.77)} \qquad \begin{cases} (\Delta_x + \omega^2) N_D^\omega(x,z) = \delta_z & \text{in } D, \\ \dfrac{\partial N_D^\omega}{\partial \nu}\Big|_{\partial D} = 0 & \text{on } \partial D. \end{cases}$$

We derive useful facts on the Neumann function. First, we note that N_D^ω has the same singularity as G^0 at 0. Then we establish the following proposition, providing a purely formal proof. We refer the reader to [**145**, Theorem 9.8] for a more rigorous one where even the case $\omega = 0$ is treated.

PROPOSITION B.24 (Spectral decomposition). *The following spectral decomposition holds pointwise:*

$$\text{(B.78)} \qquad N_D^\omega(x,z) = \sum_{j=1}^{+\infty} \frac{u_j(x)u_j(z)}{\omega^2 - \mu_j}, \quad x \neq z \in D.$$

PROOF. Consider the function

$$f(x) := \sum_{j=1}^{+\infty} a_j u_j(x), \quad x \in D.$$

If $(\Delta_x + \omega^2) f(x) = \delta_z(x)$, then we have

$$\sum_{j=1}^{+\infty} a_j(\omega^2 - \mu_j) u_j(x) = \delta_z(x).$$

Integrating both sides of the above identity against u_k over D gives

$$a_k(\omega^2 - \mu_k) = u_k(z),$$

and hence (B.78) follows. Here note that we used the orthogonality relation

$$\int_D u_j\, u_k \; \mathrm{d}x = \delta_{jk}$$

satisfied by the eigenfunctions, where δ_{jk} denotes the Kronecker symbol. □

Finally, we provide an important relation between the operator $-(1/2)\, I + \mathcal{K}_D^{\omega,*}$ and the Neumann function N_D^ω. Note that the Neumann function N_D^ω yields a solution operator for the Neumann problem for the Helmholtz equation. In fact, the function u defined by

$$u(x) := -\int_{\partial D} N_D^\omega(z,x) g(z) \; \mathrm{d}\sigma(z), \quad x \in D,$$

is the unique solution to the following Helmholtz equation:

$$\text{(B.79)} \qquad \begin{cases} \Delta u + \omega^2 u = 0 & \text{in } D, \\ \frac{\partial u}{\partial \nu} = g & \text{on } \partial D, \end{cases}$$

provided that ω^2 is not an eigenvalue of $-\Delta$ on D with Neumann boundary condition. On the other hand, under this assumption, $-(1/2)\, I + \mathcal{K}_D^{\omega,*} : L^2(\partial D) \to L^2(\partial D)$ is invertible since by Proposition B.22, $1/2$ is not an eigenvalue of the compact operator $\mathcal{K}_D^\omega$. So we can readily see that the solution to (B.79) can be represented as

$$\text{(B.80)} \qquad u(x) = \mathcal{S}_D^\omega \left(-\frac{1}{2} I + \mathcal{K}_D^{\omega,*} \right)^{-1} [g](x), \quad x \in D.$$

Therefore, assuming that ω^2 is not a Dirichlet eigenvalue of $-\Delta$ on D, we obtain that

$$\left(-\frac{1}{2} I + \mathcal{K}_D^{\omega,*} \right)^{-1} [g](x) = (\mathcal{S}_D^\omega)^{-1} \big[u|_{\partial D} \big](x), \quad x \in \partial D,$$

and hence, from (B.78) for $x, z \in \partial D$, we have the following proposition.

PROPOSITION B.25. *For ω in a small neighbourhood of* 0,

$$\left(-\frac{1}{2} I + \mathcal{K}_D^{\omega,*} \right)^{-1} = -\frac{1}{\omega^2 |D|} (\mathcal{S}_D^\omega)^{-1} \big[\chi_{\partial D}\big] \langle \chi_{\partial D}, \cdot \rangle + \mathcal{R}(\omega),$$

where $\mathcal{R}(\omega)$ is holomorphic in ω. Moreover, from (B.55), *it follows that*

$$\begin{aligned}\left(-\frac{1}{2}I+\mathcal{K}_D^{\omega,*}\right)^{-1} &= \left(-\frac{1}{\omega^2|D|}(\mathcal{S}_D^0)^{-1}[\chi_{\partial D}]\right.\\ &\left.+\frac{1}{\omega|D|}(\mathcal{S}_D^0)^{-1}\mathcal{S}_{D,1}(\mathcal{S}_D^0)^{-1}[\chi_{\partial D}]\right)\langle\chi_{\partial D},\cdot\rangle+\widetilde{\mathcal{R}}(\omega),\end{aligned} \tag{B.81}$$

where $\widetilde{\mathcal{R}}(\omega)$ is holomorphic in ω.

B.4. Quasiperiodic layer potentials

B.4.1. Quasiperiodic Green's functions. In this section we collect some notation and well known results regarding quasiperiodic layer potentials for the Helmholtz equation. We refer to [**71, 120, 127, 134, 147**] for details. These results are used for the analysis of subwavelength bandgap structures. For simplicity, here we only consider the fully periodic case (*i.e.*, the crystal case).

We let $l_1, \ldots, l_d \in \mathbb{R}^d$, for $d = 2, 3$, denote the lattice vectors generating the lattice Λ:

$$\Lambda := \{m_1 l_1 + \cdots + m_d l_d \mid m_i \in \mathbb{Z}\},$$

and denote by Y a fundamental domain of Λ:

$$Y := \{c_1 l_1 + \cdots + c_{d_l} l_d \mid 0 \le c_1, \ldots, c_d \le 1\}.$$

We recall that the dual lattice of Λ, denoted by Λ^*, is generated by $\alpha_1, \ldots, \alpha_d$ satisfying $\alpha_i \cdot l_j = 2\pi\delta_{ij}$, for $i, j = 1, \ldots, d$, and the Brillouin zone Y^* is defined as $Y^* := \mathbb{R}^d/\Lambda^*$. Moreover, for $\alpha \in Y^*$, we say that a smooth function u is α-quasiperiodic if $u(x+m) = e^{\mathrm{i}\alpha\cdot m}u(x)$ for all $x \in \mathbb{R}^d$ and $m \in \Lambda$.

We introduce the quasiperiodic Green function (or fundamental solution) $G^{\alpha,\omega}$, which satisfies

$$(\Delta+\omega^2)G^{\alpha,\omega}(x,y) = \sum_{m\in\Lambda}\delta_0(x-y-m)e^{\mathrm{i}m\cdot\alpha} \quad \text{in } \mathbb{R}^d. \tag{B.82}$$

If $\omega \neq |q+\alpha|, \forall\, q \in \Lambda^*$, then by using Poisson's summation formula

$$\frac{1}{|Y|}\sum_{q\in\Lambda^*} e^{\mathrm{i}(q+\alpha)\cdot x} = \sum_{m\in\Lambda}\delta_0(x-m)e^{\mathrm{i}m\cdot\alpha}, \tag{B.83}$$

the quasiperiodic Green function $G^{\alpha,\omega}$ can be represented as a sum of augmented plane waves over the reciprocal lattice:

$$G^{\alpha,\omega}(x,y) = \frac{1}{|Y|}\sum_{q\in\Lambda^*}\frac{e^{\mathrm{i}(q+\alpha)\cdot(x-y)}}{\omega^2-|q+\alpha|^2}. \tag{B.84}$$

Moreover, it can also be shown that $G^{\alpha,\omega}$ can be alternatively represented as a sum of images:

$$G^{\alpha,\omega}(x,y) = \sum_{m\in\Lambda} G^{\omega}(x-m-y)e^{\mathrm{i}m\cdot\alpha}, \tag{B.85}$$

where G^ω is the free-space Helmholtz Green function given by (1.7). The series in the spatial representation (B.85) of the Green function converges uniformly for x, y in compact sets of $\mathbb{R}^d$ and $\omega \neq |q+\alpha|$ for all $q \in \Lambda^*$; see, *e.g.*, [**37**, Section 2.12].

From (B.85), it follows that $G^{\alpha,\omega}(x,y) - G^0(x-y)$ is smooth for all $x, y \in Y$. A disadvantage of the form (B.84), which is often referred to as the spectral representation of the Green function, is that the singularity as $|x-y| \to 0$ is not explicit.

In all the sequel, we assume that $\omega \neq |q+\alpha|$ for all $q \in \Lambda^*$. Let $D \Subset Y$ be a domain in $\mathbb{R}^d$, with a boundary ∂D in $C^{1,s}$ with some $0 < s < 1$. Let ν denote the unit outward normal to ∂D. For $\omega > 0$ let $\mathcal{S}_D^{\alpha,\omega}$ and $\mathcal{D}_D^{\alpha,\omega}$ be the quasiperiodic single and double layer potentials associated with $G^{\alpha,\omega}$ on D; that is, for a given density $\phi \in L^2(\partial D)$,

$$\mathcal{S}_D^{\alpha,\omega}[\phi](x) = \int_{\partial D} G^{\alpha,\omega}(x,y)\phi(y)\, d\sigma(y), \quad x \in \mathbb{R}^d,$$

$$\mathcal{D}_D^{\alpha,\omega}[\phi](x) = \int_{\partial D} \frac{\partial G^{\alpha,\omega}(x,y)}{\partial \nu(y)}\phi(y)\, d\sigma(y), \quad x \in \mathbb{R}^d \setminus \partial D.$$

Then, $\mathcal{S}_D^{\alpha,\omega}[\phi]$ and $\mathcal{D}_D^{\alpha,\omega}[\phi]$ satisfy $(\Delta+\omega^2)\mathcal{S}_D^{\alpha,\omega}[\phi] = (\Delta+\omega^2)\mathcal{D}_D^{\alpha,\omega}[\phi] = 0$ in D and $Y \setminus \overline{D}$ and they are α-quasiperiodic.

The next formulas give the jump relations obeyed by the double layer potential and by the normal derivative of the single layer potential on ∂D:

$$\text{(B.86)} \qquad \frac{\partial (\mathcal{S}_D^{\alpha,\omega}[\phi])}{\partial \nu}\bigg|_{\pm}(x) = \Big(\pm \frac{1}{2} I + (\mathcal{K}_D^{-\alpha,\omega})^* \Big)[\phi](x) \quad \text{a.e. } x \in \partial D,$$

$$\text{(B.87)} \qquad (\mathcal{D}_D^{\alpha,\omega}[\phi])\Big|_{\pm}(x) = \Big(\mp \frac{1}{2} I + \mathcal{K}_D^{\alpha,\omega} \Big)[\phi](x) \quad \text{a.e. } x \in \partial D,$$

for $\phi \in L^2(\partial D)$, where $\mathcal{K}_D^{\alpha,\omega}$ is the operator on $L^2(\partial D)$ defined by

$$\text{(B.88)} \qquad \mathcal{K}_D^{\alpha,\omega}[\phi](x) = \int_{\partial D} \frac{\partial G^{\alpha,\omega}(x,y)}{\partial \nu(y)}\phi(y)\, d\sigma(y)$$

and $(\mathcal{K}_D^{-\alpha,\omega})^*$ is the L^2-adjoint operator of $\mathcal{K}_D^{-\alpha,\omega}$, which is given by

$$\text{(B.89)} \qquad (\mathcal{K}_D^{-\alpha,\omega})^*[\phi](x) = \int_{\partial D} \frac{\partial G^{\alpha,\omega}(x,y)}{\partial \nu(x)}\phi(y)\, d\sigma(y).$$

The singular integral operators $\mathcal{K}_D^{\alpha,\omega}$ and $(\mathcal{K}_D^{-\alpha,\omega})^*$ are bounded on $L^2(\partial D)$ as an immediate consequence of the fact that $G^{\alpha,\omega}(x,y) - G^0(x-y)$ is smooth for all $x, y \in Y$.

The following lemma is of use to us. In particular, it shows that $\mathcal{S}_D^{\alpha,0} : L^2(\partial D) \to H^1(\partial D)$ is invertible for $\alpha \neq 0$.

LEMMA B.26. *Suppose that $\alpha \neq 0$ and ω^2 is neither an eigenvalue of $-\Delta$ in D with the Dirichlet boundary condition on ∂D nor in $Y \setminus \overline{D}$ with the Dirichlet boundary condition on ∂D and the α-quasiperiodic condition on ∂Y. Then $\mathcal{S}_D^{\alpha,\omega} : L^2(\partial D) \to H^1(\partial D)$ is invertible.*

PROOF. Let $d = 2$. We write

$$\text{(B.90)} \qquad \mathcal{S}_D^{\alpha,\omega} = \mathcal{S}_D^0 + (\mathcal{S}_D^{\alpha,\omega} - \mathcal{S}_D^0).$$

Since $\mathcal{S}_D^0 : L^2(\partial D) \to H^1(\partial D)$ is invertible and $\mathcal{S}_D^{\alpha,\omega} - \mathcal{S}_D^0 : L^2(\partial D) \to H^1(\partial D)$ is compact, by the Fredholm alternative, injectivity of $\mathcal{S}_D^{\alpha,\omega}$ implies its invertibility.

Suppose that $\phi \in L^2(\partial D)$ satisfies $\mathcal{S}_D^{\alpha,\omega}[\phi] = 0$ on ∂D. Then $u = \mathcal{S}_D^{\alpha,\omega}[\phi]$ satisfies $(\Delta+\omega^2)u = 0$ in D and in $Y \setminus \overline{D}$. Therefore, since ω^2 is neither an eigenvalue of $-\Delta$ in D with the Dirichlet boundary condition nor in $Y \setminus \overline{D}$ with the Dirichlet boundary condition on ∂D and the quasiperiodic condition on ∂Y, it follows that $u = 0$ in Y and thus, $\phi = \partial u/\partial \nu|_+ - \partial u/\partial \nu|_- = 0$, as desired.

For $d = 2$, the proof is exactly the same after changing in (B.90) $\mathcal{S}_D^0$ with $\widetilde{\mathcal{S}}_D^0$ given by

$$\widetilde{\mathcal{S}}_D^0[\psi] = \begin{cases} \mathcal{S}_D^0[\psi] & \text{if } \psi \in L_0^2(\partial D), \\ -\chi_{\partial D} & \text{if } \psi = \varphi_0\,, \end{cases} \tag{B.91}$$

where $L_0^2(\partial D)$ is the set of L^2 functions on ∂D with mean-value zero and φ_0 is the unique normalised (in $L^2(\partial D)$) eigenfunction of $\mathcal{K}_D^{0,*}$ associated with eigenvalue 1/2. □

Define

$$G^{\alpha,0}(x,y) := -\frac{1}{|Y|}\sum_{q\in\Lambda^*}\frac{e^{\mathrm{i}(q+\alpha)\cdot(x-y)}}{|q+\alpha|^2} \quad \text{for } \alpha \neq 0,$$

and

$$G^{0,0}(x,y) := -\frac{1}{|Y|}\sum_{q\in\Lambda^*\setminus\{0\}}\frac{e^{\mathrm{i}q\cdot(x-y)}}{|q|^2}.$$

Note that $G^{\alpha,0}(x,y)$ for $\alpha \neq 0$ is a Green function of the quasiperiodic Laplacian in Y:

$$\Delta_x G^{\alpha,0}(x,y) = \delta_y(x) \quad \text{in } Y, \tag{B.92}$$

while $G^{0,0}(x,y)$ satisfies

$$\Delta_x G^{0,0}(x,y) = \delta_y(x) - \frac{1}{|Y|} \quad \text{in } Y \tag{B.93}$$

with periodic Dirichlet boundary conditions on ∂Y.

For α close to zero, we then have

$$G^{\alpha,0}(x,y) = \frac{-1}{|Y||\alpha|^2} - \frac{\mathrm{i}\alpha\cdot(x-y)}{|Y||\alpha|^2} + \frac{(\alpha\cdot(x-y))^2}{2|Y||\alpha|^2} + G^{0,0}(x,y) + \mathcal{O}(|\alpha|).$$

Consequently, for α close to zero, we have the following expansion of the single layer potential $\mathcal{S}_D^{\alpha,0}$:

$$\begin{aligned}\mathcal{S}_D^{\alpha,0}[\psi](x) = &-\frac{1}{|Y||\alpha|^2}\int_{\partial D}\psi(y)\, \mathrm{d}\sigma - \frac{\mathrm{i}}{|Y||\alpha|^2}\int_{\partial D}\alpha\cdot(x-y)\psi(y)\, \mathrm{d}\sigma \\ &+ \frac{1}{2|Y||\alpha|^2}\int_{\partial D}\big(\alpha\cdot(x-y)\big)^2\psi(y)\, \mathrm{d}\sigma + \mathcal{S}_D^{0,0}[\psi](x) + \mathcal{O}(|\alpha|).\end{aligned} \tag{B.94}$$

LEMMA B.27. *If $\mathcal{S}_D^{0,0}[\varphi] = K\chi_{\partial D}$ for some constant K and some $\varphi \in L^2(\partial D)$ satisfying $\int_{\partial D}\varphi\, \mathrm{d}\sigma = 0$, then $\varphi = 0$.*

PROOF. Let $\mathcal{D} = \sum_{m\in\Lambda}(D+m)$. For $x \in \mathbb{R}^3 \setminus \mathcal{D}$, define $V(x) := \mathcal{S}_D^{0,0}[\varphi](x)$. Then V solves the following differential problem:

$$\begin{cases} \Delta V = 0 & \text{in } \mathbb{R}^3 \setminus \mathcal{D}, \\ V|_+ = K & \text{on } \partial\mathcal{D}, \\ V(x+m) = V(x) & \text{for all } m \in \Lambda. \end{cases} \tag{B.95}$$

Moreover, using the jump relations and integration by parts, we have that

$$\int_{\partial D}\varphi\, \mathrm{d}\sigma = K\int_{Y\setminus D}|\nabla V|^2\, \mathrm{d}x = 0.$$

If $K \neq 0$, then it follows from (B.95) that $\int_{Y\setminus D} |\nabla V|^2 \, \mathrm{d}x \neq 0$, which is a contradiction. In other words, we must have $K = 0$, so that $\mathcal{S}_D^{0,0}[\varphi] = 0$ and $\int_{\partial D} \varphi \, \mathrm{d}\sigma = 0$. Therefore, we have that $\varphi = 0$. □

LEMMA B.28. *$\|(\mathcal{S}_D^{\alpha,0})^{-1}\|$, in operator norm, is bounded for α in a neighbourhood of 0.*

PROOF. To reach a contradiction, we assume that $\mathcal{S}_D^{\alpha,0}[\phi] = \mathcal{O}(|\alpha|)$ for some ϕ, which can be written as $\phi = \phi_0 + |\alpha|\phi_1$, where ϕ_0 is nonzero, does not depend on α, and $\phi_1 = \mathcal{O}(1)$ as $|\alpha| \to 0$. Also, define $v = \frac{\alpha}{|\alpha|}$. From (B.94) it follows that

$$\int_{\partial D} \phi_0 \, \mathrm{d}\sigma = 0,$$

$$\int_{\partial D} \phi_1(y) \, \mathrm{d}\sigma + \mathrm{i}\int_{\partial D} v\cdot(x-y)\phi_0(y) \, \mathrm{d}\sigma = \mathcal{O}(|\alpha|),$$

$$K(v) - \mathrm{i}v\cdot x\int_{\partial D} \phi_1(y) \, \mathrm{d}\sigma + \frac{1}{2}\int_{\partial D} \big(v\cdot(x-y)\big)^2\phi_0(y) \, \mathrm{d}\sigma + |Y|\mathcal{S}_D^{0,0}[\phi_0] = \mathcal{O}(|\alpha|),$$

where K is constant as function of x. Simplifying, we have that

$$\frac{1}{2}\int_{\partial D} \big(v\cdot(x-y)\big)^2\phi_0(y) \, \mathrm{d}\sigma = -(v\cdot x)\int_{\partial D} (v\cdot y)\phi_0(y) \, \mathrm{d}\sigma + \frac{1}{2}\int_{\partial D} (v\cdot y)^2\phi_0(y) \, \mathrm{d}\sigma.$$

In total, we get

$$\mathcal{S}_D^{0,0}[\phi_0](x) = \widetilde{K}(v) + \frac{2(v\cdot x)}{|Y|}\int_{\partial D} (v\cdot y)\phi_0(y) \, \mathrm{d}\sigma,$$

where $\widetilde{K}$ is constant in x. Observe that $\mathcal{S}_D^{0,0}[\phi_0](x)$ is independent of v. As a function of x, this function is constant for $x \in v^\perp$, and so this function is constant for all x. From Lemma B.27 we get that $\phi_0 = 0$ which proves the claim. □

The following lemma gives a complete low-frequency asymptotic expansion of $G^{\alpha,\omega}$. It is easy to prove.

LEMMA B.29. *As $\omega \to 0$, $G^{\alpha,\omega}$ can be decomposed as*

$$G^{\alpha,\omega}(x,y) = G^{\alpha,0}(x,y) - \frac{1}{|Y|}\sum_{l=1}^{+\infty}\omega^{2l}\underbrace{\sum_{q\in\Lambda^*}\frac{e^{\mathrm{i}(q+\alpha)\cdot(x-y)}}{|q+\alpha|^{2(l+1)}}}_{:=-G_l^{\alpha,0}(x,y)}, \tag{B.96}$$

for $\alpha \neq 0$, while for $\alpha = 0$, the following decomposition holds:

$$G^{0,\omega}(x,y) = \frac{1}{\omega^2} + G^{0,0}(x,y) - \frac{1}{|Y|}\sum_{l=1}^{+\infty}\omega^{2l}\underbrace{\sum_{q\in\Lambda^*\setminus\{0\}}\frac{e^{\mathrm{i}q\cdot(x-y)}}{|q|^{2(l+1)}}}_{:=-G_l^{0,0}(x,y)}. \tag{B.97}$$

The next result is also of use to us.

LEMMA B.30. *Assume that the lattice vector l_1 is aligned with the first coordinate axis. For $(\alpha_1,\ldots,\alpha_{d_l}) \in Y^*$ such that $(\alpha_2,\ldots,\alpha_{d_l}) \neq 0$, the partial derivative of the quasiperiodic Green function*

$$\frac{\partial}{\partial\alpha_1}G^{\alpha,0}(0)$$

is zero precisely when $\alpha_1 = 0$ *or* $\alpha_1 = \pi/|l_1|$.

PROOF. From the spectral form of the quasiperiodic Green function

$$G^{\alpha,0}(x) = -\frac{1}{|Y|} \sum_{q \in \Lambda^*} \frac{e^{\mathrm{i}(\alpha+q)\cdot x}}{|\alpha+q|^2},$$

it can be easily shown that

$$\nabla_\alpha G^{\alpha,0}(0) = \frac{1}{|Y|} \sum_{q \in \Lambda^*} \frac{\alpha+q}{|\alpha+q|^4}.$$

By the symmetry of the summation, we find that

$$\frac{\partial}{\partial \alpha_1} G^{\alpha,0}(0) = 0$$

if and only if $\alpha_1 = 0$ or $\alpha_1 = \pi/|l_1|$. □

Denote by $\mathcal{S}_{D,l}^{\alpha}$ and $(\mathcal{K}_{D,l}^{-\alpha})^*$, for $l \geq 0$ and $\alpha \in Y^*$, the layer potentials associated with the kernel $G_l^{\alpha,0}(x,y)$ so that

$$\text{(B.98)} \qquad \mathcal{S}_D^{\alpha,\omega} = \mathcal{S}_D^{\alpha,0} + \sum_{l=1}^{+\infty} \omega^{2l} \mathcal{S}_{D,l}^{\alpha} \quad \text{and} \quad (\mathcal{K}_D^{\alpha})^* = (\mathcal{K}_D^{\alpha,0})^* + \sum_{l=1}^{+\infty} \omega^{2l} (\mathcal{K}_{D,l}^{-\alpha})^*.$$

LEMMA B.31. *The operator* $(1/2)\, I + (\mathcal{K}_D^{-\alpha,0})^* : L^2(\partial D) \to L^2(\partial D)$ *is invertible.*

Before proving Lemma B.31, let us make a note of the following simple fact. If u and v are α-quasiperiodic smooth functions, then

$$\text{(B.99)} \qquad \int_{\partial Y} \frac{\partial u}{\partial \nu} \overline{v} \; d\sigma = 0.$$

To prove this, it is enough to see that

$$\int_{\partial Y} \frac{\partial u}{\partial \nu} \overline{v} \; d\sigma = \int_{\partial Y} \left[\frac{\partial (u e^{-\mathrm{i}\alpha\cdot x})}{\partial \nu} + \mathrm{i}\alpha \cdot \nu u e^{-\mathrm{i}\alpha\cdot x} \right] \overline{e^{-\mathrm{i}\alpha\cdot x} v} \; dx.$$

PROOF OF LEMMA B.31. Let $\phi \in L^2(\partial D)$ satisfy $\left((1/2)\, I + (\mathcal{K}_D^{-\alpha,0})^*\right)[\phi] = 0$ on ∂D. Observe that if $\alpha = 0$, then $\int_{\partial D} \phi \; d\sigma = 0$. In fact, by (B.93) and (B.99), we have

$$\mathcal{D}_D^{0,0}[1](x) = -\int_{Y\setminus\overline{D}} \Delta_y G^{0,0}(x,y) \; dy = |Y \setminus \overline{D}|, \qquad \text{for } x \in D,$$

where $|\ |$ denotes the volume, and hence

$$\text{(B.100)} \qquad (\frac{1}{2} I + \mathcal{K}_D^{0,0})[1] = |Y \setminus \overline{D}| \quad \text{on } \partial D.$$

Therefore, we have

$$|Y \setminus \overline{D}| \int_{\partial D} \phi \; d\sigma = \int_{\partial D} (\frac{1}{2} I + \mathcal{K}_D^{0,0})[1] \phi \; d\sigma = \int_{\partial D} (\frac{1}{2} I + (\mathcal{K}_D^{0,0})^*)[\phi] \; d\sigma = 0.$$

Consequently, for any $\alpha \in Y^*$, $u = \mathcal{S}_D^{\alpha,0}[\phi]$ is α-quasiperiodic and satisfies $\Delta u = 0$ in $Y \setminus \overline{D}$ with

$$\left.\frac{\partial u}{\partial \nu}\right|_+ = (\frac{1}{2} I + (\mathcal{K}_D^{-\alpha,0})^*)[\phi] = 0 \quad \text{on } \partial D.$$

Therefore, it follows from (B.99) that

$$\int_{Y\setminus\overline{D}} |\nabla u|^2 \, dx = \int_{\partial Y} \frac{\partial u}{\partial \nu}\overline{u} \, d\sigma - \int_{\partial D} \frac{\partial u}{\partial \nu}\bigg|_+ \overline{u} \, d\sigma = 0.$$

Thus, u is constant in $Y \setminus \overline{D}$ and hence in D. This implies that

$$\phi = \frac{\partial u}{\partial \nu}\bigg|_+ - \frac{\partial u}{\partial \nu}\bigg|_- = 0,$$

as desired. □

LEMMA B.32. *For $\alpha \neq 0$, let $\psi \in L^2(\partial D)$ be defined by*

$$\psi = \left(\mathcal{S}_D^{\alpha,0}\right)^{-1} [\chi_{\partial D}]. \tag{B.101}$$

Then, we have

(i)
$$\mathrm{Ker}\left(-\frac{1}{2}I + (\mathcal{K}_D^{-\alpha,0})^*\right) = \mathrm{span}\{\psi\},$$

and

$$\mathrm{Ker}\left(-\frac{1}{2}I + \mathcal{K}_D^{\alpha,0}\right) = \mathrm{span}\{\chi_{\partial D}\};$$

(ii) *For any $\phi \in L^2(\partial D)$,*

$$\int_{\partial D} \left(-\frac{1}{2}I + (\mathcal{K}_D^{-\alpha,0})^*\right)[\phi] \, d\sigma = 0; \tag{B.102}$$

(iii) *For any $\phi \in L^2(\partial D)$,*

$$\int_{\partial D} \mathcal{K}_{D,1}^{\alpha,0}[\phi](y) \, d\sigma(y) = -\int_D \mathcal{S}_D^{\alpha,0}[\phi](x) \, dx. \tag{B.103}$$

PROOF. To prove (i), we first observe that the jump relation (B.86) implies that

$$\mathrm{span}\{\psi\} \subset \mathrm{Ker}\left(-\frac{1}{2}I + (\mathcal{K}_D^{-\alpha,0})^*\right).$$

Conversely, if $\left(-\frac{1}{2}I + (\mathcal{K}_D^{-\alpha,0})^*\right)[\psi] = 0$, we define $u(x) = \mathcal{S}_D^{\alpha,0}[\psi]$ and conclude from the jump relations that

$$\begin{cases} \Delta u = 0 & \text{in } D, \\ \frac{\partial u}{\partial \nu} = 0 & \text{on } \partial D. \end{cases}$$

It follows that u is constant in D, so $u|_{\partial D} = a\chi_{\partial D}$. Therefore,

$$\mathrm{Ker}\left(-\frac{1}{2}I + (\mathcal{K}_D^{-\alpha,0})^*\right) \subset \mathrm{span}\{\psi\}.$$

The second equality of (i) follows from the first, combined with the Calderón identity

$$\mathcal{S}_D^{\alpha,0}(\mathcal{K}_D^{-\alpha,0})^* = \mathcal{K}_D^{\alpha,0}\mathcal{S}_D^{\alpha,0},$$

which can be proved analogously to (B.52).

For the proof of (ii), we use the jump relation (B.86) and integration by parts. Then

$$\int_{\partial D} \left(-\frac{1}{2}I + (\mathcal{K}_D^{-\alpha,0})^*\right)[\phi] \, d\sigma = \int_{\partial D} \frac{\partial \mathcal{S}_D^{\alpha,0}}{\partial \nu}\bigg|_- [\phi] \, d\sigma = \int_D \Delta \mathcal{S}_D^{\alpha,0}[\phi] \, dx = 0.$$

To prove (iii), we use (ii) to conclude that, on one hand,

$$\int_{\partial D}\left(-\frac{1}{2}I+(\mathcal{K}_D^{-\alpha,\omega})^*\right)[\phi]\ \mathrm{d}\sigma=\omega^2\int_{\partial D}\mathcal{K}_{D,1}^{\alpha,0}[\phi](y)\ \mathrm{d}\sigma(y)+\mathcal{O}(\omega^4\|\phi\|_{L^2(\partial D)}).$$

On the other hand, as in the proof of (ii) we have

$$\begin{aligned}\int_{\partial D}\left(-\frac{1}{2}I+(\mathcal{K}_D^{-\alpha,\omega})^*\right)[\phi]\ \mathrm{d}\sigma&=-\omega^2\int_D\mathcal{S}_D^{\alpha,k}[\phi]\ \mathrm{d}x\\&=-\omega^2\int_D\mathcal{S}_D^{\alpha,0}[\phi]\ \mathrm{d}x+\mathcal{O}(\omega^3\|\phi\|_{L^2(\partial D)}),\end{aligned}$$

where we have used the expansion (B.98). Combined, we have

$$\int_{\partial D}\mathcal{K}_{D,1}^{\alpha,0}[\phi](y)\ \mathrm{d}\sigma(y)=-\int_D\mathcal{S}_D^{\alpha,0}[\phi](x)\ \mathrm{d}x+\mathcal{O}(\omega\|\phi\|_{L^2(\partial D)}),$$

and since the leading-orders on the left- and right-hand sides are independent of ω, they must coincide. □

Finally, we derive asymptotic expansions of the α-quasiperiodic Green functions and layer potentials for α near some $\alpha^*\neq 0$. It is easy to prove that $G^{\alpha,\omega}(x)-G^{\alpha^*,\omega}(x)$ is continuously differentiable in α for $\alpha\in Y^*,\alpha\neq 0$, and bounded for $x\in Y$ and for ω in a neighbourhood of 0. We therefore have the following asymptotic expansion of $G^{\alpha,\omega}$:

(B.104)

$$\begin{aligned}G^{\alpha,\omega}(x)&=G^{\alpha^*,\omega}(x)\\&+\sum_{q\in\Lambda^*}\frac{e^{\mathrm{i}(\alpha^*+q)\cdot x}}{\omega^2-|\alpha^*+q|^2}\left(\mathrm{i}x\cdot(\alpha-\alpha^*)+2\frac{(\alpha^*+q)\cdot(\alpha-\alpha^*)}{\omega^2-|\alpha^*+q|^2}\right)+\mathcal{O}(|\alpha-\alpha^*|^2),\end{aligned}$$

uniformly for ω in a neighbourhood of 0 and for $x\in Y$. We define G_1^ω by

$$G_1^\omega(x):=\sum_{q\in\Lambda^*}\frac{e^{\mathrm{i}(\alpha^*+q)\cdot x}}{\omega^2-|\alpha^*+q|^2}\left(\mathrm{i}x+\frac{2(\alpha^*+q)}{\omega^2-|\alpha^*+q|^2}\right),$$

and the integral operators $\mathcal{S}_1^\omega$, $\mathcal{K}_1^\omega$ as

$$\mathcal{S}_1^\omega[\phi](x):=\int_{\partial D}G_1^\omega(x-y)\phi(y)\ \mathrm{d}\sigma(y),$$

and

$$\mathcal{K}_1^\omega[\phi](x):=\int_{\partial D}\frac{\partial}{\partial\nu(x)}G_1^\omega(x-y)\phi(y)\ \mathrm{d}\sigma(y).$$

We then have the expansions

$$\mathcal{S}_D^{\alpha,\omega}=\mathcal{S}_D^{\alpha^*,\omega}+\mathcal{S}_1^\omega\cdot(\alpha-\alpha^*)+\mathcal{O}(|\alpha-\alpha^*|^2),\tag{B.105}$$

$$(\mathcal{K}_D^{-\alpha,\omega})^*=(\mathcal{K}_D^{-\alpha^*,\omega})^*+\mathcal{K}_1^\omega\cdot(\alpha-\alpha^*)+\mathcal{O}(|\alpha-\alpha^*|^2),\tag{B.106}$$

uniformly for ω in a neighbourhood of 0, where the error terms are with respect to the corresponding operator norm.

B.4.2. Quasiperiodic transmission problem. With the same notation as in Appendix B.2.6, we consider the following α-quasiperiodic transmission problem for the Helmholtz equation:

$$\text{(B.107)}\qquad \begin{cases} (\Delta + k^2)u = 0 & \text{in } Y \setminus \overline{D}, \\ (\Delta + k_r^2)u = 0 & \text{in } D, \\ u|_+ = u|_- & \text{on } \partial D, \\ \displaystyle \frac{1}{\rho}\frac{\partial u}{\partial \nu}\bigg|_+ = \frac{1}{\rho_r}\frac{\partial u}{\partial \nu}\bigg|_- & \text{on } \partial D, \\ u & \text{is } \alpha\text{-quasiperiodic.} \end{cases}$$

Suppose that k_r^2 is not an eigenvalue of $-\Delta$ in $Y\setminus\overline{D}$ with the Dirichlet boundary condition on ∂D and the quasiperiodic condition on ∂Y and k^2 is not an eigenvalue of $-\Delta$ in D with the Dirichlet boundary condition. The solution u to (B.107) can be represented as

$$\text{(B.108)}\qquad u(x) = \begin{cases} \mathcal{S}_D^{\alpha,k_r}[\psi](x), & x \in D, \\ H(x) + \mathcal{S}_D^{\alpha,k}[\phi](x), & x \in Y \setminus \overline{D}, \end{cases}$$

for some densities ψ and ϕ in $L^2(\partial D)$, where the function H is given by

$$H(x) = -\mathcal{S}_Y^{\alpha,k}[\frac{\partial u}{\partial \nu}|_{\partial Y}] + \mathcal{D}_Y^{\alpha,k}[u|_{\partial Y}], \quad x \in Y.$$

Note that $(\Delta + k^2)H = 0$ in Y and hence, by integrating by parts,

$$H(x) = -\int_{\partial Y} \frac{\partial H}{\partial \nu}(y)\overline{G^{\alpha,k}}(x,y)\, dy + \int_{\partial Y} H(y)\frac{\partial G^{\alpha,k}}{\partial \nu}(x,y)\, dy, \quad x \in Y,$$

and thus, by (B.99), $H \equiv 0$. Therefore,

$$\text{(B.109)}\qquad u(x) = \begin{cases} \mathcal{S}_D^{\alpha,k_r}[\psi](x), & x \in D, \\ \mathcal{S}_D^{\alpha,k}[\phi](x), & x \in Y \setminus \overline{D}. \end{cases}$$

Now, in view of the transmission conditions in (B.107), the pair $(\phi,\psi) \in L^2(\partial D) \times L^2(\partial D)$ satisfies the following system of integral equations:

$$\text{(B.110)}\qquad \begin{cases} \mathcal{S}_D^{\alpha,k_r}[\psi] - \mathcal{S}_D^{\alpha,k}[\phi] = 0 & \text{on } \partial D, \\ \Big(-\frac{1}{2}I + (\mathcal{K}_D^{-\alpha,k_r})^*\Big)[\psi] - \frac{\rho_r}{\rho}\Big(\frac{1}{2}I + (\mathcal{K}_D^{-\alpha,k})^*\Big)[\phi] = 0 & \text{on } \partial D. \end{cases}$$

The converse is also true. If $(\phi,\psi) \in L^2(\partial D) \times L^2(\partial D)$ is a nonzero solution of (B.110), then u given by (B.109) is an eigenfunction of (B.107) associated with the eigenvalue ω^2.

THEOREM B.33. *Suppose that k_r^2 is not an eigenvalue for $-\Delta$ in $Y \setminus \overline{D}$ with Dirichlet boundary condition on ∂D and quasiperiodic boundary condition on ∂Y and assume that k^2 is not an eigenvalue for $-\Delta$ in D with Dirichlet boundary condition on ∂D. Then, for any eigenfunction u of* (B.107)*, there exists one and only one pair $(\phi,\psi) \in L^2(\partial D) \times L^2(\partial D)$ such that u has the representation* (B.109)*. Moreover, (ϕ,ψ) is the solution to the integral equation* (B.110)*. The mapping $u \longmapsto (\phi,\psi)$ from solutions of* (B.107) *to solutions of the system of integral equations* (B.110) *is one-to-one.*

In order to prove Theorem B.33, we first prove the following lemma.

LEMMA B.34. *Suppose that u is an eigenfunction of* (B.107). *Then*

$$u|_{\partial D} \perp \operatorname{Ker} \mathcal{S}_D^{-\alpha,k_r}.$$

Here, $\mathcal{S}_D^{-\alpha,k_r}$ is considered as an operator from $L^2(\partial D)$ into $H^1(\partial D)$.

PROOF. To prove this lemma, we observe that, since $(\Delta + k_r^2)u = 0$ in D,

$$u(x) = \mathcal{D}_D^{\alpha,k_r}[u|_{\partial D}](x) - \mathcal{S}_D^{\alpha,k_r}\left[\frac{\partial u}{\partial \nu}\Big|_-\right](x), \quad x \in D,$$

and consequently,

$$\frac{1}{2}u|_{\partial D} = \mathcal{K}_D^{\alpha,k_r}[u|_{\partial D}] - \mathcal{S}_D^{\alpha,k_r}\left[\frac{\partial u}{\partial \nu}\Big|_-\right].$$

Let $\phi \in \operatorname{Ker} \mathcal{S}_D^{-\alpha,k_r}$. Because of the assumption on k_r^2, we immediately deduce that $\mathcal{S}_D^{-\alpha,k_r}[\psi] = 0$ in $Y \setminus \overline{D}$, and hence

$$\text{(B.111)} \qquad \begin{cases} \mathcal{S}_D^{-\alpha,k_r}[\psi] = 0, \\ \frac{1}{2}\psi + (\mathcal{K}_D^{\alpha,k_r})^*[\psi] = 0 \end{cases} \quad \text{on } \partial D.$$

Then, we have

$$\begin{aligned} \frac{1}{2}\langle u|_{\partial D}, \phi\rangle &= \left\langle \mathcal{K}_D^{\alpha,k_r}[u|_{\partial D}], \psi\right\rangle - \left\langle \mathcal{S}_D^{\alpha,k_r}\left[\frac{\partial u}{\partial \nu}\Big|_-\right], \psi\right\rangle \\ &= \left\langle u|_{\partial D}, (\mathcal{K}_D^{\alpha,k_r})^*[\psi]\right\rangle - \left\langle \frac{\partial u}{\partial \nu}\Big|_-, \mathcal{S}_D^{-\alpha,k_r}[\psi]\right\rangle \\ &= -\frac{1}{2}\langle u|_{\partial D}, \psi\rangle - 0, \end{aligned}$$

where $\langle\cdot,\cdot\rangle$ denotes the L^2 scalar product, which proves the lemma. □

PROOF OF THEOREM B.33. We first note that the problem of finding (ϕ, ψ) is equivalent to solving the two equations

$$\text{(B.112)} \qquad \begin{cases} \mathcal{S}_D^{\alpha,k_r}[\psi] = u|_{\partial D} & \text{on } \partial D, \\ \left(-\frac{1}{2}I + (\mathcal{K}_D^{-\alpha,k_r})^*\right)[\psi] = \frac{\partial u}{\partial \nu}\Big|_- & \text{on } \partial D, \end{cases}$$

and

$$\text{(B.113)} \qquad \begin{cases} \mathcal{S}_D^{\alpha,k}[\phi] = u|_{\partial D} & \text{on } \partial D, \\ \left(\frac{1}{2}I + (\mathcal{K}_D^{-\alpha,k})^*\right)[\phi] = \frac{\partial u}{\partial \nu}\Big|_+ & \text{on } \partial D. \end{cases}$$

Here we only consider the problem of finding ψ, the solution to (B.112); the problem of finding ϕ can be solved in the same way.

From Lemma B.34 it follows that there exists $\psi_0 \in L^2(\partial D)$ such that

$$\mathcal{S}_D^{\alpha,k_r}[\psi_0 + \psi] = u|_{\partial D} \quad \text{on } \partial D, \quad \forall\, \phi \in \operatorname{Ker} \mathcal{S}_D^{\alpha,k_r}.$$

Hence, to show existence of a solution to (B.112), it suffices to prove that there exists $\psi \in \operatorname{Ker} \mathcal{S}_D^{\alpha,k_r}$ such that the second equation in (B.112) is satisfied by $\psi_0 + \psi$. Thanks to the second equation in (B.111), this equation becomes

$$\text{(B.114)} \qquad \psi = \frac{\partial \left(\mathcal{S}_D^{\alpha,k_r}[\psi_0] - u\right)}{\partial \nu}\Big|_-,$$

and then, we only need to show that

$$\frac{\partial \left(\mathcal{S}_D^{\alpha,k_r}[\psi_0]-u\right)}{\partial \nu}\Big|_- \in \operatorname{Ker} \mathcal{S}_D^{\alpha,k_r},$$

which is an immediate consequence of the fact that $\mathcal{S}_D^{\alpha,k_r}[\psi_0]-u$ is a solution to $\Delta+k_r^2$ in D with Dirichlet boundary condition. We have then proved the existence of a solution to (B.112).

Suppose now that we have two solutions ψ_1 and ψ_2 to (B.112). Then, because of the assumption on k_r^2, we have $\mathcal{S}_D^{\alpha,k_r}[\psi_1-\psi_2]=0$ in $Y\setminus\overline{D}$, and hence

$$\big(\frac{1}{2}I+(\mathcal{K}_D^{-\alpha,k_r})^*\big)[\psi_1-\psi_2]=0 \quad \text{on } \partial D.$$

By the second equation in (B.112), we have $\psi_1=\psi_2$.

So far, we have shown that there are unique ϕ and ψ satisfying (B.112) and (B.113), respectively. The jump conditions satisfied by u immediately show that the pair (ϕ,ψ) satisfies the system of integral equations (B.110).

Conversely, suppose that (ϕ,ψ) is a nontrivial solution to the system of integral equations (B.110). Then defining u by (B.109), we only need to show that u is not trivial to conclude that u is an eigenfunction of (B.107). Suppose that $u=0$ in Y. Then $\mathcal{S}_D^{\alpha,k_r}[\psi]=0$ in D, and by the assumption on k_r^2, we deduce that $\mathcal{S}_D^{\alpha,k_r}[\psi]=0$ in $Y\setminus\overline{D}$. Finally, from the jump of the normal derivative of $\mathcal{S}_D^{\alpha,k_r}[\psi]$ on ∂D, we deduce that $\psi=0$. The assumption on k^2 leads to $\phi=0$. This is in contradiction to the fact that $(\phi,\psi)\neq(0,0)$. The proof is then complete. □

B.4.3. Concept of quasiperiodic capacity.

DEFINITION B.35. *For $\alpha\in Y^*,\alpha\neq 0$ and $D\Subset Y$ with smooth boundary ∂D, we define the α-quasiperiodic capacity of ∂D as*

$$\text{(B.115)} \qquad \operatorname{Cap}_{\alpha,D}:=-\int_{\partial D}(\mathcal{S}_D^{\alpha,0})^{-1}[\chi_{\partial D}]\,\mathrm{d}\sigma.$$

If u^α is the α-quasiperiodic solution of the Laplace equation with Dirichlet boundary condition on ∂D

$$\text{(B.116)} \qquad \begin{cases}\Delta u^\alpha=0 \quad \text{in } Y\setminus\overline{D},\\ u^\alpha=1 \quad \text{on } \partial D,\\ u^\alpha \text{ is } \alpha\text{-quasiperiodic},\end{cases}$$

then $u^\alpha:=\mathcal{S}_D^{\alpha,0}[(\mathcal{S}_D^{\alpha,0})^{-1}[\chi_{\partial D}]]$ in $Y\setminus\overline{D}$.

Therefore, by the jump relation (B.86), it follows that

$$\int_{\partial D}\frac{\partial u}{\partial\nu}\Big|_+\,\mathrm{d}\sigma=\int_{\partial D}\big(\frac{1}{2}I+(\mathcal{K}_D^{-\alpha,0})^*\big)[(\mathcal{S}_D^0)^{-1}[\chi_{\partial D}]]\,\mathrm{d}\sigma=\int_{\partial D}(\mathcal{S}_D^0)^{-1}[\chi_{\partial D}]\,\mathrm{d}\sigma,$$

since, by (B.92), $\mathcal{K}_D^{\alpha,0}[1]=1/2$ for $\alpha\neq 0$. Thus, the α-quasiperiodic capacity is given by

$$\text{(B.117)} \quad \operatorname{Cap}_{\alpha,D}=-\int_{\partial D}\frac{\partial u^\alpha}{\partial\nu}\Big|_+\,\mathrm{d}\sigma=-\int_{\partial D}\frac{\partial u^\alpha}{\partial\nu}\Big|_+\overline{u^\alpha}\,\mathrm{d}\sigma=\int_{Y\setminus\overline{D}}|\nabla u^\alpha|^2\,\mathrm{d}x,$$

by multiplying (B.116) by $\overline{u^\alpha}$ and integrating by parts over $Y\setminus\overline{D}$.

For $\alpha\neq 0$, the solution u^α of (B.116) can be characterised as the solution of a minimisation problem. Let $C_\alpha^\infty(Y)$ be the set of functions in Y which can be extended to C^∞ α-quasiperiodic functions in $\mathbb{R}^d$. Let $\mathcal{H}_\alpha$ be the closure of the set $C^\infty_\alpha(Y)$ in H^1, and let $\mathcal{V}_\alpha := \{v \in \mathcal{H}_\alpha, v = 1 \text{ on } \partial D\}$. Then, by the variational principle, we can show that

$$\int_{Y\setminus\overline{D}} |\nabla u^\alpha|^2 \, \mathrm{d}x = \inf_{v\in\mathcal{V}_\alpha} \int_{Y\setminus\overline{D}} |\nabla v|^2 \, \mathrm{d}x.$$

Hence we obtain the following variational characterisation of the quasiperiodic capacity.

LEMMA B.36. *For $\alpha \neq 0$, we have*

$$\mathrm{Cap}_{\alpha,D} = \inf_{v\in\mathcal{V}_\alpha} \int_{Y\setminus\overline{D}} |\nabla v|^2 \, \mathrm{d}x.$$

Next, we state an asymptotic behaviour of $\mathrm{Cap}_{\alpha,D}$ when $D = \varepsilon B$ for a small $\varepsilon > 0$. Note that $\mathrm{Cap}_D = \mathrm{Cap}_{\varepsilon B} = \varepsilon\, \mathrm{Cap}_B$. Fix $c > 0$, the following holds.

LEMMA B.37. *For $|\alpha| > c > 0$, we have*

$$\mathrm{Cap}_{\alpha,D} = \mathrm{Cap}_D - \mathcal{R}_\alpha(0)\mathrm{Cap}_D^2 + \mathcal{O}(\varepsilon^3), \tag{B.118}$$

with

$$\mathcal{R}_\alpha(x) := G^{\alpha,0}(x) - G^0(x) = \sum_{m\in\Lambda\setminus\{0\}} G^0(x-m)e^{\mathrm{i}m\cdot\alpha},$$

where $G^{\alpha,0}$ is the quasiperiodic Green function associated to the Laplacian defined by (B.85) *for $\omega = 0$.*

PROOF. Since $\mathcal{R}_\alpha(x)$ is smooth and $\mathcal{R}_\alpha(x) = \mathcal{R}_\alpha(0) + \mathcal{O}(|x|)$ as $|x| \to 0$, we have

$$\begin{aligned}
\mathcal{S}_D^{\alpha,0}[\phi](\varepsilon x) &= \varepsilon \int_{\partial B} G^0(x-y)\widetilde{\phi}(y) \, \mathrm{d}\sigma(y) \\
&\quad + \varepsilon^2\mathcal{R}_\alpha(0) \int_{\partial B} \widetilde{\phi}(y) \, \mathrm{d}\sigma(y) + \mathcal{O}\left(\varepsilon^3\|\widetilde{\phi}\|\right) \\
&= \varepsilon\left(\mathcal{S}_B^0[\widetilde{\phi}] + \varepsilon\mathcal{R}_\alpha(0)\int_{\partial B} \widetilde{\phi} \, \mathrm{d}\sigma + \mathcal{O}\left(\varepsilon^2\|\widetilde{\phi}\|\right)\right),
\end{aligned} \tag{B.119}$$

with $\widetilde{\phi}(x) := \phi(\varepsilon x)$. Then

$$\begin{aligned}
(\mathcal{S}_D^{\alpha,0})^{-1}[\chi_{\partial D}](\varepsilon x) = \varepsilon^{-1}\Big(&(\mathcal{S}_B^0)^{-1}[\chi_{\partial B}] - \varepsilon\mathcal{R}_\alpha(0)(\mathcal{S}_B^0)^{-1}\left[\int_{\partial B}(\mathcal{S}_B^0)^{-1}[\chi_{\partial B}] \, \mathrm{d}\sigma\right] \\
&+ \mathcal{O}\left(\varepsilon^2\right)\Big),
\end{aligned}$$

and so

$$\begin{aligned}
\mathrm{Cap}_{\alpha,D} &= -\varepsilon^2 \int_{\partial B} (\mathcal{S}_D^{\alpha,0})^{-1}[\chi_{\partial D}](\varepsilon x) \, \mathrm{d}\sigma(x) \\
&= \varepsilon\left(\mathrm{Cap}_B - \varepsilon\mathcal{R}_\alpha(0)\mathrm{Cap}_B^2 + \mathcal{O}\left(\varepsilon^2\right)\right) \\
&= \mathrm{Cap}_D - \mathcal{R}_\alpha(0)\mathrm{Cap}_D^2 + \mathcal{O}(\varepsilon^3).
\end{aligned}$$

□

B.5. Helmholtz operator with an imaginary gauge potential

We provide a Green's function for the Helmholtz operator with an imaginary gauge potential $\Delta + \omega^2 + \gamma\partial_{x_1}$, where $\gamma \in \mathbb{R}, \gamma \neq 0$.

PROPOSITION B.38. *Introduce*

$$G_\gamma^\omega(x) = -\frac{e^{-\gamma x_1/2 + \mathrm{i}\sqrt{\omega^2-\gamma^2/4}|x|}}{4\pi|x|}. \tag{B.120}$$

Then, G_γ^ω satisfies

$$\Delta G_\gamma^\omega + \omega^2 G_\gamma^\omega + \gamma\partial_{x_1} G_\gamma^\omega = \delta_0 \qquad \textit{in } \mathbb{R}^3. \tag{B.121}$$

PROOF. Observe that $F_\gamma^\omega(x) = -\frac{e^{\mathrm{i}\sqrt{\omega^2-\gamma^2/4}|x|}}{4\pi|x|}$ is a Green's function of the Helmholtz operator $\Delta + \omega^2 - \gamma^2/4$. Letting

$$G_\gamma^\omega(x) = F_\gamma^\omega(x) e^{-\gamma x_1/2}, \tag{B.122}$$

we have

$$\begin{aligned}\left(\Delta + \omega^2 - \gamma^2/4\right) F_\gamma^\omega(x) &= e^{\gamma x_1/2}\left(\Delta + \omega^2 + \gamma\partial_{x_1}\right) G_\gamma^\omega(x)\\ &= \delta_0(x),\end{aligned}$$

which proves the claim. □

Observe that G_γ^ω does not satisfy the vanishing boundary condition as $|x| \to +\infty$; nevertheless, as G_γ^ω is used to represent the solution inside the resonators, the behaviour of G_γ^ω as $|x| \to +\infty$ does not affect the solution.

We now define the single layer potential associated to G_γ^ω by

$$\widetilde{\mathcal{S}}_{D,\gamma}^\omega : \phi \in L^2(\partial D) \mapsto \int_{\partial D} G_\gamma^\omega(x-y)\phi(y)\,\mathrm{d}\sigma(y). \tag{B.123}$$

We also define the Neumann-Poincaré operator associated to (B.120) by

$$\widetilde{\mathcal{K}}_{D,\gamma}^{\omega,*}[\phi](x) = \int_{\partial D} \frac{\partial}{\partial\nu(x)} G_\gamma^\omega(x-y)\phi(y)\,\mathrm{d}\sigma(y), \quad x \in \partial D, \phi \in L^2(\partial D). \tag{B.124}$$

Since G_γ^ω has the same singularity as G^ω at the origin, we can obtain the following jump conditions for the normal derivative of $\widetilde{\mathcal{S}}_{D,\gamma}^\omega$ across ∂D.

LEMMA B.39. *The Neumann-Poincaré operator $\widetilde{\mathcal{K}}_{D,\gamma}^{\omega,*}$ satisfies*

$$\frac{\partial}{\partial\nu}\widetilde{\mathcal{S}}_{D,\gamma}^\omega[\phi]\Big|_\pm = \left(\pm\frac{1}{2}I + \widetilde{\mathcal{K}}_{D,\gamma}^{\omega,*}\right)[\phi]. \tag{B.125}$$

The following result holds.

PROPOSITION B.40. *The operator $\widetilde{\mathcal{S}}_{D,\gamma}^0 : L^2(\partial D) \to L^2(\partial D)$ is injective and hence has a left inverse*

PROOF. We assume that $\widetilde{\mathcal{S}}_{D,\gamma}^0[\phi](x) = 0$ for $x \in \partial D$. For $x \in \mathbb{R}^3$, we define u and v as

$$u(x) = \widetilde{\mathcal{S}}_{D,\gamma}^0[\phi](x) \quad \text{and} \quad v(x) = e^{\gamma x_1/2}u(x).$$

Since $\Delta u + \gamma \partial_{x_1} u = 0$, we find that v satisfies the *modified* Helmholtz equation

$$\begin{cases} \Delta v - \frac{\gamma^2}{4} v = 0, \\ v|_{\partial D} = 0, \\ v(x) = \mathcal{O}\left(|x|^{-1}\right) \quad \text{as } |x| \to +\infty. \end{cases} \tag{B.126}$$

The modified Helmholtz equation is well known to have a unique solution; indeed, if we multiply (B.126) with $\overline{v}$ and integrate by parts, we have

$$\int_{\mathbb{R}^3} \left(|\nabla v|^2 + \frac{\gamma^2}{2} |v|^2 \right) \, \mathrm{d}x = 0,$$

which shows that $v(x) = 0$, and hence $u(x) = 0$. Finally, from Lemma B.39, we have that

$$\phi = \frac{\partial u}{\partial \nu}\bigg|_+ - \frac{\partial u}{\partial \nu}\bigg|_- = 0,$$

which proves the claim. □

LEMMA B.41. *If $\mathcal{D}$ consists of N connected components D_i, i.e., $\mathcal{D} = \bigcup_{i=1}^N D_i$, then the kernel of the operator $-(1/2)\,I + \widetilde{\mathcal{K}}_{\mathcal{D},\gamma}^{0,*} : L^2(\partial\mathcal{D}) \to L^2(\partial\mathcal{D})$ is N-dimensional and spanned by*

$$\psi_i = (\widetilde{\mathcal{S}}_{\mathcal{D},\gamma}^0)^{-1}[\chi_{\partial D_i}], \; i = 1, \ldots, N. \tag{B.127}$$

Here, $\chi_{\partial D_i}$ denotes the indicator function of ∂D_i.

PROOF. Let ψ_i be defined by (B.127). Introduce $u = \widetilde{\mathcal{S}}_{\mathcal{D},\gamma}^0[\psi_i]$. From (B.121), the function u satisfies the following equation:

$$\Delta u + \gamma \partial_1 u = 0 \quad \text{in } D_i. \tag{B.128}$$

Moreover, $u = 1$ on ∂D_i. Denote by $v(x) = u(x) - 1$ in D_i. Then, by integrating by parts over D_i, we have

$$\begin{aligned} 0 &= \int_{D_i} (v \Delta v + \gamma v \partial_{x_1} v) \, \mathrm{d}x \\ &= \int_{\partial D_i} v \frac{\partial v}{\partial \nu} \, \mathrm{d}\sigma - \int_{D_i} |\nabla v|^2 \, \mathrm{d}x + \frac{\gamma}{2} \int_{D_i} \partial_{x_1} v^2 \, \mathrm{d}x \\ &= \int_{\partial D_i} v \frac{\partial v}{\partial \nu} \, \mathrm{d}\sigma - \int_{D_i} |\nabla v|^2 \, \mathrm{d}x + \frac{\gamma}{2} \int_{\partial D_i} \nu_1 v^2 \, \mathrm{d}\sigma, \end{aligned} \tag{B.129}$$

where ν_1 is the first component of the outgoing normal to ∂D_i. Since v vanishes on ∂D_i, the first and third terms in (B.129) are zero. We can conclude that $\nabla v = 0$ and $v = 0$. It follows that $u = 1$ in D_i. Therefore, by the jump relation (B.125),

$$\frac{\partial u}{\partial \nu}\bigg|_- = \left(-\frac{1}{2} I + \widetilde{\mathcal{K}}_{\mathcal{D},\gamma}^{0,*} \right) [\psi_i] = 0.$$

Conversely, assume that $\psi \in L^2(\partial\mathcal{D})$ satisfies

$$\left(-\frac{1}{2} I + \widetilde{\mathcal{K}}_{\mathcal{D},\gamma}^{0,*} \right) [\psi] = 0.$$

Then, $u = \widetilde{\mathcal{S}}_{\mathcal{D},\gamma}^0[\psi]$ satisfies (B.128) in $\mathcal{D}$ together with the Neumann boundary condition $\partial u / \partial \nu|_- = 0$ on $\partial\mathcal{D}$. Let $F(x) = e^{\gamma x_1} \nabla u(x)$ for $x \in \mathcal{D}$. Then F satisfies the following curl-div system of equations in $\mathcal{D}$ with a vanishing normal component on $\partial\mathcal{D}$:

$$\begin{cases} \nabla \cdot F = 0 & \text{in } \mathcal{D}, \\ \nabla \times (e^{-\gamma x_1} F) = 0 & \text{in } \mathcal{D}, \\ F \cdot \nu = 0 & \text{on } \partial\mathcal{D}. \end{cases} \tag{B.130}$$

From the uniqueness of a solution to (B.130) (see, for instance, [**4**, Sect. 2.2]), it follows that u is constant inside each connected component D_i of $\mathcal{D}$. Therefore,

$$\psi_i = \sum_{i=1}^{N} c_i (\widetilde{\mathcal{S}}^0_{\mathcal{D},\gamma})^{-1}[\chi_{\partial D_i}],$$

for some constants c_i. □

The following result follows from Lemma B.41.

LEMMA B.42. *The kernel of the operator* $-(1/2)\, I + \widetilde{\mathcal{K}}^0_{\mathcal{D},\gamma} : L^2(\partial\mathcal{D}) \to L^2(\partial\mathcal{D})$ *is spanned by*

$$\phi_i = e^{\gamma x_1} \chi_{\partial D_i}, \ i = 1, \ldots, N. \tag{B.131}$$

PROOF. For any $\psi \in L^2(\partial\mathcal{D})$, by using the jump conditions (B.125), we have

$$\begin{aligned} \Big\langle \psi, \big(-\frac{1}{2} I + \widetilde{\mathcal{K}}^0_{\mathcal{D},\gamma}\big)[\phi_i] \Big\rangle &= \int_{\partial D_i} e^{\gamma x_1} \frac{\partial}{\partial \nu} \widetilde{\mathcal{S}}^0_{\mathcal{D},\gamma}[\psi] \Big|_- \, d\sigma \\ &= \int_{D_i} \Big(e^{\gamma x_1} \Delta \widetilde{\mathcal{S}}^0_{\mathcal{D},\gamma}[\psi] + \nabla(e^{\gamma x_1}) \cdot \nabla \widetilde{\mathcal{S}}^0_{\mathcal{D},\gamma}[\psi] \Big) \, dx \\ &= \int_{D_i} e^{\gamma x_1} \Big(\Delta \widetilde{\mathcal{S}}^0_{\mathcal{D},\gamma}[\psi] + \gamma \partial_{x_1} \widetilde{\mathcal{S}}^0_{\mathcal{D},\gamma}[\psi] \Big) \, dx \\ &= 0, \end{aligned}$$

where $\langle \ , \ \rangle$ denotes the L^2 scalar product on ∂D_i.

On the other hand, since for $\gamma = 0$, $-(1/2)\, I + \mathcal{K}^0_{\mathcal{D}} : L^2(\partial\mathcal{D}) \to L^2(\partial\mathcal{D})$ is Fredholm of index zero and $\widetilde{\mathcal{K}}^0_{\mathcal{D},\gamma} - \mathcal{K}^0_{\mathcal{D}} : L^2(\partial\mathcal{D}) \to L^2(\partial\mathcal{D})$ is compact (as G^0_γ and G^0 have the same singularity at the origin), the operator

$$-(1/2)\, I + \widetilde{\mathcal{K}}^0_{\mathcal{D},\gamma} : L^2(\partial\mathcal{D}) \to L^2(\partial\mathcal{D})$$

is also Fredholm of index zero. From Lemma B.41, it then follows that the dimension of the kernel of $-(1/2)\, I + \widetilde{\mathcal{K}}^0_{\mathcal{D},\gamma}$ is N. This completes the proof. □

B.6. Multipole expansion method

Here we derive the multipole expansion approximation of the single layer potentials $\mathcal{S}^k_D$ and $\mathcal{S}^{\alpha,k}_D$ in both two and three dimensions. The overarching principle is that when working on circular or spherical domains, the action of the single layer potential on Fourier or spherical basis functions has an explicit representation.

B.6.1. The two-dimensional case. For a radially symmetric Helmholtz equation, the cylindrical waves $J_n(kr)e^{in\theta}$ and $H_n^{(1)}(kr)e^{in\theta}$ give a basis of solutions in the polar coordinates (r,θ). When D is a circular disk of radius R, in order to derive a multipole expansion of the single layer potential $\mathcal{S}_D^k[\varphi]$, we expand the density φ into its Fourier series

$$\varphi = \sum_{n\in\mathbb{Z}} a_n e^{\mathrm{i}n\theta},$$

for some coefficients a_n. Hence, we only need to compute $u := \mathcal{S}_D^k[e^{\mathrm{i}n\theta}]$ which satisfies

$$\begin{cases} \Delta u + k^2 u = 0 & \text{in } \mathbb{R}^2 \setminus \overline{D}, \\ \Delta u + k^2 u = 0 & \text{in } D, \\ u|_+ = u|_- & \text{on } \partial D, \\ \left.\frac{\partial u}{\partial \nu}\right|_+ - \left.\frac{\partial u}{\partial \nu}\right|_- = e^{in\theta} & \text{on } \partial D, \\ |x|^{1/2}\left(\frac{\partial}{\partial |x|} - \mathrm{i}k\right) u \to 0 & \text{as } |x| \to +\infty. \end{cases} \tag{B.132}$$

The above equation can be easily solved by the separation of variables technique in polar coordinates. It gives

$$\mathcal{S}_D^k[e^{\mathrm{i}n\theta}] = \begin{cases} cJ_n(kR)H_n^{(1)}(kr)e^{in\theta}, & r \geq R, \\ cH_n^{(1)}(kR)J_n(kr)e^{in\theta}, & r < R, \end{cases} \tag{B.133}$$

where $c = -\mathrm{i}\pi R/2$.

Now, we consider the two-dimensional crystal of subwavelength resonators in Section 4.2 and compute the quasiperiodic single layer potential $\mathcal{S}_D^{\alpha,k}$ associated with the quasiperiodic Green function

$$G^{\alpha,k}(x,y) = -\frac{\mathrm{i}}{4}\sum_{m\in\mathbb{Z}^2} H_0^{(1)}(k|x-y-m|)e^{\mathrm{i}m\cdot\alpha}.$$

We have

$$\begin{aligned} \mathcal{S}_D^{\alpha,k}[e^{\mathrm{i}n\theta}] &= \mathcal{S}_D^k[e^{\mathrm{i}n\theta}] + \sum_{m\in\mathbb{Z}^2, m\neq 0} \mathcal{S}_{D+m}^k[e^{\mathrm{i}n\theta}]e^{\mathrm{i}m\cdot\alpha} \\ &= \mathcal{S}_D^k[e^{\mathrm{i}n\theta}] + cJ_n(kR)\sum_{m\in\mathbb{Z}^2, m\neq 0} H_n^{(1)}(kr_m)e^{\mathrm{i}n\theta_m}e^{\mathrm{i}m\cdot\alpha}. \end{aligned}$$

Here, $D+m$ means a translation of the disk D by m and (r_m,θ_m) is the polar coordinates with respect to the centre of $D+m$. By applying the following addition theorem:

$$H_n^{(1)}(kr_m)e^{\mathrm{i}n\theta_m} = \sum_{l\in\mathbb{Z}}(-1)^{n-l}H_{n-l}^{(1)}(k|m|)e^{\mathrm{i}n\arg(m)}J_l(kr)e^{\mathrm{i}l\theta},$$

we obtain

$$\mathcal{S}_D^{\alpha,k}[e^{\mathrm{i}n\theta}] = \mathcal{S}_D^k[e^{\mathrm{i}n\theta}] + cJ_n(kR)\sum_{l\in\mathbb{Z}}(-1)^{n-l}Q_{n-l}J_l(kr)e^{\mathrm{i}l\theta}, \tag{B.134}$$

where Q_n is the lattice sum defined by

$$Q_n := \sum_{m\in\mathbb{Z}^2, m\neq 0} H_n^{(1)}(k|m|)e^{\mathrm{i}n\arg(m)}e^{\mathrm{i}m\cdot\alpha}.$$

So, from (B.133) and (B.134), we finally obtain an explicit representation of $\mathcal{S}_D^{\alpha,k}$. For an efficient method for computing the lattice sum Q_n, we refer, for instance, to [**121**].

In the numerical computations, we should consider the truncated series

$$\sum_{m=-M}^{M} a_m \mathcal{S}_D^{\alpha,k}[e^{\mathrm{i}m\theta}]$$

instead of $\mathcal{S}_D^{\alpha,k}[\varphi] = \sum_{m\in\mathbb{Z}} a_m \mathcal{S}_D^{\alpha,k}[e^{\mathrm{i}m\theta}]$ for some large enough M. Then, using $e^{\mathrm{i}m\theta}$ as basis, we have the following matrix representation for the operator $\mathcal{S}_D^{\alpha,k}$:

$$\mathcal{S}_D^{\alpha,k}[\varphi]\big|_{\partial D} \approx \begin{pmatrix} S_{-M,-M} & S_{-M,-(M-1)} & \cdots & S_{-M,M} \\ S_{-(M-1),-M} & S_{-(M-1),-(M-1)} & \cdots & S_{-(M-1),M} \\ \vdots & \vdots & \ddots & \vdots \\ S_{M,-M} & S_{M,-(M-1)} & \cdots & S_{M,M} \end{pmatrix} \begin{pmatrix} a_{-M} \\ a_{-(M-1)} \\ \vdots \\ a_M \end{pmatrix},$$

where $S_{m,n}$ is given by

$$S_{m,n} = cJ_n(kR)H_n^{(1)}(kR)\delta_{mn} + cJ_n(kR)(-1)^{n-m}Q_{n-m}J_m(kR)$$

with δ_{mn} being the Kronecker symbol.

Similarly, we also have the following representation for $\partial \mathcal{S}_D^{\alpha,k}[\varphi]/\partial\nu|_\pm$ on ∂D:

$$\frac{\partial \mathcal{S}_D^{\alpha,k}[\varphi]}{\partial \nu}\bigg|_\pm \approx \begin{pmatrix} {S'}^{\pm}_{-M,-M} & {S'}^{\pm}_{-M,-(M-1)} & \cdots & {S'}^{\pm}_{-M,M} \\ {S'}^{\pm}_{-(M-1),-M} & {S'}^{\pm}_{-(M-1),-(M-1)} & \cdots & {S'}^{\pm}_{-(M-1),M} \\ \vdots & \vdots & \ddots & \vdots \\ {S'}^{\pm}_{M,-M} & {S'}^{\pm}_{M,-(M-1)} & \cdots & {S'}^{\pm}_{M,M} \end{pmatrix} \begin{pmatrix} a_{-M} \\ a_{-(M-1)} \\ \vdots \\ a_M \end{pmatrix},$$

where ${S'}^{\pm}_{m,n}$ is given by

$${S'}^{\pm}_{m,n} = \pm\frac{1}{2} + kc\big(J_n' H_n^{(1)} + J_n (H_n^{(1)})'\big)(kR)\delta_{mn} + kcJ_n(kR)(-1)^{n-m}Q_{n-m}J_m'(kR).$$

B.6.2. The three-dimensional case. For a radially symmetric Helmholtz equation, it is well known that the spherical waves

$$j_m(kr)Y_m^l(\theta,\phi) \quad \text{and} \quad h_m^{(1)}(kr)Y_m^l(\theta,\phi)$$

give a basis of solutions in the spherical coordinates (r,θ,ϕ). Here, $Y_m^l(\theta,\phi), m \in \mathbb{N}, l = -m,\ldots,m$, are the spherical harmonics and $j_m(kr), h_m^{(1)}(kr)$ are the spherical Bessel and Hankel functions of the first kind, respectively, defined by (B.37)–(B.39).

We begin by deriving the multipole expansion of the single layer potential $\mathcal{S}_D^k$. The spherical harmonics Y_m^l form a basis of $L^2(\partial D)$, and we seek the expansion of $\mathcal{S}_D^k$ in this basis. Define $u := \mathcal{S}_D^k[Y_m^l]$, which is the solution to

$$\text{(B.135)} \qquad \begin{cases} \Delta u + k^2 u = 0 & \text{in } \mathbb{R}^3 \setminus \overline{D}, \\ \Delta u + k^2 u = 0 & \text{in } D, \\ u|_+ = u|_- & \text{on } \partial D, \\ \frac{\partial u}{\partial \nu}\Big|_+ - \frac{\partial u}{\partial \nu}\Big|_- = Y_m^l & \text{on } \partial D, \\ |x|\left(\frac{\partial}{\partial |x|} - \mathrm{i}k\right) u \to 0 & \text{as } |x| \to +\infty. \end{cases}$$

Solving the above equation by the separation of variables technique in spherical coordinates yields

$$\mathcal{S}_D^k[Y_m^l](r,\theta,\phi) = \begin{cases} c j_m(kR) h_m^{(1)}(kr) Y_m^l(\theta,\phi), & |r| \geq R, \\ c h_m^{(1)}(kR) j_m(kr) Y_m^l(\theta,\phi), & |r| < R, \end{cases} \tag{B.136}$$

where $c = -\mathrm{i}kR^2$.

In order to handle problems posed on disjoint domains, we will need an addition theorem relating spherical waves centred around a translated origin to spherical waves around the original origin. Suppose that we have a point with coordinates $x = (r, \theta, \phi)$ in the original system and $x' = (r', \theta', \phi')$ in the translated system, with the coordinate vectors related by $x = x' + b$ for $b = (r_b, \theta_b, \phi_b)$. Moreover, we assume $r' < r_b$. Then the addition theorem reads [**82**]

$$h_m^{(1)}(kr) Y_m^l(\theta,\phi) = \sum_{m' \in \mathbb{N}, |l'| \leq m'} A_{m'l'}^{ml} j_{m'}(kr') Y_{m'}^{l'}(\theta',\phi'), \tag{B.137}$$

where the coefficients $A_{m'l'}^{ml}$ are given by

$$A_{m'l'}^{ml} = \sum_{\lambda \in \mathbb{N}, |\mu| \leq \lambda} C(m,l,m',l',\lambda,\mu) h_\lambda^{(1)}(kr_b) Y_\lambda^\mu(\theta_b,\phi_b).$$

Here, the coefficients $C(m,l,m',l',\lambda,\mu)$ are in turn given by

$$\mathrm{i}^{m'-m+\lambda}(-1)^l \sqrt{4\pi(2m+1)(2m'+1)(2\lambda+1)} \begin{pmatrix} m & m' & \lambda \\ 0 & 0 & 0 \end{pmatrix} \begin{pmatrix} m & m' & \lambda \\ -l & l' & \mu \end{pmatrix},$$

where we denote by

$$\begin{pmatrix} j_1 & j_2 & j_3 \\ l_1 & l_2 & l_3 \end{pmatrix},$$

the Wigner $3j$ symbols. To simplify these expressions slightly, we assume that the original coordinate system is aligned such that b points along the positive x_3-axis, *i.e.*, $\theta_b = 0$. In this case,

$$Y_\lambda^\mu(\theta_b,\phi_b) = \begin{cases} 0, & \mu \neq 0, \\ \sqrt{\frac{2\lambda+1}{4\pi}}, & \mu = 0. \end{cases}$$

Substituting this into the expression for $A_{m'l'}^{ml}$ gives

$$A_{m'l'}^{ml} = \sum_{\lambda \in \mathbb{N}} \sqrt{\frac{2\lambda+1}{4\pi}} C(m,l,m',l',\lambda,0) j_\lambda(kr_b).$$

Now, we consider the one-dimensional chain of resonators of period L in Section 4.4 and compute the quasiperiodic single layer potential $\mathcal{S}_D^{\alpha,k}[Y_m^l]$ in the case when D consists of a single resonator. Since

$$G^{\alpha,k}(x,y) = \sum_{n \in \mathbb{Z}} G^k(|x - y - (nL,0,0)|)\, e^{\mathrm{i}n\alpha L},$$

we have

$$\begin{aligned} \mathcal{S}_D^{\alpha,k}[Y_m^l](x) &= \mathcal{S}_D^k[Y_m^l](x) + \sum_{n \in \mathbb{Z}, n \neq 0} \mathcal{S}_{D+n}^k[Y_m^l] e^{\mathrm{i}n\alpha L} \\ &= \mathcal{S}_D^k[Y_m^l](x) + c j_m(kR) \sum_{n \in \mathbb{Z}, n \neq 0} h_m^{(1)}(kr_n') Y_m^l(\theta_n', \phi_n') e^{\mathrm{i}n\alpha}. \end{aligned}$$

Here, $D+n$ means a translation of the disk D by $(nL,0,0)$ and (r'_n,θ'_n,ϕ'_n) are the spherical coordinates with respect to the centre of $D+n$.

Using the addition theorem (B.137), we have

$$\begin{aligned}\mathcal{S}_D^{\alpha,k}[Y_m^l](x) =&\mathcal{S}_D^k[Y_m^l](x)+cj_m(kR)\\ &\times\sum_{m'\in\mathbb{N},|l'|\le m'}\left[\sum_{\lambda\in\mathbb{N},|\mu|\le\lambda}C(m,l,m',l',\lambda,\mu)Q_\lambda^\mu\right]j_{m'}(kr)Y_{m'}^{l'}(\theta,\phi)\\ :=&\mathcal{S}_D^k[Y_m^l](x)+cj_m(kR)\sum_{m'\in\mathbb{N},|l'|\le m'}B_{m'l'}^{ml}j_{m'}(kr)Y_{m'}^{l'}(\theta,\phi),\end{aligned}$$

where Q_λ^μ is the one-dimensional lattice sum in three dimensions, defined by

$$Q_\lambda^\mu=\sum_{n\in\mathbb{Z},n\neq0}h_\lambda^{(1)}(knL)Y_\lambda^\mu(\theta_n,\phi_n)e^{in\alpha L}.$$

An efficient method for computing this lattice sum is given in [**120**].

Next, we compute $\mathcal{S}_D^{\alpha,k}$ in the case when $D\Subset Y$ consists of two resonators as in Section 4.6, centred at $(-x_1,0,0)$ and $(x_1,0,0)$, respectively. By identifying $L^2(\partial D)=L^2(\partial D_1)\times L^2(\partial D_2)$ we have

$$\mathcal{S}_D^{\alpha,k}=\begin{pmatrix}\mathcal{S}_{D_1}^{\alpha,k} & \mathcal{S}_{D_2}^{\alpha,k}|_{\partial D_1}\\ \mathcal{S}_{D_1}^{\alpha,k}|_{\partial D_2} & \mathcal{S}_{D_2}^{\alpha,k}\end{pmatrix}.$$

Here the operator $\mathcal{S}_{D_i}^{\alpha,k}|_{\partial D_j}:L^2(\partial D_i)\to L^2(\partial D_j),i,j=1,2$ is the evaluation of $\mathcal{S}_{D_i}^{\alpha,k}$ on ∂D_j. To compute the multipole expansion of $\mathcal{S}_{D_1}^{\alpha,k}|_{\partial D_2}$, we again use the addition theorem. We have

$$\begin{aligned}\mathcal{S}_{D_1}^{\alpha,k}|_{\partial D_2}[Y_m^l](x')&=cj_m(kR)h_m^{(1)}(kr')Y_m^l(\theta',\phi')\\ &\quad+cj_m(kR)\sum_{m'\in\mathbb{N},|l'|\le m'}B_{m'l'}^{ml}j_{m'}(kr')Y_{m'}^{l'}(\theta',\phi')\\ &=cj_m(kR)\sum_{m''\in\mathbb{N},|l''|\le m''}\left[\sum_{\lambda\in\mathbb{N},|\mu|\le\lambda}C(m,l,m'',l'',\lambda,\mu)h_\lambda^{(1)}(kd)Y_\lambda^\mu(\theta_d,\phi_d)\right]\\ &\quad j_{m''}(kr)Y_{m''}^{l''}(\theta,\phi)+cj_m(kR)\\ &\quad\times\sum_{m''\in\mathbb{N},|l''|\le m''}\left[\sum_{\substack{m'\in\mathbb{N},|l'|\le m'\\ \lambda\in\mathbb{N},|\mu|\le\lambda}}B_{m'l'}^{ml}C(m',l',m'',l'',\lambda,\mu)j_\lambda(kd)Y_\lambda^\mu(\theta_d,\phi_d)\right]\\ &\quad\times j_{m''}(kr)Y_{m''}^{l''}(\theta,\phi).\end{aligned}$$

Finally, we perform similar computations for the operator $\mathcal{S}_D^k$ in the case where D consists of N resonators. We assume the resonators to be arranged collinearly along the x_1-axis. By identifying $L^2(\partial D) = L^2(\partial D_1) \times \ldots \times L^2(\partial D_N)$ we have

$$\mathcal{S}_D^k = \begin{pmatrix} \mathcal{S}_{D_1}^k & \mathcal{S}_{D_2}^k\big|_{\partial D_1} & \cdots & \mathcal{S}_{D_N}^k\big|_{\partial D_1} \\ \mathcal{S}_{D_1}^k\big|_{\partial D_2} & \mathcal{S}_{D_2}^k & \cdots & \mathcal{S}_{D_N}^k\big|_{\partial D_2} \\ \vdots & \vdots & \ddots & \vdots \\ \mathcal{S}_{D_1}^k\big|_{\partial D_N} & \mathcal{S}_{D_2}^k\big|_{\partial D_N} & \cdots & \mathcal{S}_{D_N}^k \end{pmatrix},$$

where, as in the quasiperiodic case, $\mathcal{S}_{D_i}^k\big|_{\partial D_j} : L^2(\partial D_i) \to L^2(\partial D_j)$ is the evaluation of $\mathcal{S}_{D_i}^k$ on ∂D_j. This relies on the addition theorem once again. The diagonal terms are easily evaluated using (B.136) directly. Away from the diagonals, the addition theorem (B.137) gives that

$$\mathcal{S}_{D_j}^k\big|_{\partial D_i}[Y_m^l](x') = c h_m^{(1)}(kR) \sum_{m' \in \mathbb{N}, |l'| \leq m'} A_{m'l'}^{ml} j_{m'}(kr') Y_{m'}^{l'}(\theta', \phi').$$

APPENDIX C

Some technical results and proofs

C.1. High-order expansions of the subwavelength resonant frequencies

Let $\omega_0 = \sqrt{\lambda_j}$, where λ_j is an eigenvalue of the capacitance matrix C and suppose that λ_j is simple. Then there exists a positive constant r such that the matrix-valued function $\omega \mapsto A(\omega, \delta)$ has exactly one characteristic value in $\overline{V_r(\omega_0)}$, where $V_r(\omega_0)$ is the disk of centre ω_0 and radius r. Here, $A(\omega, \delta)$ is defined by (1.57). This characteristic value, denoted by ω_δ, is holomorphic with respect to δ. Moreover, the following assertions hold:

(i) $\mathcal{M}(A(\omega, \delta); \partial V_r(\omega_0)) = 1$;
(ii) $A(\omega, \delta)^{-1} = (\omega - \omega_\delta)^{-1}\mathcal{L}_\delta + \mathcal{R}(\omega, \delta)$;
(iii) $\mathcal{L}_\delta : \operatorname{Ker}(A(\omega_\delta, \delta))^* \to \operatorname{Ker} A(\omega_\delta, \delta)$,

where $\mathcal{M}$ is the full multiplicity defined by (A.6), $\mathcal{R}(\omega, \delta)$ is a holomorphic function with respect to (δ, ω), and $\mathcal{L}_\delta$ is a one-dimensional operator. Note that

$$A(\omega, 0) = C - \omega^2 I.$$

We are now ready to apply the generalised argument principle in order to derive a full asymptotic expansion of the characteristic value ω_δ with respect to δ. Since ω_δ is the unique characteristic value in $V_r(\omega_0)$ of the holomorphic operator-valued function $\omega \mapsto A(\omega, \delta)$, the relevant formula will be as follows, which is an immediate consequence of Theorem A.15.

LEMMA C.1. *Let $\omega_0 = \sqrt{\lambda_j}$ and suppose that λ_j is simple. Then the characteristic value ω_δ of the matrix-value function $\omega \mapsto A(\omega, \delta)$ in $V_r(\omega_0)$ is given by*

$$\omega_\delta - \omega_0 = \frac{1}{2\pi \mathrm{i}} \operatorname{tr} \int_{\partial V_r(\omega_0)} (\omega - \omega_0) A(\omega, \delta)^{-1} \frac{\partial}{\partial \omega} A(\omega, \delta)\, \mathrm{d}\omega. \tag{C.1}$$

Substituting the expansion

$$A(\omega, \delta) = \sum_{j=0}^{+\infty} \delta^j A_j(\omega), \tag{C.2}$$

into (C.1), we obtain the following complete asymptotic expansion with respect to δ for the subwavelength resonant frequencies in the three-dimensional case.

THEOREM C.2. *The following asymptotic expansion holds:*

$$\omega_\delta - \omega_0 = \frac{1}{2\pi \mathrm{i}} \sum_{p=1}^{+\infty} \frac{1}{p} \sum_{n=p}^{+\infty} \delta^n \operatorname{tr} \int_{\partial V_r(\omega_0)} B_{n,p}(\omega)\, \mathrm{d}\omega, \tag{C.3}$$

where

$$B_{n,p}(\omega) = (-1)^p \sum_{\substack{n_1+\ldots+n_p=n \\ n_i\geq 1}} A_0(\omega)^{-1}A_{n_1}(\omega)\ldots A_0(\omega)^{-1}A_{n_p}(\omega)\omega^n.$$

PROOF. If δ is small enough, then the following Neumann series converges uniformly with respect to ω in $\partial V_r(\omega_0)$:

$$A(\omega,\delta)^{-1} = \sum_{p=0}^{+\infty} \left[A_0(\omega)^{-1}(A_0(\omega) - A(\omega,\delta))\right]^p A_0(\omega)^{-1},$$

where $A_0(\omega) = A(\omega, 0) = C - \omega^2 I$, and hence we may deduce that $\omega_\delta - \omega_0$ is equal to

$$\frac{1}{2\pi\mathrm{i}}\sum_{p=0}^{+\infty}\mathrm{tr}\int_{\partial V_r(\omega_0)} (\omega-\omega_0)\left[A_0(\omega)^{-1}(A_0(\omega) - A(\omega,\delta))\right]^p A_0(\omega)^{-1}\frac{d}{d\omega}A(\omega,\delta)\ d\omega.$$

Using the property (A.3) of the trace together with the following identity:

$$\text{(C.4)} \qquad \frac{d}{d\omega}A_0(\omega)^{-1} = -A_0(\omega)^{-1}\frac{dA_0}{d\omega}(\omega)A_0(\omega)^{-1},$$

we have

$$\begin{aligned}
&\mathrm{tr}\int_{\partial V_r(\omega_0)} (\omega-\omega_0)\frac{1}{p}\frac{\partial}{\partial\omega}\left[A_0(\omega)^{-1}(A_0(\omega) - A(\omega,\delta))\right]^p\ d\omega \\
&=\mathrm{tr}\Big[\int_{\partial V_r(\omega_0)} (\omega-\omega_0)\left[A_0(\omega)^{-1}(A_0(\omega) - A(\omega,\delta))\right]^{p-1}A_0(\omega)^{-1} \\
&\qquad\qquad \times\frac{\partial}{\partial\omega}(A_0(\omega) - A(\omega,\delta))\ d\omega \\
&\quad -\int_{\partial V_r(\omega_0)} (\omega-\omega_0)\left[A_0(\omega)^{-1}(A_0(\omega) - A(\omega,\delta))\right]^p A_0(\omega)^{-1}\frac{\partial}{\partial\omega}A_0(\omega)\ d\omega\Big],
\end{aligned}$$

and therefore,

$$\begin{aligned}
\omega_\delta - \omega_0 = &-\frac{1}{2\pi\mathrm{i}}\sum_{p=1}^{+\infty}\mathrm{tr}\int_{\partial V_r(\omega_0)} (\omega-\omega_0)\frac{1}{p}\frac{\partial}{\partial\omega}\left[A_0(\omega)^{-1}(A_0(\omega) - A(\omega,\delta))\right]^p\ d\omega \\
&+\frac{1}{2\pi\mathrm{i}}\mathrm{tr}\int_{\partial V_r(\omega_0)} (\omega-\omega_0)A_0(\omega)^{-1}\frac{d}{d\omega}A_0(\omega)\ d\omega.
\end{aligned}$$

Since ω_0 is a simple pole of $A_0(\omega)^{-1}$ and $A_0(\omega)$ is holomorphic in ω, we readily get

$$\int_{\partial V_r(\omega_0)} (\omega-\omega_0)A_0(\omega)^{-1}\frac{d}{d\omega}A_0(\omega)\ d\omega = 0.$$

Thus, it follows that

$$\omega_\delta - \omega_0 = -\frac{1}{2\pi\mathrm{i}}\sum_{p=1}^{+\infty}\mathrm{tr}\int_{\partial V_r(\omega_0)} (\omega-\omega_0)\frac{1}{p}\frac{\partial}{\partial\omega}\left[A_0(\omega)^{-1}(A_0(\omega) - A(\omega,\delta))\right]^p\ d\omega.$$

Now, a simple integration by parts yields

$$\omega_\delta - \omega_0 = \frac{1}{2\pi\mathrm{i}}\sum_{p=1}^{+\infty}\frac{1}{p}\mathrm{tr}\int_{\partial V_r(\omega_0)} \left[A_0(\omega)^{-1}(A_0(\omega) - A(\omega,\delta))\right]^p\ d\omega.$$

From the expansion (C.2), it follows that

$$\Big(A_0(\omega)^{-1}(A_0(\omega) - A(\omega,\delta))\Big)^p$$
$$= (-1)^p \sum_{n=p}^{+\infty} \delta^n \sum_{\substack{n_1+\ldots+n_p=n \\ n_i \geq 1}} A_0(\omega)^{-1} A_{n_1}(\omega) \ldots A_0(\omega)^{-1} A_{n_p}(\omega).$$

Therefore, upon inserting this into the latter formula, we arrive at the desired asymptotic expansion (C.3). □

With regards to the simplest case, the leading-order term in the asymptotic expansion of $\omega_\delta - \omega_0$ is given by

$$-\frac{1}{2\pi \mathrm{i}} \mathrm{tr} \int_{\partial V_r(\omega_0)} A_0(\omega)^{-1} A_1(\omega) \; d\omega. \tag{C.5}$$

C.2. Sensitivity enhancement using high-order exceptional points

Let $m \in L^\infty(\mathbb{R}^3)$ (possibly complex) be such that $m \equiv 1$ outside a ball B and $|m| \geq m_0 > 1$ in B for some constant m_0. Consider the volume integral operator $T_B^k : L^2(B) \to L^2(B)$ defined by

$$T_B^k[f](x) = -\int_B (m(y) - 1) G^k(x,y) f(y) \; dy, \quad x \in B, \tag{C.6}$$

where G^k is the free-space Helmholtz Green's function defined by (1.7).

In what follows, we present properties of the integral operator T_B^k. The following results are from [**54**].

LEMMA C.3. *The operator T_B^k is compact from $L^2(B)$ to $L^2(B)$. In fact, T_B^k is bounded from $L^2(B)$ to $H^2(B)$ (the usual Sobolev space of functions in $H^1(B)$ whose weak derivative is in $H^1(B)$). Moreover, T_B^k is a Hilbert-Schmidt operator.*

LEMMA C.4. *Let $\sigma(T_B^k)$ be the spectrum of T_B^k defined by* (C.6)*. We have*

(i) $\sigma(T_B^k) = \{0, \lambda_1(k), \lambda_2(k), \ldots, \lambda_n(k), \ldots\}$, *where*
$$|\lambda_1(k)| \geq |\lambda_2(k)| \geq |\lambda_3(k)| \geq \ldots$$
and $\lambda_n(k) \to 0$ *as* $n \to +\infty$;

(ii) $\{0\} = \sigma(T_B^k) \backslash \sigma_p(T_B^k)$ *with* $\sigma_p(T_B^k)$ *being the point spectrum of* T_B^k.

PROOF. We need only to prove the second assertion. Assume that

$$T_B^k[u] = -\int_B (m(y) - 1) G^k(x,y) u(y) \; dy = 0.$$

We have $0 = (\Delta + k^2) T_B^k[u] = (1 - m)u$, which shows that $u = 0$. The assertion is then proved. □

LEMMA C.5. *Let T_B^k be defined by* (C.6)*. Then, $\lambda \in \sigma(T_B^k)$ if and only if there is a nontrivial solution in $H^2_{\mathrm{loc}}(\mathbb{R}^3)$ to the following problem:*

$$\begin{cases} (\Delta + k^2)\widetilde{u}(x) = -\frac{1}{\lambda}(m(x) - 1)\widetilde{u}(x) \quad \textit{in } B, \\ (\Delta + k^2)\widetilde{u} = 0 \quad \textit{in } \mathbb{R}^3 \backslash B, \\ \widetilde{u} \textit{ satisfies the Sommerfeld radiation condition.} \end{cases} \tag{C.7}$$

PROOF. Assume that $T_B^k[u] = \lambda u$. We define

$$\widetilde{u}(x) = -\int_B G^k(x,y)(m(y)-1)u(y)\,\mathrm{d}y, \quad x \in \mathbb{R}^3.$$

Then $\widetilde{u}$ satisfies the required equations. □

LEMMA C.6. *Let $\mathcal{H}_j$ denote the generalised eigenspace of the operator T_B^k for the eigenvalue $\lambda_j(k)$. The following decomposition holds:*

$$L^2(B) = \overline{\bigcup_{j=1}^{+\infty} \mathcal{H}_j}.$$

PROOF. Let $T_B^{k,*}$ be the L^2-adjoint of T_B^k. Then we can easily check that

$$T_B^{k,*}[g](x) = (m(x)-1)T_B^{-k}[g](x), \quad x \in B, g \in L^2(B).$$

By the same argument as the one in the proof of Lemma C.4, we can then show that $\operatorname{Ker} T_B^{k,*} = \{0\}$. As a result, we have

$$\overline{T_B^k(L^2(B))} = \left(\operatorname{Ker} T_B^{k,*}\right)^{\perp} = L^2(B),$$

and the lemma is proved. □

LEMMA C.7. *There exists a basis $\{u_{j,l,q}\}$, $1 \le l \le m_j, 1 \le q \le n_{j,l}$ for $\mathcal{H}_j$ such that*

$$T_B^k(u_{j,1,1}, \dots, u_{j,m_j,n_{j,m_j}}) = (u_{j,1,1}, \dots, u_{j,m_j,n_{j,m_j}}) \begin{pmatrix} J_{j,1} & & \\ & \ddots & \\ & & J_{j,m_j} \end{pmatrix},$$

where $J_{j,l}$ is the canonical Jordan matrix of size $n_{j,l}$ in the form

$$J_{j,l} = \begin{pmatrix} \lambda_j & 1 & & \\ & \ddots & \ddots & \\ & & \lambda_j & 1 \\ & & & \lambda_j \end{pmatrix}.$$

PROOF. This follows from Jordan's theory applied to the linear operator

$$T_B^k|_{\mathcal{H}_j} : \mathcal{H}_j \to \mathcal{H}_j$$

on the finite dimensional space $\mathcal{H}_j$. □

We denote by $\Gamma = \{(j,l,q) \in \mathbb{N}\times\mathbb{N}\times\mathbb{N}; 1 \le l \le m_j, 1 \le q \le n_{j,l}\}$ the set of indices for the basis functions. We introduce a partial order on $\mathbb{N}\times\mathbb{N}\times\mathbb{N}$. Let $\gamma = (j,l,q) \in \Gamma, \gamma' = (j',l',q') \in \Gamma$, we say that $\gamma' \preceq \gamma$ if one of the following conditions is satisfied:

(i) $j > j'$;
(ii) $j = j'$, $l > l'$;
(iii) $j = j'$, $l = l'$, $q \ge q'$.

By the Gram-Schmidt orthonormalisation process, the following result is obvious.

LEMMA C.8. *There exists an orthonormal basis* $\{e_\gamma : \gamma \in \Gamma\}$ *for* $L^2(B)$ *such that*

$$e_\gamma = \sum_{\gamma' \preceq \gamma} a_{\gamma,\gamma'} u_{\gamma'},$$

where $a_{\gamma,\gamma'}$ *are constants and* $a_{\gamma,\gamma} \neq 0$.

We can regard $A = \{a_{\gamma,\gamma'}\}_{\gamma,\gamma' \in \Gamma}$ as a matrix. It is clear that A is upper-triangular and has nonzero diagonal elements. Its inverse is denoted by $B = \{b_{\gamma,\gamma'}\}_{\gamma,\gamma' \in \Gamma}$ which is also upper-triangular and has nonzero diagonal elements. We have

$$u_\gamma = \sum_{\gamma' \preceq \gamma} b_{\gamma,\gamma'} e_{\gamma'}.$$

LEMMA C.9. *The functions* $\{e_\gamma(x)\overline{e_{\gamma'}(y)}\}$ *form an orthonormal basis for the Hilbert space* $L^2(B \times B)$. *Moreover, the following completeness relation holds:*

$$\delta(x - y) = \sum_\gamma e_\gamma(x)\overline{e_\gamma(y)}.$$

Let G_m^k be the Green function of $\Delta + k^2 m$ in $\mathbb{R}^3$ subject to the Sommerfeld radiation condition.

By standard elliptic theory, we have $G_m^k(x, x_0) \in L^2(B \times B)$. Thus we have

$$G_m^k(x, x_0) = \sum_{\gamma,\gamma'} \alpha_{\gamma,\gamma'} e_\gamma(x)\overline{e_{\gamma'}}(x_0), \tag{C.8}$$

for some constants $\alpha_{\gamma,\gamma'}$ satisfying

$$\sum_{\gamma,\gamma'} |\alpha_{\gamma,\gamma'}|^2 = \|G_m^k(x, x_0)\|^2_{L^2(B\times B)} < +\infty.$$

To analyse the Green function G_m^k, we need to find the constants $\alpha_{\gamma,\gamma'}$. To do so, we first note that

$$G^k(x, x_0) = \frac{1}{(1 - m(x_0))} T_B^k[\delta(\cdot - x_0)]. \tag{C.9}$$

Then, by integrating by parts and making use of the reciprocity property (see, for instance, [**37**, Section 2.10])

$$G_m^k(x, x_0) = G_m^k(x_0, x),$$

we obtain the Lippmann-Schwinger representation

$$G_m^k(x, x_0) = G^k(x, x_0) + k^2 \int_B (1 - m(y))G_m^k(x, y)G^k(y, x_0)\, dy. \tag{C.10}$$

Thus,

$$(I - k^2 T_B^k)[G_m^k](x, x_0) = G^k(x, x_0), \quad x, x_0 \in B, \tag{C.11}$$

and hence,

$$G_m^k(x, x_0) = \big(I - k^2 T_B^k)\big)^{-1}[G^k(\cdot, x_0)](x), \tag{C.12}$$

provided that $k^2\lambda_j(k) \neq 1$ for all j, where $\lambda_j(k)$ are the eigenvalues of T_B^k. Thus, from (C.8) and (C.9), it follows that

$$G_m^k(x, x_0) = \frac{1}{(1 - m(x_0))} \sum_\gamma \big(I - k^2 T_B^k)\big)^{-1} T_B^k[e_\gamma](x)\overline{e_\gamma(x_0)}. \tag{C.13}$$

We next compute $(zI - T_B^k)^{-1} T_B^k[e_\gamma]$. For ease of notation, we define $u_{j,l,q} = 0$ for $q \leq 0$. We have

$$T_B^k[u_{j,l,q}] = \lambda_j u_{j,l,q} + u_{j,l,q-1} \quad \text{for all } j, l, q.$$

On the other hand, for $z \notin \sigma(T_B^k)$, we have

$$(zI - T_B^k)^{-1}[u_{j,l,q}] = \frac{1}{z - \lambda_j} u_{j,l,q} + \frac{1}{(z-\lambda_j)^2} u_{j,l,q-1} + \ldots + \frac{1}{(z-\lambda_j)^q} u_{j,l,1},$$

and therefore, by taking $z = 1/k^2$, it then follows that

$$\begin{aligned}(zI - T_B^k)^{-1} T_B^k[u_{j,l,q}] &= \frac{\lambda_j}{z - \lambda_j} u_{j,l,q} + \frac{\lambda_j}{(z-\lambda_j)^2} u_{j,l,q-1} + \ldots + \frac{\lambda_j}{(z-\lambda_j)^q} u_{j,l,1} \\ &\quad + \frac{1}{z - \lambda_j} u_{j,l,q-2} + \frac{1}{(z-\lambda_j)^2} u_{j,l,q-3} + \ldots + \frac{1}{(z-\lambda_j)^{q-2}} u_{j,l,1} \\ &= \sum_{\gamma'} d_{\gamma,\gamma'} u_{\gamma'},\end{aligned}$$

where the matrix $\{d_{\gamma,\gamma'}\}_{\gamma,\gamma' \in \Gamma}$ is upper-triangular and has block structure.

We arrive at the following result.

THEOREM C.10. *The following expansion holds for the Green function G_m^k:*

$$G_m^k(x, x_0) = \sum_{\gamma \in \Gamma} \sum_{\gamma''' \in \Gamma} \alpha_{\gamma,\gamma'''} \overline{e_\gamma}(x_0) e_{\gamma'''}(x), \tag{C.14}$$

where

$$\alpha_{\gamma,\gamma'''} = \frac{1}{(1 - m(x_0))} \sum_{\gamma' \preceq \gamma} \sum_{\gamma'' \preceq \gamma'} a_{\gamma,\gamma'} d_{\gamma',\gamma''} b_{\gamma'',\gamma'''}.$$

Let k^* be such that $(k^*)^2 \lambda_{j_0}(k^*) = 1$ for some j_0, *i.e.*, k^* is a characteristic value of the operator-valued function $k \mapsto I - k^2 T_B^k$. Assume that k^* is of order q_{j_0}. Theorem C.10 yields the following pole-pencil decomposition of G_m^k for k in a small neighbourhood of k^*; see [**55**] and [**74**, Theorem 4.7].

THEOREM C.11. *The following pole-pencil decomposition holds:*

$$G_m^k(x, x_0) = \sum_{l=1}^{q_{j_0}} \frac{l_l(x, x_0)}{(k^2 - (k^*)^2)^l} + r(k, x, x_0), \quad x, x_0 \in B,$$

in a neighbourhood of k^*, where l_l are in $L^2(B \times B)$ and r is holomorphic in k.

Let η be a small parameter. Assume that m is "perturbed" and the corresponding $T_{B,\eta}^k$ has the form

$$T_{B,\eta}^k = T_B^k + \eta T^{(1)} + \mathcal{O}(\eta^2),$$

in the operator norm, for some compact operator $T^{(1)} : L^2(B) \to L^2(B)$. Then the following result holds.

THEOREM C.12. *In a small neighbourhood of k^*, the perturbed system has a resonant frequency k_η with the asymptotic behaviour*

$$k_\eta - k^* = \mathcal{O}(\eta^{1/q_{j_0}}) \quad \textit{as } \eta \to 0. \tag{C.15}$$

PROOF. In a small neighbourhood $V(k^*)$ of k^*, we expand

$$\big(I - k^2(T_B^k + \eta T^{(1)} + \mathcal{O}(\eta^2))\big)^{-1}$$

in Neumann series as follows:

$$\big(I-k^2(T_B^k+\eta T^{(1)})\big)^{-1} = \big(I-k^2T_B^k\big)^{-1}+\eta k^2\big(I-k^2T_B^k\big)^{-1}T^{(1)}\big(I-k^2T_B^k\big)^{-1}+\mathcal{O}(\eta^2),$$

in the operator norm. Then, from the pole-pencil decomposition

$$\big(I - k^2T_B^k\big)^{-1} = \sum_{l=1}^{q_{j0}} \frac{L_l}{(k^2 - (k^*)^2)^l} + R(k),$$

in $V(k^*)$, where L_l are finite-dimensional operators and the operator R is holomorphic in k, it follows that the characteristic value k_η in $V(k^*)$ of the operator $I - k^2T_{B,\eta}^k$ can be approximated by

$$k_\eta^2 - (k^*)^2 = \eta^{1/q_{j0}}\Big(\frac{\langle L_{q_{j0}}T^{(1)}L_{q_{j0}}[\phi^*], \phi^*\rangle}{\langle L_{q_{j0}}[\phi^*], \phi^*\rangle}\Big)^{1/q_{j0}},$$

where $\langle\cdot,\cdot\rangle$ is the L^2 scalar product and ϕ^* is the root function associated with k^*. □

Formula (C.15) shows that the perturbed system experiences a perturbation in its characteristic values of order $\mathcal{O}(\eta^{1/q_{j0}})$. Thus, for large q_{j0}, when trying to detect small perturbations, the effect on the characteristic values will be relatively large. This offers a means to design enhanced sensing systems [**66**, **73**, **103**].

C.3. Resonant frequencies for high refractive index materials

With the notation of Appendix C.2 in hand, assume that $m(y) - 1 = \tau n(y)$, where the parameter $\tau \gg 1$ and the function $n \in L^\infty(B)$ is real-valued and satisfies $n(y) \geq n_0 > 1$ for some constant n_0.

Fix k and consider the operator

$$\widetilde{T}_B^k : f \in L^2(B) \mapsto -\int_B n(y)G^k(\cdot, y)f(y)\, dy \in L^2(B).$$

From (C.12), we have

$$G_m^k(x, x_0) = \big(I - \tau k^2\widetilde{T}_B^k)\big)^{-1}[G^k(\cdot, x_0)](x), \quad x, x_0 \in B. \tag{C.16}$$

For τ large enough, there exists j_0 such that the equation

$$\tau k^2\widetilde{\lambda}_{j_0}(k) = 1$$

has a solution k^*, where $\widetilde{\lambda}_j$ are the eigenvalues of $\widetilde{T}_B^k$. This solution can be chosen such that it corresponds to the subwavelength regime, *i.e.*, $2\pi/k^* \ll 1$. Here, k^* is near $1/\sqrt{\tau\widetilde{\lambda}_{j_0}}$.

In order to see this and to completely characterise the subwavelength resonant eigenfrequencies, we expand $\widetilde{T}_B^k$ in terms of k for k small enough as follows:

$$\widetilde{T}_B^k = \widetilde{T}_B^{(0)} + k\widetilde{T}_B^{(1)} + \ldots,$$

where

$$\widetilde{T}_B^{(0)} : f \in L^2(B) \mapsto \frac{1}{4\pi}\int_B \frac{n(y)f(y)}{|\cdot - y|}\, dy \in L^2(B),$$

and

$$\widetilde{T}_B^{(1)}[f](x) = \frac{\mathrm{i}}{4\pi}\int_B n(y)f(y)\,\mathrm{d}y, \quad x \in B, f \in L^2(B).$$

Note that $\widetilde{T}_B^{(0)}$ is compact, self-adjoint in the weighted space $L^2(B, n(x)\,\mathrm{d}x)$ and thus, its spectrum is a set of real eigenvalues, converging to 0. Let β_0 ($= \widetilde{\lambda}_{j_0}$ for some j_0) be a simple eigenvalue of $\widetilde{T}_B^{(0)}$ and let ϕ_0 be its corresponding eigenfunction. Then from the pole-pencil decomposition

$$\big(I - \tau k^2 \widetilde{T}_B^{(0)}\big)^{-1} = \frac{(\cdot, \phi_0)}{1 - \tau k^2 \beta_0} + R(k),$$

where here $(\cdot,\cdot)$ is the weighted L^2 scalar product in B and R is holomorphic in k, together with the Neumann series expansion

$$\begin{aligned}
&\big(I - \tau k^2 \widetilde{T}_B^{(0)} - \tau k^3 \widetilde{T}_B^{(1)} - \ldots\big)^{-1} \\
&\quad = \big(I - \tau k^2 \widetilde{T}_B^{(0)}\big)^{-1} + \tau k^3 \big(I - \tau k^2 \widetilde{T}_B^{(0)}\big)^{-1} \widetilde{T}_B^{(1)} \big(I - \tau k^2 \widetilde{T}_B^{(0)}\big)^{-1} + \ldots,
\end{aligned}$$

it follows that k^* approximately satisfies the equation

$$1 - \tau k^2 \beta_0 = -\tau k^3 (\widetilde{T}_B^{(1)}[\phi_0], \phi_0) = \frac{\mathrm{i}\tau}{4\pi} k^2 \Big(\int_B n(y)\phi_0(y)\,\mathrm{d}y\Big)^2,$$

or equivalently, it can be approximated by

$$k^* = \frac{1}{\sqrt{\tau \beta_0}}\Big(1 + \frac{\mathrm{i}}{8\pi}\Big(\int_B n(y)\phi_0(y)\,\mathrm{d}y\Big)^2\Big).$$

We refer the reader to [**50**] for more details.

C.4. Proof of Theorem 4.56

To prove Theorem 4.56, we need to study the characteristic values of the operator $\mathcal{M}^\varepsilon$ appearing in Lemma 4.54. Let us first analyse the operator $\int_{Y^*} (\mathcal{A}^\alpha)^{-1}\,\mathrm{d}\alpha$. From Lemma 4.3, there exists one and only one $\omega_1^\alpha = \omega_1^\alpha(\delta)$ characteristic value of $\mathcal{A}^\alpha(\omega, \delta)$ such that $\omega_1^\alpha(0) = 0$ and ω_1^α continuously depends on δ. Since ω_1^α is a simple pole of the mapping $\omega \mapsto \mathcal{A}^\alpha(\omega,\delta)^{-1}$ in a neighbourhood of ω_1^α, we can write

$$\mathcal{A}^\alpha(\omega, \delta)^{-1} = \frac{\mathcal{L}^\alpha}{\omega - \omega_1^\alpha} + \mathcal{R}^\alpha(\omega, \delta), \tag{C.17}$$

where the operator-valued function $\omega \mapsto \mathcal{R}^\alpha(\omega,\delta)$ is holomorphic in a neighbourhood of ω_1^α, and the operator $\mathcal{L}^\alpha$ maps $L^2(\partial D) \times L^2(\partial D)$ onto $\mathrm{Ker}\mathcal{A}^\alpha(\omega_1^\alpha, \delta)$.

In what follows, let us do not make explicit the dependence on δ for ease of notation. Writing

$$\mathrm{Ker}\mathcal{A}^\alpha(\omega_1^\alpha) = \mathrm{span}\{\Psi^\alpha\}, \quad \mathrm{Ker}\big(\mathcal{A}^\alpha(\omega_1^\alpha)\big)^* = \mathrm{span}\{\Phi^\alpha\},$$

we can show that

$$\mathcal{L}^\alpha = \frac{\langle \Phi^\alpha, \,\cdot\, \rangle \Psi^\alpha}{\langle \Phi^\alpha, \frac{d}{d\omega}\mathcal{A}^\alpha\big|_{\omega=\omega_1^\alpha}[\Psi^\alpha]\rangle},$$

where $\langle\,\cdot\,,\,\cdot\,\rangle$ stands for the standard inner product of $L^2(\partial D) \times L^2(\partial D)$. Hence the operator $\mathcal{M}^\varepsilon$ can be decomposed as

$$\mathcal{M}^\varepsilon = I + \frac{1}{(2\pi)^2}(\mathcal{A}^\varepsilon - \mathcal{A})\int_{Y^*} \frac{\mathcal{L}^\alpha}{\omega - \omega_1^\alpha}\,\mathrm{d}\alpha + \frac{1}{(2\pi)^2}(\mathcal{A}^\varepsilon - \mathcal{A})\int_{Y^*} \mathcal{R}^\alpha\,\mathrm{d}\alpha. \tag{C.18}$$

Note that the third term in the right-hand side of (C.18) is holomorphic with respect to ω.

Denote by $\alpha^* = (\pi, \pi)$ and $\omega_1^* = \omega_1^{(\pi,\pi)}$. In Theorem 4.7, it is proved that, using the symmetry of the square array of resonators, ω_1^α attains its maximum at $\alpha = \alpha^*$. Since we are assuming that each resonator is a circular disk, we can derive a slightly more refined result. Indeed, we can easily show that for α near α^*, we have

$$\omega_1^\alpha = \omega_1^* - \frac{1}{2} c_\delta |\alpha - \alpha^*|^2 + o(|\alpha - \alpha^*|^2)\,, \tag{C.19}$$

where c_δ is a positive constant depending on δ and scales as $c_\delta = \mathcal{O}(\sqrt{\delta})$.

In view of (C.19), we can expect that the operator $\int_{Y^*} \frac{\mathcal{L}^\alpha}{\omega - \omega_1^\alpha}$ becomes singular when ω goes to ω_1^*. Let us extract its singular part explicitly. Before doing so, we introduce some notation. We denote by

$$\mathcal{A}_* = \mathcal{A}^{(\pi,\pi)}, \Phi_* = \Phi^{(\pi,\pi)}, \quad \text{and } \mathcal{L}_* = \mathcal{L}^{(\pi,\pi)}.$$

We also define a small neighbourhood V of ω_1^* that excludes the real interval $(-\infty, \omega_1^*]$, namely,

$$V = \{|\omega - \omega_1^*| < r_*\} \setminus (-\infty, \omega_1^*]$$

for some small enough $r_* > 0$. By (C.19), we have

$$\begin{aligned}
\frac{1}{(2\pi)^2} \int_{Y^*} \frac{\mathcal{L}^\alpha}{\omega - \omega_1^\alpha} \, \mathrm{d}\alpha &= \frac{1}{(2\pi)^2} \mathcal{L}_* \left[\int_{Y^*} \frac{1}{\omega - \omega_1^* + \frac{1}{2} c_\delta |\alpha - \alpha^*|^2} \, \mathrm{d}\alpha \right] + B_1(\omega) \\
&= \frac{1}{(2\pi)^2} \mathcal{L}_* \left[\int_{|\alpha - \alpha^*| < 1} \frac{1}{\omega - \omega_1^* + \frac{1}{2} c_\delta |\alpha - \alpha^*|^2} \, \mathrm{d}\alpha \right] + B_2(\omega).
\end{aligned}$$

Hereafter, B_j means a bounded function with respect to ω in V. Then, using the polar coordinates $\alpha - \alpha^* = (r', \theta')$, we get

$$\begin{aligned}
\frac{1}{(2\pi)^2} \int_{Y^*} \frac{\mathcal{L}^\alpha}{\omega - \omega_1^\alpha} \, \mathrm{d}\alpha &= \frac{1}{(2\pi)^2} \int_0^1 \frac{2\pi r'}{\omega - \omega_1^* + \frac{1}{2} c_\delta r'^2} \, \mathrm{d}r' + B_2(\omega) \\
&= -\frac{1}{2\pi c_\delta} \ln(\omega - \omega_1^*) + B_3(\omega).
\end{aligned}$$

Here, the usual principal branch is taken for the logarithm, and so the operator $\int_{Y^*} \frac{\mathcal{L}^\alpha}{\omega - \omega_1^\alpha} \, \mathrm{d}\alpha$ has a branch cut on $(-\infty, \omega_1^*]$. We also observe that, for $\omega \in V$,

$$\ln(\omega - \omega_1^*)(\mathcal{A}^\varepsilon(\omega) - \mathcal{A}(\omega)) = \ln(\omega - \omega_1^*)(\mathcal{A}^\varepsilon(\omega_1^*) - \mathcal{A}(\omega_1^*)) + B_4(\omega).$$

Therefore, $\mathcal{M}^\varepsilon(\omega)$ in (4.96) can be rewritten as

$$\mathcal{M}^\varepsilon(\omega) = I - \frac{\ln(\omega - \omega_1^*)}{2\pi c_\delta} \mathcal{L}_* (\mathcal{A}^\varepsilon(\omega_1^*) - \mathcal{A}(\omega_1^*)) + \mathcal{R}^\varepsilon(\omega),$$

where $\mathcal{R}^\varepsilon$ is analytic and bounded with respect to ω in V and satisfies $\mathcal{R}^\varepsilon = \mathcal{O}(\varepsilon)$ in the operator norm. We first consider the principal part $\mathcal{N}^\varepsilon(\omega)$, namely,

$$\mathcal{N}^\varepsilon : \omega \mapsto \mathcal{N}^\varepsilon(\omega) = I - \frac{1}{2\pi c_\delta} \ln(\omega - \omega_1^*)(\mathcal{A}^\varepsilon(\omega_1^*) - \mathcal{A}(\omega_1^*))\mathcal{L}_*.$$

Let us find its characteristic value $\widehat{\omega}$, *i.e.*, $\widehat{\omega} \in V$ such that there exists a nontrivial function $\widehat{\Phi}$ satisfying $\mathcal{N}^\varepsilon(\widehat{\omega})[\widehat{\Phi}] = 0$. Equivalently, we have

$$\widehat{\Phi} - \frac{1}{2\pi c_\delta} \ln(\widehat{\omega} - \omega_1^*) \frac{(\mathcal{A}^\varepsilon - \mathcal{A})(\omega_1^*)[\Psi_*]}{\langle \Phi_*, \frac{d}{d\omega} \mathcal{A}_* \big|_{\omega = \omega_1^*} [\Psi_*] \rangle} \langle \Phi_*, \widehat{\Phi} \rangle = 0.$$

Then, by multiplying by Φ_*, it follows that

$$\langle \Phi_*, \widehat{\Phi}\rangle \Big(1 - \frac{1}{2\pi c_\delta} \ln(\widehat{\omega} - \omega_1^*) \frac{\langle \Phi_*, (\mathcal{A}^\varepsilon - \mathcal{A})(\omega_1^*)[\Psi_*]\rangle}{\langle \Phi_*, \frac{d}{d\omega}\mathcal{A}_*\big|_{\omega=\omega_1^*}[\Psi_*]\rangle} \Big) = 0.$$

Since $\langle \Phi_*, \widehat{\Phi}\rangle = 0$ would imply $\widehat{\Phi} = 0$, we get

$$1 - \frac{1}{2\pi c_\delta} \ln(\widehat{\omega} - \omega_1^*) \frac{\langle \Phi_*, (\mathcal{A}^\varepsilon - \mathcal{A})(\omega_1^*)[\Psi_*]\rangle}{\langle \Phi_*, \frac{d}{d\omega}\mathcal{A}_*\big|_{\omega=\omega_1^*}[\Psi_*]\rangle} = 0. \tag{C.20}$$

Before solving the above equation for ω, we need the following lemma from [**36**].

LEMMA C.13. *The following results hold:*

(i) *When $\delta \to 0$, we have*

$$\big\langle \Phi_*, \frac{d}{d\omega}\mathcal{A}_*\big|_{\omega=\omega_1^*}[\Psi_*]\big\rangle = -2\pi\omega_1^* \ln \omega_1^* R^3 + \mathcal{O}(\sqrt{\delta}),$$

which is positive for δ small enough;

(ii) *For a fixed ε, when $\delta \to 0$, we have*

$$\begin{aligned}\big\langle \Phi_*, \big(\mathcal{A}^\varepsilon(\omega_1^*) - \mathcal{A}(\omega_1^*)\big)[\Psi_*]\big\rangle &= \delta\varepsilon \ln \omega_1^* \Big(R\|\psi^{\alpha^*}\|^2_{L^2(\partial D)} - 2\mathrm{Cap}_{\alpha^*,D}\Big) \\ &\quad + \mathcal{O}(\varepsilon\delta + \varepsilon^2\delta\ln\delta), \\ &:= \varepsilon\big(\delta \ln\omega_1^* S(R)\big) + \mathcal{O}(\varepsilon\delta + \varepsilon^2\delta\ln\delta),\end{aligned}$$

where $\psi^{\alpha^*} = (\mathcal{S}_D^{\alpha^*,0})^{-1}[\chi_{\partial D}]$, $\mathrm{Cap}_{\alpha^*,D} = -\langle \psi^{\alpha^*}, \chi_{\partial D}\rangle$ and

$$S(R) := R\|\psi^{\alpha^*}\|^2_{L^2(\partial D)} - 2\mathrm{Cap}_{\alpha^*,D}.$$

For small ε and δ, the sign of $\delta \ln \omega^* S(R)$ varies with R as follows: $\delta \ln \omega_1^* S(R) > 0$ for small enough R, while $\delta \ln \omega_1^* S(R) < 0$ for R close enough to $1/2$.

In view of Lemma C.13, we have two different regimes: The dilute regime when R is small and the nondilute regime when R is close to $1/2$. For equation (C.20) to have a real solution $\widehat{\omega}$ close to ω_1^*, we need $\varepsilon < 0$ in the dilute regime and $\varepsilon > 0$ in the nondilute regime. Under this assumption we have

$$\begin{aligned}&\widehat{\omega} - \omega_1^* \\ &\quad = \exp\Big(2\pi c_\delta \big\langle \Phi_*, \frac{d}{d\omega}\mathcal{A}_*\big|_{\omega=\omega_1^*}[\Psi_*]\big\rangle \big\langle \Phi_*, \big(\mathcal{A}^\varepsilon(\omega_1^*) - \mathcal{A}(\omega_1^*)\big)[\Psi_*]\big\rangle^{-1}\Big).\end{aligned} \tag{C.21}$$

We can also see that the right-hand side in (C.21) is positive and goes to zero as ε tends to zero. In other words, $\widehat{\omega} \to \omega_1^*$ as $\varepsilon \to 0$. Now we turn to the full operator $\mathcal{M}^\varepsilon$. Recall that $\mathcal{M}^\varepsilon = \mathcal{N}^\varepsilon + \mathcal{R}^\varepsilon$ and $\mathcal{R}^\varepsilon$ is holomorphic, bounded, and satisfies $\mathcal{R}^\varepsilon = \mathcal{O}(\varepsilon)$ in V. So by the asymptotic Gohberg-Sigal theory, we can conclude that there exists a characteristic value ω^ε of the operator-valued function $\omega \mapsto \mathcal{M}^\varepsilon$ near ω_1^*. Let us denote its associated root function by Φ^ε. We choose Φ^ε so that $\langle \Phi_*, \Phi^\varepsilon\rangle = 1$. Then, as in the derivations of (C.20), we can obtain

$$1 - \frac{1}{2\pi c_\delta} \ln(\omega^\varepsilon - \omega_1^*)\gamma(\varepsilon, \delta, R) + \langle \Phi_*, \mathcal{R}^\varepsilon(\omega^\varepsilon)[\Phi^\varepsilon]\rangle = 0,$$

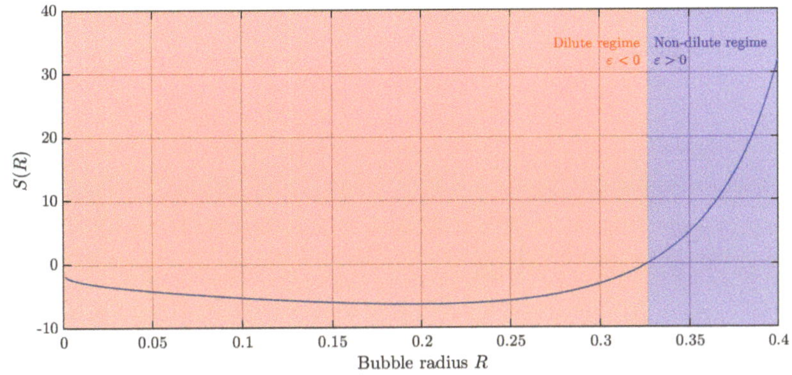

FIGURE C.1. Illustration of the sign of $S(R)$.

where

$$\gamma(\varepsilon, \delta, R) = \frac{\langle \Phi_*, (\mathcal{A}^\varepsilon - \mathcal{A})(\omega_1^*)[\Psi_*] \rangle}{\langle \Phi_*, \frac{d}{d\omega}\mathcal{A}_*\big|_{\omega=\omega_1^*}[\Psi_*] \rangle}.$$

Note that $\langle \Phi_*, \mathcal{R}^\varepsilon(\omega^\varepsilon)[\Phi^\varepsilon] \rangle = \mathcal{O}(\varepsilon)$ for small ε. We see from Lemma C.13 that

$$\gamma(\varepsilon, \delta, R) = -\frac{\delta\varepsilon \left(R\|\psi^{\alpha^*}\|^2_{L^2(\partial D)} - 2\mathrm{Cap}_{\alpha^*, D} \right)}{2\pi\omega_1^* R^3} + \mathcal{O}\Big(\varepsilon \frac{\sqrt{\delta}}{\ln \delta} + \sqrt{\delta}\varepsilon^2\Big),$$

as $\delta \to 0$ and $\varepsilon \to 0$, and the leading-order term is negative.

Note that the constant μ in Theorem 4.56 is given by

$$\mu = \frac{4\pi^2 c_\delta \omega_1^* R^3}{R\|\psi^{\alpha^*}\|^2_{L^2(\partial D)} - 2\mathrm{Cap}_{\alpha^*, D}} = \frac{4\pi^2 c_\delta \omega_1^* R^3}{S(R)}$$

with c_δ being positive. Figure C.1 illustrates the sign of $S(R)$ as a function of R. The zero $S(R_0) = 0$ is attained at $R_0 = 0.326$.

We also remark that the frequency ω^ε is a real number since the operator in the defect problem (4.85) can be considered as a real compact perturbation of a self-adjoint periodic operator.

C.5. Proof of Lemma 4.66

Recall that the quasiperiodic capacitance matrix C_{ij}^α can be written as

$$C_{ij}^\alpha = -\int_{\partial D_i} (\mathcal{S}_D^{\alpha,0})^{-1}[\chi_{\partial D_j}] \, \mathrm{d}\sigma,$$

for $\alpha \neq 0$. We shall compute the asymptotics of $(\mathcal{S}_D^{\alpha,0})^{-1}$ for small ε.

Let us decompose the quasiperiodic Green function $G^{\alpha,k}$ as

$$G^{\alpha,k}(x,y) = G^k(x,y) + \widetilde{G}^{\alpha,k}(x,y),$$

where

$$\widetilde{G}^{\alpha,k}(x,y) := \sum_{m\neq 0} e^{\mathrm{i}m\alpha L} G^k(x, y + (mL, 0, 0)).$$

Then let us define

$$\widetilde{\mathcal{S}}_D^{\alpha,0}[\varphi] = \int_{\partial D} \widetilde{G}^{\alpha,k}(x,y)\varphi(y) \, \mathrm{d}\sigma(y).$$

Note that $S_D^{\alpha,0} = S_D^0 + \widetilde{S}_D^{\alpha,0}$.

Let us write the quasiperiodic single layer potential $\mathcal{S}_D^{\alpha,0}$ in a matrix form as

$$\mathcal{S}_D^{\alpha,0} = \begin{pmatrix} \mathcal{S}_{D_1}^{\alpha,0} & \mathcal{S}_{D_2}^{\alpha,0}\big|_{\partial D_1} \\ \mathcal{S}_{D_1}^{\alpha,0}\big|_{\partial D_2} & \mathcal{S}_{D_2}^{\alpha,0} \end{pmatrix},$$

and then decompose it as

$$\begin{aligned}\mathcal{S}_D^{\alpha,0} &= \begin{pmatrix} S_{D_1}^0 & 0 \\ 0 & S_{D_2}^0 \end{pmatrix} + \begin{pmatrix} \widetilde{S}_{D_1}^{\alpha,0} & S_{D_2}^0\big|_{\partial D_1} + \widetilde{S}_{D_2}^{\alpha,0}\big|_{\partial D_1} \\ S_{D_1}^0\big|_{\partial D_2} + \widetilde{S}_{D_1}^{\alpha,0}\big|_{\partial D_2} & \widetilde{S}_{D_2}^{\alpha,0} \end{pmatrix} \\ &:= S_I + S_{II}.\end{aligned}$$

As in the proof of Lemma 1.19, in order to keep the order of the norms in $L^2(\partial D)$ and $H^1(\partial D)$ constant as $\varepsilon \to 0$, we let $\mathbb{L}$ and $\mathbb{H}$, respectively, denote the spaces $L^2(\partial D)$ and $H^1(\partial D)$ along with the inner products

$$\langle \cdot,\cdot\rangle_{\mathbb{L}} = \frac{1}{|\partial D|}\langle \cdot,\cdot\rangle_{L^2(\partial D)}, \qquad \langle \cdot,\cdot\rangle_{\mathbb{H}} = \frac{1}{|\partial D|}\langle \cdot,\cdot\rangle_{H^1(\partial D)}.$$

Next, we estimate the operator norms of S_I and S_{II}. We first recall from the proof of Lemma 1.19 that

$$\|S_I\|_{\mathcal{L}(\mathbb{L}^2,\mathbb{H}^2)} \lesssim \varepsilon, \quad \|S_I^{-1}\|_{\mathcal{L}(\mathbb{H}^2,\mathbb{L}^2)} \lesssim \varepsilon^{-1}. \tag{C.22}$$

Let us now consider S_{II}. We introduce the notation

$$z_1 = \Big(-\frac{l}{2},0,0\Big), \quad z_{1,m} = \Big(-\frac{l}{2}+mL,0,0\Big), \quad z_2 = \Big(\frac{l}{2},0,0\Big),$$

and

$$z_{2,m} = \Big(\frac{l}{2}+mL,0,0\Big).$$

We have, for small ε, that

$$\begin{aligned}\widetilde{\mathcal{S}}_{D_j}^{\alpha,0}[\varphi]\big|_{\partial D_i}(x) &= \int_{\partial D_j}\sum_{m\neq 0} e^{\mathrm{i}m\alpha L}\Big(G^0(x,z_{j,m}) + (y-z_j)\cdot\nabla_y G^0(x,y_m^*)\Big)\varphi(y)\,\mathrm{d}\sigma(y) \\ &= -\sum_{m\neq 0}\frac{e^{\mathrm{i}m\alpha L}\chi_{\partial D_i}(x)}{4\pi|z_i - z_{j,m}|}\int_{\partial D_j}\varphi(y)\,\mathrm{d}\sigma(y) \\ &\quad + \mathcal{O}\Big(\sum_{m\neq 0}\frac{\varepsilon\int_{\partial D_j}|\varphi(y)|\,\mathrm{d}\sigma(y)}{|z_i - z_{j,m}|^2}\Big).\end{aligned}$$

Here, y_m^* means a point on the line segment joining y and z_j. Note that the series in the remainder term converges. Moreover, the gradient of the remainder term is of the same order. Since

$$\int_{\partial D_j}\varphi\,\mathrm{d}\sigma = \mathcal{O}(\varepsilon^2\|\varphi\|_{\mathbb{L}}),$$

we get

$$\begin{aligned}\widetilde{\mathcal{S}}_{D_j}^{\alpha,0}[\varphi]\big|_{\partial D_i}(x) &= -\sum_{m\neq 0}\frac{e^{\mathrm{i}m\alpha L}\chi_{\partial D_i}(x)}{4\pi|z_i - z_{j,m}|}\int_{\partial D_j}\varphi(y)\,\mathrm{d}\sigma(y) + \mathcal{O}(\varepsilon^3\|\varphi\|_{\mathbb{L}}), \\ \nabla_{\partial D}\widetilde{\mathcal{S}}_{D_j}^{\alpha,0}[\varphi]\big|_{\partial D_i}(x) &= \mathcal{O}(\varepsilon^3\|\varphi\|_{\mathbb{L}}).\end{aligned} \tag{C.23}$$

Here, $\nabla_{\partial D}$ is used to denote the surface gradient on ∂D. From this we have

$$\|\widetilde{\mathcal{S}}_{D_j}^{\alpha,0}|_{\partial D_i}\|_{\mathcal{L}(\mathbb{L},\mathbb{H})} \lesssim \varepsilon^2.$$

Similarly, we can show that

$$\mathcal{S}_{D_j}^{0}[\varphi]|_{\partial D_i}(x) = -\frac{\chi_{\partial D_i}(x)}{4\pi|z_i - z_j|}\int_{\partial D_j} \varphi(y)\ \mathrm{d}\sigma(y) + \mathcal{O}(\varepsilon^3\|\varphi\|_{\mathbb{L}}), \tag{C.24}$$

$$\nabla_{\partial D}\mathcal{S}_{D_j}^{0}[\varphi]|_{\partial D_i}(x) = \mathcal{O}(\varepsilon^3\|\varphi\|_{\mathbb{L}}),$$

and $\|\mathcal{S}_{D_j}^{0}|_{\partial D_i}\|_{\mathcal{L}(\mathbb{L},\mathbb{H})} \lesssim \varepsilon^2$. These estimates imply that

$$\|S_{II}\|_{\mathcal{L}(\mathbb{L}^2,\mathbb{H}^2)} \lesssim \varepsilon^2. \tag{C.25}$$

We now compute the asymptotic behaviour of $(\mathcal{S}_D^{\alpha,0})^{-1}$. We use the definition $\psi_j := S_I^{-1}[\chi_{\partial D_j}]$ and introduce the capacitance of each individual resonator D_j as $\mathrm{Cap}_{D_j} := -\int_{\partial D_j}\psi_j\ \mathrm{d}\sigma$. Note that $\mathrm{Cap}_{D_j} = \varepsilon\mathrm{Cap}_B$. Since we know from (C.22) and (C.25) that $S_I^{-1}S_{II} = \mathcal{O}(\varepsilon)$ in the operator norm, applying the Neumann series gives

$$\begin{aligned}(\mathcal{S}_D^{\alpha,0})^{-1}[\chi_{\partial D_j}] &= (S_I + S_{II})^{-1}[\chi_{\partial D_j}]\\ &= (I + S_I^{-1}S_{II})^{-1}S_I^{-1}[\chi_{\partial D_j}]\\ &= (I - S_I^{-1}S_{II})[\psi_j] + \mathcal{O}(\varepsilon).\end{aligned} \tag{C.26}$$

We also have

$$|z_1 - z_{1,m}| = |mL|, \quad |z_1 - z_{2,m}| = |mL + l|.$$

Then, from (C.23) and (C.24), together with the fact that

$$\|\psi_j\|_{\mathbb{L}} = \|S_I^{-1}[\chi_{\partial D_j}]\|_{\mathbb{L}} \lesssim \varepsilon^{-1},\ j = 1, 2,$$

we obtain the series representations

$$S_{II}[\phi_1]|_{\partial D_1} = \sum_{m\neq 0} e^{\mathrm{i}m\alpha L}\frac{\varepsilon\mathrm{Cap}_B}{4\pi|mL|}\chi_{\partial D_1} + \mathcal{O}(\varepsilon^2),$$

$$S_{II}[\phi_2]|_{\partial D_2} = \sum_{m\in\mathbb{Z}} e^{\mathrm{i}m\alpha L}\frac{\varepsilon\mathrm{Cap}_B}{4\pi|mL + d|}\chi_{\partial D_2} + \mathcal{O}(\varepsilon^2).$$

We are now ready to compute the capacitance matrix. From (C.26), we have

$$\begin{aligned}C_{11}^{\alpha} &= -\int_{\partial D_1}(\mathcal{S}_D^{\alpha,0})^{-1}[\chi_{\partial D_1}]\ \mathrm{d}\sigma = -\int_{\partial D_1}(I - S_I^{-1}S_{II})[\psi_1]\ \mathrm{d}\sigma - \int_{\partial D_1}\mathcal{O}(\varepsilon)\ \mathrm{d}\sigma\\ &= -\int_{\partial D_1}\phi_1\ \mathrm{d}\sigma + \int_{\partial D_1} S_I^{-1}\Big[\sum_{m\neq 0} e^{\mathrm{i}m\alpha L}\frac{\varepsilon\mathrm{Cap}_B}{4\pi|mL|}\chi_{\partial D_1}\Big]\ \mathrm{d}\sigma + \mathcal{O}(\varepsilon^3)\\ &= \mathrm{Cap}_{D_1} - \sum_{m\neq 0} e^{\mathrm{i}m\alpha L}\frac{\varepsilon\mathrm{Cap}_B}{4\pi|mL|}\Big(-\int_{\partial D_1}\psi_1\ \mathrm{d}\sigma\Big) + \mathcal{O}(\varepsilon^3)\\ &= \varepsilon\mathrm{Cap}_B - \sum_{m\neq 0} e^{\mathrm{i}m\alpha L}\frac{(\varepsilon\mathrm{Cap}_B)^2}{4\pi|mL|} + \mathcal{O}(\varepsilon^3).\end{aligned}$$

The expression for C_{12}^{α} can be derived in the same way.

C.6. Proof of Lemma 4.69

Part (i): Im $C_{12}^{\alpha} > 0$ for $0 < \alpha < \pi/L$ and Im $C_{12}^{\alpha} < 0$ for $-\pi/L < \alpha < 0$. In particular, Im C_{12}^{α} is zero if and only if $\alpha \in \{0, \pi/L\}$.

Recall, from Lemma 4.66, the following expansion of $C_{12}^{\pi/L}$ for fixed $\alpha \neq 0$ in the dilute regime:

$$C_{12}^{\alpha} = -\frac{(\varepsilon \mathrm{Cap}_B)^2}{4\pi} \sum_{m=-\infty}^{+\infty} \frac{e^{\mathrm{i}m\alpha L}}{|l + mL|} + \mathcal{O}(\varepsilon^3)$$

$$:= -\frac{(\varepsilon \mathrm{Cap}_B)^2}{4\pi L} f(\alpha, l) + \mathcal{O}(\varepsilon^3), \tag{C.27}$$

where, in order to simplify notation, we have defined the function f as

$$f(\alpha, l) := \sum_{m=-\infty}^{+\infty} \frac{e^{\mathrm{i}m\alpha L}}{|l/L + m|}. \tag{C.28}$$

The imaginary part

$$\mathrm{Im}\, f(\alpha, l) = \sum_{m=-\infty}^{+\infty} \frac{\sin(m\alpha L)}{|l/L + m|}$$

converges for all $\alpha \in Y^*$, and (C.27) is valid for imaginary parts also at $\alpha = 0$.

We will express f in terms of Lerch's Transcendent function $\Phi(z, s, a)$, after having first reviewed some basic properties. For details, we refer to [**78**]. $\Phi(z, s, a)$ is defined by the power series

$$\Phi(z, s, a) = \sum_{m=0}^{+\infty} \frac{z^m}{(a + m)^s} \tag{C.29}$$

for $z \in \mathbb{C}$, where this series converges, and extended by analytic continuation. If $\mathrm{Re}\,(s) > 0, \mathrm{Re}\,(a) > 0$ and $z \in \mathbb{C} \setminus [1, +\infty)$ we have the integral representation

$$\Phi(z, s, a) = \frac{1}{\Gamma(s)} \int_0^{+\infty} \frac{t^{s-1} e^{-at}}{1 - z e^{-t}}\, \mathrm{d}t, \tag{C.30}$$

where Γ is the Gamma function.

Now, from the definition of f in (C.28), we have

$$f(\alpha, l) = \Phi(e^{\mathrm{i}\alpha L}, 1, l/L) + e^{-\mathrm{i}\alpha L} \Phi(e^{-\mathrm{i}\alpha L}, 1, 1 - l/L).$$

Using the integral representation (C.30) we get, after simplifications,

$$f(\alpha, l) = \int_0^{+\infty} \frac{e^{-\mathrm{i}\alpha L} \sinh\left(\frac{l}{L} t\right) + \sinh\left(\frac{1-l}{L} t\right)}{\cosh(t) - \cos(\alpha L)}\, \mathrm{d}t. \tag{C.31}$$

The imaginary part satisfies

$$\mathrm{Im}\, f(\alpha, l) = \sin(\alpha L) \int_0^{+\infty} \frac{\sinh\left(\frac{l}{L} t\right)}{\cos(\alpha L) - \cosh(t)}\, \mathrm{d}t. \tag{C.32}$$

At the points $\alpha = 0$ and $\alpha = \pi/L$, the functions V_1^{α} and V_2^{α} (defined by (4.60)) are real-valued and hence, Im $C_{12}^{\alpha} = 0$. We will show that, for ε small enough, this imaginary part is zero precisely when $\alpha \in \{0, \pi/L\}$. The integrand in (C.32) is positive, and hence Im $f(\alpha, l) = 0$ if and only if $\alpha \in \{0, \pi/L\}$ for

$\alpha \in Y^* = [-\pi/L, \pi/L]$. This shows that the leading-order term of Im C_{12}^α is zero precisely when $\alpha \in \{0, \pi/L\}$. Moreover, we can easily verify from (C.32) that

$$\frac{\partial}{\partial \alpha} \text{Im } f(\alpha, l)\big|_{\alpha=0} \neq 0, \qquad \frac{\partial}{\partial \alpha} \text{Im } f(\alpha, l)\big|_{\alpha=\pi/L^-} \neq 0,$$

and

$$\frac{\partial}{\partial \alpha} \text{Im } f(\alpha, l)\big|_{\alpha=-\pi/L^+} \neq 0.$$

This shows that for small enough ε, the function Im C_{12}^α will be monotonic around $\alpha = 0$ and $\alpha = \pi/L$. Since we know that $\alpha \in \{0, \pi/L\}$ are exact zeros of Im C_{12}^α, these zeros will be isolated for small enough ε. It follows that, for small enough ε, Im C_{12}^α is zero precisely when $\alpha \in \{0, \pi/L\}$. Then, from (C.32) it follows that Im $C_{12}^\alpha > 0$ for $0 < \alpha < \pi/L$ and Im $C_{12}^\alpha < 0$ for $-\pi/L < \alpha < 0$.

Part (ii): C_{12}^α is zero if and only if both $l = l'$ and $\alpha = \pi/L$.

By part (i), any zero must satisfy $\alpha = 0$ or $\alpha = \pi/L$. We begin by excluding the case $\alpha = 0$. As $\alpha \to 0$, we know from (4.12) that the quasiperiodic capacitance of a single resonator vanishes. In other words, we have, for the total capacitance of D,

$$\begin{aligned} 0 &= \int_{Y\setminus D} \nabla(V_1^0 + V_2^0) \cdot \overline{\nabla(V_1^0 + V_2^0)} \, dx \\ &= 2(C_{11}^0 + C_{12}^0), \end{aligned}$$

where the last equality follows since C_{12}^0 is real. Since $C_{11}^\alpha > 0$, we have $C_{12}^0 < 0$.

We now turn to the case $\alpha = \pi/L$. We already know from Lemma 4.68 that $C_{12}^{\pi/L} = 0$ if $l = l'$. To show that this is the only zero of $C_{12}^{\pi/L}$, we will show that $C_{12}^{\pi/L}$ is strictly monotonic as a function of $l < L$. From the definition of f we compute the derivative

$$\frac{\partial}{\partial l} f(\alpha, l)\bigg|_{\alpha=\pi/L} = -\frac{L}{l^2} + \sum_{m=1}^{+\infty} \frac{(-1)^m}{L} \left[\frac{1}{(m - l/L)^2} - \frac{1}{(m + l/L)^2}\right].$$

The sum is an alternating series, with decreasing terms and a negative first term. Hence, it converges to a negative value, and by (C.27) we have

$$\frac{\partial}{\partial l} C_{12}^{\pi/L} > 0 \tag{C.33}$$

for ε small enough. This shows that $C_{12}^{\pi/L}$ has a unique zero when $l = l'$, which completes the proof of part (ii).

Part (iii): $C_{12}^{\pi/L} < 0$ when $l < l'$ and $C_{12}^{\pi/L} > 0$ when $l > l'$. In both cases, we have $C_{12}^0 < 0$.

We already know, from the proof of part (ii), that $C_{12}^0 < 0$. The other conclusions, namely that $C_{12}^{\pi/L} < 0$ if $l < l'$ and $C_{12}^{\pi/L} > 0$ if $l > l'$, follow directly from (C.33).

C.7. Proof of Theorem 4.71

We begin by proving the following lemma.

LEMMA C.14. *For every a, b such that $-1 \leq a < 1$ and $0 < b < 1$, the following holds:*

$$\int_0^{+\infty} \frac{\sinh(bt)}{\cosh(t) - a} \, dt > -\frac{2b}{1+a} \ln\left(\frac{1}{2}(1 - a)\right). \tag{C.34}$$

PROOF. We will split the proof into the cases $a > 0$ and $a \leq 0$, and begin with $a > 0$. The right-hand side can be written as follows:

$$-\frac{2b}{1+a}\ln\left(\frac{1}{2}(1-a)\right) = \int_0^{+\infty} \frac{b\sinh(t)}{\cosh(t)(\frac{1-a}{2}\cosh(t)+\frac{1+a}{2})}\,dt. \tag{C.35}$$

Indeed, the integrand has a primitive function

$$\frac{2b}{1+a}\ln\left(\frac{\cosh(t)}{\cosh(t)+\frac{1+a}{1-a}}\right),$$

which shows (C.35). Then we have, for $a > 0$,

$$\begin{aligned}
&\int_0^{+\infty} \frac{\sinh(bt)}{\cosh(t)-a}\,dt + \frac{2b}{1+a}\ln\left(\frac{1}{2}(1-a)\right) \\
&= \int_0^{+\infty}\left(\frac{\sinh(bt)}{\cosh(t)-a} - \frac{b\sinh(t)}{\cosh(t)(\frac{1-a}{2}\cosh(t)+\frac{1+a}{2})}\right)dt \\
&> \int_0^{+\infty}\frac{1}{\cosh(t)}\left(\sinh(bt) - \frac{b\sinh(t)}{\frac{1-a}{2}\cosh(t)+\frac{1+a}{2}}\right)dt > 0,
\end{aligned}$$

where the last step follows because

$$\sinh(bt) - \frac{b\sinh(t)}{\frac{1-a}{2}\cosh(t)+\frac{1+a}{2}} > 0$$

for all $t > 0$ in the case $a > 0$. This proves the lemma in the case $a > 0$. We now turn to the case $a \leq 0$. It is easy to see that for every b,

$$\begin{aligned}
&\min_{-1\leq a\leq 0}\int_0^{+\infty}\frac{\sinh(bt)}{\cosh(t)-a}\,dt = \int_0^{+\infty}\frac{\sinh(bt)}{\cosh(t)+1}\,dt, \\
&\max_{-1\leq a\leq 0} -\frac{2b}{1+a}\ln\left(\frac{1}{2}(1-a)\right) = 2b\ln(2).
\end{aligned} \tag{C.36}$$

Moreover, we have

$$\int_0^{+\infty}\frac{\sinh(bt)}{\cosh(t)+1}\,dt > \int_0^{+\infty}\frac{bt}{\cosh(t)+1}\,dt = 2b\ln(2),$$

where we have used a known value for the integral (for example found in [**179**]). Together with (C.36), this proves the lemma in the case $a \leq 0$. □

Proof of Theorem 4.71. We will show that there is a frequency ω_0 such that

$$\max_{|\alpha|>\alpha_0}\omega_1^\alpha < \omega_0 < \min_{|\alpha|>\alpha_0}\omega_2^\alpha.$$

For sufficiently small δ, by Theorem 4.67 and Lemma 4.65, it is enough to show that there is a constant C_0 such that

$$\max_{|\alpha|>\alpha_0} C_{11}^\alpha - |C_{12}^\alpha| < C_0 < \min_{|\alpha|>\alpha_0} C_{11}^\alpha + |C_{12}^\alpha|. \tag{C.37}$$

Define C_0 as

$$C_0 = \varepsilon\mathrm{Cap}_B - \frac{(\varepsilon\mathrm{Cap}_B)^2}{4\pi L}\sum_{m\neq 0}\frac{(-1)^m}{|m|},$$

that is, C_0 is defined as the leading-order of the eigenvalues of C^α at the degenerate point $l = l', \alpha = \pi/L$. The sum appearing in the expansion of C_{11}^α can be explicitly computed as

$$\sum_{m\neq 0} \frac{e^{\mathrm{i}m\alpha L}}{|m|} = -\ln\left(2 - 2\cos(\alpha L)\right).$$

Then we have

$$\begin{aligned} C_{11}^\alpha - |C_{12}^\alpha| - C_0 &= \frac{(\varepsilon \mathrm{Cap}_B)^2}{4\pi L} \ln\left(\frac{1}{2}(1-\cos(\alpha L))\right) - |C_{12}^\alpha| + \mathcal{O}(\varepsilon^3) \\ &< 0 \end{aligned}$$

for ε small enough. This shows that, for sufficiently small ε,

$$\max_{|\alpha|>\alpha_0} C_{11}^\alpha - |C_{12}^\alpha| < C_0.$$

We now turn to the second inequality of (C.37). By (C.27) and (C.31) we have

$$\begin{aligned} C_{11}^\alpha + |C_{12}^\alpha| - C_0 = \frac{(\varepsilon \mathrm{Cap}_B)^2}{4\pi L} \Bigg(&\ln\left(\frac{1}{2}(1-\cos(\alpha L))\right) \\ &+ \left| \int_0^{+\infty} \frac{e^{\mathrm{i}\alpha L} \sinh\left(\frac{l}{L}t\right) + \sinh\left(\frac{1-l}{L}t\right)}{\cosh(t) - \cos(\alpha L)} \,\mathrm{d}t \right| \Bigg) + \mathcal{O}(\varepsilon^3). \end{aligned} \tag{C.38}$$

Recall that $0 < l < L$, so we can apply (C.34) with $b = l/L$ or with $b = (1-l)/L$ and with $a = \cos(\alpha L)$. Expanding the absolute value and applying (C.34), we find, after simplifications, that

$$\begin{aligned} \left| \int_0^{+\infty} \frac{e^{\mathrm{i}\alpha L} \sinh\left(\frac{l}{L}t\right) + \sinh\left(\frac{1-l}{L}t\right)}{\cosh(t) - \cos(\alpha L)} \,\mathrm{d}t \right|^2 &> \ln\left(\frac{1}{2}(1-\cos(\alpha L))\right)^2 \\ &\quad \times \frac{4\left(l^2 + 2l(1-l)\cos(\alpha L) + (1-l)^2\right)}{L^2\left(1+\cos(\alpha L)\right)^2} \\ &> \ln\left(\frac{1}{2}(1-\cos(\alpha L))\right)^2. \end{aligned}$$

Together with (C.38), this shows that

$$\min_{|\alpha|>\alpha_0} C_{11}^\alpha + |C_{12}^\alpha| > C_0$$

for ε small enough. We have thus proved (C.37), from which the theorem follows.

C.8. Integer unit length dislocation in a chain of dimers

With the notation of Section 4.6.3 in hand, we study edge modes in dislocated chains of dimers when the dislocation d is an integer number of unit cell lengths. This is equivalent to the case where an integer multiple of dimers are removed from the original periodic structure, thus creating a cavity. We model this defect cavity problem using the fictitious source method described in Section 4.5.1. For simplicity, we assume that the unit cell length $L = 1$. The results in this appendix are from [**22**].

C.8.1. Fictitious sources for a removed resonator. Here, we describe the method of fictitious sources when a single dimer $D = D_1 \cup D_2$ is removed. We assume that we have a reference solution u satisfying

$$\Delta u + \omega^2 u = 0 \qquad \text{in } Y.$$

Let $\widetilde{u}$ satisfy the fictitious source problem

$$\text{(C.39)} \qquad \begin{cases} \Delta \widetilde{u} + \omega^2 \widetilde{u} = 0 & \text{in } Y \setminus \partial D, \\ \widetilde{u}|_+ - \widetilde{u}|_- = 0 & \text{on } \partial D, \\ \delta \frac{\partial \widetilde{u}}{\partial \nu}\Big|_+ - \frac{\partial \widetilde{u}}{\partial \nu}\Big|_- = g & \text{on } \partial D. \end{cases}$$

We want to determine the fictitious source g such that $u = \widetilde{u}$ inside Y. Any solution $\widetilde{u}$ can be represented as

$$\text{(C.40)} \qquad \widetilde{u} = \begin{cases} \mathcal{S}_D^\omega[\widetilde{\psi}] & \text{in } D, \\ \mathcal{S}_D^\omega[\widetilde{\phi}] + H & \text{in } Y \setminus D, \end{cases}$$

for some H satisfying $\Delta H + \omega^2 H = 0$ in Y. Imposing $\widetilde{u} = u$ in Y is equivalent to

$$\widetilde{\psi} = (\mathcal{S}_D^\omega)^{-1}[u|_{\partial D}], \qquad \widetilde{\phi} = 0, \qquad H = u.$$

Moreover, using the jump conditions in (C.39), we find the following expression of g.

PROPOSITION C.15. *The fictitious source g and the layer density ψ satisfy*

$$\text{(C.41)} \qquad g = B_D(\omega, \delta)[\widetilde{\psi}], \quad B_D(\omega, \delta) := (\delta - 1)\left(-\frac{1}{2}I + \mathcal{K}_D^{\omega,*}\right).$$

Conversely, by the unique continuation property of the Helmholtz equation, if g satisfies Proposition C.15, then $\widetilde{u} = u$ in Y.

C.8.2. Integral equation for the dislocated system. We now assume that $2N$ resonators are removed (see Figure C.2), so that u satisfies

$$\text{(C.42)} \qquad \begin{cases} \Delta u + \omega^2 u = 0 & \text{in } \mathbb{R}^3 \setminus \partial \mathcal{D}_{\text{disloc}}, \\ u|_+ - u|_- = 0 & \text{on } \partial \mathcal{D}_{\text{disloc}}, \\ \delta \frac{\partial u}{\partial \nu}\Big|_+ - \frac{\partial u}{\partial \nu}\Big|_- = 0 & \text{on } \partial \mathcal{D}_{\text{disloc}}, \\ u(x_1, x_2, x_3) \text{ satisfies the outgoing radiation condition} \\ \text{as } \sqrt{x_2^2 + x_3^2} \to +\infty, \end{cases}$$

with

$$\text{(C.43)} \qquad \mathcal{D}_{\text{disloc}} = \mathcal{D} \setminus \left(\bigcup_{m=0}^{N-1} D^m\right), \qquad D^m = D + (m, 0, 0).$$

To model this using the fictitious source method, we put g_m as in Proposition C.15 for all removed resonators.

Assume that we have a nonzero solution u to (C.42). Inside

$$Y^m := Y + (m, 0, 0), m = 0, 1, \ldots, N - 1,$$

we can define the layer density $\widetilde{\psi}_m$ as

$$\text{(C.44)} \qquad \widetilde{\psi}_m = \left(\mathcal{S}_{D^m}^\omega\right)^{-1}[u|_{\partial D^m}].$$

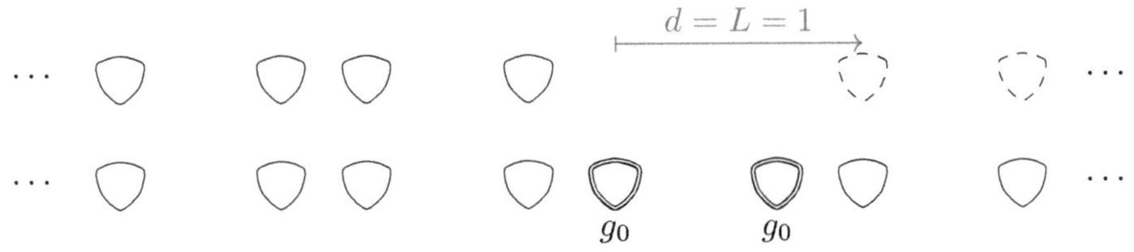

FIGURE C.2. The dislocated system with dislocation equal to a multiple of the length of the unit cell (that is, $d = NL$) is equivalent to the original array with the addition of so-called fictitious sources g_m on the boundary of D^m for $m = 0, \ldots, N-1$. The case $N = 1$ is depicted here. Legend: — untouched resonator, ═ resonator with fictitious sources, - - - dislocated resonator.

We then set the fictitious sources as

$$g_m = B_{D^m}[\widetilde{\psi}_m], \quad 0 < m < N-1, \tag{C.45}$$

and $g_m = 0$ otherwise. Then, putting $\widetilde{\phi} = 0$ and $H = u$, we obtain a solution $\widetilde{u}$ defined by (C.40) (with D, Y replaced with D^m, Y^m), which coincides with u on $Y^m \setminus D^m$.

Conversely, if we have a nonzero solution $\widetilde{u}$ to

$$\begin{cases} \Delta \widetilde{u} + \omega^2 \widetilde{u} = 0 & \text{in } \mathbb{R}^3 \setminus \partial \mathcal{D}, \\ \widetilde{u}|_+ - \widetilde{u}|_- = 0 & \text{on } \partial D^m, m \in \mathbb{N}, \\ \delta \dfrac{\partial \widetilde{u}}{\partial \nu}\Big|_+ - \dfrac{\partial \widetilde{u}}{\partial \nu}\Big|_- = g_m & \text{on } \partial D^m, m \in \mathbb{N}, \\ \widetilde{u}(x_1, x_2, x_3) \text{ satisfies the outgoing radiation condition} \\ \text{ as } \sqrt{x_2^2 + x_3^2} \to +\infty, \end{cases} \tag{C.46}$$

represented as (C.40) in Y^m and with sources satisfying (C.45), then we can define a nonzero solution $u = \widetilde{u}$ to (C.42) coinciding with $\widetilde{u}$ on $Y^m \setminus D^m$.

We introduce the extended operator on $(L^2(\partial D))^2$,

$$B = \begin{pmatrix} 0 & 0 \\ B_D & 0 \end{pmatrix}, \tag{C.47}$$

where B_D is given by (C.41).

For $\alpha \in Y^*$, define $\mathcal{B}^\alpha : (L^2(\partial D))^{2N} \to (L^2(\partial D))^2$ blockwise as

$$\mathcal{B}^\alpha = \begin{pmatrix} B & e^{-\mathrm{i}\alpha} B & \cdots & e^{-(N-1)\mathrm{i}\alpha} B \end{pmatrix},$$

and define $E^\alpha : (L^2(\partial D))^2 \to (L^2(\partial D))^{2N}$ blockwise as

$$E^\alpha = \begin{pmatrix} I \\ e^{\mathrm{i}\alpha} I \\ e^{2\mathrm{i}\alpha} I \\ \vdots \\ e^{(N-1)\mathrm{i}\alpha} I \end{pmatrix}.$$

Next, we derive an integral equation for the dislocated system. By taking the Floquet transform, we obtain

$$\begin{cases} \Delta u^\alpha + \omega^2 u^\alpha = 0 & \text{in } Y \setminus \partial D, \\ u^\alpha|_+ - u^\alpha|_- = 0 & \text{on } \partial D, \\ \delta \dfrac{\partial u^\alpha}{\partial \nu}\Big|_+ - \dfrac{\partial u^\alpha}{\partial \nu}\Big|_- = g^\alpha & \text{on } \partial D, \\ e^{-\mathrm{i}\alpha x_1} u^\alpha(x_1, x_2, x_3) \text{ is periodic in } x_1, \\ u^\alpha(x_1, x_2, x_3) \text{ satisfies the } \alpha\text{-quasiperiodic} \\ \text{ outgoing radiation condition as } \sqrt{x_2^2 + x_3^2} \to +\infty, \end{cases} \tag{C.48}$$

where

$$g^\alpha = \sum_{m \in \mathbb{Z}} g_m e^{-\mathrm{i}\alpha m}, \tag{C.49}$$

together with the relation

$$\widetilde{\psi}_m = \frac{1}{2\pi} \int_{Y^*} e^{\mathrm{i}\alpha m} \widetilde{\psi}^\alpha \, \mathrm{d}\alpha, \tag{C.50}$$

for $\widetilde{\psi}_m$. Putting

$$\widetilde{\Phi}_N = \begin{pmatrix} \widetilde{\psi}_0 \\ 0 \\ \vdots \\ \widetilde{\psi}_{N-1} \\ 0 \end{pmatrix},$$

we then obtain the following result.

PROPOSITION C.16. *For $\mathcal{D}_{\text{disloc}}$ as in* (C.43)*, the midgap frequencies of* (C.42) *are precisely the values ω such that there is a nonzero solution $\widetilde{\Phi}_N \in \big(L^2(\partial D)\big)^{2N}$ to the equation*

$$\widetilde{\Phi}_N = -\frac{1}{2\pi} \left(\int_{Y^*} E^\alpha \big(\mathcal{A}^\alpha(\omega, \delta)\big)^{-1} \mathcal{B}^\alpha \, \mathrm{d}\alpha \right) \widetilde{\Phi}_N, \tag{C.51}$$

where

$$\mathcal{A}^\alpha(\omega, \delta) = \begin{pmatrix} \mathcal{S}_D^\omega & -\mathcal{S}_D^{\alpha,\omega} \\ -\frac{1}{2} I + \mathcal{K}_D^{\omega,*} & -\delta \left(\frac{1}{2} I + \left(\mathcal{K}_D^{-\alpha,\omega} \right)^* \right) \end{pmatrix}. \tag{C.52}$$

In order to analyse (C.51), we will need the following lemma, which is an immediate consequence of the structure of B. Before stating the lemma, we introduce the following notation. We let $\omega_j^\alpha, j = 1, 2$, be the subwavelength resonant frequencies of the operator $\mathcal{A}^\alpha(\omega, \delta)$ defined as in Theorem 4.67. We let C^α be the quasiperiodic capacitance matrix associated with the dimer D. For α such that $C_{12}^\alpha \neq 0$, we let θ_α be as in Lemma 4.65 and introduce the basis $\{u_1, u_2\}$ of $\operatorname{Ker}\Big(-\frac{1}{2} I + \widehat{\mathcal{K}}_D^{0,*}\Big) \subset L^2(\partial D)$ as

$$u_1 = \frac{1}{\sqrt{2}} \left(-e^{\mathrm{i}\theta_\alpha} \psi_1 + \psi_2 \right), \qquad u_2 = \frac{1}{\sqrt{2}} \left(e^{\mathrm{i}\theta_\alpha} \psi_1 + \psi_2 \right),$$

where $\psi_j = (\mathcal{S}_{D_j}^0)^{-1}[\chi_{\partial D_j}], j = 1, 2$.

Moreover, we define the bases $\{u_1^\alpha, u_2^\alpha\}$, $\{\chi_1^\alpha, \chi_2^\alpha\}$ of $\mathrm{Ker}\Big(-\frac{1}{2}I + \big(\mathcal{K}_D^{-\alpha,0}\big)^*\Big)$ and $\mathrm{Ker}\Big(-\frac{1}{2}I + \mathcal{K}_D^{\alpha,0}\Big)$, respectively, as

$$u_1^\alpha = \frac{1}{\sqrt{2}}\left(-e^{\mathrm{i}\theta_\alpha}\psi_1^\alpha + \psi_2^\alpha\right), \qquad u_2^\alpha = \frac{1}{\sqrt{2}}\left(e^{\mathrm{i}\theta_\alpha}\psi_1^\alpha + \psi_2^\alpha\right),$$
$$\chi_1^\alpha = \frac{1}{\sqrt{2}}\left(-e^{\mathrm{i}\theta_\alpha}\chi_{\partial D_1} + \chi_{\partial D_2}\right), \qquad \chi_2^\alpha = \frac{1}{\sqrt{2}}\left(e^{\mathrm{i}\theta_\alpha}\chi_{\partial D_1} + \chi_{\partial D_2}\right),$$

where $\psi_j^\alpha = (\mathcal{S}_D^{\alpha,0})^{-1}[\chi_{\partial D_j}], j = 1, 2$.

LEMMA C.17. *We have*

$$\big(\mathcal{A}^\alpha(\omega,\delta)\big)^{-1} B = \begin{pmatrix} A_{12}B_D & 0 \\ A_{22}B_D & 0 \end{pmatrix}. \tag{C.53}$$

Moreover, as $\delta \to 0$ and $\omega = \mathcal{O}(\sqrt{\delta})$, the operators A_{12} and A_{22} can be approximated respectively by

$$\begin{aligned} A_{12} = &-\frac{\langle \chi_{\partial D_1}, \cdot\rangle\psi_1 + \langle \chi_{\partial D_2}, \cdot\rangle\psi_2}{\omega^2|D_1|} - \frac{\langle \chi_1^\alpha, \cdot\rangle u_1}{\omega^2|D_1|}\left(\frac{(\omega_1^\alpha)^2}{\omega^2 - (\omega_1^\alpha)^2}\right) \\ &- \frac{\langle \chi_2^\alpha, \cdot\rangle u_2}{\omega^2|D_1|}\left(\frac{(\omega_2^\alpha)^2}{\omega^2 - (\omega_2^\alpha)^2}\right) + \mathcal{O}\left(\frac{1}{\omega}\right), \end{aligned} \tag{C.54}$$

and

$$A_{22} = -\frac{\langle \chi_1^\alpha, \cdot\rangle u_1^\alpha}{\omega^2|D_1|}\left(\frac{\omega^2}{\omega^2 - (\omega_1^\alpha)^2}\right) - \frac{\langle \chi_2^\alpha, \cdot\rangle u_2^\alpha}{\omega^2|D_1|}\left(\frac{\omega^2}{\omega^2 - (\omega_2^\alpha)^2}\right) + \mathcal{O}\left(\frac{1}{\omega}\right), \tag{C.55}$$

with respect to the norm in $\mathcal{L}(L^2(\partial D))$.

Defining

$$\Psi_j^\alpha = \begin{pmatrix} u_j \\ u_j^\alpha \end{pmatrix} \qquad \Phi_j^\alpha = \begin{pmatrix} -\delta u_j^\alpha \\ \chi_j^\alpha \end{pmatrix}, \quad j = 1, 2,$$

the above computations imply the following result.

PROPOSITION C.18. *As $\omega \to \omega_j^\alpha, j = 1, 2$, we have*

$$\big(\mathcal{A}^\alpha(\omega,\delta)\big)^{-1} = -\frac{1}{2\omega_j^\alpha|D_1|}\frac{\langle \Phi_j^\alpha, \cdot\rangle \Psi_j^\alpha}{\omega - \omega_j^\alpha} + \mathcal{R}_j^\alpha(\omega),$$

where $\omega \mapsto \mathcal{R}_j^\alpha(\omega)$ is holomorphic for ω in a neighbourhood of ω_j^α.

Due to the zero column in (C.53), it is clear that (C.51) reduces to a system of $N \times N$ equations for $\widetilde{\psi}_0, \dots, \widetilde{\psi}_{N-1}$. In fact, from (C.51), it follows that

$$\widetilde{\Psi}_N = -\frac{1}{2\pi}\int_{Y^*}\begin{pmatrix} 1 & e^{\mathrm{i}\alpha} & \cdots & e^{-(N-1)\mathrm{i}\alpha} \\ e^{\mathrm{i}\alpha} & 1 & \cdots & e^{-(N-2)\mathrm{i}\alpha} \\ \vdots & \vdots & \ddots & \vdots \\ e^{(N-1)\mathrm{i}\alpha} & e^{(N-2)\mathrm{i}\alpha} & \cdots & 1 \end{pmatrix}\begin{pmatrix} A_{12}B_D & 0 & \cdots & 0 \\ 0 & A_{12}B_D & \cdots & 0 \\ \vdots & \vdots & \ddots & \vdots \\ 0 & 0 & \cdots & A_{12}B_D \end{pmatrix}\widetilde{\Psi}_N \mathrm{d}\alpha, \tag{C.56}$$

where

$$\widetilde{\Psi}_N = \begin{pmatrix} \widetilde{\psi}_0 \\ \widetilde{\psi}_1 \\ \vdots \\ \widetilde{\psi}_{N-1} \end{pmatrix}.$$

From Lemma C.17, we obtain that, to leading-order in $\omega = \mathcal{O}(\delta^{1/2})$, $\widetilde{\psi}_m$ is a linear combination of ψ_1 and ψ_2 defined as in Lemma 1.4:

$$\widetilde{\psi}_m = c_m\psi_1 + d_m\psi_2 + \mathcal{O}(\omega),$$

with respect to the $L^2(\partial D)$-norm. Define, for $j = 1, 2$,

$$t_{i,j}^m = \frac{1}{2\pi}\int_{Y^*} e^{\mathrm{i}m\alpha}\left\langle \chi_{\partial D_i}, (I + A_{12}B_D)[\psi_j]\right\rangle \mathrm{d}\alpha.$$

Then, taking inner products $\langle \chi_{\partial D_i}, \cdot\rangle$ in equation (C.56), we find that

$$\frac{1}{2\pi}\int_{Y^*} e^{\mathrm{i}m\alpha}\begin{pmatrix}\left\langle \chi_{\partial D_1}, I + A_{12}B_D[\widetilde{\psi}_n]\right\rangle\\ \left\langle \chi_{\partial D_2}, I + A_{12}B_D[\widetilde{\psi}_n]\right\rangle\end{pmatrix}\mathrm{d}\alpha = T_m\begin{pmatrix}c_n\\ d_n\end{pmatrix},$$

where T_m denotes the 2×2 matrix $\left(t_{i,j}^m\right)$. We thus have

$$\mathcal{T}_N(\omega)C_N = 0, \tag{C.57}$$

where we have defined

$$\mathcal{T}_N(\omega) = \begin{pmatrix} T_0 & T_{-1} & \cdots & T_{-(N-1)}\\ T_1 & T_0 & \cdots & T_{-(N-2)}\\ \vdots & \vdots & \ddots & \vdots\\ T_{N-1} & T_{N-2} & \cdots & T_0\end{pmatrix}, \qquad C_N = \begin{pmatrix} c_0\\ d_0\\ c_1\\ \vdots\\ d_{N-1}\end{pmatrix}.$$

Observe that $\mathcal{T}_N$ is a block Toeplitz matrix generated by the symbol φ given by

$$\varphi = \varphi(\alpha) = \begin{pmatrix}\varphi_{11} & \varphi_{12}\\ \varphi_{21} & \varphi_{22}\end{pmatrix}, \qquad \varphi_{ij} = \left\langle \chi_{\partial D_i}, (I + A_{12}B_D)[\psi_j]\right\rangle, \quad i,j = 1,2.$$

In the following lemma, we compute the symbol φ.

LEMMA C.19. *We have*

$$\varphi(\alpha) = -\frac{\mathrm{Cap}_{D_1}}{2}\begin{pmatrix}\eta_1+\eta_2 & -e^{\mathrm{i}\theta_\alpha}(\eta_1-\eta_2)\\ -e^{-\mathrm{i}\theta_\alpha}(\eta_1-\eta_2) & \eta_1+\eta_2\end{pmatrix},$$

and

$$\det\varphi(\alpha) = \left(\mathrm{Cap}_{D_1}\right)^2\eta_1\eta_2,$$

where

$$\eta_j = \frac{\left(\omega_j^\alpha\right)^2}{\omega^2 - \left(\omega_j^\alpha\right)^2}, \quad j = 1, 2.$$

PROOF. From Lemma B.13, we have

$$\left\langle \chi_{\partial D_j}, \left(-\frac{1}{2}I + \mathcal{K}_{D_j}^{\omega,*}\right)[\psi_j]\right\rangle = -\omega^2|D_1| + \mathcal{O}(\omega^3),$$

and therefore,

$$\langle \chi_{\partial D_i}, B_D[\psi_j]\rangle = \omega^2|D_1|\delta_{ij} + \mathcal{O}(\omega^3),\ i,j = 1,2.$$

From this, using Lemma C.17 and (C.54), we find that

$$A_{12}B_D[\psi_1] = \left(-1 + \frac{\eta_1+\eta_2}{2}\right)\psi_1 - e^{-\mathrm{i}\theta_\alpha}\frac{\eta_1-\eta_2}{2}\psi_2 + \mathcal{O}(\omega),$$

$$A_{12}B_D[\psi_2] = -e^{\mathrm{i}\theta_\alpha}\frac{\eta_1-\eta_2}{2}\psi_1 + \left(-1 + \frac{\eta_1+\eta_2}{2}\right)\psi_2 + \mathcal{O}(\omega),$$

with respect to the $L^2(\partial D)$-norm. The result now follows from the facts that $\langle \chi_{\partial D_i}, \psi_j \rangle = -\mathrm{Cap}_{D_i}\delta_{ij}$ and $\mathrm{Cap}_{D_1} = \mathrm{Cap}_{D_2}$. □

Observe, in particular, that φ is a Hermitian matrix and, therefore, the Toeplitz matrices $\mathcal{T}_N$ are also Hermitian. We define the "exchange" matrix $J_m \in \mathbb{R}^{2m}, m \in \mathbb{N}$, as

$$J_m = \begin{pmatrix} 0 & \cdots & 0 & 1 \\ 0 & \cdots & 1 & 0 \\ \vdots & \iddots & \vdots & \vdots \\ 1 & \cdots & 0 & 0 \end{pmatrix}.$$

The following lemma describes the centrosymmetry property of Hermitian Toeplitz matrices.

LEMMA C.20. *We have*

$$T_m = J_1 \overline{T}_{-m} J_1, \qquad \mathcal{T}_N = J_N \overline{\mathcal{T}_N} J_N.$$

PROOF. We have $J_1 \varphi J_1 = \overline{\varphi}$ and therefore,

$$\begin{aligned} T_m &= \frac{1}{2\pi}\int_{Y^*} e^{im\alpha}\varphi(\alpha)\,\mathrm{d}\alpha \\ &= \frac{1}{2\pi}\int_{Y^*} J_1\overline{e^{-im\alpha}\varphi(\alpha)}J_1\,\mathrm{d}\alpha \\ &= J_1\overline{T}_{-m}J_1. \end{aligned}$$

The second equality of the statement follows from the first one together with the Toeplitz structure of $\mathcal{T}_N$. □

We will study the solutions to (C.57) in the two cases $N = 1$ and $N \to +\infty$. The following proposition characterises the solutions in the case $N = 1$, corresponding to two removed resonators.

PROPOSITION C.21. *If $N = 1$, equation* (C.57) *has a nonzero solution if and only if ω is a solution to one of the two equations*

$$\text{(C.58)} \qquad \frac{1}{2\pi}\int_{Y^*} \Big(\eta_1\left(1 \pm e^{\mathrm{i}\theta_\alpha}\right) + \eta_2\left(1 \mp e^{\mathrm{i}\theta_\alpha}\right)\Big)\,\mathrm{d}\alpha = 0.$$

If $l_0 < 1/2$, there are no solutions to equations (C.58)*, while if $l_0 > 1/2$, each equation has exactly one solution. Here, l_0 is the separation of the resonators D_1 and D_2 (assuming that the unit cell length $L = 1$).*

PROOF. In the case $N = 1$, equation (C.57) reads

$$T_0\begin{pmatrix} c_0 \\ d_0 \end{pmatrix} = 0,$$

which has a nonzero solution if and only if $\det T_0 = 0$. We have

$$\det T_0 = \frac{\left(\mathrm{Cap}_{D_1}\right)^2}{4}\left((I_1)^2 - |I_2|^2\right),$$

where

$$I_1 = \frac{1}{2\pi}\int_{Y^*} (\eta_1 + \eta_2)\,\mathrm{d}\alpha, \qquad I_2 = \frac{1}{2\pi}\int_{Y^*} e^{\mathrm{i}\theta_\alpha}(\eta_1 - \eta_2)\,\mathrm{d}\alpha.$$

By time-reversal symmetry, we have $\omega_j^{-\alpha} = \omega_j^{\alpha}, j = 1, 2$, which implies $I_2 \in \mathbb{R}$. Hence, $\det T_0 = 0$ is equivalent to

$$I_1 - I_2 = 0 \text{ or } I_1 + I_2 = 0.$$

In [**22**, Appendix B], it is shown that each of these equations has a unique solution in the case $l_0 > 1/2$, while no solutions in the case $l_0 < 1/2$. □

Denote by $\mathcal{T}(\omega)$ the infinite Toeplitz matrix corresponding to $\mathcal{T}_N(\omega)$, that is,

$$\mathcal{T}(\omega) = \begin{pmatrix} T_0 & T_{-1} & T_{-2} & \cdots \\ T_1 & T_0 & T_{-1} & \cdots \\ T_2 & T_1 & T_0 & \cdots \\ \vdots & \vdots & \vdots & \ddots \end{pmatrix},$$

which defines a bounded operator on the space l_2^2 of sequences of two-dimensional vectors. More precisely, l_2^2 consists of sequences $\{x_n\}_{n=0}^{+\infty}$ of vectors $x_n \in \mathbb{R}^2$ such that

$$\left(\sum_{n=0}^{+\infty} \|x_n\|^2 \right)^{1/2} < +\infty,$$

where $\| \cdot \|$ denotes the Euclidean norm.

Proposition C.22. *Given ω_∞ inside the bandgap such that $\mathcal{T}(\omega_\infty)$ has eigenvalue 0, there are two frequencies $\omega_1(N), \omega_2(N) \to \omega_\infty$ as $N \to +\infty$, such that $\mathcal{T}_N$ is not invertible at $\omega_1(N), \omega_2(N)$.*

Proof. Let $X = \{x_n\}_{n=0}^{+\infty} \in l_2^2$ be an eigenvector with $\mathcal{T}(\omega_\infty)X = 0$ and let $x \in \mathbb{R}^{2N}$ be a truncation of X. Since $\mathcal{T}(\omega_\infty)X = 0$, we have

$$\sum_{n=0}^{+\infty} T_{k-n} x_n = 0 \tag{C.59}$$

for all $k \in \mathbb{N}$. Define $z_1, z_2 \in \mathbb{R}^{4N}$,

$$z_1 = \begin{pmatrix} x \\ J_N \overline{x} \end{pmatrix}, \qquad z_2 = \begin{pmatrix} x \\ -J_N \overline{x} \end{pmatrix}.$$

Then, using Lemma C.20, we have

$$\mathcal{T}_{2N}(\omega_\infty) z_1 = \begin{pmatrix} \vdots \\ \sum_{n=0}^{N-1} T_{k-n} x_n + \sum_{n=0}^{N-1} T_{k-N-n} J_N \overline{x}_{N-1-n} \\ \vdots \end{pmatrix}$$
$$= \begin{pmatrix} \vdots \\ \sum_{n=0}^{N-1} T_{k-n} x_n + J_N \overline{\sum_{n=0}^{N-1} T_{2N-1-k-n} x_n} \\ \vdots \end{pmatrix}$$

for $k = 0, \ldots, 2N-1$.

In view of (C.59), given $\varepsilon > 0$, we can choose N such that

$$\|\mathcal{T}_{2N} z_1\| < \varepsilon,$$

which implies that 0 is in the ε-pseudospectrum of $\mathcal{T}_{2N}(\omega_\infty)$ (see Appendix C.10 for a discussion on the definition and properties of pseudospectra of Toeplitz matrices).

Since $\mathcal{T}_{2N}(\omega_\infty)$ is Hermitian, it follows that there is an eigenvalue μ_1 of $\mathcal{T}_{2N}(\omega_\infty)$ with $|\mu_1| < \varepsilon$. From this, it follows that there is a value ω_1 such that $\mathcal{T}_{2N}(\omega_1)$ is not invertible, satisfying $|\omega_1 - \omega_\infty| < K\varepsilon$ for some K independent of ε.

In the same way, we can show that given $\varepsilon > 0$ we can choose N such that

$$\|\mathcal{T}_{2N} z_2\| < \varepsilon,$$

and therefore there is a value ω_2 such that $\mathcal{T}_{2N}(\omega_2)$ is not invertible, satisfying $|\omega_2 - \omega_\infty| < K\varepsilon$ for some K independent of ε.

The above argument shows that $\omega_i(2N) \to \omega_\infty$ as $N \to +\infty$. The case of the sequence $\omega_i(2N-1)$, corresponding to odd indices, follows similarly by choosing the truncation $x \in \mathbb{R}^{2N-1}$ and constructing z_1, z_2 analogously. □

REMARK C.23. *The values of the nonzero solutions C_N to (C.57) correspond to the values attained by the midgap modes inside the dislocation region. The two pseudomodes z_1 and z_2 can be interpreted as approximations of the monopole and dipole modes, respectively, arising from the hybridisation of the two semi-infinite half-structures. As the dislocation increases, that is, as $N \to +\infty$, the strength of the hybridisation decreases and the frequencies corresponding to these modes converge to the same value ω_∞.*

REMARK C.24. *This section shows the intimate connection between localised edge modes and the fact that block Toeplitz matrices with sufficiently smooth symbols have eigenvectors that are exponentially localised to the "edges" (that is, the first and last entries) of the vector.*

C.9. Periodic capacitance matrix

Let $D = D_1 \cup D_2$ be a parity symmetric dimer of resonators, *i.e.*, satisfying the inversion symmetry assumption (4.127). In this appendix, we give a proof of the independence of the periodic capacitance matrix C^0 based on its integral representation (4.73) given in Definition 4.42. Now, in order to prove that C^0 is independent of α_0, it suffices to show that ψ_i^0 is independent of α_0. Let

$$\mathbf{c}_i = \int_{\partial D} y \psi_i^0(y) \, \mathrm{d}\sigma(y). \tag{C.60}$$

To emphasise the role of α_0, we will use the notation $\psi_i^0 = \psi_i^{0,\alpha_0}$ and $\mathbf{c}_i = \mathbf{c}_i^{\alpha_0}$. Expanding the orders of ω in (4.71), we find that

$$\int_{\partial D} \psi_j^{0,\alpha_0} \, \mathrm{d}\sigma = 0,$$

$$\mathcal{S}_D^{0,0}[\psi_j^{0,\alpha_0}] + \frac{1}{2\mathrm{i}w_0|Y_l|} \int_{\partial D} \widehat{\psi}_j^{1,\alpha_0} \, \mathrm{d}\sigma - \int_{\partial D} \frac{\alpha_0 \cdot y}{2w_0|Y_l|} \psi_j^{0,\alpha_0}(y) \, \mathrm{d}\sigma(y) = \chi_{\partial D_j}, \tag{C.61}$$

and

$$\mathcal{S}_D^{0,0}[\widehat{\psi}_j^{1,\alpha_0}] + \frac{\alpha_0 \cdot x}{2w_0|Y_l|} \int_{\partial D} \widehat{\psi}_j^{1,\alpha_0} \, \mathrm{d}\sigma + K\chi_{\partial D} = 0$$

for some constant K. Under the parity symmetry assumption (4.127), we have

$$\psi_1^{0,\alpha_0}(y) = \psi_2^{0,-\alpha_0}(\mathcal{P}y) = -\psi_1^{0,-\alpha_0}(\mathcal{P}y), \qquad \widehat{\psi}_1^{1,\alpha_0}(y) = \widehat{\psi}_2^{1,-\alpha_0}(\mathcal{P}y), \tag{C.62}$$

and, in particular, $\mathbf{c}_i^{\alpha_0} = \mathbf{c}_i^{-\alpha_0}$. Then, from (C.61), we find that

$$\mathcal{S}_D^{0,0}[\psi_1^{0,\alpha_0} + \psi_2^{0,-\alpha_0}] + \frac{1}{\mathrm{i}w_0|Y_l|}\int_{\partial D} \widehat{\psi}_1^{1,\alpha_0}\, \mathrm{d}\sigma - \frac{\alpha_0 \cdot \mathbf{c}_1^{\alpha_0}}{w_0|Y_l|} = \chi_{\partial D},$$

and that

$$\mathcal{S}_D^{0,0}[\psi_1^{0,\alpha_0} - \psi_2^{0,-\alpha_0}] = \chi_{\partial D_1} - \chi_{\partial D_2}. \tag{C.63}$$

In other words, for some constant K, we have

$$\mathcal{S}_D^{0,0}[\psi_1^{0,\alpha_0} + \psi_2^{0,-\alpha_0}] = K\chi_{\partial D},$$

and since $\int_{\partial D} \psi_i^{0,\alpha_0} = 0$, we have from Lemma 4.39 that $\psi_1^{0,\alpha_0} = -\psi_2^{0,-\alpha_0}$. Then, from (C.63), we find that

$$\mathcal{S}_D^{0,0}[\psi_1^{0,\alpha_0}] = \frac{1}{2}\chi_{\partial D_1} - \frac{1}{2}\chi_{\partial D_2}.$$

Since $\mathcal{S}_D^{0,0}$ and the right-hand side are independent of α_0 and since $\mathcal{S}_D^{0,0}$ is injective on $L_0^2(\partial D)$, we obtain that ψ_i^{0,α_0} is independent of α_0.

C.10. Spectra and pseudospectra of Toeplitz matrices and operators

In this section, we first recall some basic results on the spectra and pseudospectra of Toeplitz matrices, Toeplitz operators, and Laurent operators. Then we characterise the pseudospectra of banded Toeplitz matrices with perturbations. This characterisation from [**7**] is useful for studying the gauge capacitance matrix in Section 5.2.

C.10.1. Spectra and pseudospectra of Toeplitz matrices.

An $N \times N$ Toeplitz matrix is a matrix whose entries are constant along diagonals:

$$\mathbf{A}_N = \begin{pmatrix} a_0 & a_1 & & \cdots & a_{N-1} \\ a_{-1} & a_0 & & & \vdots \\ & \ddots & \ddots & \ddots & \\ \vdots & & & a_0 & a_1 \\ a_{-(N-1)} & \cdots & & a_{-1} & a_0 \end{pmatrix}. \tag{C.64}$$

A semi-infinite matrix of the same form is known as a Toeplitz operator, and a doubly infinite matrix of this kind is a Laurent operator and we denote them by $\mathbf{A}$. The symbol of a Toeplitz matrix, Toeplitz operator, or Laurent operator is the function

$$f(z) = \sum_k a_k z^k$$

with $\sum_k |a_k| < +\infty$. This symbol is important for deriving many properties of the Toeplitz matrices, Toeplitz operators, and Laurent operators. For example, it fully determines the spectrum Λ of a Toeplitz or a Laurent operator $\mathbf{A}$. Define $\mathbb{T} := \{z \in \mathbb{C} : |z| = 1\}$ and $I(f(\mathbb{T}), \lambda)$ as the winding number of $f(\mathbb{T})$ about λ in the usual positive (counterclockwise) sense. We have the following result.

THEOREM C.25. *Let* $\mathbf{A}$ *and* f *be as described above. The following holds:*

(i) *If* $\mathbf{A}$ *is a Laurent operator, then* $\Lambda(\mathbf{A}) = f(\mathbb{T})$;
(ii) *If* $\mathbf{A}$ *is a Toeplitz operator, then* $\Lambda(\mathbf{A}) = f(\mathbb{T}) \cup \{\lambda \in \mathbb{C} : I(f(\mathbb{T}), \lambda) \neq 0\}$.

Note that the winding number $I(f(\mathbb{T}),\lambda)$ or index of the continuous curve $f(\mathbb{T})$ about the point $\lambda \in \mathbb{C}$ is also given by

$$I(f(\mathbb{T}),\lambda) = \frac{1}{2\pi\mathrm{i}} \int_{\mathbb{T}} \frac{f'(z)}{f(z)-\lambda} dz, \quad \lambda \notin f(\mathbb{T}).$$

Consequently, if $\lambda \in f(\mathbb{T})$, then $I(f(\mathbb{T}),\lambda)$ is undefined.

A specific property of the Toeplitz operator $\mathbf{A}$ is that, for the eigenvectors of $\mathbf{A}$ corresponding to the eigenvalue λ with $I(f(\mathbb{T}),\lambda) \neq 0$, the amplitude of its entries a_j's decreases or increases as $j \to +\infty$. As stated in [**144**, Eq. (3.4)]:

- If $I(f(S),\lambda) > 0$, then the right eigenvectors of $\mathbf{A}$ are geometrically decreasing;
- If $I(f(S),\lambda) < 0$, then the right eigenvectors of $\mathbf{A}$ are geometrically increasing.

This is known to be the topological origin of the skin effect in non-Hermitian physical systems with complex gauge transforms; see Section 5.2 and [**5**].

For characterising the spectrum of Toeplitz matrices, no results similar to those in Theorem C.25 are available in the literature. Moreover, a detailed characterisation of general Toeplitz matrices seems to be out of reach. The only known results are asymptotic characterisations of eigenpairs of some Hermitian Toeplitz matrices and Toeplitz matrices with specific symbols; see [**61**] for a survey in this regard. We also refer the reader to some representations of eigenpairs of tridiagonal Toeplitz matrices [**68, 96, 174**], which led to our recent demonstration of the skin effect for finitely periodic systems of subwavelength resonators with complex gauge transformations in the one-dimensional case [**5, 11**].

On the other hand, there are some results for the pseudospectra of banded and semi-banded Toeplitz matrices. In order to state them, we introduce the concept of pseudo-eigenvalue.

DEFINITION C.26. *Given $\varepsilon > 0$, the number $\lambda \in \mathbb{C}$ is a pseudo-eigenvalue of a matrix $\mathbf{A}$ if any of the following equivalent conditions is satisfied:*

(i) *λ is an eigenvalue of $\mathbf{A}+\mathbf{E}$ for some $\mathbf{E} \in \mathbb{C}^{N\times N}$ with $||\mathbf{E}|| \leqslant \varepsilon$;*
(ii) *$\exists\, u \in \mathbb{C}^N, ||u|| = 1$, such that $||(\mathbf{A}-\lambda I)u|| \leqslant \varepsilon$;*
(iii) *$||(\lambda I - \mathbf{A})^{-1}|| \geqslant \varepsilon^{-1}$;*
(iv) *$\sigma_{\min}(\lambda I - \mathbf{A}) \leqslant \varepsilon$, where $\sigma_{\min}(\lambda I - \mathbf{A})$ is the smallest singular value of $\lambda I - \mathbf{A}$.*

Given $\varepsilon > 0$, the set of all pseudo-eigenvalues of $\mathbf{A}$, the pseudospectrum, is denoted by $\Lambda_\varepsilon(\mathbf{A})$ or simply by Λ_ε.

An l-banded Toeplitz matrix is defined as $\mathbf{A}_N$ in (C.64) with $a_j = 0$ for $|j| > l$ and is denoted by $\mathbf{A}_{N,l}$. To state the result on its pseudospectrum, we need to introduce the sets Ω_r and $\Omega^{1/r}$ defined by

$$\text{(C.65)} \quad \Omega_r = \{z \in \mathbb{C} : I\left(f\left(\mathbb{T}_r\right), z\right) > 0\}, \quad \Omega^R = \left\{z \in \mathbb{C} : I\left(f\left(\mathbb{T}_{1/r}\right), z\right) < 0\right\},$$

where $\mathbb{T}_r := \{z \in \mathbb{C} : |z| = r\}$ and $\mathbb{T}_{1/r}$ is defined analogously. The following result ([**144**, Theorem 3.2]) characterises the pseudo-eigenvalues of an l-banded Toeplitz matrix.

THEOREM C.27. *Let $\mathbf{A}$ be a banded Toeplitz operator with bandwidth l,* i.e., *$a_k = 0$ for $|k| > l$. Let $f(z)$ be the symbol of $\mathbf{A}$, and let $\mathbf{A}_N$ denote the $N \times N$*

Toeplitz matrix defined by (C.64). *Then, for any* $r < 1$ *and* $\rho > r$, *we have*

$$\Omega_r \cup \Omega^{1/r} \cup (\Lambda(\mathbf{A}) + \Delta_\varepsilon) \subseteq \Lambda_\varepsilon(\mathbf{A}_{N,l}) \quad \text{for} \quad \varepsilon = C\rho^N,$$

where $\Delta_\varepsilon = \{z \in \mathbb{C} : |z| < \varepsilon\}$ *and* C *is a constant independent of* N.

C.10.2. Pseudospectra of banded Toeplitz matrices with perturbations. Note that our gauge capacitance matrix is a Toeplitz matrix with perturbations on the nonzero elements. Here, we consider and analyse the pseudospectrum of an l-banded Toeplitz matrix with perturbations on the corner blocks. More precisely, we define the l-banded Toeplitz matrix with the first and last l rows perturbed as follows:

$$\widehat{\mathbf{A}}_{N,l} = \begin{pmatrix} B_1 \\ B_2 \\ B_3 \end{pmatrix}, \tag{C.66}$$

where B_1, B_2, and B_3 are, respectively, of size $(l-1) \times N$, $(N-2l+2) \times N$, and $(l-1) \times N$ and are given by

$$\begin{pmatrix} a_0+b_{0,1} & \cdots & a_{l-1}+b_{l-1,1} & 0 & \cdots & \cdots & \cdots & \cdots \\ a_{-1}+b_{-1,2} & \cdots & a_{l-2}+b_{l-2,2} & a_{l-1}+b_{l-1,2} & 0 & \ddots & \ddots & \vdots \\ \vdots & \ddots & \ddots & \ddots & \ddots & \ddots & \ddots & \vdots \\ a_{2-l}+b_{2-l,l-1} & \cdots & a_0+b_{0,l-1} & a_1+b_{1,l-1} & \cdots & a_{l-1}+b_{l-1,l-1} & 0 & \cdots \end{pmatrix},$$

$$\begin{pmatrix} a_{-l+1} & a_{-l+2} & \cdots & a_{l-1} & 0 & \cdots & 0 \\ 0 & a_{-l+1} & a_{-l+2} & \cdots & a_{l-1} & 0 & \vdots \\ \vdots & \ddots & \ddots & \ddots & \ddots & \ddots & \vdots \\ 0 & \cdots & 0 & a_{-l+1} & a_{-l+2} & \cdots & a_{l-1} \end{pmatrix},$$

and

$$\begin{pmatrix} 0 & \cdots & 0 & a_{1-l}+b_{1-l,N-l+2} & \cdots & \cdots & \cdots & a_{l-3}+b_{l-3,N-l+2} & a_{l-2}+b_{l-2,N-l+2} \\ 0 & \cdots & 0 & 0 & \ddots & \ddots & \ddots & \cdots & a_{l-3}++b_{l-3,N-l+3} \\ \vdots & \ddots & \vdots & \ddots & \ddots & \ddots & \ddots & \ddots & \vdots \\ 0 & \cdots & 0 & 0 & \cdots & 0 & a_{1-l}+b_{1-l,N} & \cdots & a_0+b_{0,N} \end{pmatrix}.$$

We generalise Theorem C.27 to the perturbed l-banded Toeplitz $\widehat{\mathbf{A}}_{N,l}$ as follows.

THEOREM C.28. *Let* $f(z)$ *be the symbol defined by* $\sum_{q=1-l}^{l-1} a_q z^q$, *and let* $\widehat{\mathbf{A}}_{N,l}$ *be the* $N \times N$ l*-banded perturbed Toeplitz matrix in* (C.66). *Then, for any* $r < 1$ *and* $\rho > r$, *we have*

$$\Omega_r \cup \Omega^{1/r} \subseteq \Lambda_\varepsilon\left(\widehat{\mathbf{A}}_{N,l}\right) \quad \text{for} \quad \varepsilon = \max\left(C_1, C_2 N^{l-1}\right) \rho^{N-2l+2}, \tag{C.67}$$

where $\Omega_r, \Omega^{\frac{1}{r}}$ *are defined as in* (C.65) *and* C_1, C_2 *are constants independent of* N. *Moreover, there exist nonzero pseudo-eigenvectors* $\mathbf{v}^{(N)}$ *satisfying*

$$\frac{\left|\left|\left(\widehat{\mathbf{A}}_{N,l} - \lambda I\right)\mathbf{v}^{(N)}\right|\right|_2}{\left|\left|\mathbf{v}^{(N)}\right|\right|_2} \leq \max\left(C_1, C_2 N^{l-1}\right) \rho^{N-2l+2}, \tag{C.68}$$

such that

$$\frac{\left|\left(\mathbf{v}^{(N)}\right)_j\right|}{\max_j \left|\left(\mathbf{v}^{(N)}\right)_j\right|} \leq \begin{cases} C_3 \rho^{j-1}, & \text{if } \lambda \in \Omega_r, \\ C_3 \rho^{N-j}, & \text{if } \lambda \in \Omega^{\frac{1}{r}}, \end{cases} \quad 1 \leq j \leq N, \tag{C.69}$$

where C_3 *is independent of* N.

PROOF. By symmetry, $\Omega^{1/r}$ must satisfy an estimate of type (C.67) if Ω_r does. Thus, we only need to prove that $\Omega_r \subseteq \Lambda_\varepsilon\left(\widehat{\mathbf{A}}_{N,l}\right)$.

The idea is to construct geometrically decreasing pseudo-eigenvectors. Given any $r < 1$, let $\lambda \in \Omega_r$ be arbitrary. Assume without loss of generality that

$$a_{-l+1} \neq 0.$$

The symbol is

$$f(z) = \sum_{q=-l+1}^{l-1} a_q z^q.$$

Then $f(z)$ has a pole of order exactly $l-1$ at $z=0$. Since $I\left(f\left(\mathbb{T}_r\right), \lambda\right) \geqslant 1$ (by the definition of Ω_r), it follows by the argument principle that the equation $f(z) = \lambda$ has at least l solutions in $\{z \in \mathbb{C} : |z| < r\}$, counted with their multiplicities. Let $z_1, \ldots, z_l$ be any l of these solutions. Assume for the moment that the z_j's are distinct. Corresponding to each z_j is a vector $u_j = \left(1, z_j^1, z_j^2, \ldots, z_j^{(N-1)}\right)^\top$. Since

$$\lambda = f(z_j) = \sum_{q=-l+1}^{l-1} a_q z_j^q,$$

we can compute that

$$\left((\lambda I - \widehat{\mathbf{A}}_{N,l})u_j\right)_q = 0, \quad l \leq q \leq N-l,$$

$$\left((\lambda I - \widehat{\mathbf{A}}_{N,l})u_j\right)_q = \sum_{k=-l+1}^{2-q} a_k z_j^k z_j^{q-1} + \sum_{k=1-q}^{l-1} -b_{k,q} z_j^k z_j^{q-1},$$

for $1 \leq q \leq l-1$, and

$$\left((\lambda I - \widehat{\mathbf{A}}_{N,l})u_j\right)_q = \sum_{k=N-q+1}^{l-1} a_k z_j^k z_j^{q-1} + \sum_{k=1-l}^{N-q} -b_{k,q} z_j^k z_j^{q-1},$$

for $N-l+2 \leq q \leq N$. Thus, we have

$$(\lambda I - \widehat{\mathbf{A}}_{N,l})u_j = z_j^{N-2l+2} L u_j + z_j^{-N+2l-2} R u_j, \tag{C.70}$$

where L is given by

$$\begin{pmatrix} 0 & \cdots & 0 & 0 & 0 & 0 & \cdots & 0 & 0 & \cdots & \cdots & 0 \\ & & \vdots & \vdots & \vdots & \vdots & \vdots & \vdots & \vdots & & & \\ 0 & \cdots & & & & & & & & \cdots & \cdots & 0 \\ 0 & \cdots & 0 & 0 & 0 & 0 & \cdots & 0 & 0 & \cdots & \cdots & 0 \\ -b_{1-l,\zeta} & \cdots & \cdots & \cdots & \cdots & -b_{l-2,\zeta} & a_{l-1} & \cdots & 0 & \cdots & \cdots & 0 \\ & \ddots & \ddots & \ddots & \ddots & \vdots & \vdots & \ddots & \ddots & \ddots & \ddots & \\ 0 & & & & & & & & & & & 0 \\ 0 & \cdots & -b_{1-l,N-1} & \cdots & -b_{0,N-1} & -b_{1,N-1} & a_2 & \cdots & a_{l-1} & 0 & \cdots & 0 \\ 0 & \cdots & \cdots & -b_{1-l,N} & \cdots & -b_{0,N} & a_1 & \cdots & \cdots & a_{l-1} & \cdots & 0 \end{pmatrix}$$

with $\zeta = N-l+2$ and R is given by

$$\begin{pmatrix} 0 & \cdots & a_{1-l} & \cdots & \cdots & a_{-1} & -b_{0,1} & -b_{1,1} & \cdots & -b_{l-1,1} & \cdots & \cdots & 0 \\ 0 & \cdots & 0 & a_{1-l} & \cdots & a_{-2} & -b_{-1,2} & -b_{0,2} & -b_{1,2} & \cdots & -b_{l-1,2} & \cdots & 0 \\ 0 & \cdots & 0 & \ddots & \ddots & \vdots & \vdots & \ddots & \ddots & \ddots & \ddots & \ddots & 0 \\ 0 & \cdots & 0 & \ddots & \ddots & a_{1-l} & -b_{2-l,l-1} & \cdots & \cdots & \cdots & \cdots & \cdots & -b_{l-1,l-1} \\ 0 & \cdots & 0 & 0 & \cdots & 0 & 0 & \cdots & 0 & 0 & \cdots & \cdots & 0 \\ \vdots & \cdots & \vdots & \vdots & \vdots & \vdots & \vdots & \vdots & \vdots & \vdots & \cdots & \cdots & \vdots \\ 0 & \cdots & 0 & 0 & \cdots & 0 & 0 & \cdots & 0 & 0 & \cdots & \cdots & 0 \end{pmatrix}.$$

Let $R_{1:l-1}$ be the matrix consisting of the first to the $(l-1)^{\text{th}}$ row of R. Since there are l linearly independent vectors $R_{1:l-1}z_j^{-N+2l-2}u_j$ of length $l-1$, we can find l complex numbers c_j so that

$$\sum_{j=1}^{l} c_j R_{1:l-1} z_j^{-N+2l-2} u_j = \mathbf{0}, \tag{C.71}$$

and let

$$\mathbf{v} = \sum_{j=1}^{l} c_j u_j.$$

Then $z_j^{-N+2l-2}Ru_j = \mathbf{0}$ in (C.70) and we have

$$\left(\lambda I - \widehat{\mathbf{A}}_{N,l}\right)\mathbf{v} = L\sum_{j=0}^{l} z_j^{N-2l+2} c_j u_j. \tag{C.72}$$

We now need to relate the norm of the right-hand side of this equation to $||\mathbf{v}||_2$. To do this, we write $\mathbf{v} = Uc$, where U is the $N \times l$ Vandermonde matrix whose columns are the vectors u_j, and $c = (c_1, \dots, c_l)^\top$. We let D be the diagonal matrix with elements $z_1^{N-2l+2}, \dots, z_l^{N-2l+2}$. Then, we have

$$||UDc||_2 = \left|\left|UDU^\dagger Uc\right|\right|_2 = \left|\left|UDU^\dagger \mathbf{v}\right|\right|_2 \leqslant \kappa(U)\,||D||_2\,||\mathbf{v}||_2\,,$$

where $U^\dagger$ denotes the pseudo-inverse of U and $\kappa(U) = \sigma_1/\sigma_l$ is its condition number. Since

$$||D||_2 \leqslant r^{N-2l+2},$$

it follows that (C.72) implies

$$\frac{\left|\left|\left(\lambda I - \widehat{\mathbf{A}}_{N,l}\right)\mathbf{v}\right|\right|_2}{||\mathbf{v}||_2} \leqslant r^{N-2l+2}\kappa(U)\,||L||_2\,. \tag{C.73}$$

This completes the proof under the assumption that the roots $z_1, \dots, z_l$ are distinct. We can handle the case of multiple roots as follows. If some of the roots $z_0, \dots, z_l$ are confluent at some points $\lambda \in \overline{\Omega}_r$, then we analyse the problem using a confluent Vandermonde matrix. More precisely, assuming that the root z_j has multiplicity k, we have

$$f(z) - \lambda = P(z)(z - z_j)^k,$$

which yields

$$f^{(1)}(z_j) = 0,\ f^{(2)}(z_j) = 0,\ \dots,\ f^{(k-1)}(z_j) = 0. \tag{C.74}$$

Now, we consider $g(z) = f(z)z^s, s = l-1, l, l+1, \dots$ By (C.74), we have

$$\begin{aligned} g^{(1)}(z_j) &= f^{(1)}(z_j)z_j^s + f(z_j)sz_j^{s-1} = f(z_j)sz_j^{s-1} = \lambda s z_j^{s-1} \\ g^{(2)}(z_j) &= \cdots = \lambda s(s-1)z_j^{s-2} \\ &\vdots \\ g^{(k-1)}(z_j) &= \cdots = \lambda s(s-1)\dots(s-k+2)z_j^{s-(k-1)}. \end{aligned}$$

Since $f(z)z^s = \sum_{q=-l+1}^{l-1} a_q z^{q+s}$, for $s \geq l-1$, the above equalities give

$$\begin{aligned}
\lambda s z_j^{s-1} &= \sum_{q=-l+1}^{l-1} a_q (q+s) z_j^{q+s-1} \\
\lambda s(s-1) z_j^{s-2} &= \sum_{q=-l+1}^{l-1} a_q (q+s)(q+s-1) z_j^{q+s-2} \\
&\vdots \\
\lambda s(s-1)\dots(s-k+2) z_j^{s-(k-1)} &= \sum_{q=-l+1}^{l-1} a_q (q+s)(q+s-1)\dots \\
&\qquad \times (q+s-k+2) z_j^{q+s-(k-1)}.
\end{aligned}$$

This implies that we can use

$$\begin{pmatrix} 1 \\ z_j \\ z_j^2 \\ \vdots \\ \vdots \\ z_j^{(N-1)} \end{pmatrix}, \begin{pmatrix} 0 \\ 1 \\ 2z_j \\ 3z_j^2 \\ \vdots \\ \vdots \\ (N-1)z_j^{(N-2)} \end{pmatrix}, \dots, \begin{pmatrix} 0 \\ 0 \\ \vdots \\ (k-1)! \\ k(k-1)\cdots 2z_j \\ \vdots \\ (N-1)\cdots(N-(k-1))z_j^{N-k-2} \end{pmatrix},$$

and other vectors corresponding to other roots z_j's to construct the pseudo-eigenvector. Instead, we consider to use

(C.75)

$$\begin{pmatrix} 1 \\ z_j \\ z_j^2 \\ \vdots \\ \vdots \\ z_j^{(N-1)} \end{pmatrix}, \begin{pmatrix} 0 \\ z_j \\ 2z_j^2 \\ 3z_j^3 \\ \vdots \\ \vdots \\ (N-1)z_j^{(N-1)} \end{pmatrix}, \dots, \begin{pmatrix} 0 \\ 0 \\ \vdots \\ (k-1)!z_j^{k-1} \\ k(k-1)\cdots 2z_j^k \\ \vdots \\ (N-1)\cdots(N-(k-1))z_j^{N-1} \end{pmatrix}$$

to construct the pseudo-eigenvector.

The above discussion shows that by considering the confluent Vandermonde matrix, we can always find such l linearly independent vectors like the ones in (C.75) to construct the pseudo-eigenvector. Then an analysis involving confluent Vandermonde matrices yields a bound analogous to (C.73) except with r^{N-2l+2} replaced by an algebraically growing factor at worst as $N^l r^{N-2l+2}$. This proves (C.68).

Now, we prove (C.69). As the pseudo-eigenvector is constructed by $\mathbf{v} = \sum_{j=1}^{l} c_j u_j$ with c_j's satisfying (C.71) and u_j's from (C.75), by the form of vectors in (C.75), we obtain

$$\frac{|\mathbf{v}_j|}{\max_j |\mathbf{v}_j|} \leq C_3 \rho^{j-1}, \ 1 \leq j \leq N, \text{ if } \lambda \in \Omega_r,$$

with C_3 being independent of N. □

C.11. Proof of Theorem 5.22

We decompose the k-banded matrix $\mathcal{C}^{\gamma}_{N,k}$ as

$$\mathcal{C}^{\gamma}_{N,k} = A + M,$$

where the entries of the matrices A and M are, respectively, given by

$$A_{ij} = \begin{cases} \left(\mathcal{C}^{\gamma}_{N,k}\right)_{ij}, & 1 \le i \le k, \text{ or } N-(k-1) \le i \le N, \\ \mathcal{C}^{\gamma}_{ij}, & |i-j| < k, k < i < N-(k-1), \\ 0, & |i-j| \ge k, \end{cases} \tag{C.76}$$

and

$$M_{ij} = \begin{cases} 0, & 1 \le i \le k, \text{ or } N-(k-1) \le i \le N, \\ \left(\mathcal{C}^{\gamma}_{N,k}\right)_{ij} - \mathcal{C}^{\gamma}_{ij}, & |i-j| < k, k < i < N-(k-1), \\ 0, & |i-j| \ge k. \end{cases} \tag{C.77}$$

Note that, based on the definition of $\mathcal{C}^{\gamma}_{ij}$ in (5.19), A is a k-banded perturbed Toeplitz matrix. By Theorem C.28, we have that for any complex number λ with $I(f(\mathbb{T}), \lambda) \neq 0$, where f is defined by (5.36) and for some $\rho < 1$ and sufficiently large $N > 4k$, there exist nonzero pseudo-eigenvectors $\mathbf{v}^{(N)}$ with $||\mathbf{v}^{(N)}||_2 = 1$ satisfying

$$\left|\left|(A-\lambda)\mathbf{v}^{(N)}\right|\right|_2 \le \max\left(C_1, C_2 N^{k-1}\right)\rho^{N-2k+2},$$

such that

$$\frac{\left|\left(\mathbf{v}^{(N)}\right)_j\right|}{\max_j \left|\left(\mathbf{v}^{(N)}\right)_j\right|} \le \begin{cases} C_3 \rho^{j-1}, & \text{if } I(f(\mathbb{T}), \lambda) > 0, \\ C_3 \rho^{N-j}, & \text{if } I(f(\mathbb{T}), \lambda) < 0, \end{cases} \quad 1 \le j \le N, \tag{C.78}$$

where C_1, C_2, and C_3 are independent of N. To prove (5.38), we estimate $||M\mathbf{v}^{(N)}||_2$. By the definition of M, we have

$$\left(M\mathbf{v}^{(N)}\right)_j = 0 \quad \text{for } 1 \le j \le k, N-(k-1) \le j \le N.$$

For $k+1 \le j \le \lfloor \frac{N}{2} \rfloor - k$, we have

$$\begin{aligned} \left|\left(M\mathbf{v}^{(N)}\right)_j\right| &= \left|\sum_{q=j+1-k}^{j+k-1} M_{j,q}\left(\mathbf{v}^{(N)}\right)_q\right| \le \sum_{q=j+1-k}^{j+k-1} |M_{j,q}| \left|\left(\mathbf{v}^{(N)}\right)_q\right| \\ &\le \sum_{q=j+1-k}^{j+k-1} \frac{\delta K}{(1+j)(1+q)} \\ &\qquad \left(\text{by Lemma 5.18, } \left|\left|\mathbf{v}^{(N)}\right|\right|_2 = 1, \text{ and } N > 4k\right) \\ &= \frac{1}{1+j}\sum_{q=j+1-k}^{j+k-1} \frac{\delta K}{1+q} = \frac{\delta K}{1+j}\xi(j+2-k, j+k), \end{aligned}$$

where $\xi(j+2-k, j+k)$ is defined as in (5.37). Therefore,

$$\sum_{j=k+1}^{\lfloor \frac{N}{2} \rfloor - k} \left| \left(M\mathbf{v}^{(N)} \right)_j \right|^2 \leq (\delta K)^2 \sum_{j=k+1}^{\lfloor \frac{N}{2} \rfloor} \frac{\xi(j+2-k, j+k)^2}{(1+j)^2} < (\delta K)^2 \varepsilon,$$

where the last inequality is from the condition on k. Similarly, we can also prove that

$$\sum_{j=\lfloor \frac{N}{2} \rfloor + k}^{N-k} \left| \left(M\mathbf{v}^{(N)} \right)_j \right|^2 < (\delta K)^2 \varepsilon.$$

For $\lfloor \frac{N}{2} \rfloor - k + 1 \leq j \leq \lfloor \frac{N}{2} \rfloor + k - 1$, we have

$$\begin{aligned} \left| \left(M\mathbf{v}^{(N)} \right)_j \right| &= \left| \sum_{q=j+1-k}^{j+k-1} M_{j,q} \left(\mathbf{v}^{(N)} \right)_q \right| \leq \sum_{q=j+1-k}^{j+k-1} |M_{j,q}| \left| \left(\mathbf{v}^{(N)} \right)_q \right| \\ &\leq \sum_{q=j+1-k}^{j+k-1} \frac{\delta K}{(1 + \lfloor \frac{N}{2} \rfloor - k + 1)(1 + \lfloor \frac{N}{2} \rfloor - k + 1)} \\ &\qquad \left(\text{by Lemma 5.18, } \left\| \mathbf{v}^{(N)} \right\|_2 = 1, N > 4k \right) \\ &\leq \frac{2\delta K}{k} \quad (\text{since } N > 4k). \end{aligned}$$

It follows that

$$\sum_{j=\lfloor \frac{N}{2} \rfloor - k + 1}^{\lfloor \frac{N}{2} \rfloor + k - 1} \left| \left(M\mathbf{v}^{(N)} \right)_j \right|^2 \leq \frac{8(\delta K)^2}{k} \leq 8(\delta K)^2 \varepsilon.$$

Combining the above estimates, we have

$$\left\| M\mathbf{v}^{(N)} \right\|_2 < \delta K \sqrt{10\varepsilon},$$

and

$$\begin{aligned} \left\| \left(\mathcal{C}_{N,k}^{\gamma} - \lambda \right) \mathbf{v}^{(N)} \right\|_2 &\leq \left\| (A - \lambda) \mathbf{v}^{(N)} \right\|_2 + \left\| M\mathbf{v}^{(N)} \right\|_2 \\ &< \max \left(C_1, C_2 N^{k-1} \right) \rho^{N-2k+2} + \delta K \sqrt{10\varepsilon}. \end{aligned}$$

C.12. Proof of Theorem 5.23

By Proposition 5.21, we have

$$|(\mathcal{C}_N^{\gamma})_{ij}| \leq \frac{\delta K}{|i-j|}, \tag{C.79}$$

for some constant K. We decompose $\mathcal{C}_N^{\gamma}$ as follows:

$$\mathcal{C}_N^{\gamma} = \mathcal{C}_{N,k}^{\gamma} + L + R,$$

where the matrices L and R are defined respectively by

$$(L)_{ij} = \begin{cases} (\mathcal{C}_N^{\gamma})_{ij}, & i-j \geq k, \\ 0, & i-j < k, \end{cases}$$

and

$$(R)_{ij} = \begin{cases} (\mathcal{C}_N^{\gamma})_{ij}, & i-j \leq -k, \\ 0, & i-j > -k. \end{cases}$$

In order to prove (5.42), we estimate $||L\mathbf{v}^{(N)}||_2$ and $||R\mathbf{v}^{(N)}||_2$. Since $||\mathbf{v}^{(N)}||_2 = 1$ implies

$$\max_j \left|\left(\mathbf{v}^{(N)}\right)_j\right| \leq 1,$$

(5.41) gives

$$\left|\left(\mathbf{v}^{(N)}\right)_j\right| \leq C\rho^{j-1}, 1 \leq j \leq N, \text{ or } \left|\left(\mathbf{v}^{(N)}\right)_j\right| \leq C\rho^{N-j}, 1 \leq j \leq N. \tag{C.80}$$

First, for $\mathbf{v}^N$ satisfying

$$\left|\left(\mathbf{v}^{(N)}\right)_j\right| \leq C\rho^{j-1}, \quad 1 \leq j \leq N, \tag{C.81}$$

we calculate $||L\mathbf{v}^{(N)}||_2$. Note that the entries of $L\mathbf{v}^{(N)}$ are given by

$$(L\mathbf{v}^{(N)})_j = \begin{cases} 0, & j = 1, \dots, k, \\ \sum_{q=k+1}^{j} (\mathcal{C}_N^\gamma)_{j,q-k} \left(\mathbf{v}^{(N)}\right)_{q-k}, & j = k+1, \dots, N. \end{cases} \tag{C.82}$$

Define

$$s := \min_{p \geq k} \min_{q=1,2,\dots} \left(\frac{p}{p+q}\right)^{\frac{1}{q}}. \tag{C.83}$$

It is not hard to see that

$$1 - \varepsilon_1 < \frac{k}{k+1} \leq s \leq 1. \tag{C.84}$$

By (C.79) and (C.81), we have for $j = k+1, \dots, N$,

$$\begin{aligned}
\left|(L\mathbf{v}^{(N)})_j\right| =& \left|\sum_{q=k+1}^{j} (\mathcal{C}_N^\gamma)_{j,q-k} \left(\mathbf{v}^{(N)}\right)_{q-k}\right| \leq \sum_{q=k+1}^{j} \left|(\mathcal{C}_N^\gamma)_{j,q-k}\right| \left|\left(\mathbf{v}^{(N)}\right)_{q-k}\right| \\
\leq& \delta KC \sum_{q=k+1}^{j} \frac{1}{j-q+k}\rho^{q-k-1} = \delta KC \sum_{q=k+1}^{j} \frac{1}{q-1}\rho^{j-q} \\
=& \delta KC \sum_{q=k+1}^{j} \frac{1}{q-1} s^{j-q} (\rho/s)^{j-q} \\
=& \frac{\delta KC}{j-1} \left(\sum_{q=k+1}^{j} \frac{j-1}{q-1} s^{j-q} (\rho/s)^{j-q}\right) \\
=& \frac{\delta KC}{j-1} \left(\sum_{q=k+1}^{j} \frac{q-1+j-q}{q-1} s^{j-q} (\rho/s)^{j-q}\right) \\
\leq& \frac{\delta KC}{j-1} \left(\sum_{q=k+1}^{j} (\rho/s)^{j-q}\right) \quad \text{(by (C.83))} \\
\leq& \frac{\delta KC}{j-1} \sum_{q=0}^{+\infty} \left(\frac{\rho}{1-\varepsilon_1}\right)^q \quad \text{(by (C.84))} \\
=& \frac{\delta KC}{j-1} \frac{1-\varepsilon_1}{1-\varepsilon_1-\rho}.
\end{aligned}$$

Thus, combining the above estimates, we obtain

$$\left|\left|L\mathbf{v}^{(N)}\right|\right|_2 \leq \delta K C \frac{1-\varepsilon_1}{1-\varepsilon_1-\rho}\sqrt{\sum_{j=k+1}^{N}\frac{1}{(j-1)^2}} \leq \delta K C \frac{1-\varepsilon_1}{1-\varepsilon_1-\rho}\sqrt{\varepsilon_1}, \tag{C.85}$$

where the last inequality is from the condition on k in the theorem.

Next, for $\mathbf{v}^{(N)}$ satisfying

$$\left|\left(\mathbf{v}^{(N)}\right)_j\right| \leq C\rho^{N-j}, \quad 1 \leq j \leq N, \tag{C.86}$$

we estimate $\left|\left|L\mathbf{v}^{(N)}\right|\right|_2$. The entries of $L\mathbf{v}^{(N)}$ are given by (C.82). By (C.79) and (C.86), we similarly have, for $j = k+1, \ldots, N$, that

$$\begin{aligned}\left|(L\mathbf{v}^{(N)})_j\right| &= \left|\sum_{q=k+1}^{j}(\mathcal{C}_N^{\gamma})_{j,q-k}\left(\mathbf{v}^{(N)}\right)_{q-k}\right| \leq \sum_{q=k+1}^{j}\left|(\mathcal{C}_N^{\gamma})_{j,q-k}\right|\left|\left(\mathbf{v}^{(N)}\right)_{q-k}\right| \\ &\leq \delta K C \sum_{q=k+1}^{j}\frac{1}{j-q+k}\rho^{N-(q-k)} = \delta K C \sum_{q=k+1}^{j}\frac{1}{q-1}\rho^{N-1-(j-q)}.\end{aligned}$$

Further estimating $\left|(L\mathbf{v}^{(N)})_j\right|$, we obtain

$$\begin{aligned}\left|(L\mathbf{v}^{(N)})_j\right| &\leq \delta K C \sum_{q=k+1}^{j}\frac{1}{q-1}\rho^{N-1-(j-q)} \\ &\leq \frac{\delta K C}{k}\rho^{N-1-j+(k+1)}\left(\sum_{q=k+1}^{j}\frac{k}{q-1}\rho^{q-(k+1)}\right) \\ &\leq \frac{\delta K C}{k}\rho^{N-1-j+(k+1)}\left(\sum_{q=k+1}^{j}\rho^{q-(k+1)}\right) \\ &\leq \frac{\delta K C}{k}\rho^{N-1-j+(k+1)}\frac{1}{1-\rho}.\end{aligned}$$

Therefore, combining the above estimates yields

$$\begin{aligned}\left|\left|L\mathbf{v}^{(N)}\right|\right|_2 &\leq \frac{\delta K C}{k}\frac{1}{1-\rho}\sqrt{\sum_{j=k+1}^{N}\rho^{2(N-1-j+(k+1))}} \\ &\leq \frac{\delta K C}{k}\rho^k\frac{1}{1-\rho}\sqrt{\frac{1}{1-\rho^2}} \\ &\leq \delta K C \varepsilon_1 \rho^k \frac{1}{1-\rho}\sqrt{\frac{1}{1-\rho^2}},\end{aligned} \tag{C.87}$$

where the last inequality is from the condition on k in the theorem.

In the same fashion, we can prove that for $\mathbf{v}^{(N)}$ satisfying (C.81), we have

$$\left|\left|R\mathbf{v}^{(N)}\right|\right|_2 \leq \delta K C \varepsilon_1 \rho^k \frac{1}{1-\rho}\sqrt{\frac{1}{1-\rho^2}},$$

and for $\mathbf{v}^{(N)}$ satisfying (C.86), we have

$$\left\|R\mathbf{v}^{(N)}\right\|_2 \leq \delta KC \frac{1-\varepsilon_1}{1-\varepsilon_1-\rho}\sqrt{\varepsilon_1}.$$

Therefore, we arrive at

$$\begin{aligned}\left\|(\mathcal{C}_N^\gamma - \lambda I)\mathbf{v}^{(N)}\right\|_2 \leq& \left\|\left(\mathcal{C}_{N,k}^\gamma - \lambda I\right)\mathbf{v}^{(N)}\right\|_2 + \left\|L\mathbf{v}^{(N)}\right\|_2 + \left\|R\mathbf{v}^{(N)}\right\|_2 \\ \leq& \varepsilon_2 + \delta KC\left(\frac{1-\varepsilon_1}{1-\varepsilon_1-\rho}\sqrt{\varepsilon_1} + \varepsilon_1\rho^k\frac{1}{1-\rho}\sqrt{\frac{1}{1-\rho^2}}\right),\end{aligned}$$

which completes the proof.

C.13. Asymptotic Floquet theory for differential equations

Let $A(t)$ be a T-periodic $N \times N$ complex matrix function, and consider the ordinary differential equation

$$\frac{\mathrm{d}x}{\mathrm{d}t}(t) = A(t)x(t). \tag{C.88}$$

Recall that the matrix-valued fundamental solution of (C.88) is an $N \times N$ matrix with linearly independent column vectors, which solves (C.88). The following theorem is classical.

THEOREM C.29. *(Floquet's theorem) Denote by $X(t)$ the matrix-valued fundamental solution with initial value $X(0) = \mathrm{I}$, where I is the identity matrix. There exists a constant matrix F and a T-periodic matrix function $P(t)$ such that*

$$X(t) = P(t)e^{Ft}. \tag{C.89}$$

For each eigenvalue $\lambda := e^{\mathrm{i}\omega}$ of e^F, there is a Bloch solution $x(t)$ which is ω-quasiperiodic, *i.e.*,

$$x(t+T) = e^{\mathrm{i}\omega T}x(t).$$

Observe that ω is defined modulo $\Omega := 2\pi/T$. Therefore, we define the time Brillouin zone as $Y_t^* := \mathbb{C}/(\Omega\mathbb{Z})$.

REMARK C.30. *In some literature,* e.g. **[171]**, *$e^{\mathrm{i}\omega T}$ is called a characteristic multiplier. We refer to ω as a quasifrequency, while $\mathrm{i}\omega$ is a Floquet exponent.*

If A is time-independent, the solution to (C.88) can be written as $x(t) = e^{At}x(0)$. The Floquet exponents are then given by the eigenvalues of A. Since the Floquet exponent is defined modulo $\mathrm{i}\Omega$, we need the following definition.

DEFINITION C.31 (Folding number). *Let ω_A be the imaginary part of an eigenvalue of the time-independent matrix A, we can uniquely write $\omega_A = \omega_0 + m\Omega$, where $\omega_0 \in [-\Omega/2, \Omega/2)$. The integer m is called the folding number.*

C.13.1. Floquet matrix elements.

We now describe the asymptotic Floquet analysis in a general setting. We consider the perturbed system of linear ordinary differential equations

$$\frac{\mathrm{d}y}{\mathrm{d}t}(t) = A_\varepsilon(t)y(t), \tag{C.90}$$

where ε is a small parameter, and assume that the T-periodic continuous matrix $A_\varepsilon(t)$ is an analytic function of ε at $\varepsilon = 0$ and has the following expansion:

$$A_\varepsilon(t) = A_0 + \varepsilon A_1(t) + \varepsilon^2 A_2(t) + \ldots + \varepsilon^n A_n(t) + \ldots, \tag{C.91}$$

as $\varepsilon \to 0$. By Floquet's theorem (Theorem C.29), the fundamental solution of (C.90) can be written as

$$X_\varepsilon(t) = P_\varepsilon(t) e^{F_\varepsilon t}, \tag{C.92}$$

where $P_\varepsilon(t)$ is T-periodic with $P_\varepsilon(0) = \mathrm{I}$ and F_ε is constant in time. Crucially, we assume that

(i) A_0 is constant in time and diagonal;
(ii) $F_0 := F_{\varepsilon=0}$ has no distinct eigenvalues which are congruent modulo $2\pi\mathrm{i}/T$;
(iii) the series in (C.91) is convergent for $|\varepsilon| < r_0$, where r_0 is independent of t.

Under these assumptions, it follows from [**171**] that the matrices P_ε and F_ε are analytic functions of ε at $\varepsilon = 0$ and therefore they can be expanded as follows:

$$P_\varepsilon(t) = P_0(t) + \varepsilon P_1(t) + \varepsilon^2 P_2(t) + \dots \quad \text{and} \quad F_\varepsilon = F_0 + \varepsilon F_1 + \varepsilon^2 F_2 + \dots. \tag{C.93}$$

By inserting (C.92) into (C.90), we derive the following systems of ordinary differential equations:

$$\begin{cases} \frac{\mathrm{d}P_0}{\mathrm{d}t}(t) = A_0 P_0(t) - P_0(t) F_0, \\ \frac{\mathrm{d}P_n}{\mathrm{d}t}(t) = A_0 P_n(t) - P_n(t) F_0 + \sum_{i=1}^n (A_i(t) P_{n-i}(t) - P_{n-i}(t) F_i), \ n \geq 1, \end{cases} \tag{C.94}$$

with the initial conditions $P_0(0) = \mathrm{I}$ and $P_n(0) = 0$ if $n \geq 1$.

Note that assumption (ii) is crucial for the continuity of the map $\varepsilon \mapsto F_\varepsilon$. Nevertheless, in order to satisfy Assumption (ii) above, we choose F_0 so that all eigenvalues are inside the first Brillouin zone. In other words, $\mathrm{Im}\,(\sigma(F_0)) \subset [-\Omega/2, \Omega/2)$, where $\sigma(F_0)$ denotes the set of eigenvalues of F_0. Since F_0 is defined modulo $2\pi\mathrm{i}/T$, this choice is always possible and is described in the following result from [**42**].

LEMMA C.32. *$A_0 - F_0$ takes the diagonal form* $\frac{2\pi\mathrm{i}}{T}\operatorname{diag}(m_1, m_2, \dots, m_N)$, *where m_i is the folding number of $(A_0)_{ii}$ as in Definition* C.31.

We now vectorise (C.94) by first vectorising P_0 in the basis $\{E_{kj}\}_{\{k,j=1,\dots,N\}}$, where the $(kj)^{\text{th}}$ entry E_{kj} is 1 and 0 otherwise. We denote the vectorised quantity by $\operatorname{vect}(P_0) = p_0$ with

$$p_0(0) = \operatorname{vect}(\mathrm{I}) = \sum_{k=1}^N e_{(k-1)N+k},$$

where $e_{(k-1)N+k}$ is an $N^2 \times 1$ vector with the $((k-1)N+k)^{\text{th}}$ entry being 1 and 0 otherwise. Equation (C.94) reads

$$\begin{cases} \frac{\mathrm{d}p_0}{\mathrm{d}t}(t) = (1 \otimes A_0 - F_0 \otimes 1) p_0, \\ \frac{\mathrm{d}p_n}{\mathrm{d}t}(t) = (1 \otimes A_0 - F_0 \otimes 1) p_n + \sum_{i=1}^n (1 \otimes A_i - F_i^\top \otimes 1) p_{n-1} \quad \text{for } n \geq 1. \end{cases} \tag{C.95}$$

Here, we have used the following tensor notation:

$$1 \otimes A = \begin{pmatrix} A & & & \\ & A & & \\ & & \ddots & \\ & & & A \end{pmatrix}, \qquad B \otimes 1 = \begin{pmatrix} B_{11}\,\mathrm{I} & \cdots & B_{1N}\,\mathrm{I} \\ \vdots & \ddots & \vdots \\ B_{N1}\,\mathrm{I} & \cdots & B_{NN}\,\mathrm{I} \end{pmatrix}.$$

The following result from [**16**] holds.

LEMMA C.33. *We have the following expansion for p_0:*

$$p_0 = \sum_{k=1}^{N} e^{\mathrm{i}\Omega m_k t} e_{(k-1)N+k}. \tag{C.96}$$

PROOF. First, we write

$$1 \otimes A_0 - F_0 \otimes 1 = \begin{pmatrix} A_0 & & & \\ & A_0 & & \\ & & \ddots & \\ & & & A_0 \end{pmatrix} - \begin{pmatrix} (F_0)_{11}\,\mathrm{I} & & & \\ & (F_0)_{22}\,\mathrm{I} & & \\ & & \ddots & \\ & & & (F_0)_{NN}\,\mathrm{I} \end{pmatrix}.$$

We have

$$\begin{aligned} p_0 &= e^{(1\otimes A_0 - F_0 \otimes 1)t} \sum_{k=1}^{N} e_{(k-1)N+k} \\ &= \sum_{k=1}^{N} e^{(A_0 - F_0)_{kk} t} e_{(k-1)N+k} \\ &= \sum_{k=1}^{N} e^{\mathrm{i}\Omega m_k t} e_{(k-1)N+k}. \end{aligned}$$

□

Furthermore, from (C.94) for $n = 1$ it follows that

$$\begin{aligned} \frac{\mathrm{d}p_1}{\mathrm{d}t}(t) &= (1 \otimes A_0 - F_0 \otimes 1)p_1 + (1 \otimes A_1 - F_1^\top \otimes 1)p_0 \\ &= (1 \otimes A_0 - F_0 \otimes 1)p_1 + (1 \otimes A_1 - F_1^\top \otimes 1) \sum_{k=1}^{N} e^{\mathrm{i}\Omega m_k t} e_{(k-1)N+k}. \end{aligned} \tag{C.97}$$

We insert the following Fourier series expansions:

$$p_1(t) = \sum_{m \in \mathbb{Z}} e^{\mathrm{i}\Omega m t} p_1^{(m)}$$

and

$$A_1(t) = \sum_{m \in \mathbb{Z}} e^{\mathrm{i}\Omega m t} A_1^{(m)}$$

into (C.97) to obtain that

$$\begin{aligned} &\sum_{m \in \mathbb{Z}} \mathrm{i}\Omega m\, e^{\mathrm{i}\Omega m t} p_1^{(m)} \\ &\quad = \sum_{m \in \mathbb{Z}} (1 \otimes A_0 - F_0 \otimes 1) e^{\mathrm{i}\Omega m t} p_1^{(m)} + (1 \otimes A_1 - F_1^\top \otimes 1) \sum_{k=1}^{N} e^{\mathrm{i}\Omega m_k t} e_{(k-1)N+k}. \end{aligned}$$

Then, it follows that

$$\begin{aligned} &\sum_{m \in \mathbb{Z}} \left(\mathrm{i}\Omega m - (1 \otimes A_0 - F_0 \otimes 1)\right) e^{\mathrm{i}\Omega m t} p_1^{(m)} \\ &\quad = \left(\sum_{m \in \mathbb{Z}} e^{\mathrm{i}\Omega m t} 1 \otimes A_1^{(m)} - F_1^\top \otimes 1 \right) \sum_{k=1}^{N} e^{\mathrm{i}\Omega m_k t} e_{(k-1)N+k}. \end{aligned} \tag{C.98}$$

By comparing the coefficients in (C.98), we conclude that for $m \in \mathbb{Z}$,

$$(\mathrm{i}\Omega m - (1 \otimes A_0 - F_0 \otimes 1))p_1^{(m)} = \sum_{k=1}^{N} \left(1 \otimes A_1^{(m-m_k)}\right) e_{(k-1)N+k} - \sum_{k=1}^{N} (\delta_{mm_k} F_1^\top \otimes 1) e_{(k-1)N+k}, \tag{C.99}$$

where δ_{mm_k} is the Kronecker symbol.

LEMMA C.34. *For every $j = 1, \ldots, N$, we have $(F_1)_{jj} = (A_1^{(0)})_{jj}$.*

PROOF. For $j = 1, \ldots, N$, we consider (C.99) with $m = m_j$. The $((j-1)N + j)^{\text{th}}$ entry of the left-hand side is 0. Multiplying by $e_{(j-1)N+j}^\top$ gives us the $((j-1)N + j)^{\text{th}}$ entry of the right-hand side as well:

$$\begin{aligned} e_{(j-1)N+j}^\top & \left(\sum_{k=1}^{N} 1 \otimes A_1^{(m_j - m_k)} e_{(k-1)N+k} - \sum_{k=1}^{N} (\delta_{m_j m_k} F_1^\top \otimes 1) e_{(k-1)N+k} \right) \\ &= \sum_{k=1}^{N} \delta_{jk} \left(\left(A_1^{(m_j - m_k)}\right)_{jk} - (F_1^\top)_{jk} \right). \end{aligned}$$

This means precisely that $(A_1^{(0)})_{jj} = (F_1)_{jj}$. □

LEMMA C.35. *If $(F_0)_{ll} = (F_0)_{jj}$ for some $l \neq j$, then $(F_1)_{jl} = (A_1^{(m_l - m_j)})_{jl}$.*

PROOF. Consider equation (C.99) with $m = m_l$. If $(F_0)_{ll} = (F_0)_{jj}$ for some $l \neq j$, then the $((l-1)N + j)^{\text{th}}$ entry of the left-hand side is 0. Multiplying by $e_{(l-1)N+j}^\top$ also gives us the $((l-1)N + j)^{\text{th}}$ entry of the right-hand side:

$$\begin{aligned} e_{(l-1)N+j}^\top & \left(\sum_{k=1}^{N} 1 \otimes A_1^{(m_l - m_k)} e_{(k-1)N+k} - \sum_{k=1}^{N} (\delta_{m_l m_k} F_1^\top \otimes 1) e_{(k-1)N+k} \right) \\ &= \sum_{k=1}^{N} \delta_{lk} \left(A_1^{(m_l - m_k)}\right)_{jk} - \sum_{k=1}^{N} \delta_{m_l m_k} (F_1^\top)_{lk} \delta_{kj} \\ &= \left(A_1^{(m_l - m_j)}\right)_{jl} - (F_1)_{jl}. \end{aligned}$$

□

C.13.2. Asymptotic analysis of quasifrequency perturbations. Assume that $F = F_0 + \varepsilon F_1 + \varepsilon^2 F_2 + \ldots$ and F_0 is diagonal with respect to the basis vectors $w_1, \ldots, w_N$.

DEFINITION C.36 (Degenerate point). *Suppose that f_0 is a multiple eigenvalue of F_0 of multiplicity $r \geq 2$. Then f_0 is called a* degenerate point. *If f_0 is simple, then we call it a* nondegenerate point.

We now focus on the perturbation of degenerate points. Let f_0 be a degenerate point of multiplicity r and let $w_1, \ldots, w_r$ be its associated eigenvectors. Without loss of generality, we assume that $(F_0)_{ii} = f_0$ for $i = 1, \ldots, r$, *i.e.*, the diagonal entries of A_0 are permuted to make the first r diagonal entries of F_0 coincide. In this setting, there are standard expansions for the eigenvalue perturbation of f_0 (see, for instance, [**95**] and [**16**, Appendix A]). We only need the upper left $r \times r$ block of F_1, which is given by Lemmas C.34 and C.35, to derive the first-order perturbation at degenerate points or eigenvalues of F_0. The following result from [**16**] gives an asymptotic expansion of the quasifrequencies in terms of ε.

THEOREM C.37. *Let f_0 be a degenerate point with multiplicity r. Then F has associated eigenvalues given by*

$$f_0 + \varepsilon f_i + \mathcal{O}(\varepsilon^2),$$

where f_i, for $i = 1, \ldots, r$, are the eigenvalues of the $r \times r$ upper-left block of F_1, whose entries are given by

$$(F_1)_{lk} = \left(A_1^{(m_l - m_k)}\right)_{lk} \quad \text{for } l, k = 1, \ldots, r, \tag{C.100}$$

with m_l and m_k being the folding numbers of the lth and k^{th} eigenvalues of A_0.

PROOF. Formula (C.100) follows from Lemmas C.34 and C.35 together with a standard eigenvalue perturbation argument. □

Typically, it is sufficient to consider degenerate points of order $r = 2$. Moreover, it is natural to assume that A_1 has no constant part; *i.e.*, $A_1^{(0)} = 0$. In this setting, the ε-perturbations are given by the eigenvalues of the matrix

$$\begin{pmatrix} 0 & (F_1)_{12} \\ (F_1)_{21} & 0 \end{pmatrix}. \tag{C.101}$$

Thus, the eigenvalues f of F associated with f_0 are given by

$$f = f_0 \pm \varepsilon \sqrt{(F_1)_{12}(F_1)_{21}} + \mathcal{O}(\varepsilon^2). \tag{C.102}$$

Bibliography

[1] M. J. Ablowitz, C. W. Curtis, and Y. Zhu, *On tight-binding approximations in optical lattices*, Stud. Appl. Math. **129** (2012), no. 4, 362–388, DOI 10.1111/j.1467-9590.2012.00558.x. MR2993124

[2] K. Alexopoulos and B. Davies, *Topologically protected modes in dispersive materials: the case of undamped systems*, Multiscale Model. Simul. **23** (2025), no. 1, 611–639, DOI 10.1137/24M1661674. MR4871808

[3] K. Alexopoulos, B. Davies, and E. O. Hiltunen, *Topological interface modes in systems with damping*, SIAM J. Appl. Math. **85** (2025), no. 4, 1500–1518, DOI 10.1137/24M1710371. MR4941896

[4] A. Alonso Rodríguez, E. Bertolazzi, and A. Valli, *The curl-div system: theory and finite element approximation*, Maxwell's equations—analysis and numerics, Radon Ser. Comput. Appl. Math., vol. 24, De Gruyter, Berlin, 2019, pp. 1–43. MR4398312

[5] H. Ammari, S. Barandun, J. Cao, B. Davies, and E. O. Hiltunen, *Mathematical foundations of the non-Hermitian skin effect*, Arch. Ration. Mech. Anal. **248** (2024), no. 3, Paper No. 33, 34, DOI 10.1007/s00205-024-01976-y. MR4728767

[6] H. Ammari, S. Barandun, J. Cao, B. Davies, and E. O. Hiltunen, *Mathematical foundations of the non-Hermitian skin effect*, Arch. Ration. Mech. Anal. **248** (2024), no. 3, Paper No. 33, 34, DOI 10.1007/s00205-024-01976-y. MR4728767

[7] H. Ammari, S. Barandun, B. Davies, E. O. Hiltunen, and P. Liu, *Stability of the non-Hermitian skin effect in one dimension*, SIAM J. Appl. Math. **84** (2024), no. 4, 1697–1717, DOI 10.1137/23M1610537. MR4781658

[8] H. Ammari, S. Barandun, B. Davies, E. O. Hiltunen, and A. Uhlmann, *Universal estimates for the density of states for aperiodic block subwavelength resonator systems*, J. Phys. A **58** (2025), no. 41, Paper No. 415201, 29. MR4975220

[9] H. Ammari, S. Barandun, B. Davies, E. O. Hiltunen, T. Kosche, and P. Liu, *Exponentially localized interface eigenmodes in finite chains of resonators*, Stud. Appl. Math. **153** (2024), no. 4, Paper No. e12765, 25, DOI 10.1111/sapm.12765. MR4830980

[10] H. Ammari, S. Barandun, B. Davies, E. O. Hiltunen, and P. Liu, *Stability of the non-Hermitian skin effect in one dimension*, SIAM J. Appl. Math. **84** (2024), no. 4, 1697–1717, DOI 10.1137/23M1610537. MR4781658

[11] H. Ammari, S. Barandun, and P. Liu, *Perturbed block Toeplitz matrices and the non-Hermitian skin effect in dimer systems of subwavelength resonators* (English, with English and French summaries), J. Math. Pures Appl. (9) **195** (2025), Paper No. 103658, 28, DOI 10.1016/j.matpur.2025.103658. MR4853452

[12] H. Ammari, S. Barandun, P. Liu, and A. Uhlmann, *Generalized Brillouin zone for non-reciprocal systems*, Proc. A **481** (2025), no. 2310, Paper No. 20240643, 15. MR4994984

[13] H. Ammari, S. Barandun, C. Thalhammer, and A. Uhlmann. Competing edge and bulk localisation in non-reciprocal disordered systems. *Multiscale Model. Simul., to appear (arXiv:2507.02096)*, 2025.

[14] H. Ammari, S. Barandun, and A. Uhlmann, *Subwavelength localization in disordered systems*, Proc. A **481** (2025), no. 2319, Paper No. 20250407, 28. MR4965050

[15] H. Ammari, S. Barandun, and A. Uhlmann, *Truncated Floquet-Bloch transform for computing the spectral properties of large finite systems of resonators*, Trans. London Math. Soc. **12** (2025), no. 1, Paper No. e70009, 26, DOI 10.1112/tlm3.70009. MR4949664

[16] H. Ammari, J. Cao, and E. O. Hiltunen, *Nonreciprocal wave propagation in space-time modulated media*, Multiscale Model. Simul. **20** (2022), no. 4, 1228–1250, DOI 10.1137/21M1449427. MR4500526

[17] H. Ammari, J. Cao, E. O. Hiltunen, and L. Rueff, *Effective Medium Theory for Time-Modulated Subwavelength Resonators*, SIAM J. Appl. Math. **86** (2026), no. 1, 305–326, DOI 10.1137/24M1719426. MR5020403

[18] H. Ammari, J. Cao, and X. Zeng, *Transmission properties of space-time modulated metamaterials*, Stud. Appl. Math. **150** (2023), no. 2, 558–581, DOI 10.1111/sapm.12549. MR4563238

[19] H. Ammari, A. Dabrowski, B. Fitzpatrick, P. Millien, and M. Sini, *Subwavelength resonant dielectric nanoparticles with high refractive indices*, Math. Methods Appl. Sci. **42** (2019), no. 18, 6567–6579, DOI 10.1002/mma.5760. MR4037920

[20] H. Ammari and B. Davies, *A fully coupled subwavelength resonance approach to filtering auditory signals*, Proc. A **475** (2019), no. 2228, 20190049, 17. MR4008362

[21] H. Ammari and B. Davies, *Asymptotic links between signal processing, acoustic metamaterials, and biology*, SIAM J. Imaging Sci. **16** (2023), no. 1, 64–88, DOI 10.1137/22M1510352. MR4536585

[22] H. Ammari, B. Davies, and E. O. Hiltunen, *Robust edge modes in dislocated systems of subwavelength resonators*, J. Lond. Math. Soc. (2) **106** (2022), no. 3, 2075–2135, DOI 10.1112/jlms.12619. MR4498550

[23] H. Ammari, B. Davies, and E. O. Hiltunen, *Convergence rates for defect modes in large finite resonator arrays*, SIAM J. Math. Anal. **55** (2023), no. 6, 7616–7634, DOI 10.1137/23M1575937. MR4665040

[24] H. Ammari, B. Davies, and E. O. Hiltunen, *Anderson localization in the subwavelength regime*, Comm. Math. Phys. **405** (2024), no. 1, Paper No. 1, 20, DOI 10.1007/s00220-023-04880-w. MR4688698

[25] H. Ammari, B. Davies, and E. O. Hiltunen, *Spectral convergence in large finite resonator arrays: the essential spectrum and band structure*, Bull. Lond. Math. Soc. **57** (2025), no. 3, 730–747, DOI 10.1112/blms.13225. MR4877527

[26] H. Ammari, B. Davies, E. O. Hiltunen, H. Lee, and S. Yu, *Bound states in the continuum and Fano resonances in subwavelength resonator arrays*, J. Math. Phys. **62** (2021), no. 10, Paper No. 101506, 24, DOI 10.1063/5.0051292. MR4323026

[27] H. Ammari, B. Davies, E. O. Hiltunen, H. Lee, and S. Yu, *High-order exceptional points and enhanced sensing in subwavelength resonator arrays*, Stud. Appl. Math. **146** (2021), no. 2, 440–462, DOI 10.1111/sapm.12349. MR4206386

[28] H. Ammari, B. Davies, E. O. Hiltunen, H. Lee, and S. Yu, *Exceptional points in parity-time-symmetric subwavelength metamaterials*, SIAM J. Math. Anal. **54** (2022), no. 6, 6223–6253, DOI 10.1137/22M1469821. MR4512271

[29] H. Ammari, B. Davies, E. O. Hiltunen, and S. Yu, *Topologically protected edge modes in one-dimensional chains of subwavelength resonators* (English, with English and French summaries), J. Math. Pures Appl. (9) **144** (2020), 17–49, DOI 10.1016/j.matpur.2020.08.007. MR4175443

[30] H. Ammari, B. Davies, and S. Yu, *Close-to-touching acoustic subwavelength resonators: eigenfrequency separation and gradient blow-up*, Multiscale Model. Simul. **18** (2020), no. 3, 1299–1317, DOI 10.1137/20M1313350. MR4128998

[31] H. Ammari, F. Fiorani, and E. O. Hiltunen, *On the validity of the tight-binding method for describing systems of subwavelength resonators*, SIAM J. Appl. Math. **82** (2022), no. 4, 1611–1634, DOI 10.1137/21M1449804. MR4470470

[32] H. Ammari, B. Fitzpatrick, D. Gontier, H. Lee, and H. Zhang, *A mathematical and numerical framework for bubble meta-screens*, SIAM J. Appl. Math. **77** (2017), no. 5, 1827–1850, DOI 10.1137/16M1090235. MR3717823

[33] H. Ammari, B. Fitzpatrick, D. Gontier, H. Lee, and H. Zhang, *Sub-wavelength focusing of acoustic waves in bubbly media*, Proc. A **473** (2017), no. 2208, 20170469, 17, DOI 10.1098/rspa.2017.0469. MR3749985

[34] H. Ammari, B. Fitzpatrick, D. Gontier, H. Lee, and H. Zhang, *Minnaert resonances for acoustic waves in bubbly media*, Ann. Inst. H. Poincaré C Anal. Non Linéaire **35** (2018), no. 7, 1975–1998, DOI 10.1016/j.anihpc.2018.03.007. MR3906861

[35] H. Ammari, B. Fitzpatrick, E. O. Hiltunen, H. Lee, and S. Yu, *Honeycomb-lattice Minnaert bubbles*, SIAM J. Math. Anal. **52** (2020), no. 6, 5441–5466, DOI 10.1137/19M1281782. MR4169751

[36] H. Ammari, B. Fitzpatrick, E. O. Hiltunen, and S. Yu, *Subwavelength localized modes for acoustic waves in bubbly crystals with a defect*, SIAM J. Appl. Math. **78** (2018), no. 6, 3316–3335, DOI 10.1137/18M1178748. MR3885766

[37] H. Ammari, B. Fitzpatrick, H. Kang, M. Ruiz, S. Yu, and H. Zhang, *Mathematical and computational methods in photonics and phononics*, Mathematical Surveys and Monographs, vol. 235, American Mathematical Society, Providence, RI, 2018, DOI 10.1090/surv/235. MR3837172

[38] H. Ammari, B. Fitzpatrick, H. Lee, S. Yu, and H. Zhang, *Subwavelength phononic bandgap opening in bubbly media*, J. Differential Equations **263** (2017), no. 9, 5610–5629, DOI 10.1016/j.jde.2017.06.025. MR3688425

[39] H. Ammari, B. Fitzpatrick, H. Lee, S. Yu, and H. Zhang, *Double-negative acoustic metamaterials*, Quart. Appl. Math. **77** (2019), no. 4, 767–791, DOI 10.1090/qam/1543. MR4009331

[40] H. Ammari and E. O. Hiltunen, *Edge modes in active systems of subwavelength resonators*, `arXiv:2006.05719`, 2020.

[41] H. Ammari and E. O. Hiltunen, *Time-dependent high-contrast subwavelength resonators*, J. Comput. Phys. **445** (2021), Paper No. 110594, 18, DOI 10.1016/j.jcp.2021.110594. MR4302658

[42] H. Ammari, E. O. Hiltunen, and T. Kosche, *Asymptotic Floquet theory for first order ODEs with finite Fourier series perturbation and its applications to Floquet metamaterials*, J. Differential Equations **319** (2022), 227–287, DOI 10.1016/j.jde.2022.02.047. MR4389752

[43] H. Ammari, E. O. Hiltunen, and T. Kosche, *Design of defected non-Hermitian chains of resonator dimers for spatial and spatio-temporal localizations*, SIAM J. Appl. Math. **83** (2023), no. 6, 2443–2468, DOI 10.1137/23M1573896. MR4673678

[44] H. Ammari, E. O. Hiltunen, and L. Rueff, *Space-time wave localization in systems of subwavelength resonators*, Proc. A **481** (2025), no. 2307, Paper No. 20240698, 23. MR4882249

[45] H. Ammari, E. O. Hiltunen, and S. Yu, *A high-frequency homogenization approach near the Dirac points in bubbly honeycomb crystals*, Arch. Ration. Mech. Anal. **238** (2020), no. 3, 1559–1583, DOI 10.1007/s00205-020-01572-w. MR4160806

[46] H. Ammari, E. O. Hiltunen, and S. Yu, *Subwavelength guided modes for acoustic waves in bubbly crystals with a line defect*, J. Eur. Math. Soc. (JEMS) **24** (2022), no. 7, 2279–2313, DOI 10.4171/jems/1126. MR4413767

[47] H. Ammari and H. Kang, *Polarization and moment tensors*, Applied Mathematical Sciences, vol. 162, Springer, New York, 2007. With applications to inverse problems and effective medium theory. MR2327884

[48] H. Ammari and T. Kosche, *Nonlinear subwavelength resonances in three dimensions*, Stud. Appl. Math. **154** (2025), no. 3, Paper No. e70036, 14, DOI 10.1111/sapm.70036. MR4881245

[49] H. Ammari, H. Lee, and H. Zhang, *Bloch waves in bubbly crystal near the first band gap: a high-frequency homogenization approach*, SIAM J. Math. Anal. **51** (2019), no. 1, 45–59, DOI 10.1137/18M116722X. MR3895120

[50] H. Ammari, B. Li, and J. Zou, *Mathematical analysis of electromagnetic scattering by dielectric nanoparticles with high refractive indices*, Trans. Amer. Math. Soc. **376** (2023), no. 1, 39–90, DOI 10.1090/tran/8641. MR4510105

[51] H. Ammari, P. Millien, M. Ruiz, and H. Zhang, *Mathematical analysis of plasmonic nanoparticles: the scalar case*, Arch. Ration. Mech. Anal. **224** (2017), no. 2, 597–658, DOI 10.1007/s00205-017-1084-5. MR3614756

[52] H. Ammari and J. Qiu, *Analysis of nonlinear resonances in resonator crystals: Tight-binding approximation and existence of subwavelength soliton-like solutions*, `arXiv:2509.04184`, 2025.

[53] H. Ammari, C. Thalhammer, and A. Uhlmann, *Uniform hyperbolicity, bandgaps and edge modes in aperiodic systems of subwavelength resonators*, `arXiv:2509.22417`, 2025.

[54] H. Ammari and H. Zhang, *A mathematical theory of super-resolution by using a system of sub-wavelength Helmholtz resonators*, Comm. Math. Phys. **337** (2015), no. 1, 379–428, DOI 10.1007/s00220-015-2301-4. MR3324165

[55] H. Ammari and H. Zhang, *Super-resolution in high-contrast media*, Proc. A **471** (2015), no. 2178, 20140946, 11, DOI 10.1098/rspa.2014.0946. MR3367310

[56] H. Ammari and H. Zhang, *Effective medium theory for acoustic waves in bubbly fluids near Minnaert resonant frequency*, SIAM J. Math. Anal. **49** (2017), no. 4, 3252–3276, DOI 10.1137/16M1078574. MR3690650

[57] P. W. Anderson. Absence of diffusion in certain random lattices. *Phys. Rev.*, 109(5):1492, 1958.

[58] A. Aspect and M. Inguscio. Anderson localization of ultracold atoms. *Physics Today*, 62:30–35, 2009.

[59] B. A. Bernevig, *Topological insulators and topological superconductors*, Princeton University Press, Princeton, NJ, 2013. With Taylor L. Hughes, DOI 10.1515/9781400846733. MR3185492

[60] D. S. Borgnia, A. J. Kruchkov, and R.-J. Slager, *Non-Hermitian boundary modes and topology*, Phys. Rev. Lett. **124** (2020), no. 5, 056802, 6, DOI 10.1103/physrevlett.124.056802. MR4068223

[61] A. Bëttcher, Ĭ. M. Bogoya, S. M. Grudskiĭ, and E. A. Maksimenko, *Asymptotics of the eigenvalues and eigenvectors of Toeplitz matrices* (Russian, with Russian summary), Mat. Sb. **208** (2017), no. 11, 4–28, DOI 10.4213/sm8865; English transl., Sb. Math. **208** (2017), no. 11, 1578–1601. MR3717195

[62] A. Bourget and T. M. Goode, *Density of states of Jacobi matrices with periodic and asymptotically periodic coefficients*, Proc. Amer. Math. Soc. **143** (2015), no. 12, 5293–5306, DOI 10.1090/proc/12658. MR3411147

[63] T. A. Brody, J. Flores, J. B. French, P. A. Mello, A. Pandey, and S. S. M. Wong, *Random-matrix physics: spectrum and strength fluctuations*, Rev. Modern Phys. **53** (1981), no. 3, 385–479, DOI 10.1103/RevModPhys.53.385. MR619406

[64] Y. Cao and Y. Zhu, *Double Dirac cones in band structures of periodic Schroedinger operators*, Multiscale Model. Simul. **21** (2023), no. 3, 1147–1169, DOI 10.1137/22M1539824. MR4628924

[65] P. Cheben, R. Halir, J. H. Schmid, H. A. Atwater, and D. R. Smith. Subwavelength integrated photonics. *Nature*, 560:565–572, 2018.

[66] W. Chen, Ş. K. Özdemir, G. Zhao, J. Wiersig, and L. Yang. Exceptional points enhance sensing in an optical microcavity. *Nature*, 548(7666):192–196, 2017.

[67] R. V. Craster and B. Davies, *Asymptotic characterization of localized defect modes: Su-Schrieffer-Heeger and related models*, Multiscale Model. Simul. **21** (2023), no. 3, 827–848, DOI 10.1137/22M1511217. MR4613767

[68] C. M. da Fonseca, *The characteristic polynomial of some perturbed tridiagonal k-Toeplitz matrices*, Appl. Math. Sci. (Ruse) **1** (2007), no. 1-4, 59–67. MR2300976

[69] B. Davies and L. Herren, *Robustness of subwavelength devices: a case study of cochlea-inspired rainbow sensors*, Proc. A **478** (2022), no. 2262, Paper No. 20210765, 21. MR4456637

[70] Y. De Bruijn and E. O. Hiltunen, *Complex band structure for subwavelength evanescent waves*, Stud. Appl. Math. **154** (2025), no. 2, Paper No. e70022, 17, DOI 10.1111/sapm.70022. MR4863326

[71] A. Dienstfrey, F. Hang, and J. Huang, *Lattice sums and the two-dimensional, periodic Green's function for the Helmholtz equation*, R. Soc. Lond. Proc. Ser. A Math. Phys. Eng. Sci. **457** (2001), no. 2005, 67–85, DOI 10.1098/rspa.2000.0656. MR1843935

[72] A. Drouot, C. L. Fefferman, and M. I. Weinstein, *Defect modes for dislocated periodic media*, Comm. Math. Phys. **377** (2020), no. 3, 1637–1680, DOI 10.1007/s00220-020-03787-0. MR4121609

[73] R. Duggan, S. A. Mann, and A. Alù. Limitations of sensing at an exceptional point. *ACS Photonics*, 9(5):1554–1566, 2022.

[74] S. Dyatlov and M. Zworski, *Mathematical theory of scattering resonances*, Graduate Studies in Mathematics, vol. 200, American Mathematical Society, Providence, RI, 2019, DOI 10.1090/gsm/200. MR3969938

[75] E. N. Economou, *Green's functions in quantum physics*, 3rd ed., Springer Series in Solid-State Sciences, vol. 7, Springer-Verlag, Berlin, 2006, DOI 10.1007/3-540-28841-4. MR2247484

[76] J. T. Edwards and D. J. Thouless. Numerical studies of localization in disordered systems. *J. Phys. C: Solid State Phys.*, 5(8):807, 1972.

[77] J. C. Eilbeck, P. S. Lomdahl, and A. C. Scott, *The discrete self-trapping equation*, Phys. D **16** (1985), no. 3, 318–338, DOI 10.1016/0167-2789(85)90012-0. MR805707

[78] A. Erdélyi, W. Magnus, F. Oberhettinger, and F. G. Tricomi. Higher transcendental functions vol. i, 1953.

[79] C. L. Fefferman, J. P. Lee-Thorp, and M. I. Weinstein, *Topologically protected states in one-dimensional systems*, Mem. Amer. Math. Soc. **247** (2017), no. 1173, vii+118, DOI 10.1090/memo/1173. MR3633291

[80] C. L. Fefferman, J. P. Lee-Thorp, and M. I. Weinstein, *Honeycomb Schrödinger operators in the strong binding regime*, Comm. Pure Appl. Math. **71** (2018), no. 6, 1178–1270, DOI 10.1002/cpa.21735. MR3794531

[81] C. L. Fefferman and M. I. Weinstein, *Wave packets in honeycomb structures and two-dimensional Dirac equations*, Comm. Math. Phys. **326** (2014), no. 1, 251–286, DOI 10.1007/s00220-013-1847-2. MR3162492

[82] B. U. Felderhof and R. B. Jones, *Addition theorems for spherical wave solutions of the vector Helmholtz equation*, J. Math. Phys. **28** (1987), no. 4, 836–839, DOI 10.1063/1.527572. MR880310

[83] F. Feppon and H. Ammari, *Analysis of a Monte-Carlo Nystrom method*, SIAM J. Numer. Anal. **60** (2022), no. 3, 1226–1250, DOI 10.1137/21M1432338. MR4431376

[84] F. Feppon and H. Ammari, *Modal decompositions and point scatterer approximations near the Minnaert resonance frequencies*, Stud. Appl. Math. **149** (2022), no. 1, 164–229, DOI 10.1111/sapm.12493. MR4472790

[85] F. Feppon and H. Ammari, *Homogenization of sound-soft and high-contrast acoustic metamaterials in subcritical regimes*, ESAIM Math. Model. Numer. Anal. **57** (2023), no. 2, 491–543, DOI 10.1051/m2an/2022098. MR4564294

[86] F. Feppon and H. Ammari, *Subwavelength resonant acoustic scattering in fast time-modulated media* (English, with English and French summaries), J. Math. Pures Appl. (9) **187** (2024), 233–293, DOI 10.1016/j.matpur.2024.05.012. MR4756020

[87] M. Filoche, S. Mayboroda, and T. Tao, *The effective potential of an M-matrix*, J. Math. Phys. **62** (2021), no. 4, Paper No. 041902, 15, DOI 10.1063/5.0042629. MR4246780

[88] S. Franca, V. Könye, F. Hassler, J. van den Brink, and C. Fulga, *Non-Hermitian physics without gain or loss: the skin effect of reflected waves*, Phys. Rev. Lett. **129** (2022), no. 8, Paper No. 086601, 6, DOI 10.1103/physrevlett.129.086601. MR4474436

[89] H. Gazzah, P. A. Regalia, and J.-P. Delmas, *Asymptotic eigenvalue distribution of block Toeplitz matrices and application to blind SIMO channel identification*, IEEE Trans. Inform. Theory **47** (2001), no. 3, 1243–1251, DOI 10.1109/18.915697. MR1830075

[90] J. S. Geronimo, E. M. Harrell II, and W. Van Assche, *On the asymptotic distribution of eigenvalues of banded matrices*, Constr. Approx. **4** (1988), no. 4, 403–417, DOI 10.1007/BF02075470. MR956176

[91] A. Ghatak, M. Brandenbourger, J. Van Wezel, and C. Coulais. Observation of non-Hermitian topology and its bulk–edge correspondence in an active mechanical metamaterial. *Proc. Natl. Acad. Sci. USA*, 117(47):29561–29568, 2020.

[92] I. C. Gohberg and E. I. Sigal, *An operator generalization of the logarithmic residue theorem and Rouché's theorem* (Russian), Mat. Sb. (N.S.) **84(126)** (1971), 607–629. MR313856

[93] D. Gontier, *Edge states in ordinary differential equations for dislocations*, J. Math. Phys. **61** (2020), no. 4, 043507, 21, DOI 10.1063/1.5128886. MR4090721

[94] D. Gontier, *Edge states for second order elliptic operators in a channel*, J. Spectr. Theory **12** (2022), no. 3, 1155–1202, DOI 10.4171/jst/430. MR4578347

[95] K. Gottfried and T.-M. Yan, *Quantum mechanics: fundamentals*, 2nd ed., Springer, New York, 2004. MR2174511

[96] M. J. C. Gover, *The eigenproblem of a tridiagonal 2-Toeplitz matrix*, Linear Algebra Appl. **197/198** (1994), 63–78, DOI 10.1016/0024-3795(94)90481-2. Second Conference of the International Linear Algebra Society (ILAS) (Lisbon, 1992). MR1275608

[97] G. M. Graf and J. Shapiro, *The bulk-edge correspondence for disordered chiral chains*, Comm. Math. Phys. **363** (2018), no. 3, 829–846, DOI 10.1007/s00220-018-3247-0. MR3858823

[98] R. M. Gray, *On the asymptotic eigenvalue distribution of Toeplitz matrices*, IEEE Trans. Inform. Theory **IT-18** (1972), 725–730, DOI 10.1109/tit.1972.1054924. MR453214

[99] H. Guo, M. Zhang, and Y. Zhu, *Threefold Weyl points for the Schrödinger operator with periodic potentials*, SIAM J. Math. Anal. **54** (2022), no. 3, 3654–3695, DOI 10.1137/21M1410464. MR4442440

[100] F. Haldane and S. Raghu. Possible realization of directional optical waveguides in photonic crystals with broken time-reversal symmetry. *Phys. Rev. Lett.*, 100(1):013904, 2008.

[101] N. Hatano and D. R. Nelson. Localization transitions in non-Hermitian quantum mechanics. *Phys. Rev. Lett.*, 77:570–573, 1996.

[102] W. D. Heiss, *The physics of exceptional points*, J. Phys. A **45** (2012), no. 44, 444016, 11, DOI 10.1088/1751-8113/45/44/444016. MR2991883

[103] H. Hodaei, A. U. Hassan, S. Wittek, H. Garcia-Gracia, R. El-Ganainy, D. N. Christodoulides, and M. Khajavikhan. Enhanced sensitivity at higher-order exceptional points. *Nature*, 548(7666):187–191, 2017.

[104] M. Hussein, M. Leamy, and M. Ruzzene. Dynamics of phononic materials and structures: Historical origins, recent progress, and future outlook. *Applied Mechanics Reviews*, 66:040802, 2014.

[105] Y. Katznelson, *An introduction to harmonic analysis*, 3rd ed., Cambridge Mathematical Library, Cambridge University Press, Cambridge, 2004, DOI 10.1017/CBO9781139165372. MR2039503

[106] P. Kuchment, *Floquet theory for partial differential equations*, Operator Theory: Advances and Applications, vol. 60, Birkhäuser Verlag, Basel, 1993, DOI 10.1007/978-3-0348-8573-7. MR1232660

[107] P. Kuchment, *An overview of periodic elliptic operators*, Bull. Amer. Math. Soc. (N.S.) **53** (2016), no. 3, 343–414, DOI 10.1090/bull/1528. MR3501794

[108] A. Lagendijk, B. Van Tiggelen, and D. S. Wiersma. Fifty years of Anderson localization. *Phys. Today*, 62(8):24–29, 2009.

[109] P. D. Lax, *Functional analysis*, Pure and Applied Mathematics (New York), Wiley-Interscience [John Wiley & Sons], New York, 2002. MR1892228

[110] N. N. Lebedev, *Special functions and their applications*, Dover Publications, Inc., New York, 1972. Revised edition, translated from the Russian and edited by Richard A. Silverman; Unabridged and corrected republication. MR350075

[111] D. Leykam, K. Y. Bliokh, C. Huang, Y. D. Chong, and F. Nori, *Edge modes, degeneracies, and toplogical numbers in non-Hermitian systems*, Phys. Rev. Lett. **118** (2017), no. 4, 040401, 6, DOI 10.1103/PhysRevLett.118.040401. MR3669319

[112] W. Li, J. Lin, J. Qiu, and H. Zhang, *Interface modes in honeycomb topological photonic structures with broken reflection symmetry*, `arXiv:2405.03238`, 2024.

[113] W. Li, J. Lin, and H. Zhang, *Dirac points for the honeycomb lattice with impenetrable obstacles*, SIAM J. Appl. Math. **83** (2023), no. 4, 1546–1571, DOI 10.1137/22M1505116. MR4620431

[114] J. Lin, *A perturbation approach for near bound-state resonances of photonic crystal with defect*, European J. Appl. Math. **27** (2016), no. 1, 66–86, DOI 10.1017/S0956792515000340. MR3436036

[115] J. Lin and F. Santosa, *Resonances of a finite one-dimensional photonic crystal with a defect*, SIAM J. Appl. Math. **73** (2013), no. 2, 1002–1019, DOI 10.1137/120897304. MR3045667

[116] J. Lin and F. Santosa, *Scattering resonances for a two-dimensional potential well with a thick barrier*, SIAM J. Math. Anal. **47** (2015), no. 2, 1458–1488, DOI 10.1137/140952053. MR3336303

[117] J. Lin and H. Zhang, *Fano resonance in metallic grating via strongly coupled subwavelength resonators*, European J. Appl. Math. **32** (2021), no. 2, 370–394, DOI 10.1017/s0956792520000200. MR4228266

[118] J. Lin and H. Zhang, *Mathematical theory for topological photonic materials in one dimension*, J. Phys. A **55** (2022), no. 49, Paper No. 495203, 45. MR4523797

[119] R. Lin, T. Tai, L. Li, and C. H. Lee. Topological non-Hermitian skin effect. *Front. Phys.*, 18:53605, 2023.

[120] C. M. Linton, *The Green's function for the two-dimensional Helmholtz equation in periodic domains*, J. Engrg. Math. **33** (1998), no. 4, 377–402, DOI 10.1023/A:1004377501747. MR1638860

[121] C. M. Linton, *Lattice sums for the Helmholtz equation*, SIAM Rev. **52** (2010), no. 4, 630–674, DOI 10.1137/09075130X. MR2736967

[122] S. Longhi, D. Gatti, and G. Valle. Robust light transport in non-Hermitian photonic lattices. *Scientific Reports*, 5:13376, 2015.

[123] J. Lu, J. L. Marzuola, and A. B. Watson, *Defect resonances of truncated crystal structures*, SIAM J. Appl. Math. **82** (2022), no. 1, 49–74, DOI 10.1137/21M1415601. MR4359277

[124] M. L. Lyra, S. Mayboroda, and M. Filoche. Dual landscapes in Anderson localization on discrete lattices. *Europhysics Letters*, 109(4):47001, 2015.

[125] K. Matsushimaa and T. Yamada. Non-Bloch band theory for time-modulated discrete mechanical systems. *Journal of Sound and Vibration*, 595:118757, 2024.

[126] W. McLean, *Strongly elliptic systems and boundary integral equations*, Cambridge University Press, Cambridge, 2000. MR1742312

[127] R. C. McPhedran, N. A. Nicorovici, L. C. Botten, and K.-D. Bao, *Green's function, lattice sums and Rayleigh's identity for a dynamic scattering problem*, Wave propagation in complex media (Minneapolis, MN, 1994), IMA Vol. Math. Appl., vol. 96, Springer, New York, 1998, pp. 155–186, DOI 10.1007/978-1-4612-1678-0_8. MR1489749

[128] M.-A. Miri and A. Alù, *Exceptional points in optics and photonics*, Science **363** (2019), no. 6422, 42, DOI 10.1126/science.aar7709. MR3889190

[129] A. L. Mironov and V. L. Oleĭnik, *Limits on the applicability of the tight-binding approximation method* (Russian, with English and Russian summaries), Teoret. Mat. Fiz. **99** (1994), no. 1, 103–120, DOI 10.1007/BF01018800; English transl., Theoret. and Math. Phys. **99** (1994), no. 1, 457–469. MR1302462

[130] A. L. Mironov and V. L. Oleĭnik, *On the limits of applicability of the tight binding approximation method for a complex-valued potential* (Russian, with English and Russian summaries), Teoret. Mat. Fiz. **112** (1997), no. 3, 448–466, DOI 10.1007/BF02583047; English transl., Theoret. and Math. Phys. **112** (1997), no. 3, 1157–1171 (1998). MR1486800

[131] S. Mukherjee and M. Rechtsman. Observation of floquet solitons in a topological bandgap. *Science*, 368:856–859, 2020.

[132] N. C. Murphy, R. Wortis, and W. A. Atkinson. Generalized inverse participation ratio as a possible measure of localization for interacting systems. *Phys. Rev. B*, 83:184206, 2011.

[133] J.-C. Nédélec, *Acoustic and electromagnetic equations*, Applied Mathematical Sciences, vol. 144, Springer-Verlag, New York, 2001. Integral representations for harmonic problems, DOI 10.1007/978-1-4757-4393-7. MR1822275

[134] N. A. Nicorovici, R. C. McPhedran, and L. C. Botten. Photonic band gaps for arrays of perfectly conducting cylinders. *Phys. Rev. E*, 52:1135–1145, 1995.

[135] N. Okuma, K. Kawabata, K. Shiozaki, and M. Sato, *Topological origin of non-Hermitian skin effects*, Phys. Rev. Lett. **124** (2020), no. 8, 086801, 7, DOI 10.1103/physrevlett.124.086801. MR4071786

[136] S. Ozawa, *Point interaction potential approximation for* $(-\Delta+U)^{-1}$ *and eigenvalues of the Laplacian on wildly perturbed domain*, Osaka J. Math. **20** (1983), no. 4, 923–937. MR727440

[137] R. Ozdemir and J. Lin, *A rigorous mathematical theory for topological phases and edge modes in spring-mass mechanical systems*, Proc. A **480** (2024), no. 2287, Paper No. 20230910, 23. MR4740026

[138] G. C. Papanicolaou, *Diffusion in random media*, Surveys in applied mathematics, Vol. 1, Surveys Appl. Math., vol. 1, Plenum, New York, 1995, pp. 205–253. MR1366209

[139] R. J. Plemmons, *M-matrix characterizations. I. Nonsingular M-matrices*, Linear Algebra Appl. **18** (1977), no. 2, 175–188, DOI 10.1016/0024-3795(77)90073-8. MR444681

[140] G. Poole and T. Boullion, *A survey on M-matrices*, SIAM Rev. **16** (1974), 419–427, DOI 10.1137/1016079. MR352146

[141] J. Qiu, J. Lin, P. Xie, and H. Zhang, *Mathematical theory for the interface mode in an acoustic waveguide bifurcated from a Dirac point*, Commun. Math. Sci. **24** (2026), no. 3, 749–796, DOI 10.4310/cms.260210003613. MR5030060

[142] S. Raghu and F. Haldane. Analogs of quantum-Hall-effect edge states in photonic crystals. *Phys. Rev. A*, 78:033834, 2008.

[143] M. C. Rechtsman. Applied physics: Optical sensing gets exceptional. *Nature*, 548(7666):161–162, 2017.

[144] L. Reichel and L. N. Trefethen, *Eigenvalues and pseudo-eigenvalues of Toeplitz matrices*, Linear Algebra Appl. **162/164** (1992), 153–185, DOI 10.1016/0024-3795(92)90374-J. Directions in matrix theory (Auburn, AL, 1990). MR1148398

[145] G. F. Roach, *Green's functions*, 2nd ed., Cambridge University Press, Cambridge-New York, 1982. MR660842

[146] S. Shankar, A. Souslov, M. Bowick, M. Marchetti, and V. Vitelli. Topological active matter. *Nature Reviews Physics*, 4:380–398, 2022.

[147] S. P. Shipman and S. Venakides, *Resonance and bound states in photonic crystal slabs*, SIAM J. Appl. Math. **64** (2003), no. 1, 322–342, DOI 10.1137/S0036139902411120. MR2029138

[148] S. Steinberg, *Meromorphic families of compact operators*, Arch. Rational Mech. Anal. **31** (1968/69), 372–379, DOI 10.1007/BF00251419. MR233240

[149] J. Stoer and R. Bulirsch, *Introduction to numerical analysis*, 2nd ed., Texts in Applied Mathematics, vol. 12, Springer-Verlag, New York, 1993. Translated from the German by R. Bartels, W. Gautschi and C. Witzgall, DOI 10.1007/978-1-4757-2272-7. MR1295246

[150] W. P. Su, J. R. Schrieffer, and A. J. Heeger. Solitons in polyacetylene. *Phys. Rev. Lett.*, 42:1698–1701, Jun 1979.

[151] A. Szameit and M. Rechtsman. Discrete nonlinear topological photonics. *Nature Physics*, 20:905–912, 2024.

[152] M. E. Taylor, *Partial differential equations. II*, Applied Mathematical Sciences, vol. 116, Springer-Verlag, New York, 1996. Qualitative studies of linear equations, DOI 10.1007/978-1-4757-4187-2. MR1395149

[153] G. C. Thiang and H. Zhang, *Bulk-interface correspondences for one-dimensional topological materials with inversion symmetry*, Proc. A **479** (2023), no. 2270, Paper No. 20220675, 22. MR4556004

[154] E. L. Thomas. Bubbly but quiet. *Nature*, 462(7276):990–991, 2009.

[155] D. J. Thouless. Electrons in disordered systems and the theory of localization. *Phys. Rep.*, 13(3):93–142, 1974.

[156] P. Tilli, *On the asymptotic spectrum of Hermitian block Toeplitz matrices with Toeplitz blocks*, Math. Comp. **66** (1997), no. 219, 1147–1159, DOI 10.1090/S0025-5718-97-00840-5. MR1408378

[157] P. Tilli, *A note on the spectral distribution of Toeplitz matrices*, Linear and Multilinear Algebra **45** (1998), no. 2-3, 147–159, DOI 10.1080/03081089808818584. MR1671591

[158] E. E. Tyrtyshnikov and N. L. Zamarashkin, *Spectra of multilevel Toeplitz matrices: advanced theory via simple matrix relationships*, Linear Algebra Appl. **270** (1998), 15–27. MR1484073

[159] L. Van Hove, *The occurrence of singularities in the elastic frequency distribution of a crystal*, Phys. Rev. (2) **89** (1953), 1189–1193. MR56512

[160] J. Veenstra, O. Gamayun, X. Guo, A. Sarvi, C. Meinersen, and C. Coulais. Non-reciprocal topological solitons in active metamaterials. *Nature*, 627:528–533, 2024.

[161] F. Vollmer and S. Arnold. Whispering-gallery-mode biosensing: label-free detection down to single molecules. *Nat. Methods*, 5(7):591, 2008.

[162] F. Vollmer, S. Arnold, and D. Keng. Single virus detection from the reactive shift of a whispering-gallery mode. *Proc. Natl. Acad. Sci. U.S.A.*, 105(52):20701–20704, 2008.

[163] P. R. Wallace. The band theory of graphite. *Phys. Rev.*, 71(9):622, 1947.

[164] Q. Wang and Y. Chong. Non-Hermitian photonic lattices: tutorial. *Journal of the Optical Society of America B*, 40:1443–1466, 2023.

[165] W. Wang, X. Wang, and G. Ma. Non-Hermitian morphing of topological modes. *Nature*, 608:50–55, 2022.

[166] X. Wen, X. Zhu, A. Fan, W. Y. Tam, J. Zhu, H. W. Wu, F. Lemoult, M. Fink, and J. Li. Unidirectional amplification with acoustic non-Hermitian space-time varying metamaterial. *Commun. Phys.*, 5, 2022.

[167] J. Wiersig. Sensors operating at exceptional points: general theory. *Phys. Rev. A*, 93(3):033809, 2016.

[168] G. Windisch, *M-matrices in numerical analysis*, Teubner-Texte zur Mathematik [Teubner Texts in Mathematics], vol. 115, BSB B. G. Teubner Verlagsgesellschaft, Leipzig, 1989. With German, French and Russian summaries, DOI 10.1007/978-3-663-10818-4. MR1059459

[169] P. Xie and Y. Zhu, *Wave packet dynamics in slowly modulated photonic graphene*, J. Differential Equations **267** (2019), no. 10, 5775–5808, DOI 10.1016/j.jde.2019.06.006. MR3996785

[170] P. Xie and Y. Zhu, *Wave packets in the fractional nonlinear Schrödinger equation with a honeycomb potential*, Multiscale Model. Simul. **19** (2021), no. 2, 951–979, DOI 10.1137/20M1340757. MR4264639

[171] V. A. Yakubovich and V. M. Starzhinskii, *Linear differential equations with periodic coefficients. 1, 2*, Halsted Press [John Wiley & Sons], New York-Toronto; Israel Program for Scientific Translations, Jerusalem-London, 1975. Translated from Russian by D. Louvish. MR364740